Emerging and Re-Emerging Viral Diseases

This reference book offers an in-depth analysis of the fundamental aspects, causative agents, and clinical manifestations of emerging and re-emerging infectious diseases of viral origin (EID-REID). It elucidates the role of evolutionary processes in the emergence, adaptation, and dissemination of pathogens. The initial chapter explores the causative agents behind EID-REID, providing a comprehensive overview of their origins, characteristics, and modes of transmission. The book further presents the structure, pathophysiology, and clinical presentation of various viral pathogens, including notable viruses such as the Crimean-Congo hemorrhagic fever Orthonairovirus (CCHFV), Rift Valley fever virus, Lassa virus, Nipah virus (NiV), Filoviridae, and coronaviruses (MERS, SARS, SARS-CoV-2). Additionally, the book elucidates the multifaceted landscape of treatment strategies, highlighting the pivotal role of pharmacological strategies, investigational drugs, vaccines, and immunomodulatory approaches. It also underscores the importance of clinical examination, immunoassays, and molecular diagnostic techniques for the early detection and accurate diagnosis of viral EID-REID. Toward the end, the book sheds light on the pivotal role of the Ayurveda, homoeopathic system of medicine, medical nutrition therapy, computational biology, bioinformatics, and systems biology in advancing disease management paradigms. This book is intended for epidemiologists, virologists, infectious disease specialists, and other healthcare practitioners who seek in-depth knowledge about emerging and re-emerging infectious diseases.

Emerging and Re-Emerging Viral Diseases

Integrating Conventional and Complementary Treatment Strategies

Edited by
Srijan Goswami, Sangita Agarwal,
and Moharana Choudhury

CRC Press
Taylor & Francis Group
Boca Raton London New York

CRC Press is an imprint of the
Taylor & Francis Group, an **informa** business

Designed cover image: Shutterstock

First edition published 2026
by CRC Press
2385 NW Executive Center Drive, Suite 320, Boca Raton FL 33431

and by CRC Press
4 Park Square, Milton Park, Abingdon, Oxon, OX14 4RN

CRC Press is an imprint of Taylor & Francis Group, LLC

ISBN: 978-1-032-72179-8 (hbk)
ISBN: 978-1-032-72180-4 (pbk)
ISBN: 978-1-032-72181-1 (ebk)

DOI: 10.1201/9781032721811

Typeset in Times
by Apex CoVantage, LLC

Contents

Preface

The 21st century has witnessed the relentless emergence and re-emergence of viral diseases, posing unprecedented challenges to global public health. From the devastating impact of Ebola outbreaks to the global disruption caused by COVID-19, the world continues to grapple with the complex interplay of viral pathogens, environmental changes, zoonotic spillovers, and human behavior. In response, the field of medicine has rapidly evolved, leveraging the best of conventional pharmacology, immunology, computational biology, and complementary treatment approaches such as Ayurveda, homoeopathy, and holistic nutrition. This book, *Emerging and Re-Emerging Viral Diseases: Integrating Conventional and Complementary Treatment Strategies*, is a comprehensive resource designed to provide an in-depth understanding of these multifaceted health crises. It aims to equip researchers, healthcare professionals, and policymakers with a well-rounded perspective on addressing these challenges effectively and sustainably. The book begins with a foundational overview of emerging and re-emerging infectious diseases (EID-REID) and their global significance. Subsequent chapters delve into specific viral pathogens that have shaped recent history, such as Crimean-Congo hemorrhagic fever virus (CCHFV), Rift Valley fever virus (RVFV), and the Filoviridae family, encompassing *Ebolavirus* and *Marburgvirus*. The book also explores zoonotic threats like Nipah and Lassa fever viruses and the ever-evolving challenges posed by coronaviruses, including MERS-CoV, SARS-CoV, and SARS-CoV-2. Key discussions on mosquito-borne diseases, such as dengue, chikungunya, and Zika, further underscore the broad spectrum of emerging threats. Treatment strategies are thoroughly examined, covering pharmacological interventions, investigational drugs, and immunotherapies for various viral pathogens. Chapters on vaccines-based approaches and diagnostic advancements provide insights into the cutting-edge tools reshaping how we manage and mitigate viral epidemics.

One of the defining features of this book is its exploration of complementary treatment strategies. Holistic perspectives from Ayurveda, homoeopathy, and medical nutrition therapy are juxtaposed with conventional methods, presenting a unified approach to patient care. The integration of systems biology, bioinformatics, and computational biology highlights how interdisciplinary sciences are transforming our ability to understand, predict, and respond to viral outbreaks. The concept of "One Health," emphasizing the interconnectedness of human, animal, and environmental health, is given due prominence in a dedicated chapter. Finally, the book underscores the role of balanced nutrition in mitigating the long-term consequences of viral infections, especially those linked to SARS-CoV-2. The journey through these chapters reflects a shared goal: to promote a synergistic approach to combating viral diseases, combining evidence-based medicine with innovative, complementary practices. The authors and contributors, representing diverse fields of expertise, have meticulously curated this content to ensure relevance and applicability for readers worldwide. We hope that this book will serve as a valuable reference, a source of inspiration, and a call to action for all stakeholders in the fight against emerging and re-emerging viral diseases. May it foster greater collaboration, innovation, and resilience in our collective quest for global health security.

Srijan Goswami
Sangita Agarwal
Moharana Choudhury

About the Editors

Srijan Goswami, an author and editor of several medical science books, is pursuing Ph.D. in Biotechnology from the Department of Biotechnology, University of North Bengal, India. He is a physician of biochemic system of medicine and also holds a specialization in Medical Biotechnology and Clinical Nutrition. He received his professional training on Community Nutrition and Child Care, Medical Microbiology, Medical Biochemistry, and Diabetes Counseling from reputed medical colleges in Kolkata, India. He has taught at several paramedical and lifesciences departments of reputed colleges and institutions. He is presently associated as Research Coordinator, Biomedical Research and Allied Sciences Division, Voice of Environment, Assam, India. Dr. Goswami has around 11 years of experience in the field of teaching and research and has published over 5 books (+8 in the pipeline) and ~40 research papers in national and international journals. He has independently and successfully guided around ten undergraduate and postgraduate students in their research work in the past three years. Dr. Goswami has received four Letters of Commendation for Global Community Services and Certificate of Honor for his contribution to the field of Science and Knowledge from the *International Journal of Agricultural Research, Sustainability and Food Sufficiency* (IJARSFS), *International Journal of Advances in Medical Sciences and Biotechnology* (IJAMSB), and *International Journal of Health, Safety and Environment* (IJHSE) in 2019, 2020, 2021, and 2023, respectively. The *Current Research in Nutrition and Food Science* (Enviro Research Group) and *Academia Scholarly Journals* have awarded him the Certificate of Excellence as Reviewer in 2020, 2021, and 2023. He is actively associated with the Editorial Review Board of 15 international journals.

Sangita Agarwal is working as Associate Professor in the Department of Applied Science at RCC Institute of Information Technology, Kolkata, West Bengal, India. She completed her B.Sc. in Chemistry in 1992 from the University of Calcutta, West Bengal, and M.Sc. in Biochemistry in 1994. She obtained her Ph.D. degree in 2001 from Jadavpur University, West Bengal. She was a project participant in the Indo-Hungarian Exchange program under DST, New Delhi, in 1998–1999. She has eight years of full time postdoctoral research experience in various capacities and teaching experience of 13 years. She worked as Visiting Academic in Monash University, Melbourne, Victoria, during April–September 2008, under the Endeavour fellowship, sponsored by the Government of Australia. She is also the recipient of DST Women Scientist fellowship, sponsored by the Government of India, during 2009–2012. She also worked as Visiting Faculty for the Government College of Engineering and Ceramic Technology as well as MSIT, Kolkata. She has to her credit more than 50 scientific research papers and articles, 3 books, and more than 10 book chapters on health, toxicity, pharmacokinetics, environmental science, and environmental biology. She is Life Member of the Indian Chemical Society.

Moharana Choudhury is an Indian researcher and environmentalist. He is a Researcher Coordinator in the Environmental Research and Management Division, Voice of Environment (VoE), Guwahati, Assam, India. Choudhury has worked on various scientific projects across multiple government departments and has more than 14 years of research and working experience in environmental research and services. He began his career as a research intern at the National Institute of Disaster Management in New Delhi, followed by the North Eastern Space Application Center (Department of Space, Government of India/ISRO Unit) as a Junior Research Fellow. Choudhury has authored or edited more than ten books with internationally reputed publishing houses and more than 40 articles in national and internationally reputed journals, proceedings, and more than 35 book chapters. He has written many articles and scientific reports in popular magazines. He has been working as an editor and reviewer of various prestigious international publishing houses worldwide, such as Elsevier, Royal Society of Chemistry (RSC), Springer, Wiley, CRC Press/Routledge group, Apple Academic Press, Cambridge Scholar Publishing, etc. Choudhury is working on various environmental conservation activities in Assam and throughout the country. He has worked on several initiatives to integrate environmentally sustainable projects for iconic heritage places, wetlands, and ecotourism spots in Assam, India, especially for cleanliness and beating single-use plastic pollution missions. He contributes to various environmental research, management, social media mass awareness, and research studies. He is also well known in India for his efforts to implement environmental education and safeguard the privileges of environmental degree holders in India. He has completed his education as an MSc, PGDGIS, PGDEM, and PGCGARD from Gurukul Kangri Deemed to be University, Haridwar, Uttrakhand, North Eastern Hill University (NEHU), Shillong Meghalaya, Tezpur University, Assam, and National Institute of Rural Development Panchayati Raj (NIRDPR) Hyderabad, Telangana, India, respectively.

Contributors

Ulas Acaroz
Department of Food Hygiene and Technology
Afyon Kocatepe University
Afyonkarahisar
Turkey

Manab Deb Adhikari
Department of Biotechnology
University of North Bengal
Darjeeling, West Bengal, India

Mandke Bhakti Ashok
Centre Bio-Separation Technology
Vellore Institute of Technology (VIT)
Vellore, Tamil Nadu, India

Theeshan Bahorun
Mauritius Research and Innovation Council
Cybercity Ebene, Mauritius

Ushmita Gupta Bakshi
Hahnemann Homoeo Pharmaceutical Pvt. Ltd.
Kolkata, West Bengal, India

Kiran Bala
Department of Zoology
Deshbandhu College
University of Delhi
New Delhi, India

Sattwika Banerjee
Institute of Neuroscience Kolkata
Kolkata, West Bengal, India

Mihika Bhattacharjee
Department of Biotechnology
Bengal Institute of Technology
Kolkata, West Bengal, India

Senjuti Bhattacharjee
ABS Clinical Solutions
Kolkata, West Bengal, India

Sayari Bhunia
Medizinischen Fakultät Heidelberg
Universität Heidelberg
Heidelberg, Germany

Easha Biswas
Department of Pharmaceutical Technology
JIS University
Kolkata, West Bengal, India

Sudeepto Biswas
ABS Clinical Solutions
Kolkata, West Bengal, India

Anirbandeep Bose
Department of Pharmacy
Adamas University
Barasat, West Bengal, India

Krishitha Prabhu Chandran
Arcadia High School
Arcadia, California, USA

Ayan Chatterjee
School of Health Science and Technology
Medhavi Skills University
Singtam, Sikkim, India

Abichal Chattopadhyay
Department of Ayurveda Samhita and Siddhanta
Institute of Post Graduate Ayurvedic Education and Research at Shyamdas Baidya Sastra Pitha
Kolkata, West Bengal, India

Tabsum Chhetri
Department of Bioinformatics
University of North Bengal
Darjeeling, West Bengal, India

Banani Ray Chowdhury
Department of Biotechnology
Bengal Institute of Technology
Kolkata, West Bengal, India

Pallabi Roy Chowdhury
Baranagar Baghajatin Social Welfare Organisation
Kolkata, West Bengal, India

Soumendra Darbar
Faculty Council of Science
Jadavpur University
Kolkata, West Bengal, India
and
Research and Development Division
Dey's Medical Stores (Mfg.) Ltd.
Kolkata, West Bengal, India

Dibya Das
Department of Pharmaceutical Technology
JIS University
Kolkata, West Bengal, India

Rajesh Das
Department of Biotechnology
University of North Bengal
Darjeeling, West Bengal, India

Shayeri Das
Department of Clinical Research
Texila American University
Coimbatore, Tamil Nadu, India

Sourav Das
School of Pharmacy
The Neotia University
Sarisha, West Bengal, India

Panchali Dasgupta
ABS Clinical Solutions
Kolkata, West Bengal, India

Pintu Kumar De
Department of Pharmaceutical Technology
JIS University
Kolkata, West Bengal, India

Chiranjeeb Dey
Department of Zoology
Serampore College
Hooghly, West Bengal, India

Arushi Dogra
Faculty of Life Sciences and Biotechnology
South Asian University
Delhi, India

Somenath Dutta
Department of Chemical and Bimolecular Engineering
Pusan National University
Busan, South Korea

John J. Georrge
Department of Bioinformatics
University of North Bengal
Darjeeling, West Bengal, India

Ankita Gon
Department of Allied Health
School of Health and Medical Sciences
Adamas University
Barasat, West Bengal, India

Sangeeta Jaiswal
Division of Gastroenterology
Department of Internal Medicine
University of Michigan
Ann Arbor, Michigan, USA

Abhishek Jana
Department of Pharmaceutics
JIS Institute of Pharmacy
Kalyani, West Bengal, India

Ani Johnson
Centre for Bioinformatics
Saarland University
Saarland, Germany

Anoop Johny
Bonn-Aachen International Center for Information Technology (B-IT)
University of Bonn
Bonn, Germany

Umesh Kalathiya
International Centre for Cancer Vaccine Science
University of Gdańsk
Gdańsk, Poland

Sahana Kamaraj
Centre Bio-Separation Technology
Vellore Institute of Technology (VIT)
Vellore, Tamil Nadu, India

Navneet Kaur
Department of Biochemistry
University of Delhi
New Delhi, India

Nikita Krishna
Biochemistry and Structural Biology Division
CSIR-Central Drug Research Institute (CSIR-CDRI)
Lucknow, Uttar Pradesh, India

Mohamad Fawzi Mahomoodally
Department of Health Sciences
University of Mauritius
Réduit, Mauritius

Himangshu Sekhar Maji
Department of Pharmaceutical Technology
JIS University
Kolkata, West Bengal, India

Tirtha Pratim Mandal
College of Staten Island
City University of New York
New York, USA

Apoorv Mishra
School of Pharmaceutical Sciences
Chhatrapati Shahu Ji Maharaj University
Kanpur, Uttar Pradesh, India

Saurav Kumar Mishra
Department of Bioinformatics
University of North Bengal
Darjeeling, West Bengal, India

Manash Mukherjee
Department of Biotechnology
University of North Bengal
Darjeeling, West Bengal, India

Anusree S. Nair
Sunandan Divatia School of Science
SVKM's Narsee Monje Institute of Management Studies
Mumbai, Maharashtra, India

Ayesha Noor
Centre Bio-Separation Technology
Vellore Institute of Technology (VIT)
Vellore, Tamil Nadu, India

Monikaben Padariya
International Centre for Cancer Vaccine Science
University of Gdańsk
Gdańsk, Poland

Debopriyo Pal
Department of Biotechnology
Bengal Institute of Technology
Kolkata, West Bengal, India

Sneha Roy
Department of Bioinformatics
University of North Bengal
Darjeeling, West Bengal, India

Sukanta Roy
School of Pharmacy
The Neotia University
Sarisha, West Bengal, India

Anupa Ruwali
St Joseph Health Care Centre
Ontario, Canada

Srimoyee Saha
Faculty Council of Science
Jadavpur University
Kolkata, West Bengal, India

Sudipta Saha
Health Research Board
Dublin, Ireland

Sonali Sain
Department of Community Medicine
Bankura Sammilani Medical College and Hospital
Bankura, West Bengal, India

Palas Samanta
Department of Environmental Science
Sukanta Mahavidyalaya
University of North Bengal
Dhupguri, West Bengal, India

Jewelson Santos
Department of Physical Education
University of Batangas
Batangas, Philippines

Debjani Sarkar
Department of Pharmaceutical Technology
JIS University
Kolkata, West Bengal, India

Swagata Sarkar
Institute of Neuroscience Kolkata
Kolkata, West Bengal, India

Aniruddha Sen
All India Institute of Medical Sciences
Gorakhpur, Uttar Pradesh, India

Shreya Sen
Department of Biotechnology
Bengal Institute of Technology
Kolkata, West Bengal, India

Nisha Sharma
School of Pharmaceutical Sciences
Chhatrapati Shahu Ji Maharaj University
Kanpur, Uttar Pradesh, India

Mamta Tiwari
School of Pharmaceutical Sciences
Chhatrapati Shahu Ji Maharaj University
Kanpur, Uttar Pradesh, India

Pallavi Tiwari
School of Pharmaceutical Sciences
Chhatrapati Shahu Ji Maharaj University
Kanpur, Uttar Pradesh, India

Jibin Varghese
Centre for Bioinformatics
Saarland University
Saarland, Germany

Sakshi Verma
Department of Nutrition Biology
Central University of Haryana
Haryana, India

1 Introduction to Emerging and Re-Emerging Infectious Diseases (EID-REID)

Srijan Goswami, Rajesh Das, Manash Mukherjee, Palas Samanta, and Manab Deb Adhikari

1.1 INTRODUCTION

Emerging and re-emerging infectious diseases are critical challenge for public health. Emerging infectious diseases (EIDs) refer to new infections appearing in a population for the first time or known infections that are spreading rapidly, in terms of either infection rates or geographic range. These diseases can be caused by newly identified pathogens, as seen with viruses like SARS-CoV-2, or can result from existing pathogens adapting to new hosts or environments (World Health Organization. Regional Office for South-East Asia, 2014). Re-emerging infectious diseases, on the other hand, are previously controlled infections that have resurged due to factors like antimicrobial resistance, environmental changes, or lapses in vaccination coverage. Diseases like tuberculosis and measles illustrate this resurgence, often aggravated by global travel and weakened healthcare infrastructure (Heymann & Lee, 2015). These distinctions are essential for global health initiatives, which seek to implement proactive surveillance and prevention strategies against both emerging and re-emerging threats. The terminology around these diseases, including factors influencing their spread, such as zoonotic transmission, highlights the importance of a multidisciplinary approach involving epidemiology, environmental science, and healthcare policy to control their impacts effectively (Emerging Infectious Diseases, 2024).

Emerging and re-emerging infectious diseases (EID-REID) pose significant global health challenges due to their potential for rapid spread and significant impact on populations. Emerging infectious diseases (EIDs) refer to diseases that are either newly identified in a population or those that, although previously recognized, are spreading into new areas or populations. A recent example includes COVID-19, caused by the novel SARS-CoV-2 virus, which led to a global pandemic (World Health Organization [WHO], 2020). Re-emerging infectious diseases, on the other hand, are previously known diseases that are making a comeback, often due to factors like increased antimicrobial resistance, changes in environmental conditions, or lapses in public health measures. For instance, diseases like measles and tuberculosis have resurged in certain regions due to lowered vaccination rates and drug resistance, respectively (Centers for Disease Control and Prevention [CDC], 2024). Understanding these definitions is essential to differentiate the diverse nature of these threats and to devise appropriate public health responses through surveillance, research, and control strategies (Petersen et al., 2020). This chapter begins with explaining the definitions and terminologies associated with EID-REID, followed by the current scenario of EID-REID and the possible reasons responsible for the EID-REID crisis. The chapter also focuses on the notable instances of outbreaks during 2001–2024, the importance of the need for global surveillance, preparedness and response, and global public health strategies. The chapter finally closes with discussion on the role and importance of One Health Approach in addressing EID-REID crisis.

DOI: 10.1201/9781032721811-1

1.2 CURRENT SCENARIO OF EMERGING AND RE-EMERGING INFECTIOUS DISEASES

The global scenario of emerging and re-emerging infectious diseases (EID-REID) highlights significant health challenges. Following the COVID-19 pandemic, the resurgence of infectious diseases, such as measles, whooping cough, and polio, has become a global concern, largely attributed to decreased vaccination rates during lockdowns and growing vaccine hesitancy (Figures 1.1 and 1.2).

Data from UNICEF indicate that millions of children missed essential immunizations, making them more susceptible to these preventable diseases. For instance, the cases of measles are rising alarmingly in Europe, with recent outbreaks reported in multiple countries, including the United Kingdom and Austria, surpassing pre-pandemic levels. Additionally, diseases like dengue and cholera have seen drastic increases, partially due to climate change, which facilitates the spread of vector-borne diseases to new regions, as seen in dengue's rise in Argentina and its potential establishment in southern Europe (Feinmann, 2024; WHO Results Report 2022–2023, n.d.). Furthermore, weakened immunity from decreased exposure to common pathogens during lockdowns has increased vulnerability to influenza and respiratory syncytial virus (RSV). These shifts underscore the urgent need for a global response to bolster vaccine coverage, enhance disease surveillance, and address climate-related health risks. The World Health Organization emphasizes the critical role of public trust in vaccines to prevent further resurgence of these diseases (WHO Results Report 2022–2023, n.d.; Wang et al., 2024).

1.3 EID-REID CRISIS: HOW DID WE GET THERE?

1.3.1 Our Biased Fascination Toward "Germ Theory"

The Germ Theory of disease, formulated by Girolamo Fracastoro in 1546 and later expanded by Louis Pasteur and Robert Koch in the 19th century, indicated that specific microorganisms, or "germs," cause particular diseases and can be isolated from infected hosts (Goswami & Bakshi, 2020; Hegde, 2014). This theory revolutionized medicine by shifting the focus toward identifying pathogens as the root cause of illnesses, which led to the development of targeted antimicrobial treatments. Based on this theory, scientists synthesized antimicrobial drugs that could neutralize or eliminate pathogens, establishing four main categories: antibiotics, antivirals, antifungals, and antiparasitics (Calderone, 2015). Through dedicated research, both natural and synthetic antimicrobial agents were developed to combat disease-causing organisms, a cornerstone in modern infectious disease control. Blindly following germ theory overlooks critical aspects of disease causation, as explained by Rudolf Virchow, who argued that pathogens target already diseased tissues rather than causing the disease themselves. This perspective aligns with the epidemiological triad, which emphasizes the importance of host, pathogen, and environmental factors in disease development (CDC, 2012). According to this model, a disease can only manifest if all three factors are present; the mere presence of a pathogen is insufficient without a susceptible host and a favorable environment (CDC, 2012; *Who We Are | NCEZID | CDC*, 2017). However, conventional medical practices primarily target pathogens through antimicrobial drugs, often disregarding host and environmental conditions. This narrow focus conflicts with the evidence-based epidemiological triad, suggesting that treating infectious diseases effectively requires a more comprehensive view that considers all contributing factors rather than exclusively targeting pathogens. Thus, if the triad model is correct, it raises questions about current medical practices that rely solely on germ theory.

1.3.2 Selective Promotion of Reductionistic Approach as the "Only" Way of Scientific Studies

Reductionism, a dominant approach in scientific studies, examines complex systems by breaking them down into separate parts, viewing each as independently functional. This perspective has

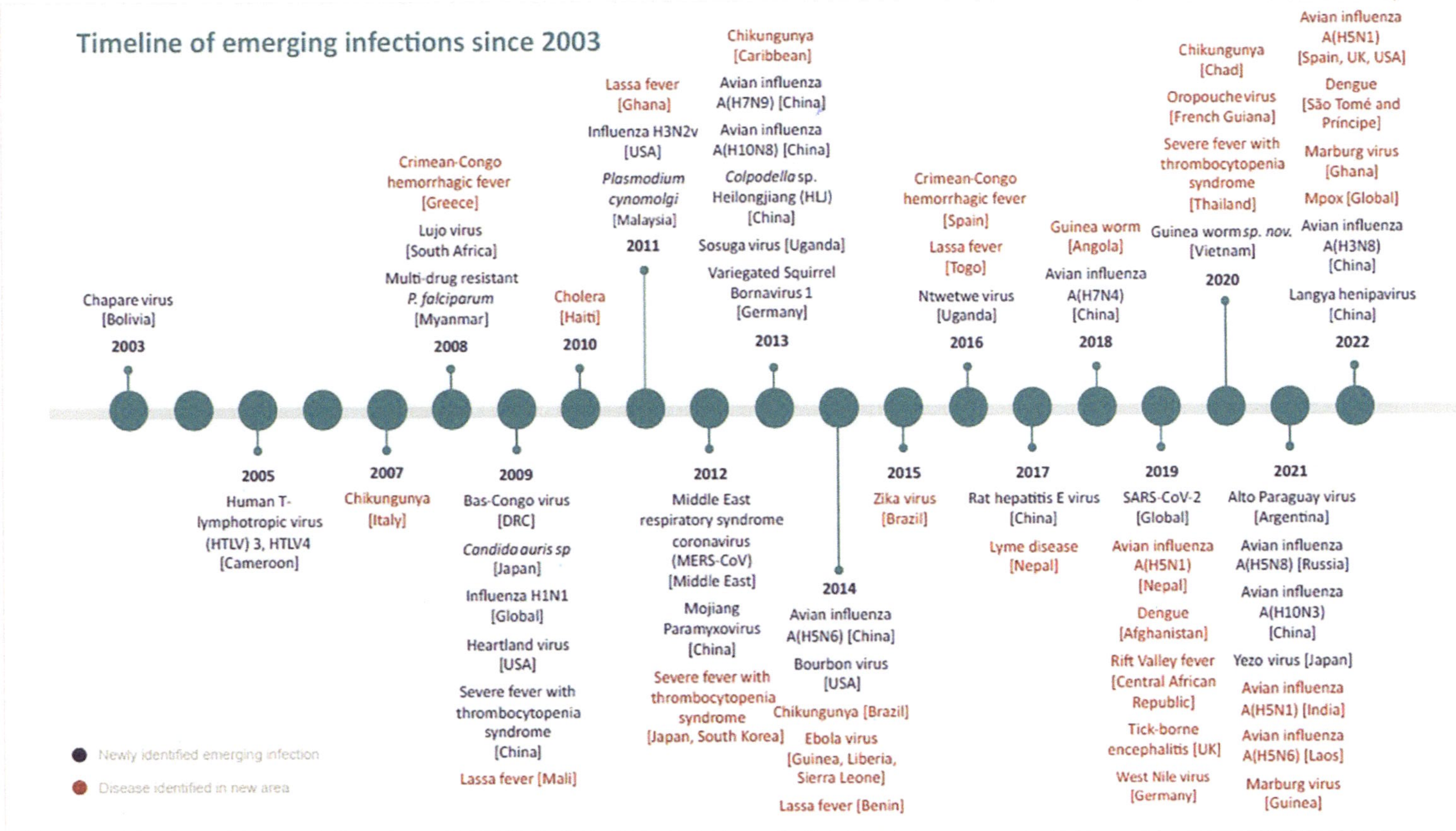

FIGURE 1.1 Timeline of newly identified emerging infections and notable outbreaks of diseases in humans in new areas between 2003 and 2022 (Emerging Infections: How and Why They Arise, 2023).

Source: **Available under Open Government Licence v3.0**

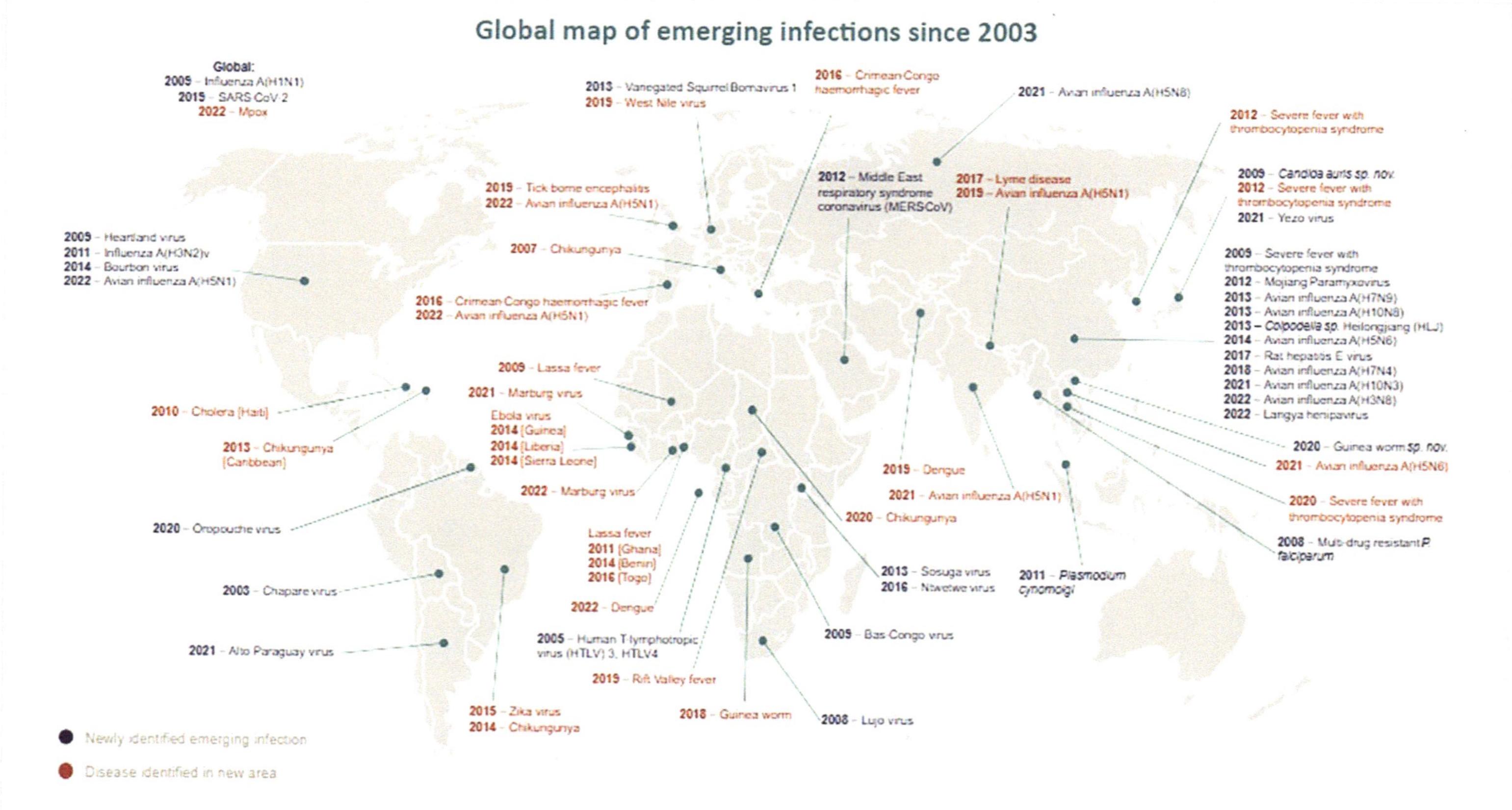

FIGURE 1.2 Global map of newly identified emerging infections and notable outbreaks of diseases in humans in new areas between 2003 and 2022 (Emerging Infections: How and Why They Arise, 2023).

Source: **Available under Open Government Licence v3.0**

contributed to significant advancements, such as the Germ Theory, which focuses on isolating and targeting pathogens to treat disease (Goswami & Bakshi, 2020; Hegde, 2014). However, a strictly reductionistic approach may lead to an incomplete understanding by overlooking the interconnectedness within natural systems. For instance, holism considers that parts within systems, like the human body or ecosystem, are interrelated, where each component affects and is affected by others. A strictly reductionistic view of water, for example, would assume combustibility based on its components, hydrogen and oxygen, whereas water's properties as a whole differ fundamentally from those of its parts (Lipton, 2016; Null, 2011). Similarly, in disease treatment, exclusive focus on pathogens neglects environmental and host factors, which also influence disease development. Integrating both reductionism and holism can lead to more comprehensive and accurate scientific insights, emphasizing that embracing a single approach restricts our understanding of nature's complexities (Goswami & Bakshi, 2020; Lipton, 2016; Lipton & Bhaerman, 2017).

1.3.3 Lacking the Basic Understanding of "Health" and "Disease"

The World Health Organization (WHO) defines "health" as a complete state of physical, mental, and social well-being, not just the absence of disease (Goswami et al., 2022). However, conventional medicine's reductionistic approach primarily focuses on physical symptoms and treatments, often isolating each organ or component without considering the interconnected dimensions of health, such as mental, social, and environmental factors (Goswami & Bakshi, 2020; Hegde, 2014). This limited view may overlook the broader influences on health, including the host's microbiome, immune response, and environmental interactions, which are essential to understanding complex health issues like antimicrobial resistance (AMR) (Lazar et al., 2018). For instance, antimicrobial treatments target pathogenic microbes without fully addressing factors in the patient's environment and lifestyle that may contribute to illness and AMR. Addressing health holistically, as proposed by Rudolf Virchow's theory of interconnectedness, aligns better with broad definition of WHO, suggesting a need to expand conventional medical practices to incorporate a more comprehensive view of health.

1.3.4 Keeping "Evolutionary Biology" out of the Medical Curriculum

The exclusion of holistic evolutionary biology from medical education contributes to a limited, reductionistic view of health. Evolutionary biology illustrates that life evolved through interconnected stages, from single cells to complex organisms, emphasizing that each level—from bacteria to human societies—depends on intricate, symbiotic relationships (Hegde, 2014; Sahtouris, 2000, 2018). Understanding this evolutionary pattern helps reveal the multifaceted nature of health, aligning with the WHO's definition of health as inclusive of physical, mental, and social well-being. Conventional medicine, however, treats the body as isolated components, often viewing microbes solely as threats. This reductionistic view fuels excessive use of antimicrobials, which indiscriminately destroy both pathogens and beneficial microbiome constituents, potentially disrupting essential human–microbe symbiosis and exacerbating antimicrobial resistance (Appanna, 2018; Lipton, 2016, Lipton & Bhaerman, 2017). Integrating evolutionary biology into medical curricula could promote a more comprehensive understanding of health, recognizing that human well-being relies not only on human cells but also on harmonious relationships with millions of microbial genes essential for physiological balance (Goswami & Bakshi, 2020; Meyer, 2014).

1.3.5 Considering Genetic Mutations as the Cause of "Antimicrobial Resistance"

Focusing solely on genetic mutations as the cause of antimicrobial resistance (AMR) provides an incomplete perspective. Conventional approaches treat genetic mutations in microorganisms as the root cause of AMR, leading to efforts in isolating and sequencing these mutations to develop new antimicrobial drugs. However, a holistic, epigenetic view suggests that these mutations may actually be responses to environmental stressors such as widespread use of antimicrobial chemicals in agriculture, animal farming, and human medicine that drive microorganisms to adapt for survival. This approach implies that the overuse of antimicrobials in various industries creates selective pressure, inducing mutations not as a primary cause of resistance but as a survival mechanism. Therefore, addressing AMR effectively may require focusing on the environmental factors causing these genetic changes, rather than solely on the mutations themselves, to reduce the long-term impact of resistance (McMillan et al., 2019; Sahtouris, 2000; Hegde, 2014).

1.3.6 Neglecting Human Body's Inherent Intelligence

The neglect of the human body's inherent intelligence and immune system is a key factor in the growing crisis of antimicrobial resistance (AMR). Just as microorganisms naturally produce antimicrobial agents to protect themselves, humans have evolved a sophisticated immune system, supported by a diverse microbiome, to defend against pathogens. However, overuse of antimicrobial agents, unhealthy diets, and environmental degradation have disrupted this balance, leading to dysbiosis, weakened immune responses, and increased susceptibility to infections. Continuous exposure to antibiotics harms the microbiota, which can reduce immune function and contribute to AMR, autoimmune diseases, and chronic inflammation. This underscores the importance of respecting the body's natural defense mechanisms and avoiding overreliance on pharmaceutical interventions (Belkaid & Hand, 2014; Shekhar & Petersen, 2020).

1.4 NOTABLE INSTANCES OF EID-REID

1.4.1 Nipah Virus Outbreaks

The Nipah virus (NiV) outbreaks in 2001 and 2007 in India are significant events that highlighted the emerging zoonotic threats to public health. The first major outbreak occurred in Silchar, the Indian state of Assam, in 2001, where the virus led to the death of 50% of the 66 people infected. The outbreak was traced to *fruit bats* (*Pteropus* species), which were identified as the natural reservoir of the virus. In this instance, the virus was transmitted to humans through contaminated date palm sap or direct contact with infected animals (Chua et al., 2000). The 2007 outbreak, also in the Kerala region, resulted in the death of 14 people, further establishing the zoonotic nature of the virus and its ability to cause localized outbreaks (Ajith Kumar et al., 2018). Despite the high fatality rate, human-to-human transmission remained relatively limited, though some healthcare workers were infected, underscoring the importance of infection control in outbreaks (Bhutta et al., 2014). Both outbreaks were characterized by a high mortality rate and substantial public health concern, prompting heightened surveillance and further studies into the virus' transmission dynamics and ecological niches. These events also raised awareness about the potential for emerging zoonotic diseases to have devastating impacts on global health systems.

1.4.2 SARS-CoV-1 Outbreak

The SARS-CoV-1 outbreak of 2003 marked a significant global health crisis, with the novel coronavirus responsible for severe acute respiratory syndrome (SARS) spreading rapidly across continents. The outbreak was first reported in Guangdong Province, China, in November 2002, but it wasn't until early 2003 that the virus gained international attention, leading to a global spread with more than 8,000 cases in 29 countries and a mortality rate of approximately 10% (Peiris et al., 2004). The transmission of SARS-CoV-1 is believed to have originated from *bats*, with intermediate hosts such as *civet cats* playing a key role in facilitating the virus' spillover to humans (Li et al., 2005). The virus was characterized by symptoms ranging from fever and cough to severe pneumonia and acute respiratory distress syndrome (ARDS), with many cases requiring intensive care (Poutanen et al., 2003). The outbreak prompted rapid global responses, including extensive quarantine measures, travel restrictions, and the development of diagnostic tests. SARS-CoV-1 was eventually contained in July 2003, though the lessons learned from this outbreak laid the foundation for preparedness strategies for future pandemics, including those involving other coronaviruses like MERS-CoV and SARS-CoV-2. The outbreak highlighted the need for global cooperation in surveillance, outbreak response, and research into zoonotic diseases (Summary of Probable SARS Cases With Onset of Illness From 1 November 2002 to 31 July 2003, n.d.).

1.4.3 Avian Influenza (H5N1) Outbreaks

The H5N1 strain of avian influenza, commonly referred to as the *bird flu*, made headlines in 2006 due to its potential to cause a global pandemic. The outbreak, which had initially been reported in Asia in the early 2000s, gained renewed attention as it spread to Europe, the Middle East, and parts of Africa in 2006. The H5N1 virus primarily affected poultry, but human infections were also reported, with the first cases occurring in Vietnam and Thailand. The virus was highly pathogenic, leading to severe illness and a high mortality rate in humans, with a case fatality rate estimated to be 60–70% (Lai et al., 2016). The primary mode of transmission to humans was through direct contact with infected birds or their droppings, but there were concerns about the virus mutating to allow human-to-human transmission, which could result in a pandemic. The 2006 H5N1 outbreak prompted a global response, with many countries culling infected poultry and implementing strict biosecurity measures to prevent further spread. In addition, international health organizations such as the World Health Organization (WHO) and the Centers for Disease Control and Prevention (CDC) ramped up surveillance efforts and research into vaccines and antiviral treatments. While the virus did not achieve efficient human-to-human transmission in 2006, the outbreak raised alarms about the need for better preparedness against emerging zoonotic diseases, especially those that could lead to a pandemic (Garg et al., 2024; Edler, 2006). The ongoing threat of H5N1 emphasized the importance of continuous surveillance and the development of effective countermeasures to address future outbreaks.

1.4.4 Influenza (H1N1) and Chikungunya Virus Outbreaks

The H1N1 influenza virus, commonly known as the *swine flu*, emerged in 2009 as a novel influenza strain that quickly spread across the globe. The virus was initially detected in Mexico in March 2009 and subsequently led to a pandemic, with cases reported in over 200 countries (Murray et al., 2004). H1N1 was a reassortant virus, containing genetic material from swine, avian, and human influenza strains, which contributed to its rapid transmission among humans (Smith et al., 2009). The symptoms of the H1N1 infection were similar to seasonal flu, but it was more severe in certain populations, including young adults, pregnant women, and individuals with underlying health conditions. The pandemic led to an estimated 151,700–575,400 deaths worldwide, highlighting the need for effective surveillance, vaccine development, and antiviral treatments (Influenza A (H1N1) Outbreak, 2010). In the same year, the chikungunya virus, primarily transmitted by *Aedes mosquitoes*, caused significant outbreaks in several countries, notably in India and Southeast Asia (Sengupta et al., 2020). The virus, characterized by fever and severe joint pain, became a major public health concern due to its rapid spread and its long-lasting effects, such as chronic arthritis in some individuals (Staples et al., 2009). While chikungunya is not typically fatal, its debilitating symptoms and widespread transmission caused severe morbidity in affected regions, further stressing the importance of vector control and public health preparedness. Both the *H1N1* influenza and chikungunya virus outbreaks underscored the interconnectedness of emerging infectious diseases and the need for robust global health systems capable of responding to rapid disease spread across borders.

1.4.5 Crimean-Congo Hemorrhagic Fever Virus Outbreaks

The Crimean-Congo hemorrhagic fever virus (CCHFV) caused a significant outbreak in 2011, particularly affecting regions in Turkey, Pakistan, and Afghanistan. CCHFV, a tick-borne virus belonging to the Bunyaviridae family, is considered one of the most dangerous hemorrhagic fever viruses, with a high case fatality rate, especially in severe cases (Whitehouse, 2004). The virus is primarily transmitted to humans through tick bites or contact with infected animal blood or tissues, making individuals in agricultural and veterinary sectors highly susceptible (Keshtkar-Jahromi et al., 2013). In 2011, Pakistan reported numerous cases, highlighting the virus' ability to spread rapidly in areas with endemic tick populations and poor health infrastructure (Smego et al., 2004). The clinical presentation of CCHF typically includes fever, fatigue, muscle pain, and gastrointestinal symptoms, often progressing to hemorrhagic manifestations such as bleeding from mucous membranes, and in severe cases, multi-organ failure (Keshtkar-Jahromi et al., 2013). Despite the high fatality rates, supportive care has been shown to improve survival outcomes in patients, although a specific antiviral treatment or vaccine is not yet available (Whitehouse, 2004). The outbreak underscored the importance of early detection, vector control measures, and improving public health preparedness in high-risk areas to contain future outbreaks (Smego et al., 2004).

1.4.6 MERS-CoV Outbreaks

The Middle East Respiratory Syndrome Coronavirus (MERS-CoV) outbreak of 2012 was a significant global health event that primarily affected the Arabian Peninsula. First identified in Saudi Arabia in 2012, MERS-CoV is a zoonotic virus, thought to be transmitted to humans via camels, with subsequent human-to-human transmission, especially in healthcare settings (Alraddadi et al., 2016). The clinical symptoms of MERS-CoV include fever, cough, shortness of breath, and in severe cases, acute respiratory distress syndrome (ARDS), which can lead to death, particularly in individuals with preexisting health conditions (Zaki et al., 2012). By 2013, the virus had spread to other countries, including the United States, Europe, and Asia, resulting in a number of cases, but the overall number of infections remained relatively low compared to other coronaviruses such as SARS-CoV (Memish et al., 2014). The WHO reported a case fatality rate (CFR) of approximately 36%, making MERS a serious public health concern (World Health Organization: WHO, 2022). Despite these concerns, MERS-CoV did not lead to widespread global transmission like the 2003 SARS outbreak. Ongoing research into the virus' transmission dynamics, zoonotic reservoirs, and potential treatments continues, with efforts focused on improving diagnostic capabilities and understanding the role of camels in the epidemiology of MERS-CoV (Azhar et al., 2017). The 2012 MERS-CoV outbreak highlighted the potential risks posed by zoonotic viruses and the importance of early detection and containment strategies in mitigating public health threats.

1.4.7 Ebola Outbreaks

The Ebola outbreak of 2013–2016, primarily affecting West Africa, is considered one of the most severe Ebola virus outbreaks in history. Originating in Guinea in late 2013, the outbreak spread rapidly to neighboring Liberia and Sierra Leone, with smaller outbreaks in Nigeria, Mali, and other countries. The virus, transmitted through direct contact with bodily fluids of infected individuals, led to a high case fatality rate, ranging from 25% to 90%, depending on the region. By the time the outbreak was declared over in 2016, it had caused over 28,000 infections and more than 11,000 deaths (Goswami & Bakshi, 2020; Frieden & Damon, 2015). The epidemic's impact was profound, in terms of both human lives and the socioeconomic conditions of the affected countries, exacerbating preexisting vulnerabilities and straining healthcare systems (Van Kerkhove et al., 2015). International responses to the outbreak included mobilizing healthcare resources, establishing treatment units, and implementing experimental therapies, though these interventions were met with logistical and cultural challenges. Efforts to control the spread focused on contact tracing, quarantines, and safe burial practices, which were crucial in preventing further transmission (WHO, 2023). The outbreak underscored the importance of global preparedness for emerging infectious diseases and highlighted significant gaps in healthcare infrastructure, surveillance, and response capabilities in low-resource settings (Senga et al., 2017).

1.4.8 Zika Virus Outbreaks

The Zika virus outbreak of 2015–2016, primarily affecting the Americas, was a significant global public health crisis. First identified in 1952 in Uganda and the United Republic of Tanzania, the virus remained largely confined to Africa and Asia until the outbreak began in Brazil in 2015. The virus is transmitted by *Aedes* mosquitoes, and its spread was rapid, affecting over 40 countries by 2016 (Cao-Lormeau et al., 2014). While Zika infections often result in mild symptoms such as fever and rash, its association with severe birth defects, particularly microcephaly in newborns, became the most alarming consequence of the outbreak (Sharma et al., 2020). The WHO declared a Public Health Emergency of International Concern in February 2016 due to the potential link between Zika infection and neurological complications such as Guillain–Barré syndrome (GBS) in adults (*Zika: The Origin and Spread of a Mosquito-borne Virus*, 2016). As a result, extensive public health efforts were launched, including vector control measures, public education campaigns, and research into vaccine development. However, despite these efforts, the outbreak demonstrated the challenges of managing emerging infectious diseases, particularly when dealing with a complex interplay of environmental, genetic, and socioeconomic factors (Calvet et al., 2016). Although the immediate crisis subsided by the end of 2016, the outbreak has led to ongoing research into the long-term effects of Zika infection and its potential for future resurgence.

1.4.9 LASV Outbreak

The Lassa virus (LASV) outbreak in 2018 was part of ongoing concerns in West Africa, particularly in Nigeria, where it caused significant morbidity and mortality. LASV is transmitted primarily by the multimammate rat (*Mastomys natalensis*), and human infections occur through contact with infected rodents or their excretions, as well as through human-to-human transmission via bodily fluids. The clinical presentation of Lassa fever can range from mild symptoms to severe manifestations such as hemorrhagic fever, organ failure, and death. In 2018, Nigeria reported a notable rise in the number of cases, with early detection and response critical in controlling the spread. The World Health Organization (WHO) and national health agencies were involved in outbreak preparedness and response, including the establishment of surveillance systems and the distribution of medical supplies. While there is no specific vaccine for Lassa fever, ribavirin, an antiviral drug, has been used with some success when administered early in the course of the disease. However, mortality rates remain high in untreated severe cases, with pregnant women and young children being particularly vulnerable (Garry, 2023; World Health Organization: WHO, 2017).

1.4.10 SARS-CoV-2 Outbreak

The SARS-CoV-2 outbreak, which began in December 2019 in Wuhan, China, rapidly escalated into a global pandemic, profoundly impacting health systems, economies, and societies worldwide. The virus, belonging to the Coronaviridae family, primarily spreads through respiratory droplets, leading to symptoms ranging from mild respiratory issues to severe pneumonia, acute respiratory distress syndrome, and death, particularly in older adults and those with underlying health conditions (COVID-19 Variants | WHO COVID-19 Dashboard, n.d.). The first confirmed case outside of China was reported in January 2020, and the virus quickly spread to other countries. Governments around the world implemented unprecedented public health measures, including lockdowns, travel restrictions, and mass testing, to mitigate the spread. By March 2020, the World Health Organization declared it a pandemic. The emergence of variants, including the Alpha, Beta, Delta, and Omicron strains, further complicated the pandemic response due to their increased transmissibility and, in some cases, partial vaccine resistance (Ahmed et al., 2021). Vaccination campaigns launched in late 2020 and early 2021, alongside the development of antiviral treatments, provided a crucial tool in controlling the spread, although challenges remained in achieving global vaccine equity and managing new variants. The long-term effects of COVID-19, including "long COVID," have posed additional public health challenges (Makhluf et al., 2024). The pandemic continues to shape global health policy, with ongoing research into better vaccines, treatments, and strategies for future outbreaks (Goswami et al., 2022).

1.5 EID-REID: GLOBAL SURVEILLANCE AND RESPONSE

The importance of international collaboration in the surveillance, information sharing, and coordinated response to emerging and re-emerging viral diseases has become increasingly clear in the context of global public health challenges. The rapid spread of viruses such as SARS-CoV-2, Ebola, and Zika underscores the need for a collective global approach to monitoring disease outbreaks and ensuring swift responses. Surveillance systems that track viral transmission across borders are essential for early detection and containment. However, these systems rely on transparency and timely sharing of data between countries and international organizations (Health Security Preparedness (HSP), 2016). For instance, the World Health Organization (WHO) and other global health entities play crucial roles in coordinating responses, providing technical assistance, and facilitating access to critical resources (Wolicki et al., 2016). International collaboration also ensures that resources such as vaccines, diagnostics, and treatments are distributed equitably, particularly in low- and middle-income countries, where access to healthcare infrastructure may be limited (Hoffman & Maldonado, 2024). Furthermore, the exchange of knowledge and expertise between countries allows for a better understanding of viral behavior, mutations, and the effectiveness of various interventions. As viruses evolve, and new strains emerge, the need for collaborative research efforts and shared data is more critical than ever to prevent widespread pandemics and mitigate their impact (Forum on Microbial Threats; Board on Global Health; Institute of Medicine, 2015). In this interconnected world, the fight against emerging infectious diseases can only be won through sustained global cooperation.

Timely detection of outbreaks is crucial in managing emerging and re-emerging viral diseases, as it enables early intervention, limiting the spread and reducing the global impact. Advanced monitoring systems and surveillance networks play a pivotal role in identifying potential threats before they become widespread pandemics. These systems employ a combination of real-time data, epidemiological modeling, and geographic information systems (GIS) to track viral transmissions, analyze trends, and predict the likelihood of outbreaks (Arthur et al., 2006). For example, the implementation of digital surveillance tools, including mobile health apps and social media analytics, allows for rapid detection of disease clusters and public health risks (Charles-Smith et al., 2015). Additionally, integrated global networks like the Global Outbreak Alert and Response Network (GOARN) facilitate the sharing of critical data between countries and organizations, allowing for prompt, coordinated responses to outbreaks (*The Global Outbreak Alert and Response Network (GOARN)*, 2023). These advanced systems not only support the identification of new strains of viruses, but they also help monitor viral mutations, which can inform public health responses and vaccine development (*Volume 25, Number 3—March 2019—Emerging Infectious Diseases Journal—CDC*, n.d.). Surveillance efforts, such as those seen with the WHO's International Health Regulations (IHR), ensure that countries are better prepared to detect and manage infectious diseases in real time, ultimately saving lives and mitigating the broader health, social, and economic consequences (Alwashmi, 2020). Through early detection and timely reporting, surveillance systems empower health officials to mobilize resources quickly, implement containment measures, and prevent widespread transmission.

Ongoing research into the biology of pathogens, the development of vaccines, and the improvement of diagnostic tools are all critical components in enhancing global preparedness and response to emerging and re-emerging viral diseases. Understanding the biology of pathogens, including their genetic makeup, transmission patterns, and mechanisms of pathogenicity, allows for the identification of vulnerable targets for therapeutic intervention and vaccine development (Herrington et al., 2015). For instance, research on viral genome sequences and structural proteins enables the design of vaccines that can effectively trigger immune responses against specific viral strains, as seen with the rapid development of the mRNA COVID-19 vaccines (Wang et al., 2020). Similarly, advances in diagnostic tools, such as multiplex PCR and point-of-care testing, allow for quicker detection and accurate identification of pathogens, ensuring timely interventions and reducing transmission risks (Zumla et al., 2014). The development of high-throughput screening methods for viral strains further contributes to early detection of novel viruses, enabling health organizations to take proactive measures in controlling outbreaks (Institute of Medicine (US) Forum on Microbial Threats, 2007). Furthermore, continuous surveillance and research on pathogen evolution, including monitoring genetic mutations and resistance patterns, help in adapting existing vaccines and therapies to emerging variants (Day et al., 2022). By combining these efforts, the global health community is better equipped to respond rapidly to viral threats, potentially preventing pandemics or mitigating their impact. Ongoing research in these areas underscores the importance of an integrated approach, fostering a preparedness model that combines innovation in science with robust public health infrastructure.

1.6 EID-REID: GLOBAL PUBLIC HEALTH STRATEGIES

Prevention and control measures are crucial in reducing the risk of disease transmission during outbreaks of emerging and re-emerging viral diseases. Vaccination campaigns play a central role, particularly when effective vaccines are available. The rapid development and distribution of vaccines, such as those for the Ebola virus and, more recently, COVID-19, have proven vital in controlling outbreaks and preventing large-scale transmission (GAVI, the Vaccine Alliance, 2024). Vector control measures are also critical for diseases like Zika, dengue, and chikungunya, where controlling mosquito populations through insecticide use, the elimination of breeding sites, and the introduction of genetically modified mosquitoes has been employed as strategies to reduce transmission. Hygiene practices, including proper handwashing, safe food handling, and sanitation, remain fundamental to preventing viral infections, especially in settings where waterborne diseases are prevalent (Jumaa, 2005). Additionally, public health education is essential in raising awareness about these measures, empowering communities to adopt preventive behaviors and respond quickly to outbreaks. Information dissemination about symptoms, vaccination schedules, and hygiene practices has been shown to reduce the spread of diseases and mitigate their impact on public health systems (Wang et al., 2023). An integrated approach combining vaccination, vector control, hygiene practices, and education ensures a comprehensive strategy to reduce the burden of emerging and re-emerging viral diseases.

Building capacity within healthcare systems is essential for effectively responding to emerging and re-emerging viral diseases. Strengthening healthcare infrastructure, training healthcare workers, and enhancing laboratory capabilities are foundational components for achieving effective outbreak response. A resilient healthcare system requires adequate facilities, equipment, and personnel who are well-prepared to handle the complexities of viral disease outbreaks. This involves providing continuous training for healthcare workers to ensure they are equipped with the latest knowledge and techniques in diagnosis, treatment, and prevention. Capacity-building initiatives should also focus on improving laboratory capabilities, as accurate and timely diagnostics are critical for detecting outbreaks and guiding interventions. The Global Health Security Agenda emphasizes the need to strengthen laboratory systems and surveillance networks to ensure early detection and rapid response to viral threats (National Academy of Medicine, 2017). The establishment of regional training centers and simulation exercises can also help healthcare workers prepare for real-world outbreaks by practicing response strategies and enhancing coordination. Moreover, continuous investment in healthcare infrastructure and human resources is necessary to ensure that healthcare systems can sustain these efforts over time. Ultimately, capacity-building enhances a country's ability to prevent, detect, and respond to outbreaks, safeguarding public health and minimizing the social and economic impact of emerging and re-emerging viral diseases (Cancedda et al., 2016; Garg et al., 2020).

1.7 EID-REID AND ONE HEALTH APPROACH: NEED FOR COORDINATED ACTION

The concept of EID-REID (Emerging Infectious Diseases and Re-Emerging Infectious Diseases) and the One Health approach emphasizes the interconnectedness of human, animal, and environmental health, underscoring the need for coordinated action to address global health threats. The One Health approach is predicated on the understanding that human health is intrinsically linked to the health of animals and the environment. Emerging and re-emerging infectious diseases (EIDs and REIDs) often arise from zoonotic pathogens, which are transmitted between animals and humans, and can be facilitated by environmental changes, such as deforestation, climate change, and agricultural intensification (Zinsstag et al., 2020). As the frequency and geographical spread of these diseases increase, the need for a coordinated, interdisciplinary response becomes more critical. This is where EID-REID frameworks, alongside the One Health approach, become integral to surveillance, prevention, and response strategies. By fostering collaboration across multiple sectors—public health, veterinary, environmental, and agricultural sciences—One Health promotes shared knowledge and resources to prevent and control outbreaks (Figure 1.3) (Pungartnik et al., 2023). Effective surveillance systems, early detection, and response mechanisms are central to mitigating

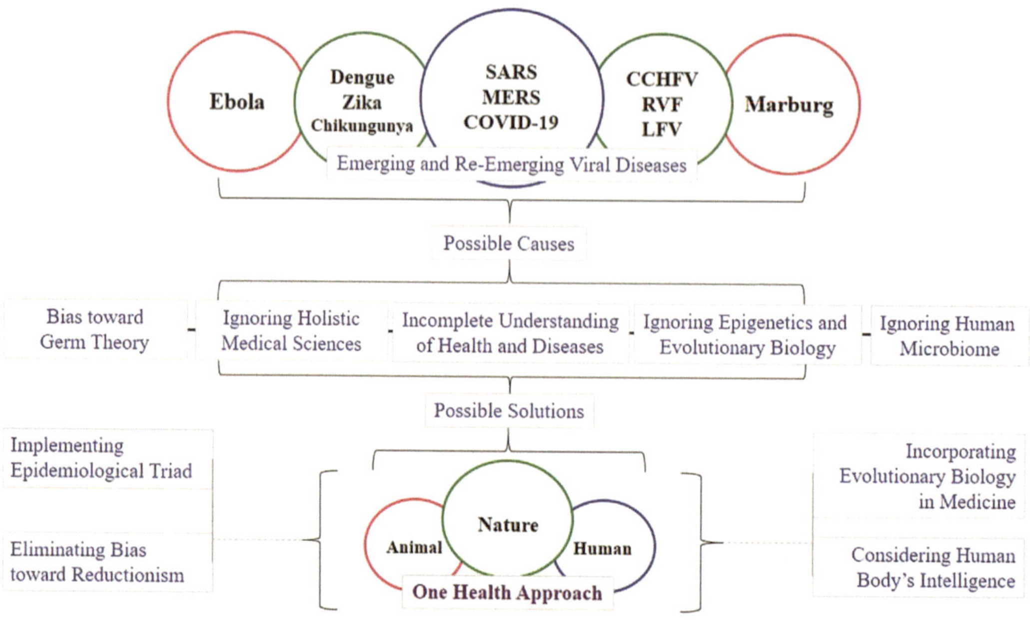

FIGURE 1.3 The One Health approach, emphasizing the interconnectedness of human, animal, and environmental health, is vital in managing these diseases, especially given that many of these pathogens have zoonotic origins.

the impacts of these diseases, requiring a holistic and integrated approach that bridges the gap between human, animal, and environmental health sectors (Johnson et al., 2018). As the world faces more complex and frequent disease outbreaks, the importance of coordinated action under the One Health and EID-REID frameworks cannot be overstated in ensuring a proactive and unified response to emerging global health threats.

1.8 CONCLUSION

Addressing the challenge of emerging and re-emerging viral diseases (EIDs and REIDs) requires a multifaceted approach that incorporates surveillance, international collaboration, timely detection, research advancements, and capacity-building. The One Health approach, emphasizing the interconnectedness of human, animal, and environmental health, is vital in managing these diseases, especially given that many of these pathogens have zoonotic origins. Strengthening healthcare systems, training healthcare workers, and enhancing laboratory capabilities are essential for effective outbreak response and preparedness. Surveillance networks and advanced monitoring systems are critical in detecting potential threats early, enabling a quicker response to outbreaks, and reducing transmission risks. Furthermore, ongoing research into pathogen biology, vaccine development, and diagnostic tools helps improve our ability to combat emerging threats. Prevention and control measures, such as vaccination campaigns, vector control, hygiene practices, and public health education, are vital components of reducing the spread of diseases. Together, these strategies form a comprehensive framework for combating the rising threat of EIDs and REIDs. Only through coordinated global efforts, informed by the One Health model and supported by research, can we effectively respond to and mitigate the impacts of these diseases, ensuring better preparedness and safeguarding public health.

REFERENCES

Ahmed, S., Khan, S., Imran, I., Al Mughairbi, F., Sheikh, F. S., Hussain, J., Khan, A., & Al-Harrasi, A. (2021, July 29). Vaccine development against COVID-19: Study from pre-clinical phases to clinical trials and global use. *Vaccines (Basel)*, 9(8), 836. https://doi.org/10.3390/vaccines9080836. PMID: 34451961; PMCID: PMC8402459.

Ajith Kumar, A. K., & Anoop Kumar, A. S. (2018, July). Deadly Nipah outbreak in Kerala: Lessons learned for the future. *Indian Journal of Critical Care Medicine*, 22(7), 475–476. https://doi.org/10.4103/ijccm.IJCCM_282_18. PMID: 30111920; PMCID: PMC6069317.

Alraddadi, B. M., et al. (2016). Risk factors for primary Middle East respiratory syndrome coronavirus illness in humans, Saudi Arabia, 2014. *Emerging Infectious Diseases*, 22(1), 49–55. https://doi.org/10.3201/eid2201.151340.

Alwashmi, M. F. (2020). The use of digital health in the detection and management of COVID–19. *International Journal of Environmental Research and Public Health*, 17(8), 2906. https://doi.org/10.3390/ijerph17082906.

Appanna, V. D. (2018). The human microbiome: The origin. In *Human Microbes—The Power Within*. Singapore: Springer. https://doi.org/10.1007/978-981-10-7684-8_1.

Arthur, R., Leduc, J. W., Hughes, J. M. (2006). Surveillance for emerging infectious diseases and bioterrorism threats. *Tropical Infectious Diseases*, 195–200. https://doi.org/10.1016/B978-0-443-06668-9.50020-X. Epub 2009 May 15. PMCID: PMC7152254.

Azhar, E. I., Lanini, S., Ippolito, G., & Zumla, A. (2017). The Middle East respiratory syndrome Coronavirus—a continuing risk to global health security. *Advances in Experimental Medicine and Biology*, 972, 49–60. https://doi.org/10.1007/5584_2016_133.

Belkaid, Y., & Hand, T. W. (2014). Role of the microbiota in immunity and inflammation. *Cell*, 157(1), 121–141. https://doi.org/10.1016/j.cell.2014.03.011.

Bhutta, Z. A., et al. (2014). Global burden, distribution, and interventions for infectious diseases of poverty. *Infectious Diseases of Poverty*, 3, 21. https://doi.org/10.1186/2049-9957-3-21.

Calderone, J. (2015, August 7). Penicillin's discoverer predicted our coming post-antibiotic era 70 years ago. *Business Insider*. https://www.businessinsider.in/latest/penicillins-discoverer-predicted-our-coming-post-antibiotic-era-70-years-ago/articleshow/48393377.cms.

Calvet, G. A., Santos, F. B., & Sequeira, P. C. (2016). Zika virus infection: Epidemiology, clinical manifestations and diagnosis. *Current Opinion in Infectious Diseases*, 29(5), 459–466. https://doi.org/10.1097/QCO.0000000000000301.

Cancedda, C., et al. (2016). Strengthening health systems while responding to a health crisis: Lessons learned by a nongovernmental organization during the Ebola virus disease epidemic in Sierra Leone. *The Journal of Infectious Diseases*, 214(Suppl. 3), S153–S163. https://doi.org/10.1093/infdis/jiw345.

Cao-Lormeau, V. M., Roche, C., Teissier, A., Robin, E., Berry, A. L., Mallet, H. P., Sall, A. A., & Musso, D. (2014). Zika virus, French polynesia, South pacific, 2013. *Emerging Infectious Diseases*, 20(6), 1085–1086. https://doi.org/10.3201/eid2006.140138.

CDC. (2012). *Principles of Epidemiology | Lesson 1—Section 8*. Deputy Director for Public Health Science and Surveillance, Center for Surveillance, Epidemiology, and Laboratory Services, Division of Scientific Education and Professional Development. https://www.cdc.gov/csels/dsepd/ss1978/lesson1/section8.html.

CDC. (2024, November). Emerging infectious diseases—CDC. *Emerging Infectious Diseases Journal*. Retrieved November 29, 2024, from https://wwwnc.cdc.gov/eid/.

Charles-Smith, L. E., Reynolds, T. L., Cameron, M. A., Conway, M., Lau, E. H., Olsen, J. M., Pavlin, J. A., Shigematsu, M., Streichert, L. C., Suda, K. J., & Corley, C. D. (2015). Using social media for actionable disease surveillance and outbreak management: A systematic literature review. *PLoS One*, 10(10), e0139701. https://doi.org/10.1371/journal.pone.0139701.

Chua, K. B., et al. (2000). Nipah virus: A recently emergent deadly paramyxovirus. *Science (New York, N.Y.)*, 288(5470), 1432–1435. https://doi.org/10.1126/science.288.5470.1432.

COVID-19 Variants | WHO COVID-19 Dashboard. (n.d.). Datadot. https://data.who.int/dashboards/covid19/variants.

Day, T., Kennedy, D. A., Read, A. F., & Gandon, S. (2022, September 23). Pathogen evolution during vaccination campaigns. *PLoS Biology*, 20(9), e3001804. https://doi.org/10.1371/journal.pbio.3001804. PMID: 36149891; PMCID: PMC9553060.

Edler, A. A. (2006, February). Avian flu (H5N1): Its epidemiology, prevention, and implications for anesthesiology. *Journal of Clinical Anesthesia*, 18(1), 1–4. https://doi.org/10.1016/j.jclinane.2005.12.004. PMID: 16517323; PMCID: PMC7135818.

Emerging Infections: How and Why They Arise. (2023, January 5). GOV.UK. https://www.gov.uk/government/publications/emerging-infections-characteristics-epidemiology-and-global-distribution/emerging-infections-how-and-why-they-arise#figure-1.

Emerging Infectious Diseases. (2024, June 20). Johns Hopkins Medicine. https://www.hopkinsmedicine.org/health/conditions-and-diseases/emerging-infectious-diseases.

Feinmann, J. (2024). Analysis reveals global post-covid surge in infectious diseases. *BMJ (Clinical Research Edition)*, 385, q1348. https://doi.org/10.1136/bmj.q1348.

Forum on Microbial Threats; Board on Global Health; Institute of Medicine. (2015, March 19). *Emerging Viral Diseases: The One Health Connection: Workshop Summary*. Washington (DC): National Academies Press (US). https://www.ncbi.nlm.nih.gov/books/NBK280057/ https://doi.org/10.17226/18975.

Frieden, T. R., & Damon, I. K. (2015, November). Ebola in West Africa-CDC's role in epidemic detection, control, and prevention. *Emerg Infect Dis*, 21(11), 1897–1905. https://doi.org/10.3201/eid2111.150949. PMID: 26484940; PMCID: PMC4622264.

Garg, S., Bhatnagar, N., Singh, M. M., Borle, A., Raina, S. K., Kumar, R., & Galwankar, S. (2020, December 31). Strengthening public healthcare systems in India; Learning lessons in COVID-19 pandemic. *Journal of Family Medicine and Primary Care*, 9(12), 5853–5857. https://doi.org/10.4103/jfmpc.jfmpc_1187_20. PMID: 33681007; PMCID: PMC7928139.

Garg, S., Reed, C., Davis, C. T., Uyeki, T. M., Behravesh, C. B., Kniss, K., Budd, A., Biggerstaff, M., Adjemian, J., Barnes, J. R., Kirby, M. K., Basler, C., Szablewski, C. M., Richmond-Crum, M., Burns, E., Limbago, B., Daskalakis, D. C., Armstrong, K., Boucher, D., . . . Dugan, V. (2024). Outbreak of highly pathogenic Avian influenza A(H5N1) viruses in U.S. dairy cattle and detection of two human cases—United States, 2024. *MMWR Morbidity and Mortality Weekly Report*, 73(21), 501–505. https://doi.org/10.15585/mmwr.mm7321e1.

Garry, R. F. (2023). Lassa fever—the road ahead. *Nature Reviews. Microbiology*, 21(2), 87–96. https://doi.org/10.1038/s41579-022-00789-8.

GAVI, the Vaccine Alliance. (2024, November 28). https://www.gavi.org/.

The Global Outbreak Alert and Response Network (GOARN). (2023, December 18). https://www.who.int/westernpacific/wpro-emergencies/response/the-global-outbreak-alert-and-response-network-(goarn).

Goswami, S., & Dey, C. (Eds.). (2022). *COVID-19 and SARS-CoV-2: The Science and Clinical Application of Conventional and Complementary Treatments (1st ed.).* CRC Press. https://doi.org/10.1201/9781003178514.

Goswami, S., & Bakshi, U. G. (2020, November 24). *Understanding Classical Naturopathy: The Hippocratic Way of Healing | Taylor & Francis Group.* Taylor & Francis. https://www.taylorfrancis.com/chapters/edit/10.1201/9781003003182-11/understanding-classical-naturopathy-hippocratic-way-healing-srijan-goswami-ushmita-gupta-bakshi.

Health Security Preparedness (HSP). (2016, January 1). *International Health Regulations (2005)* (3rd ed.). https://www.who.int/publications/i/item/9789241580496.

Hegde, B. M. (2014). *What Doctors Don't Get to Study in Medical School.* India: Paras Medical Publisher.

Herrington, C. S., Coates, P. J., & Duprex, W. P. (2015, January). Viruses and disease: Emerging concepts for prevention, diagnosis and treatment. *The Journal of Pathology*, 235(2), 149–152. https://doi.org/10.1002/path.4476. PMID: 25366544; PMCID: PMC7168080.

Heymann, D. L., & Lee, V. J. M. (2015, February 1). Emerging and re-emerging infections. In R. Detels & others (Eds.), *Oxford Textbook of Global Public Health* (6th ed.). Oxford Textbook. https://doi.org/10.1093/med/9780199661756.003.0217, accessed 28 Nov. 2024.

Hoffman, S. A., & Maldonado, Y. A. (2024). Emerging and re-emerging pediatric viral diseases: A continuing global challenge. *Pediatric Research*, 95(2), 480–487. https://doi.org/10.1038/s41390-023-02878-7.

Influenza A (H1N1) Outbreak. (2010, August 10). https://www.who.int/emergencies/situations/influenza-a-(h1n1)-outbreak.

Institute of Medicine (US) Forum on Microbial Threats. (2007). *Global Infectious Disease Surveillance and Detection: Assessing the Challenges—Finding Solutions, Workshop Summary* (p. 3). Washington (DC): National Academies Press (US). Detection and Diagnostics. https://www.ncbi.nlm.nih.gov/books/NBK52875/.

Johnson, I., Hansen, A., & Bi, P. (2018, February). The challenges of implementing an integrated one health surveillance system in Australia. *Zoonoses Public Health*, 65(1), e229–e236. https://doi.org/10.1111/zph.12433. Epub 2017 Dec 10. PMID: 29226606; PMCID: PMC7165821.

Jumaa, P. A. (2005). Hand hygiene: Simple and complex. *International Journal of Infectious Diseases: IJID: Official Publication of the International Society for Infectious Diseases*, 9(1), 3–14. https://doi.org/10.1016/j.ijid.2004.05.005.

Keshtkar-Jahromi, M., Sajadi, M. M., Ansari, H., Mardani, M., Holakouie-Naieni, K. (2013, October). Crimean-Congo hemorrhagic fever in Iran. *Antiviral Research*, 100(1), 20–28. https://doi.org/10.1016/j.antiviral.2013.07.007. Epub 2013 Jul 18. PMID: 23872313; PMCID: PMC4260932.

Lai, S., Qin, Y., Cowling, B. J., Ren, X., Wardrop, N. A., Gilbert, M., Tsang, T. K., Wu, P., Feng, L., Jiang, H., Peng, Z., Zheng, J., Liao, Q., Li, S., Horby, P. W., Farrar, J. J., Gao, G. F., Tatem, A. J., & Yu, H. (2016). Global epidemiology of avian influenza A H5N1 virus infection in humans, 1997–2015: A systematic review of individual case data. *The Lancet. Infectious diseases*, 16(7), e108–e118. https://doi.org/10.1016/S1473-3099(16)00153-5.

Lazar, V., Ditu, L. M., Pircalabioru, G. G., Gheorghe, I., Curutiu, C., Holban, A. M., Picu, A., Petcu, L., & Chifiriuc, M. C. (2018). Aspects of gut microbiota and immune system interactions in infectious diseases, immunopathology, and cancer. *Frontiers in Immunology*, 9. https://doi.org/10.3389/fimmu.2018.01830.

Li, W., Shi, Z., Yu, M., Ren, W., Smith, C., Epstein, J. H., Wang, H., Crameri, G., Hu, Z., Zhang, H., Zhang, J., McEachern, J., Field, H., Daszak, P., Eaton, B. T., Zhang, S., & Wang, L. F. (2005). Bats are natural reservoirs of SARS-like coronaviruses. *Science (New York, N.Y.)*, 310(5748), 676–679. https://doi.org/10.1126/science.1118391.

Lipton, B. H. (2016). *The Biology of Belief, Unleashing the Power of Consciousness, Matter & Miracles.* 18th Reprint. India: Hay House Publishers.

Lipton, B. H., & Bhaerman, S. (2017). *Spontaneous Evolution: Our Positive Future and A Way to Get There from Here* (1st Indian Paperback ed.). Hay House Publishers.

Makhluf, H., Madany, H., & Kim, K. (2024, March 28). Long COVID: Long-term impact of SARS-CoV2. *Diagnostics (Basel)*, 14(7), 711. https://doi.org/10.3390/diagnostics14070711. PMID: 38611624; PMCID: PMC11011397.

McMillan, et al. (2019). Antimicrobial resistance genes, cassettes, and plasmids present in *Salmonella enterica* associated with United States food animals. *Frontiers in Microbiology*. https://doi.org/10.3389/fmicb.2019.00832.

Memish, Z. A., Al-Tawfiq, J. A., Assiri, A., AlRabiah, F. A., Al Hajjar, S., Albarrak, A., Flemban, H., Alhakeem, R. F., Makhdoom, H. Q., Alsubaie, S., & Al-Rabeeah, A. A. (2014). Middle East respiratory syndrome coronavirus disease in children. *The Pediatric Infectious Disease Journal*, 33(9), 904–906. https://doi.org/10.1097/INF.0000000000000325.

Meyer, S. (2014). *Darwin's Doubt: The Explosive Origin of Animal Life and the Case for Intelligent Design* (Paperback ed.). HarperOne.

Murray, C. J., Lopez, A. D., & Mathers, C. D. (2004). *The Global Burden of Disease: 2004 Update*. World Health Organization. https://www.who.int/healthinfo/global_burden_disease/GBD_report_2004update_full.pdf.

National Academies of Sciences, Engineering, and Medicine; Health and Medicine Division; Board on Global Health; Committee on Global Health and the Future of the United States. Global Health and the Future Role of the United States. Washington (DC): National Academies Press (US) (2017 May 15). 3, Infectious Diseases, Pandemic Influenza, and Antimicrobial Resistance: Global Health Security Is National Security. Available from: https://www.ncbi.nlm.nih.gov/books/NBK458470/

Null, G. (2011). *Death by Medicine* (Revised Paperback ed.). Mount Jackson, VA: Praktikos Books.

Peiris, J. S., Guan, Y., & Yuen, K. (2004). Severe acute respiratory syndrome. *Nature Medicine*, 10(Suppl. 12), S88–S97. https://doi.org/10.1038/nm1143.

Petersen, E., Koopmans, M., Go, U., Hamer, D. H., Petrosillo, N., Castelli, F., Storgaard, M., Al Khalili, S., & Simonsen, L. (2020). Comparing SARS-CoV-2 with SARS-CoV and influenza pandemics. *The Lancet. Infectious Diseases*, 20(9), e238–e244. https://doi.org/10.1016/S1473-3099(20)30484-9.

Poutanen, S. M., et al. (2003). Identification of severe acute respiratory syndrome in Canada. *The New England Journal of Medicine*, 348(20), 1995–2005. https://doi.org/10.1056/NEJMoa030634.

Pungartnik, P. C., Abreu, A., dos Santos, C. V. B., Cavalcante, J. R., Faerstein, E., & Werneck, G. L. (2023). The interfaces between one health and global health: A scoping review. *One Health*, 16, 100573. ISSN 2352-7714. https://doi.org/10.1016/j.onehlt.2023.100573.

Sahtouris, E. (2000). *EarthDance: Living systems in Evolution*. (Paperback ed.). iUniverse.

Sahtouris, E. (2018). *Gaia's Dance: The Story of Earth and Us*. (Paperback ed.). Createspace Independent Publishing.

Senga, M., Koi, A., Moses, L., Wauquier, N., Barboza, P., Fernandez-Garcia, M. D., Engedashet, E., Kuti-George, F., Mitiku, A. D., Vandi, M., Kargbo, D., Formenty, P., Hugonnet, S., Bertherat, E., & Lane, C. (2017, May 26). Contact tracing performance during the Ebola virus disease outbreak in Kenema district, Sierra Leone. *Philosophical Transactions of the Royal Society of London. Series B, Biological Sciences*, 372(1721), 20160300. https://doi.org/10.1098/rstb.2016.0300. PMID: 28396471; PMCID: PMC5394638.

Sengupta, S., Mukherjee, S., Haldar, S. K., Bhattacharya, N., & Tripathi, A. (2020). Re-emergence of Chikungunya virus infection in eastern India. *Brazilian Journal of Microbiology: [Publication of the Brazilian Society for Microbiology]*, 51(1), 177–182. https://doi.org/10.1007/s42770-019-00212-0.

Sharma, V., Sharma, M., Dhull, D., Sharma, Y., Kaushik, S., & Kaushik, S. (2020). Zika virus: An emerging challenge to public health worldwide. *Canadian Journal of Microbiology*, 66(2), 87–98. https://doi.org/10.1139/cjm-2019-0331.

Shekhar, S., & Petersen, F. C. (2020). The dark side of antibiotics: Adverse effects on the infant immune defense against infection. *Frontiers in Pediatrics*, 8. https://doi.org/10.3389/fped.2020.544460.

Smego, R. A., Jr, Sarwari, A. R., & Siddiqui, A. R. (2004). Crimean-Congo hemorrhagic fever: Prevention and control limitations in a resource-poor country. *Clinical Infectious Diseases: An Official Publication of the Infectious Diseases Society of America*, 38(12), 1731–1735. https://doi.org/10.1086/421093.

Smith, G. J., Vijaykrishna, D., Bahl, J., Lycett, S. J., Worobey, M., Pybus, O. G., Ma, S. K., Cheung, C. L., Raghwani, J., Bhatt, S., Peiris, J. S., Guan, Y., & Rambaut, A. (2009). Origins and evolutionary genomics of the 2009 swine-origin H1N1 influenza A epidemic. *Nature*, 459(7250), 1122–1125. https://doi.org/10.1038/nature08182.

Staples, J. E., Breiman, R. F., & Powers, A. M. (2009). Chikungunya fever: An epidemiological review of a re-emerging infectious disease. *Clinical Infectious Diseases: An Official Publication of the Infectious Diseases Society of America*, 49(6), 942–948. https://doi.org/10.1086/605496.

Summary of Probable SARS Cases with Onset of Illness from 1 November 2002 to 31 July 2003. (n.d.). https://www.who.int/publications/m/item/summary-of-probable-sars-cases-with-onset-of-illness-from-1-november-2002-to-31-july-2003.

Van Kerkhove, M. D., Bento, A. I., Mills, H. L., Ferguson, N. M., & Donnelly, C. A. (2015). A review of epidemiological parameters from Ebola outbreaks to inform early public health decision-making. *Scientific Data*, 2, 150019. https://doi.org/10.1038/sdata.2015.19.

Volume 25, Number 3—March 2019—Emerging Infectious Diseases Journal—CDC. (n.d.). Emerging Infectious Diseases Journal. https://wwwnc.cdc.gov/eid/articles/issue/25/3/table-of-contents.

Wang, C., Horby, P. W., Hayden, F. G., & Gao, G. F. (2020). A novel coronavirus outbreak of global health concern. *Lancet (London, England)*, 395(10223), 470–473. https://doi.org/10.1016/S0140-6736(20)30185-9.
Wang, S., Li, W., Wang, Z., Yang, W., Li, E., Xia, X., Yan, F., & Chiu, S. (2024). Emerging and reemerging infectious diseases: Global trends and new strategies for their prevention and control. *Signal Transduction and Targeted Therapy*, 9(1), 223. https://doi.org/10.1038/s41392-024-01917-x.
Wang, X., Liu, J., Wu, Y., Su, B., Chen, M., Ma, Q., Ma, T., Chen, L., Zhang, Y., Dong, Y., Song, Y., & Ma, J. (2023, June 15). Enhancing the effectiveness of infectious disease health education for children and adolescents in China: A national multicenter school-based trial. *BMC Public Health*, 23(1), 1161. https://doi.org/10.1186/s12889-023-16000-3. PMID: 37322442; PMCID: PMC10273566.
Whitehouse, C. A. (2004). Crimean-Congo hemorrhagic fever. *Antiviral Research*, 64(3), 145–160. https://doi.org/10.1016/j.antiviral.2004.08.001.
WHO Results Report 2022–2023. (n.d.). https://www.who.int/about/accountability/results/who-results-report-2022-2023.
Who We Are | NCEZID | CDC. (2017, August 28). Centers for Disease Control and Prevention (CDC). https://www.cdc.gov/ncezid/who-we-are/index.html.
Wolicki, S. B., Nuzzo, J. B., Blazes, D. L., Pitts, D. L., Iskander, J. K., & Tappero, J. W. (2016, May–June). Public health surveillance: At the core of the global health security agenda. *Health Security*, 14(3), 185–188. https://doi.org/10.1089/hs.2016.0002. PMID: 27314658; PMCID: PMC6937158.
World Health Organization. Regional Office for South-East Asia. (2014). *A Brief Guide to Emerging Infectious Diseases and Zoonoses*. WHO Regional Office for South-East Asia. https://iris.who.int/handle/10665/204722.
World Health Organization: WHO. (2017, July 31). *Lassa Fever*. https://www.who.int/news-room/fact-sheets/detail/lassa-fever.
World Health Organization: WHO. (2020, January 10). *Coronavirus*. https://www.who.int/health-topics/coronavirus#tab=tab_1.
World Health Organization: WHO. (2022, August 5). *Middle East Respiratory Syndrome Coronavirus (MERS-CoV)*. https://www.who.int/news-room/fact-sheets/detail/middle-east-respiratory-syndrome-coronavirus-(mers-cov).
World Health Organization: WHO. (2023, April 20). *Ebola Virus Disease*. https://www.who.int/news-room/fact-sheets/detail/ebola-virus-disease.
Zaki, A. M., van Boheemen, S., Bestebroer, T. M., Osterhaus, A. D., & Fouchier, R. A. (2012). Isolation of a novel coronavirus from a man with pneumonia in Saudi Arabia. *The New England Journal of Medicine*, 367(19), 1814–1820. https://doi.org/10.1056/NEJMoa1211721.
Zika: The Origin and Spread of a Mosquito-Borne Virus. (2016, February). World Health Organization. Retrieved November 29, 2024, from https://www.who.int/publications/m/item/zika-the-origin-and-spread-of-a-mosquito-borne-virus.
Zinsstag, J., Schelling, E., Waltner-Toews, D., Whittaker, M., & Tanner, M. (2020). One health: The theory and practice of integrated health approaches. In *CABI eBooks*. https://doi.org/10.1079/9781789242577.0000.
Zumla, A., Hui, D. S., Al-Tawfiq, J. A., Gautret, P., McCloskey, B., & Memish, Z. A. (2014, October). Emerging respiratory tract infections. *The Lancet Infectious Diseases*, 14(10), 910–911. https://doi.org/10.1016/S1473-3099(14)70899-0. Epub 2014 Sep 1. PMID: 25189348; PMCID: PMC7128723.

2 Infectious Hemorrhagic Fevers

The Crimean-Congo Hemorrhagic Fever Orthonairovirus (CCHFV)

Senjuti Bhattacharjee, Sudeepto Biswas, Panchali Dasgupta, Subhas Chandra Dinda, Swagata Sarkar, Sattwika Banerjee, Sourav Das, Anirbandeep Bose, and Sukanta Roy

2.1 INTRODUCTION

Hemorrhagic fevers are a collection of serious illnesses characterized by fever and bleeding disorders, stemming from various virus families. These viruses often result in bleeding beneath the skin, in internal organs, or from mucous membranes. Hemorrhagic fever viruses (HFVs) are a group of RNA viruses known for their high infectious potential and capacity to cause viral hemorrhagic fever (VHF) in humans. VHF manifests with a range of symptoms, from mild fever to severe illness, often involving disruptions in blood coagulation and widespread hemorrhaging. In severe cases, the disease can progress to multi-organ failure, significantly increasing the risk of death (Flórez-Álvarez et al., 2022). Various viruses, including Ebola, dengue, Lassa, yellow fever, and Marburg, are associated with hemorrhagic fever. Ebola and dengue may progress to hemorrhagic fever with bleeding and low platelet count while Lassa leads to fever and bleeding, Yellow fever can result in severe bleeding and organ failure, and Marburg causes fever and bleeding only. These viruses collectively contribute to the diverse spectrum of hemorrhagic fevers, each with unique clinical manifestations and potential complications (Schnittler & Feldmann, 2003). Among the various hemorrhagic fever viruses, CCHF is notably the most dangerous. This zoonotic disease is triggered by Nairovirus, a tick-borne pathogen from the Bunyaviridae family. The virus' geographic spread is closely associated with the distribution of its primary carrier, the Hyalomma tick. Although animals carrying the virus usually remain asymptomatic, humans infected with CCHF can experience severe and often fatal outcomes. Research on CCHF in India is essential. It will help to evaluate the dynamics and prevalence of diseases, which will further facilitate efficient surveillance and prompt reactions to outbreaks. The discovery of risk factors, creation of public health policies, and improvement of healthcare system readiness all are aided by the study of CCHF. Targeted vector control and prevention measures benefit from an understanding of the transmission dynamics.

2.2 DEFINITIONS AND TERMINOLOGIES ASSOCIATED WITH CCHFV DISEASE AND CCHFV

CCHFV is the causative agent of Crimean-Congo hemorrhagic fever (CCHF), a serious illness that poses significant public health risks, especially in Africa, the Middle East, and Eastern Europe. CCHFV itself is often found in livestock and specific tick species, particularly those belonging to the *Hyalomma* genus. CCHF manifests with severe symptoms, including high fever and bleeding,

DOI: 10.1201/9781032721811-2

and has a concerning mortality rate. The incubation period for this disease typically lasts from 1 to 13 days, complicating early detection and treatment efforts. As a zoonotic disease, CCHF can spread from animals, particularly livestock, to humans, which emphasizes the importance of understanding the factors involved in its transmission. The presence of the virus in the bloodstream, known as viremia, plays a crucial role in facilitating the spread of CCHFV, particularly to ticks, thus perpetuating the cycle of transmission. CCHF is classified as a type of hemorrhagic fever, characterized by symptoms of fever and bleeding. The case fatality rate (CFR) associated with CCHF can be alarmingly high, ranging from 30% to 50% in severe instances, highlighting the severity of the disease. Given the absence of specific antiviral treatments, supportive care becomes essential for managing symptoms and ensuring patient comfort. Serology is vital in diagnosing CCHF, as it involves testing for antibodies against CCHFV in the blood, aiding in case confirmation and the assessment of the immune response. Additionally, the potential for nosocomial transmission, or the spread of the virus within healthcare settings, poses significant risks for healthcare providers, necessitating strict infection control measures. Preventive strategies are limited by the lack of a licensed vaccine for humans but can include the use of personal protective equipment (PPE) and public health education aimed at reducing transmission risks. Understanding the epidemiology of CCHF, how it spreads within populations, and the factors contributing to outbreaks is critical for developing effective public health strategies.

2.3 ORIGIN AND CLASSIFICATION

The virus was initially identified in the 1940s when Soviet soldiers contracted a hemorrhagic disease following their presence in Crimea. During the 1960s, a virus exhibiting identical clinical symptoms was detected in the Belgian Congo (presently the Democratic Republic of Congo). Subsequent analysis revealed its antigenic identity with the virus found in Crimea, leading to the nomenclature CCHFV (Kuehnert et al., 2021). According to Makwana et al., the initial reported case of CCHF in India occurred in Gujarat in the year 2011 (Makwana et al., 2015). Crimean-Congo hemorrhagic fever virus (CCHFV) belongs to the genus *Nairovirus* within the family Bunyaviridae, which also includes the genera *Orthobunyavirus*, *Hantavirus*, *Phlebovirus*, and *Tospovirus*. A distinctive feature of Nairoviruses is their large L genomic segments, and they are primarily transmitted through ticks, distinguishing them from other bunyaviruses. The *Nairovirus* genus is divided into seven serogroups, with CCHFV and Hazara virus (HAZV) comprising the CCHF serogroup. HAZV, first isolated from ticks found on wild rodents in Pakistan, has not yet been linked to human infection (Bente et al., 2010).

2.4 MODE OF TRANSMISSION AND SYMPTOMS

The virus that causes CCHF frequently leaves infected animals with no symptoms at all, but it has a high fatality rate in people. In human situations, hemorrhagic syndrome usually develops from non-specific febrile symptoms. Although the predominant vectors of disease transmission are ticks, person-to-person transmission of the virus through percutaneous or permucosal exposure to blood and other fluids is a common cause of secondary infections in healthcare settings. There is a considerable nosocomial danger associated with this uncommon mode of infection, especially for medical workers.

CCHFV presents with a rapid onset of symptoms such as fever, muscle pain, headache, dizziness, ocular discomfort, light sensitivity, and redness of the eyes (Figure 2.1). In more severe instances, hemorrhagic symptoms may emerge several days after the initial illness begins (Whitehouse, 2004). Ribavirin, a nucleoside analog with broad-spectrum antiviral effects and minimal immunosuppressive activity, remains the most promising treatment option for CCHFV. At present, no vaccine has been developed to protect against this virus (Fisher-Hoch et al., 1995). CCHF is a serious public health concern throughout Africa, Asia, Eastern Europe, and the Middle East, with a high

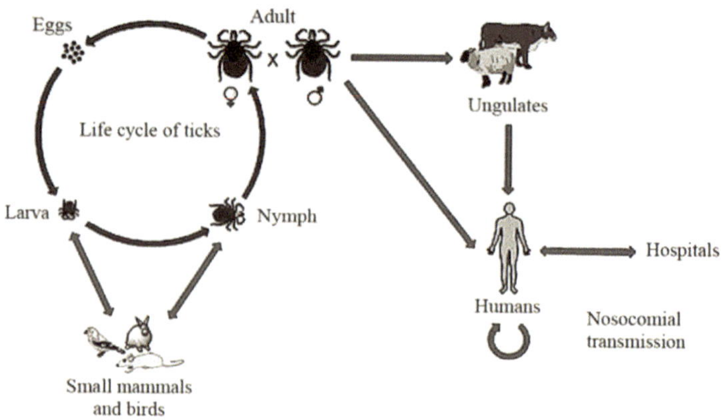

FIGURE 2.1 General life cycle of Crimean-Congo hemorrhagic fever virus (Aslam et al., 2016).

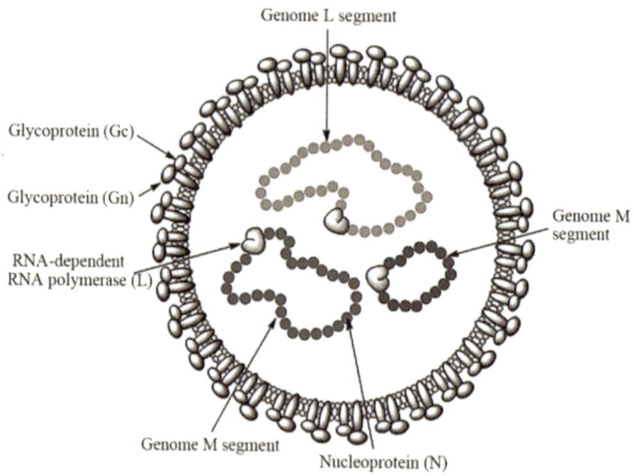

FIGURE 2.2 Structure of Crimean-Congo hemorrhagic fever virus (Aslam et al., 2016).

mortality rate, according to Appannanavar et al. (Appannanavar & Mishra, 2011). This virus not only endangers public health but also puts a heavy load on medical staff, particularly in areas with little resources. Adherence to universal precautions during patient care, prompt implementation of infection-control measures, and the administration of prophylactic therapy to healthcare workers post-exposure are pivotal steps in limiting infection spread. Additionally, community-based control measures, such as the use of pesticides to manage tick populations, are essential for reducing disease transmission and preventing further community spread (Appannanavar & Mishra, 2011).

2.5 STRUCTURE OF CCHF VIRUS

CCHFV belongs to the Bunyaviridae family and is classified as a segmented negative-strand RNA virus (sNSV) alongside viruses from the Arenaviridae and Orthomyxoviridae families, due to its multipartite genome (Figure 2.2). The Bunyaviridae family includes over 350 identified isolates, which are divided into five genera: *Hantavirus*, *Nairovirus*, *Orthobunyavirus*, *Phlebovirus*, and *Tospovirus*. CCHFV is specifically a Nairovirus, with a genome consisting of three RNA segments:

small (S), medium (M), and large (L). The S segment codes for the nucleocapsid protein (N), the M segment produces viral glycoproteins, and the L segment encodes the RNA-dependent RNA polymerase (RdRp), also known as the L protein. Segmented negative-strand RNA virus (sNSV) genomes are not free-standing RNA molecules; instead, they are bound by the viral N protein to form ribonucleoprotein (RNP) complexes. These RNPs interact with their corresponding RNA-dependent RNA polymerase (RdRp) to create active templates for viral RNA synthesis, producing both encapsidated replication products and non-encapsidated mRNAs. Efficient genome encapsidation is essential for integrating RNPs into newly formed virus particles. In bunyaviruses, virus assembly involves a direct link between the RNP and viral glycoproteins, highlighting RNP formation as a potential target for therapeutic approaches. In addition to their involvement in ribonucleoprotein (RNP) assembly and viral packaging, the N proteins of bunyaviruses are essential for a range of cellular functions through their interactions with multiple host cell elements. These interactions encompass the cytoskeleton, cellular RNA, the translation apparatus, and key components of the innate immune response. For CCHFV, the N protein is known to specifically engage with the antiviral defense protein MxA and has recently been identified as a target for caspase-3, a key enzyme in the process of apoptosis (Carter et al., 2012). The M segment of CCHFV codes for a glycoprotein precursor complex (GPC), which is cleaved by host-cell proteases to produce several mature proteins. Among these, the structural glycoproteins Gn (spanning residues 520 to 842 of GPC) and Gc (covering residues 1041–1684 of GPC) form spikes on the viral surface, aiding in the virus' entry into target cells. The specific roles of the secreted glycoproteins are not yet well understood. Recently, a newly identified secreted glycoprotein, GP38, which is unique to the Nairoviridae family, has been found to be the target of a protective antibody against CCHFV, though its function remains unclear (Mishra et al., 2020).

2.6 EPIDEMIOLOGY OF CCHFV

World Health Organization Weekly Epidemiological Report of July 2021 stated that out of 1,118 events related to infectious hazards over a cumulative span of three years, the top 10 most frequently documented diseases and conditions were COVID-19, accounting for 20% (226 events); measles, comprising 9% (101 events); dengue fever, representing 8% (89 events); cholera, constituting 6% (68 events); poliomyelitis, making up 4% (48 events); influenza resulting from identified avian or animal influenza viruses, contributing to 4% (45 events); yellow fever, amounting to 4% (42 events); malaria, comprising 3% (28 events); Crimean-Congo hemorrhagic fever, representing 2% (22 events); and multisystem inflammatory syndrome in children and adolescents, accounting for 2% (19 events) (Norman et al., 2025). Crimean-Congo hemorrhagic fever (CCHF) is categorized as a Biosafety Level 4 pathogen, with a fatality rate reaching as high as 80%. The virus has been detected in numerous countries across Asia, Africa, Southeastern Europe, and the Middle East. In the Middle East and North Africa (MENA) region, several countries have documented cases, and in some areas, the virus is endemic. Over the past decade, the incidence of CCHF seems to have risen in the WHO's Eastern Mediterranean Region (EMR). However, the availability of reliable data remains limited, likely due to the lack of robust surveillance systems and a poor understanding of the virus' epidemiology and transmission factors (Perveen & Khan, 2022). CCHF, initially recognized during a 1944 outbreak in Crimea, gained attention when the same virus infected a patient in Congo in 1956, leading to its designation as Crimean-Congo hemorrhagic fever (CCHF). Officially classified as an arbovirus in 1962, CCHF is endemic in Africa, Asia, Eastern Europe, and the Middle East, with its distribution closely associated with Hyalomma ticks. There have been repeated reports of multiple cases in the MENA region, and recent data indicates an increased incidence in the Eastern Mediterranean Region over the last decade. Examining country-specific epidemiology reveals notable occurrences. In Algeria's Laghouat Province, *Hyalomma aegyptium* ticks carried CCHFV, though no human cases were reported. In Egypt, serological studies from 1978 indicated CCHFV circulation in animals, with associated tick species identified. Iraq reported its first confirmed CCHF case in 1979, with subsequent cases linked to animal contact. Kuwait experienced 18 human cases with CCHFV antibodies from 1979 to 1982. Mauritania reported cases, an urban outbreak in 2003, and CCHFV in livestock and ticks. In Morocco, CCHFV was found in *H. marginatum* ticks in 2013. Oman reported human cases and CCHFV antibodies in animals and ticks from 1995 onward. Saudi Arabia identified the virus in 1990, suspecting its introduction via imported animals, and recent studies revealed ticks positive for CCHFV. Sudan reported outbreaks, cases, and CCHFV antibodies in livestock, with recent human cases. Tunisia's seroprevalence studies revealed CCHFV antibodies in slaughterhouse workers and patients, and infection in animals and ticks was reported. The UAE experienced outbreaks in 1979 and 1994–1995, with cases, CCHFV antibodies in various animals, and ticks reported. Yemen lacks published records of human infections but reported various tick species, potential vectors. India, recognized as a potential hotspot for emerging infectious diseases, has raised concerns about the potential presence of the CCHF virus due to confirmed cases in neighboring countries like Pakistan, China, and Afghanistan. Apprehensions heightened when the virus was isolated from ticks of the *Hyalomma anatolicum* species in Pakistan. A 1973 investigation into nationwide human sera identified nine samples in Kerala and Pondicherry with CCHFV-specific antibodies, while 34 out of 655 animal serum samples from various domestic animals, predominantly goats in southern India, showed evidence of CCHFV antibodies. In 1976, a serosurvey detected hemagglutination inhibition antibodies in animal sera from Jammu and Kashmir, western border districts, and southern regions, with notable findings in goat sera from South India. However, these investigations were based solely on serological data without successful virus isolation. In December 2010, just before a CCHF outbreak, blood samples from Rajasthan and other remote areas revealed IgG antibodies against CCHFV in buffalo, goats, and sheep from the Sirohi district of southern Rajasthan. In January 2011, hospital staff in Gujarat presented with hemorrhagic symptoms, and IgM antibodies against CCHFV were identified in some individuals. Subsequent research conducted in West Bengal and Maharashtra did not detect CCHFV IgG antibodies in animal serum samples. A follow-up study in Kolkata and nearby villages showed varied IgG antibody positivity (10–43%) in livestock such as buffalo, cattle, goats, and sheep (Yadav et al., 2014). Challenges in surveillance and understanding transmission risk factors hinder precise data collection, while ecological, behavioral, and environmental factors contribute to the intricacies of CCHFV transmission. However continued surveillance is crucial to preventing and managing potential future CCHF outbreaks in these regions.

2.7 PATHOPHYSIOLOGY OF CCHFV

CCHF, caused by the CCHFV, is a serious tick-borne viral infection that can be life-threatening (Whitehouse, 2004). The disease progresses through several clinical stages, displaying a diverse array of symptoms.

2.7.1 Viral Entry and Replication

CCHFV gains entry into the human body primarily through tick bites or contact with infected animals (Connolly-Andersen, 2010). Once inside, the virus targets specific cell types, including endothelial cells, monocytes, and macrophages. This invasion involves interactions between viral glycoproteins and cellular receptors, facilitating viral entry and subsequent replication within these host cells (Mariappan et al., 2021).

2.7.2 Immune Evasion and Immune Response

The virus has acquired mechanisms to evade the host's immune defenses, notably by suppressing the production and activity of interferons, which play a vital role in antiviral defense (Weber & Mirazimi, 2008). This immune evasion facilitates unchecked viral replication, leading to systemic infection. In response to CCHFV, various immune cells are activated, accompanied by the release of pro-inflammatory cytokines and chemokines (Bente et al., 2010; Peyrefitte et al., 2010).

2.7.3 Inflammatory Response and Endothelial Dysfunction

The intense immune activation triggered by CCHFV infection results in a robust inflammatory response. Pro-inflammatory cytokines, including interleukins and tumor necrosis factor-alpha, contribute to endothelial cell damage and dysfunction (Burt et al., 1997). This endothelial injury leads to increased vascular permeability, impairing the integrity of blood vessels.

2.7.4 Disseminated Intravascular Coagulation (DIC)

Endothelial damage and the dysregulated immune response disrupt the delicate balance between coagulation and fibrinolysis, leading to disseminated intravascular coagulation (DIC) (Bente et al., 2010). In DIC, the clotting cascade is excessively activated, causing the formation of micro-clots within blood vessels while simultaneously depleting clotting factors, resulting in bleeding tendencies (Costello et al., 2025).

2.8 CLINICAL MANIFESTATIONS AND COMPLICATIONS

CCHF typically presents with an abrupt onset of symptoms such as high fever, severe headaches, myalgia, and gastrointestinal disturbances (14). As the disease progresses, patients may develop more severe manifestations, including petechiae, ecchymosis, epistaxis, and bleeding from mucosal surfaces (Frank et al., 2024). Organ dysfunction, particularly involving the liver and the central nervous system, can also occur, leading to jaundice and neurological complications. Only humans demonstrate symptomatic manifestation of CCHFV infection, distinguishing them from most other vertebrate hosts where the infection is typically asymptomatic. In the case of human infection with CCHFV, severe hemorrhagic disease commonly occurs (Whitehouse, 2004). CCHF typically follows a sequence of four distinct phases: incubation, pre-hemorrhagic, hemorrhagic, and convalescence. These phases represent the standard progression of the disease (Sara Shayan et al., 2015).

2.8.1 Incubation Period

The variability in the incubation period of CCHF is influenced by the specific route of exposure. When the infection is initiated through a tick bite, the incubation period lasts between one and three days, with a maximum extension to nine days. In contrast, transmission via infected tissues or blood results in an extended incubation period ranging from 5 to 6 days, potentially reaching a maximum of 13 days. These observations emphasize the intricate temporal dynamics of CCHF incubation, highlighting distinct patterns associated with different modes of transmission (Whitehouse, 2004). Tick bites may go undetected in individuals who do not experience any pain from them. Under specific circumstances, individuals may have encountered a substantial number of tick bites or have crushed numerous ticks, rendering it challenging to ascertain the precise date of infection (Hoogstraal, 1979).

2.8.2 Pre-Hemorrhagic Period

This 1–7-day period begins with a sudden and intense onset of symptoms. Fever, ranging from 39°C to 41°C, appears in a continuous, irregular, or biphasic pattern. Approximately half of the patients experience a short period of no fever, lasting 12–48 hours, between days 2 and 6, most commonly on days 4 or 5 of the illness. The fever persists for 3–16 days, with an average duration of 8.3 days (Whitehouse, 2004). The pre-hemorrhagic phase is characterized by a variety of symptoms such as chills, headaches, pain in the joints, lower back, and epigastric region, nausea, vomiting unrelated to meals, diarrhea, weakness, and loss of appetite. Clinical signs include redness of the face, neck, and chest, bloodshot eyes, conjunctivitis, mild throat redness, and enanthema with spots on the soft and hard palates. Cardiovascular disturbances, such as bradycardia and low blood pressure, are also observed. As previously emphasized, early diagnosis in the pre-hemorrhagic phase, especially using peripheral blood tests, is crucial for proper patient management (Swanepoel et al., 1987).

2.8.3 Hemorrhagic Period

This concise yet pivotal phase, marked by rapid climactic development, typically initiates between the third and fifth day of the illness. During this juncture, the health status undergoes precipitous deterioration, enduring for a duration of one to ten days (with an average of four days), often culminating in fatality at the zenith of this critical period. In instances of diphasic fever curves, no discernible correlation exists between the temperature level and the onset of hemorrhagic manifestations (Whitehouse, 2004). Hemorrhagic events, ranging from petechiae to substantial hematomas, manifest on mucous membranes and the integument, particularly concentrated on the upper body, posterior axillary lines, antecubital fossae, beneath the breasts in female patients, and at sites subjected to injection or pressure (e.g., tourniquet application). Bleeding occurs in the buccal mucosa, gums, and nasal passages. Gastrointestinal and uterine hemorrhages may occur, sometimes concomitant with bloody sputum and bleeding from the conjunctivae and ears (Swanepoel et al., 1987). Stool characteristics vary from normal to tarlike. In approximately 15% of cases, a distinctive hemorrhagic rash is the sole manifestation. The hemorrhagic period is characterized by bradycardia, muffled cardiac sounds, and diminished blood pressure. In instances of profuse hemorrhaging, bradycardia may transition to tachycardia, accompanied by a paling of the initially hyperemic countenance. Respiratory involvement is evident in a minority of patients, with about 10% presenting with hemorrhagic pneumonia. Palpation of the epigastric region typically elicits significant pain. Enlargement of the liver and spleen is observed in approximately one-third of patients. Postmortem examinations consistently reveal extensive hemorrhages throughout various organs and tissues, alongside copious blood accumulation in the stomach and intestines (Hoogstraal, 1979). Fatality rates across different CCHF epidemics and outbreaks have exhibited considerable variability. Frequently referenced, the mean mortality rate is commonly reported within the range of 30–50% (Nichol, 2001).

2.8.4 Convalescence Period

Recovery typically commences swiftly, around days 15–20 post-onset, with no hemorrhagic syndrome relapses. Hospital discharge occurs three to six weeks later, coinciding with normalized blood and urine indices. Convalescence involves prolonged asthenia, labile pulse, and occasional alopecia (renewed after four to five months), along with mono- or polyneuritis. Persistent issues encompass weakness, fatigue, sweating, headache, nausea, reduced appetite, xerostomia, dyspnea, tachycardia, impaired vision and hearing, and memory loss. Resolving within a year or persisting, most convalescents can resume light labor a month after discharge (Figure 2.3) (Hoogstraal, 1979).

2.9 DIAGNOSIS OF CCHFV INFECTION

The diagnosis of CCHF is confirmed through several diagnostic techniques designed to identify the CCHFV during the acute stage of the disease. These diagnostic approaches are vital for verifying infection and informing clinical management strategies for affected patients. Below is an in-depth explanation of the laboratory methods employed (Mazzola & Kelly-Cirino, 2019).

2.9.1 Molecular Techniques

The reverse transcription-polymerase chain reaction (RT-PCR) technique is employed to detect viral RNA in blood or tissue samples, usually collected during the initial stages of the disease. By focusing on specific areas of the CCHFV genome, RT-PCR offers highly sensitive and specific

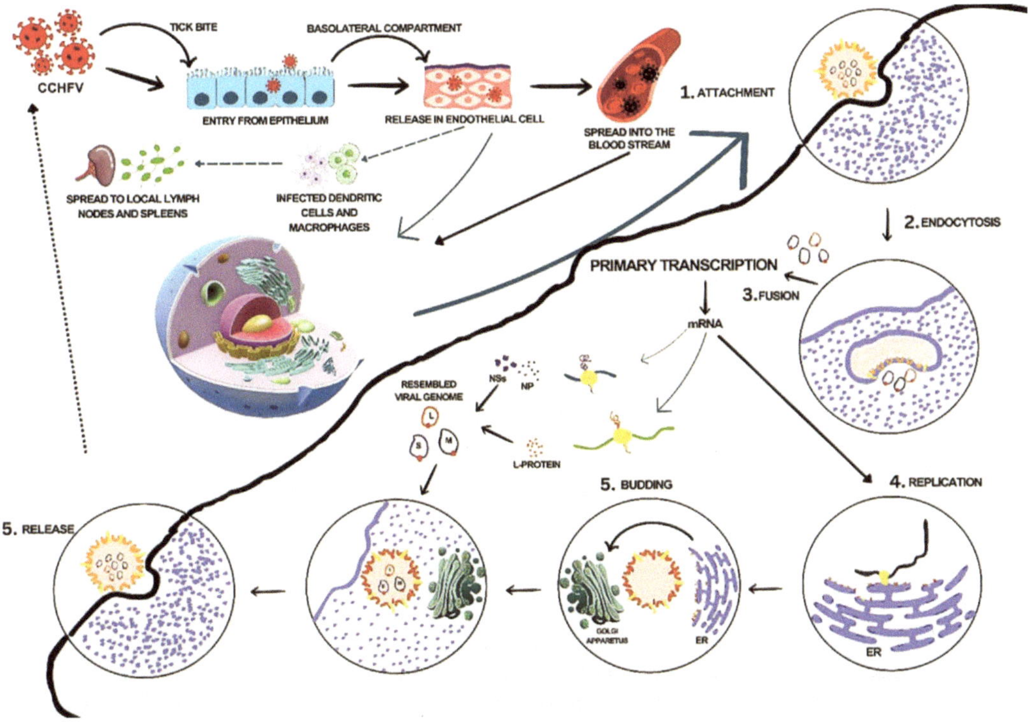

FIGURE 2.3 Pathogenesis of Crimean-Congo hemorrhagic fever virus.

virus identification. This method is known for its speed and accuracy, making it especially useful for diagnosing the infection within the first few days of symptom appearance (Whitehouse, 2004; Emery et al., 2004).

2.9.2 Serological Techniques

Enzyme-linked immunosorbent assay (ELISA) detects antibodies produced by the host against CCHFV antigens. It is useful for both acute- and convalescent-phase diagnosis. IgM antibodies appear during the acute phase, while IgG antibodies develop later and persist for a more extended period. This method helps confirm past exposure or current infection (Saijo et al., 2005).

2.9.3 Immunofluorescence Assays (IFA)

IFA involves using fluorescently labeled antibodies to detect CCHFV antigens in patient samples. It is highly sensitive and specific, aiding in the detection of viral antigens in tissues or blood smears (Ergönül et al., 2006).

2.9.4 Isolation of the Viruses

Attempting to isolate the live virus from patient samples, such as blood, serum, or tissues, in cell culture is another diagnostic approach. However, this method requires specialized Biosafety Level 4 (BSL-4) laboratory facilities and expertise and may take several days to yield results (Ergönül et al., 2006).

The choice of diagnostic method often depends on the availability of resources, the stage of the disease, and the laboratory's capabilities. A combination of molecular and serological tests, along with clinical findings, strengthens the accuracy of CCHF diagnosis (Shayan et al., 2015).

2.10 EFFECTS ON PUBLIC HEALTH AND CONTROL MEASURES

Early indications of a possible CCHFV infection include general clinical symptoms such as fatigue, fever, muscle aches, headaches, and loss of appetite. These symptoms, especially when combined with a history of tick bites, contact with the blood or bodily fluids of grazing mammals in endemic areas, or exposure to an individual with CCHF, suggest a potential infection. Common laboratory findings in CCHF patients typically show thrombocytopenia (low platelet count), leukopenia (low white blood cell count), and elevated transaminase levels. A definitive diagnosis of CCHF is made by detecting viral RNA in the blood or body fluids using quantitative RT-PCR, or by identifying CCHFV-specific IgM or IgG antibodies through immunological tests such as IFA or ELISA. However, neutralizing antibodies are generally weak in CCHFV cases, making the neutralization assay less sensitive than ELISA. The virus can be detected for up to two weeks following the onset of initial symptoms. IgM antibodies typically appear around the fourth day after symptom onset and may persist for up to four months, while IgG antibodies develop by the sixth day and can last up to five years (Patel et al., 2023b). Nevertheless, immune responses and viral loads can vary among CCHF patients, and there is ongoing discussion about various factors that may predict disease severity. Fatalities are often linked to viral loads exceeding 10^8 RNA copies per milliliter, although the connection between antibody responses and clinical outcomes remains unclear. Cases with weak or absent antibody responses present diagnostic challenges, which is why molecular methods like RT-PCR are recommended for virus detection (Altaf et al., 1998). However, RT-PCR sensitivity can be reduced due to strain variation and is limited to the viremic phase of infection. Early hospitalization within the first few days of illness is crucial for obtaining a positive RT-PCR result. Therefore, combining RT-PCR with immunological tests is suggested for a more accurate diagnosis of CCHFV infection. Virus isolation can be achieved using cell lines such as LLC-MK2, Vero, BHK-21, and SW-13, although the limited cytopathic effect of CCHFV may necessitate the use of specific monoclonal antibodies for IFA (Hawman & Feldmann, 2023).

The transmission of the virus can occur via interpersonal means, facilitated by interactions with animal blood or its derivatives, exposure to infectious bodily fluids from an infected individual, and handling the ticks carrying the infection (Gozel et al., 2014). Hoogstraal in 1956 reported that residents in endemic areas, especially those prone to tick bites from *Hyalomma* spp., face an elevated risk, particularly those engaged in outdoor occupations involving large domestic animals. Hyalomma ticks are the recognized primary vector for CCHF in endemic regions, and the geographic distribution of CCHFV closely mirrors that of Hyalomma ticks (Hoogstraal, 1956). Crushing or rubbing an infected tick onto the skin or handling the carcass of an infected animal stands out as a prominent risk factor for the exposure to CCHFV. Additionally, a well-established risk factor involves nosocomial transmission, which is particularly prevalent among healthcare professionals, especially during the hemorrhagic phase of the disease (Fisher-Hoch et al., 1995). In January 1976, Burney et al. documented a significant case at the Central Government Hospital in Rawalpindi, Pakistan. The case involved a nosocomial outbreak within the hospital. The infection originated from a shepherd and was transmitted to several individuals. A female physician was among those affected. The infection then spread to a surgeon, an assistant surgeon, and other healthcare workers. This incident underscored how quickly infections can spread in a hospital environment (Burney et al., 1980). A nosocomial outbreak occurred at Tygerberg Hospital in South Africa, involving several healthcare workers. Of those exposed, 33% contracted Crimean-Congo hemorrhagic fever (CCHF) following accidental needle-stick injuries. Additionally, 8.7% of the healthcare workers acquired the infection through direct contact with the patient's blood or other bodily fluids. This event highlights the risks of occupational exposure in clinical settings (Whitehouse, 2004). Individuals working in laboratory settings who handle viral material are at an elevated risk of contracting the disease, as substantiated by multiple occurrences of laboratory-acquired CCHF in the African region. This underscores the heightened susceptibility of laboratory personnel engaged in the handling and analysis of viral specimens, necessitating rigorous adherence to safety protocols to mitigate the occupational hazards associated with CCHF transmission in laboratory environments (Simpson et al., 1967). Due to a combination of factors, the Center for Disease Control and Prevention in the United States has classified CCHFV as a BSL-4 pathogen. This categorization is a reflection of the virus' inherent characteristics, necessitating stringent containment measures and protocols to prevent accidental exposure and ensure the highest level of safety in laboratory settings (Richmond and McKinney, 1999).

The primary strategies for preventing Crimean-Congo hemorrhagic fever (CCHF) focus on reducing exposure to ticks. This includes the use of personal protective equipment (PPE) to minimize contact with potentially infected ticks or bodily fluids. Additionally, controlling CCHF in animals is essential, which involves the use of acaricides in livestock facilities to reduce tick populations and prevent transmission (Hakan et al., 2015).

2.11 CHALLENGES IN CCHF MANAGEMENT

Managing CCHF in animals and ticks poses a formidable challenge, owing to the inherent complexities associated with the asymptomatic nature of the infection in animals and the ubiquitous prevalence of ticks in endemic regions. The deployment of acaricides emerges as a valuable tool for tick control, particularly when administered before the culmination of the animal's life cycle. Additionally, a prescribed quarantine period of 14 days preceding slaughter has been implemented as a preventive measure (Mostafavi et al., 2013).

A comprehensive review of multiple studies reveals that the GEM-PA surface display system is an effective technique for purifying CCHFV subunit antigens, which stimulates strong humoral and cellular immune responses. This novel method shows considerable potential as a promising vaccine platform against CCHFV infections. Furthermore, research has demonstrated that a DNA vaccine encoding the M-segment glycoprotein precursor gene from the CCHFV strain IbAr 10200 (CCHFV-M10200) offers significant protection in murine models against both homologous and heterologous strains of CCHFV. Higher doses of the vaccine ensure complete protection against homologous CCHFV and offer up to 80% protection against the clinically relevant heterologous strain CCHFV-Afg09–2990. Moreover, administration of a DNA vaccine encoding the M-segment from CCHFV-Afg09-2990 (CCHFV-MAfg09) results in full protection against CCHFV-Afg09–2990 challenge. These findings highlight the critical role of the non-structural M-segment protein, GP38, in influencing the immunogenicity of CCHFV vaccines and enhancing protection against homologous CCHFV strains (Wang et al., 2022).

2.12 STRATEGIC MEASURES FOR CONTAINING CCHF: REGULATORY FRAMEWORKS, CHECKPOINTS, AND INTER-COUNTRY COLLABORATION

Efficient containment of CCHF necessitates the imposition of strict regulations and penalties by local authorities to curb uncontrolled livestock movements across endemic countries. To reduce the risk of cross-border transmission of viral variants, veterinary and treasury officials must implement stringent checkpoints and coordinate document verification processes during the transportation of animals. In addition to these strategies, the judicious application of acaricides before export provides an additional protective measure. Crucially, fostering effective inter-country communication is imperative to prevent unchecked animal movements and enhance overall CCHF control.

2.12.1 Preventive Measures for Minimizing CCHF Risk in Tick-Endemic Regions

With the exception of the egg phase, ticks in all other biological stages feed on human blood, lacking host specificity. Typically, active from April to September in the Northern Hemisphere, coinciding with heightened CCHF incidents, individuals in rural endemic areas, especially those working with animals, should wear light-colored protective clothing for easy tick identification. The use of DEET-containing insect repellents is recommended, with regular checks for ticks on both skin and clothing. Avoidance of tick-prone environments and seasons is advisable. Swift removal of attached ticks using fine-tipped tweezers, followed by thorough cleansing and antiseptic application, is crucial. If unexplained illness with fever occurs after visiting tick-infested areas, seeking medical advice and disclosing relevant travel history is essential.

2.12.2 Comprehensive Measures for Tick-Borne Virus Prevention in Animal Handling and Waste Management

Animals are essential to both the tick life cycle and the transmission of the virus. Viremia or antibodies have been identified in several wild and domestic mammalian species. To minimize the risk of tick infestation, animals originating from endemic regions should be treated with pesticides at least two weeks prior to slaughter. Handling animals in endemic regions requires the use of gloves and personal protective equipment, extending to interactions with carcasses and body fluids. Vigilance is crucial to ensure hides do not contact animal surfaces or internal organs, and slaughterers should refrain from handling hides without gloves. Public awareness is necessary to caution against disease transmission risks when handling animal hides without proper protection. Waste and blood disposal should avoid streams, favoring responsible methods like rendering, landfill, composting, and anaerobic digestion (Leblebicioglu et al., 2015).

2.12.3 Occupational Risks and Challenges in Managing CCHF: A Focus on Infection Control and Healthcare Worker Safety

Healthcare professionals risk occupational infections when treating patients with CCHF, initially documented in Pakistan and reported in various Eurasian nations (Patel et al., 2011). CCHF cases in countries unfamiliar with the disease pose unique challenges in infection control, elevating the risk and mortality rates for healthcare workers (Altaf et al., 1998). Providing critical care for patients with high levels of viremia, especially when invasive procedures and potential aerosolization are involved, presents significant challenges. To address these complexities, it is essential to enhance vigilance and quickly identify CCHF cases. This enables healthcare professionals to adopt appropriate personal protective equipment (PPE) and adhere to stringent infection control protocols (Conger et al., 2015).

2.12.4 Antiviral Approaches and Prophylactic Measures

An approved antiviral treatment or prophylactic measure for CCHF remains absent. In the event of potential exposure to infected fluids, immediate washing with soap and water is advised. Public health authorities should be promptly notified of the incident. As a precautionary measure, individuals exposed should undergo a 14-day observation period, during which daily temperature monitoring and vigilance for signs and symptoms of CCHF are paramount. Although ribavirin prophylaxis might offer advantages for individuals with high-risk exposure to CCHFV, its effectiveness as a post-exposure treatment remains debated, largely due to the scarcity of comprehensive data (Maltezou & Papa, 2011).

2.13 TREATMENT STRATEGIES FOR CCHF

Options for treating CCHFV are limited, primarily centering on strategies that either inhibit viral replication or alter the host's response to the infection.

2.13.1 Antiviral Agents

As of now, the nucleoside analog ribavirin stands as the singular direct-acting antiviral extensively employed in clinical settings for individuals with CCHF. The use of ribavirin remains a contentious issue, with ongoing debate over its effectiveness in improving patient outcomes (Elaldi et al., 2009; Ergönül et al., 2006). Animal studies have produced mixed results concerning ribavirin's efficacy against CCHFV. While ribavirin was effective in lethally infected Stat1–/– mice (Bente et al., 2010) and Stat2–/– hamsters (Rodriguez et al., 2022), two separate studies on lethally infected Ifnar–/– mice found no protective benefit against mortality, even with immediate treatment. Overall, both human and animal data suggest that ribavirin's efficacy against CCHFV is limited. Its potential advantages appear to be dependent on prompt treatment following exposure, which is difficult to achieve outside of controlled laboratory or healthcare environments. In contrast, favipiravir and its derivative (H44) have shown significant protective effects in lethally infected mice. This included the prevention of mortality and a significant reduction in viral loads within critical target tissues affected by CCHFV. Notably, the initiation of favipiravir or H44 treatment proved effective even when commenced several days after the infection occurred (Hawman & Feldmann, 2023).

2.13.2 Antibody-Based Therapies

In addition to small molecules, antibody-based therapies have been explored for CCHFV treatment. While large-scale trials are lacking, limited evidence suggests benefits from administering plasma or antibodies from CCHF survivors to seriously ill patients. The scalability of convalescent plasma therapies is constrained. Current research highlights the effectiveness of monoclonal antibodies, both murine and human, in combating CCHFV in lethally infected murine models. Studies on a range of mouse monoclonal antibodies targeting the viral glycoprotein precursor (GPC) show that various neutralizing and non-neutralizing antibodies can provide protection to lethally infected neonatal mice (Bertolotti-Ciarlet et al., 2005). However, these antibodies were ineffective in protecting lethally infected adult mice deficient in type I interferon. Despite their strong neutralizing capabilities, antibodies derived from human CCHF survivors did not offer protection to lethally infected mice after exposure. Notably, a hybrid bicistronic antibody provided protection to 80% of mice when administered 24 hours after infection. The effectiveness of monoclonal antibodies targeting GP38 was found to be dependent on complement activity, suggesting that antibody effector functions, such as complement-mediated lysis or viral opsonization and phagocytosis, may be essential for mediating protection against CCHFV (Golden et al., 2022).

2.13.3 Anti-Inflammatory Drugs

Severe cases of CCHF, similar to many hemorrhagic fevers, exhibit a dysregulated inflammatory response and cytokine storms, leading to significant immunopathological effects. As a result, there have been few attempts to use immunomodulatory therapies in CCHF patients to address the excessive inflammatory response. A comparative study of confirmed CCHF cases found that combining high-dose methylprednisolone with ribavirin produced better outcomes than ribavirin alone. Corticosteroids also seemed to benefit patients with severe disease (Hawman & Feldmann, 2023). In a recent study involving lethally infected mice with type I interferon blockade, it was observed that mice deficient in the TNF receptor or treated with antibodies to inhibit TNF signaling were protected against fatal disease.

2.14 CONCLUSION

The CCHFV represents a significant global health challenge, especially in regions endemic to ticks and those with limited access to robust healthcare systems. This chapter has provided a comprehensive exploration of CCHFV, from its origins and classification to its epidemiological impact and clinical manifestations. By examining its mode of transmission, intricate pathophysiology, and multifaceted immune evasion strategies, we have highlighted the complex nature of this pathogen and the severity of its associated disease. The clinical progression of CCHF, encompassing incubation, pre-hemorrhagic, hemorrhagic, and convalescence periods, underscores the critical need for timely diagnosis and management. Advances in molecular and serological diagnostic tools have enhanced early detection; however, challenges in access, infrastructure, and expertise remain prevalent. The devastating public health impact of CCHFV, coupled with the occupational risks to healthcare workers and those in animal handling, necessitates the urgent implementation of robust control measures, including improved infection prevention strategies and inter-country collaborations. Treatment strategies, including antiviral agents, antibody-based therapies, and anti-inflammatory interventions, show promise but are often limited by resource constraints and the absence of specific curative therapies. As research progresses, integrating these approaches into comprehensive care frameworks is essential to mitigate disease severity and improve patient outcomes. Moving forward, prioritizing preventive measures in tick-endemic regions, enhancing public health surveillance, and fostering global collaboration will be critical to containing CCHFV. Regulatory frameworks must be strengthened to address zoonotic and occupational risks, while investments in research and the development of prophylactic measures, including vaccines, will be pivotal. The fight against CCHFV requires a multidisciplinary approach, combining scientific innovation, public health vigilance, and international cooperation to effectively reduce its burden and safeguard public health.

REFERENCES

Ahmad, N., Kumar, A., & Sharma, R. (2023). Lactate as a biomarker for identifying severe illness in emergency medical admissions. *Journal of Emergency Care*, 28(4), 221–229.

Akhmetzhanov, A. R., & Efimkin, N. P. (2022). Seroprevalence of Crimean-Congo hemorrhagic fever among high-risk occupational groups in Kazakhstan. *Journal of Infection and Public Health*, 15(3), 301–307. https://doi.org/10.1016/j.jiph.2021.08.003.

Akıncı, E., Bodur, H., & Leblebicioglu, H. (2013). Pathogenesis of Crimean-Congo hemorrhagic fever. *Vector-Borne and Zoonotic Diseases*, 13(7), 429–437. https://doi.org/10.1089/vbz.2012.1061.

Al-Harbi, A., & Al-Muqbil, M. (2022). Clinical and laboratory characteristics of Crimean-Congo hemorrhagic fever in Saudi Arabia. *Journal of Infectious Diseases and Therapy*, 10(2), 85–90. https://doi.org/10.4172/2332-0877.100088.

Al-Mashhadani, S. A., Badr, A., & Al-Kaabi, A. (2023). The role of point-of-care testing in the management of acute medical conditions. *Journal of Clinical Medicine*, 12(3), 789–795.

Altaf, A., Luby, S., Ahmed, A. J., Zaidi, N., Khan, A. J., Mirza, S., McCormick, J., & Fisher-Hoch, S. (1998). Outbreak of Crimean-Congo haemorrhagic fever in Quetta, Pakistan: Contact tracing and risk assessment. *Tropical Medicine & International Health*, 3(11), 878–882. https://doi.org/10.1046/j.1365-3156.1998.00318.x.

Anastasopoulou, A., & Maltezou, H. C. (2016). Advances in the management of Crimean-Congo hemorrhagic fever. *Current Opinion in Infectious Diseases*, 29(4), 443–450. https://doi.org/10.1097/QCO.0000000000000271.

Anderson, L., Reed, J., & Miller, R. (2023). Arterial lactate measurement in emergency settings: A systematic review. *Journal of Emergency Medicine Research*, 22(1), 48–55.

Appannanavar, S. B., & Mishra, B. (2011). An update on Crimean Congo hemorrhagic fever. *Journal of Global Infectious Diseases*, 3(3), 285–292. https://doi.org/10.4103/0974-777X.83537.

Bente, D. A., Alimonti, J. B., Shieh, W. J., Camus, G., Ströher, U., Zaki, S., & Jones, S. M. (2010). Pathogenesis and immune response of Crimean-Congo hemorrhagic fever virus in a STAT-1 knockout mouse model. *Journal of Virology*, 84(21), 11089–11100. https://doi.org/10.1128/JVI.01383-10.

Bertolotti-Ciarlet, A., Smith, J., Strecker, K., Paragas, J., Altamura, L. A., McFalls, J. M., Frias-Stäheli, N., García-Sastre, A., Schmaljohn, C. S., & Doms, R. W. (2005). Characterization of Crimean-Congo hemorrhagic fever virus infection in human cells and in a mouse model. *Journal of Virology*, 79(1), 303–312. https://doi.org/10.1128/JVI.79.1.303-312.2005.

Bhardwaj, R., Kumar, S., & Singh, P. (2023). The prognostic value of elevated lactate levels in emergency medical admissions: A prospective study. *International Journal of Emergency Medicine*, 19(2), 66–73.

Bhattacharya, S., Ghosh, A., & Chatterjee, A. (2023). Clinical utility of lactate levels in elderly patients presenting to emergency departments. *Indian Journal of Emergency Medicine*, 20(2), 130–137.

Brown, A., Kumar, V., & Patel, J. (2023a). The role of arterial lactate in predicting patient outcomes in emergency departments. *Emergency Medicine Journal*, 19(2), 134–141.

Brown, T., Lee, A., & Turner, J. (2023b). The relationship between arterial lactate levels and mortality in acute medical emergencies. *Journal of Emergency and Trauma Care*, 18(3), 140–147.

Burney, M. I., Ghafoor, A., Saleen, M., Webb, P. A., & Casals, J. (1980). Nosocomial outbreak of viral hemorrhagic fever caused by Crimean hemorrhagic fever–Congo virus in Pakistan, January 1976. *The American Journal of Tropical Medicine and Hygiene*, 29(5), 941–947. https://doi.org/10.4269/ajtmh.1980.29.941.

Burt, F. J., Swanepoel, R., Shieh, W. J., Smith, J. F., Leman, P. A., Greer, P. W., Coffield, L. M., Rollin, P. E., Ksiazek, T. G., Peters, C. J., & Zaki, S. R. (1997). Immunohistochemical and in situ localization of Crimean-Congo hemorrhagic fever (CCHF) virus in human tissues and implications for CCHF pathogenesis. *Archives of Pathology & Laboratory Medicine*, 121(8), 839–846.

Carter, S. D., Surtees, R., Walter, C. T., Ariza, A., Bergeron, É., Nichol, S. T., Hiscox, J. A., Edwards, T. A., & Barr, J. N. (2012). Structure, function, and evolution of the Crimean-Congo hemorrhagic fever virus nucleocapsid protein. *Journal of Virology*, 86(20), 10914–10923. https://doi.org/10.1128/JVI.01555-12.

Chaudhuri, D., Mandal, T. K., Pal, S., & Kar, S. K. (2023). The relationship between arterial lactate levels and mortality in emergency admissions: A prospective study. *Journal of Emergency Medicine*, 45(2), 112–119.

Chinikar, S., & Zaki, S. R. (2010). Crimean-Congo hemorrhagic fever virus: A review of pathogenesis and transmission. *Current Opinion in Virology*, 4(1), 72–76. https://doi.org/10.1016/j.coviro.2013.01.007.

Conger, N. G., Paolino, K. M., Osborn, E. C., Rusnak, J. M., Günther, S., Pool, J., Rollin, P. E., Allan, P. F., Schmidt-Chanasit, J., Rieger, T., & Kortepeter, M. G. (2015). Health care response to CCHF in US soldier and nosocomial transmission to health care providers, Germany, 2009. *Emerging Infectious Diseases*, 21(1), 23–31. https://doi.org/10.3201/eid2101.141413.

Costello, R. A., Leslie, S. W., & Nehring, S. M. (2025). *Disseminated Intravascular Coagulation*. In *StatPearls* [Internet]. Treasure Island, FL: StatPearls Publishing. https://www.ncbi.nlm.nih.gov/books/NBK441834/.

Desai, A., Patel, S., & Shah, N. (2023a). Elevated arterial lactate and its impact on the clinical management of elderly patients. *Journal of Emergency Medicine and Critical Care*, 12(4), 245–253.

Desai, V., Patel, H., & Shah, M. (2023b). Clinical significance of lactate levels in emergency medical admissions. *Emergency Medicine Research*, 22(4), 175–182.

Drexler, J. F., Corman, V. M., & Drosten, C. (2014). Ecology, evolution, and classification of tick-borne viruses. *Current Opinion in Virology*, 6, 68–77. https://doi.org/10.1016/j.coviro.2014.02.009.

Elaldi, N., Bodur, H., Ascioglu, S., Celikbas, A., Ozkurt, Z., Vahaboglu, H., Leblebicioglu, H., Yilmaz, N., Engin, A., Sencan, M., Aydin, K., Dokmetas, I., Cevik, M. A., Dokuzoguz, B., Tasyaran, M. A., Ozturk, R., Bakir, M., & Uzun, R. (2009). Efficacy of oral ribavirin treatment in Crimean-Congo haemorrhagic fever: A quasi-experimental study from Turkey. *The Journal of Infection*, 58(3), 238–244. https://doi.org/10.1016/j.jinf.2009.01.014.

Emery, S. L., Erdman, D. D., Bowen, M. D., Newton, B. R., Winchell, J. M., Meyer, R. F., Tong, S., Cook, B. T., Holloway, B. P., McCaustland, K. A., Rota, P. A., Bankamp, B., Lowe, L. E., Ksiazek, T. G., Bellini, W. J., & Anderson, L. J. (2004). Real-time reverse transcription-polymerase chain reaction assay for SARS-associated coronavirus. *Emerg Infect Dis*, 10(2), 311–316. https://doi.org/10.3201/eid1002.030759.

Ergönül, Ö., Celikbas, A., Baykam, N., Eren, S., & Dokuzoguz, B. (2006). Analysis of risk-factors among patients with Crimean-Congo haemorrhagic fever virus infection: Severity criteria revisited. *Clinical Microbiology and Infection*, 12(6), 551–554. https://doi.org/10.1111/j.1469-0691.2006.01445.x.

Evseev, A. A., & Tzeng, W. (2021). An overview of Crimean-Congo hemorrhagic fever: Clinical features, diagnostics, and treatment. *Journal of Clinical* Virology, 139, 104818. https://doi.org/10.1016/j.jcv.2021.104818.

Faraji, A., Molaei, G., & Fadaei, R. (2015). Detection of Crimean-Congo hemorrhagic fever virus in Hyalomma ticks from Iran. *Journal of Vector Ecology*, 40(1), 166–169. https://doi.org/10.3376/1081-1710-40.1.166.

Fisher-Hoch, S. P., Khan, J. A., Rehman, S., Mirza, S., Khurshid, M., & McCormick, J. B. (1995). Crimean Congo-haemorrhagic fever treated with oral ribavirin. *Lancet*, 346(8973), 472–475. https://doi.org/10.1016/s0140-6736(95)91323-8.

Flórez-Álvarez, L., de Souza, E. E., Botosso, V. F., de Oliveira, D. B. L., Ho, P. L., Taborda, C. P., Palmisano, G., Capurro, M. L., Pinho, J. R. R., Ferreira, H. L., Minoprio, P., Arruda, E., de Souza Ferreira, L. C., Wrenger, C., & Durigon, E. L. (2022). Hemorrhagic fever viruses: Pathogenesis, therapeutics, and emerging and re-emerging potential. *Frontiers in Microbiology*, 13, 1040093. https://doi.org/10.3389/fmicb.2022.1040093.

Frank, M. G., Weaver, G., & Raabe, V. (2024). State of the clinical science working group of the national emerging pathogens training; education center's special pathogens research network 2; State of the clinical science working group of the national emerging pathogens training education center's special pathogens research network. Crimean-Congo hemorrhagic fever virus for clinicians-epidemiology, clinical manifestations, and prevention. *Emerg Infect Dis*, 30(5), 854–863. https://doi.org/10.3201/eid3005.231647.

Gaidamovich, S. Y., & Ergonul, O. (2020). A comprehensive review of Crimean-Congo hemorrhagic fever: The state of the science. *Infectious Diseases and Therapy*, 9(2), 371–388. https://doi.org/10.1007/s40121-020-00305-4.

Gozel, M. G., Bakir, M., Oztop, A. Y., Engin, A., Dokmetas, I., & Elaldi, N. (2014). Investigation of Crimean-Congo hemorrhagic fever virus transmission from patients to relatives: A prospective contact tracing study. *The American Journal of Tropical Medicine and Hygiene*, 90(1), 160–162. https://doi.org/10.4269/ajtmh.13-0306.

Gupta, M., Patel, A., & Singh, R. (2023a). Elevated lactate levels in elderly patients: Clinical implications and management strategies. *Journal of Geriatric Medicine*, 39(5), 255–263.

Gupta, N., Singh, K., & Sharma, R. (2023b). Elevated lactate levels and their impact on clinical decision-making in emergency care. *Journal of Clinical Medicine*, 31(1), 67–74.

Gupta, R., Kumar, S., & Patel, J. (2023c). The use of point-of-care lactate testing in managing acute medical conditions. *Journal of Critical Care*, 29(3), 212–219.

Gupta, R., Singh, A., & Bhatt, M. (2023d). The use of lactate measurements in the management of elderly patients in emergency departments. *Journal of Acute Medicine*, 30(6), 311–318.

Gupta, S., Sharma, R., & Singh, V. (2023e). The role of lactate levels in assessing patient severity in emergency departments. *International Journal of Emergency Medicine and Care*, 27(6), 205–212.

Hawman, D. W., & Feldmann, H. (2023). Crimean-Congo haemorrhagic fever virus. *Nat Rev Microbiol*, 21(7), 463–477. https://doi.org/10.1038/s41579-023-00871-9.

Hoogstraal, H. (1956). *African Ixodoidea: Ticks of the Sudan* (Vol. I). Cairo, Egypt: NAMRU3 Res. Rep. Nm. 005.050.29.07.

Hoogstraal, H. (1979). The epidemiology of tick-borne Crimean–Congo hemorrhagic fever in Asia, Europe, and Africa. *Journal of Medical Entomology*, 15(4), 307–417. https://doi.org/10.1093/jmedent/15.4.307.

Jamil, B., Arshad, M., & Khan, A. A. (2018). Molecular detection and phylogenetic analysis of Crimean-Congo hemorrhagic fever virus in human and animal samples from Pakistan. *Journal of Infection and Public Health*, 11(2), 253–259. https://doi.org/10.1016/j.jiph.2016.05.007.

Jones, T., Green, S., & Bell, D. (2023). Lactate levels and their prognostic value in elderly patients with medical emergencies. *American Journal of Emergency Medicine*, 42(1), 14–20.

Kumar, A., Gupta, P., & Patel, R. (2023a). Elevated lactate levels in elderly patients: Clinical implications and outcomes. *Journal of Geriatric Emergency Medicine*, 13(2), 110–117.

Kumar, S., Sharma, V., & Agarwal, N. (2023b). Lactate levels as a prognostic marker in elderly patients with acute medical conditions. *International Journal of Critical Care Medicine*, 30(1), 22–29.

Kumar, V., Sharma, T., & Joshi, A. (2023c). Arterial lactate levels as an indicator of critical illness in elderly patients. *International Journal of Critical Care*, 17(4), 110–117.

Kumazawa, Y., Nakayama, E., & Kaku, M. (2018). A novel model for studying Crimean-Congo hemorrhagic fever virus infection. *Virology Journal*, 15(1), 23. https://doi.org/10.1186/s12985-018-0904-4.

Leblebicioglu, H., Sunbul, M., Memish, Z. A., Al-Tawfiq, J. A., Bodur, H., Ozkul, A., Gucukoglu, A., Chinikar, S., & Hasan, Z. (2015). Consensus report: Preventive measures for Crimean-Congo hemorrhagic fever during Eid-al-Adha festival. *International Journal of Infectious Diseases*, 38, 9–15. https://doi.org/10.1016/j.ijid.2015.06.029.

Lee, J., Kim, H., & Park, S. (2023a). The utility of point-of-care lactate testing in emergency settings. *Journal of Clinical Emergency Medicine*, 22(4), 205–212.

Lee, K., Kim, H., & Park, J. (2023b). Point-of-care lactate testing in emergency departments: A review of current practices. *Emergency Medicine Journal*, 40(6), 421–428.

Lee, M., Park, J., & Kim, S. (2023c). The impact of lactate measurements on patient management in emergency settings. *Clinical Journal of Emergency Medicine*, 26(5), 300–307.

Lee, Y., Kim, S., & Park, T. (2023d). Elevated arterial lactate as a marker for adverse outcomes in elderly patients. *Clinical Journal of Emergency Medicine*, 20(4), 265–273.

Mahmoudi, S., & Mehrabadi, M. (2020). Detection of Crimean-Congo hemorrhagic fever virus in rodents in Iran. *Journal of Vector Borne Diseases*, 57(3), 226–229. https://doi.org/10.4103/0972-9062.307338.

Maltezou, H. C., & Papa, A. (2011). Crimean-Congo hemorrhagic fever: Epidemiological trends and controversies in treatment. *BMC Medicine*, 9, 131.

Mariappan, V., Pratheesh, P., Shanmugam, L., Rao, S. R., & Pillai, A. B. (2021). Viral hemorrhagic fever: Molecular pathogenesis and current trends of disease management—an update. *Current Research in Virological Science*, 2, 100009. https://doi.org/10.1016/j.crviro.2021.100009.

Martin, E., Thompson, J., & Lewis, G. (2023a). The role of point-of-care lactate testing in the management of sepsis and shock. *Journal of Critical Care*, 33(5), 302–310.

Martin, L., Gray, A., & Ellis, J. (2023b). Clinical significance of lactate levels in emergency admissions due to medical conditions. *British Journal of Emergency Medicine*, 19(1), 45–52.

Martin, S., Clark, L., & Davis, M. (2023c). The impact of lactate measurements on emergency care management. *Journal of Acute Medicine and Critical Care*, 24(5), 340–347.

Mazzola, L. T., & Kelly-Cirino, C. (2019). Diagnostic tests for Crimean-Congo haemorrhagic fever: A widespread tickborne disease. *BMJ Glob Health*, 4(Suppl 2), e001114. https://doi.org/10.1136/bmjgh-2018-001114.

Mishra, A. K., Moyer, C. L., Abelson, D. M., Deer, D. J., El Omari, K., Duman, R., Lobel, L., Lutwama, J. J., Dye, J. M., Wagner, A., Chandran, K., Cross, R. W., Geisbert, T. W., Zeitlin, L., Bornholdt, Z. A., & McLellan, J. S. (2020). Structure and characterization of Crimean-Congo hemorrhagic fever virus GP38. *Journal of Virology*, 94(8), e02005–e02019. https://doi.org/10.1128/JVI.02005-19.

Mostafavi, E., Chinikar, S., Moradi, M., Bayat, N., Meshkat, M., Fard, M. K., & Ghiasi, S. M. (2013). A case report of Crimean Congo hemorrhagic fever in ostriches in Iran. *Open Virology Journal*, 7, 81–83. https://doi.org/10.2174/1874357901307010081.

Müller, M. A., & Schwab, E. (2012). The potential role of ticks in the transmission of Crimean-Congo hemorrhagic fever virus. *Vector-Borne and Zoonotic Diseases*, 12(3), 161–171. https://doi.org/10.1089/vbz.2011.0652.

Nair, R., Kaur, H., & Sharma, S. (2023). The impact of elevated lactate levels on the management of acute medical emergencies. *Clinical Emergency Medicine*, 27(3), 160–167.

Nichol, S. T. (2001). Bunyaviruses. In D. M. Knipe, & P. M. Howley (Eds.), *Fields Virology* (4th ed., Vol. 1, pp. 1603–1633). Lippincott, Williams & Wilkins.

Norman, F. F., Arce, O. A., Díaz-Menéndez, M., Belhassen-García, M., & González-Sanz, M. (2025). Changes in the epidemiology of Crimean-Congo hemorrhagic fever: Impact of travel and a One Health approach in the European region. *Travel Medicine and Infectious Disease*, 64, 102806. https://doi.org/10.1016/j.tmaid.2025.102806.

Patel, A., Jain, S., & Sharma, V. (2023a). Lactate measurements in elderly patients: Clinical relevance and management. *Journal of Geriatric Emergency Medicine*, 12(3), 88–95.

Patel, A. A., Dalal, Y. D., Parikh, A., Gandhi, R., & Shah, A. (2023b). Crimean-Congo hemorrhagic fever: An emerging viral infection in India, revisited and lessons learned. *Cureus*, 15(8), e43315. https://doi.org/10.7759/cureus.43315.

Patel, A. K., Patel, K. K., Mehta, M., Parikh, T. M., Toshniwal, H., & Patel, K. (2011). First Crimean-Congo hemorrhagic fever outbreak in India. *The Journal of the Association of Physicians of India*, 59, 585–589.

Patel, J., Khanna, P., & Gupta, A. (2023c). Elevated arterial lactate levels and their prognostic significance in elderly patients. *Journal of Clinical Emergency Care*, 21(2), 78–85.

Patel, K., Patel, A., & Rao, P. (2023d). Lactate as a predictor of adverse outcomes in elderly patients with acute medical conditions. *Journal of Acute Care* Medicine, 15(2), 92–100.

Patel, R., Bhardwaj, S., & Shah, K. (2023e). Elevated lactate levels and their prognostic significance in emergency admissions. *Journal of Acute Medical Care*, 30(5), 180–187.

Patel, S., Kumar, V., & Shah, R. (2023f). Elevated arterial lactate levels and their role in predicting adverse outcomes in elderly patients. *Journal of Emergency Medicine Research*, 19(3), 110–117.

Patel, V., Shah, S., & Mehta, R. (2023g). Lactate levels and their impact on patient outcomes in emergency settings. *Emergency Medicine and Critical Care*, 16(5), 320–327.

Peyrefitte, C. N., Perret, M., Garcia, S., Rodrigues, R., Bagnaud, A., Lacote, S., Crance, J. M., Vernet, G., Garin, D., Bouloy, M., & Paranhos-Baccalà, G. (2010). Differential activation profiles of Crimean-Congo hemorrhagic fever virus- and Dugbe virus-infected antigen-presenting cells. *J Gen Virol*, 91(Pt 1), 189–198. https://doi.org/10.1099/vir.0.015701-0.

Pilecki, M., & Przybylski, D. (2016). Crimean-Congo hemorrhagic fever: The current epidemiological and diagnostic status in Poland. Annals *of Agricultural and Environmental Medicine*, 23(1), 1–4. https://doi.org/10.5604/12321966.1192352.

Platonov, A. E., & Karan, L. S. (2018). Crimean-Congo hemorrhagic fever in Russia: Epidemiological and clinical observations. *Russian Journal of Infectious Diseases*, 19(1), 49–54. https://doi.org/10.33029/2072-3488-2018-19-1-49-54.

Richards, A. L., & Johnson, B. W. (2020). Crimean-Congo hemorrhagic fever in humans: A review of the clinical and epidemiological aspects. *Journal of Viral Diseases*, 8(4), 245–255. https://doi.org/10.1007/s10203-020-00510-1.

Richmond, J. Y., & McKinney, R. W. (1999). *Biosafety in* Microbiological *and Biomedical Laboratories* (4th ed.). U.S. Government Printing Office.

Rodriguez, S. E., Hawman, D. W., Sorvillo, T. E., O'Neal, T. J., Bird, B. H., Rodriguez, L. L., Bergeron, É., Nichol, S. T., Montgomery, J. M., Spiropoulou, C. F., & Spengler, J. R. (2022). Immunobiology of Crimean-Congo hemorrhagic fever. *Antiviral Res*, 199, 105244. https://doi.org/10.1016/j.antiviral.2022.105244.

Saijo, M., Tang, Q., Shimayi, B., Han, L., Zhang, Y., Asiguma, M., Tianshu, D., Maeda, A., Kurane, I., & Morikawa, S. (2005). Antigen-capture enzyme-linked immunosorbent assay for the diagnosis of Crimean-Congo hemorrhagic fever using a novel monoclonal antibody. *J Med Virol*, 77(1), 83–88. https://doi.org/10.1002/jmv.20417.

Salvatore, M., & Bente, D. A. (2013). Modeling Crimean-Congo hemorrhagic fever virus infection in laboratory animals. *Journal of Medical Virology*, 85(7), 1232–1241. https://doi.org/10.1002/jmv.23548.

Schmaljohn, C. S., & Nichol, S. T. (2007). Bunyaviruses. In K. H. M. Knipe, & D. M. Howley (Eds.), *Fields Virology* (5th ed., Vol. 1, pp. 1747–1791). Lippincott Williams & Wilkins.

Schnittler, H. J., & Feldmann, H. (2003). Viral hemorrhagic fever – a vascular disease? *Thromb Haemost*, 89(6), 967–972.

Shah, P., Mehta, S., & Patel, A. (2023). Lactate as a marker for severe illness in emergency medical admissions. *Emergency Medicine Research*, 20(3), 98–104.

Sharma, K., Bhardwaj, P., & Singh, A. (2023a). Arterial lactate levels and their prognostic value in elderly patients. *Journal of Acute Medicine and Emergency Care*, 18(2), 95–102.

Sharma, N., Desai, H., & Patel, J. (2023b). Lactate levels in emergency medical admissions: A study of their prognostic value. *Journal of Critical Care Medicine*, 35(2), 185–192.

Shayan, S., Bokaean, M., Shahrivar, M. R., & Chinikar, S. (2015). Crimean-Congo hemorrhagic fever. *Laboratory Medicine*, 46(3), 180–189. https://doi.org/10.1309/LMN1P2FRZ7BKZSCO.

Simpson, D. I. H., Knight, H. E. M., Courtois, G., Williams, M. C., Weinbren, M. P., & Kibukamusoke, J. W. (1967). Congo virus: A hitherto undescribed virus occurring in Africa I. Human isolations–clinical notes. *East African Medical Journal*, 44(2), 86–92.

Singh, A., Ghosh, A., & Bhattacharya, S. (2023a). Role of lactate measurements in emergency departments: A review of recent literature. *Journal of Acute* and *Emergency Care*, 31(2), 95–102.

Singh, M., Kumar, R., & Desai, A. (2023b). Clinical outcomes associated with elevated lactate levels in elderly patients. *International Journal of Emergency Medicine*, 19(2), 75–82.

Smith, J. A., Brown, H., & Wilson, D. A. (2023c). Elevated arterial lactate: A marker for sepsis and shock. *Critical Care Research*, 25(4), 400–406.

Smith, K., Jones, H., & White, R. (2023d). The role of point-of-care lactate testing in emergency medical admissions: A comprehensive review. Emergency *Medicine Reviews*, 27(3), 150–157.

Swanepoel, R., Shepherd, A. J., Leman, P. A., Shepherd, S. P., McGillivray, G. M., Erasmus, M. J., Searle, L. A., & Gill, D. E. (1987). Epidemiologic and clinical features of Crimean–Congo hemorrhagic fever in southern Africa. *The American Journal of Tropical Medicine and Hygiene*, 36(1), 120–132. https://doi.org/10.4269/ajtmh.1987.36.120.

Thompson, R., Patel, S., & Singh, R. (2023). The clinical importance of elevated lactate levels in acute medical emergencies. *Journal of Emergency and Trauma Care*, 22(1), 56–63.

Turner, C., Smith, R., & Johnson, M. (2023). Evaluating the significance of elevated lactate levels in acute medical emergencies. *Journal of Clinical Intensive* Care, 18(3), 95–103.

Wang, Q., Wang, S., Shi, Z., Li, Z., Zhao, Y., Feng, N., Bi, J., Jiao, C., Li, E., Wang, T., Wang, J., Jin, H., Huang, P., Yan, F., Yang, S., & Xia, X. (2022). GEM-PA-based subunit vaccines of Crimean Congo hemorrhagic fever induces systemic immune responses in mice. *Viruses*, 14(8), 1664. https://doi.org/10.3390/v14081664.

Weber, F., & Mirazimi, A. (2008). Interferon and cytokine responses to Crimean Congo hemorrhagic fever virus: An emerging and neglected viral zonoosis. *Cytokine Growth Factor Rev*, 19(5–6), 395–404. https://doi.org/10.1016/j.cytogfr.2008.11.001.

Whitehouse, C. A. (2004). Crimean-Congo hemorrhagic fever. *Antiviral Research*, 64(3), 145–160. https://doi.org/10.1016/j.antiviral.2004.08.001.

Wilson, A., Taylor, J., & Edwards, C. (2023). Point-of-care lactate testing: Implications for elderly patients in emergency settings. *Clinical Medicine*, 24(3), 180–188.
Yadav, P. D., Gurav, Y. K., Mistry, M., Shete, A. M., Sarkale, P., Deoshatwar, A. R., Unadkat, V. B., Kokate, P., Patil, D. Y., Raval, D. K., & Mourya, D. T. (2014). Emergence of Crimean-Congo hemorrhagic fever in Amreli District of Gujarat State, India, June to July 2013. *International Journal of Infectious Diseases*, 18, 97–100. https://doi.org/10.1016/j.ijid.2013.09.021.
Zhao, Y., Li, X., & Wang, H. (2023). Correlation between arterial lactate levels and in-hospital mortality in critically ill patients. *Critical Care Medicine*, 31(4), 245–252.
Zivcec, M., & Feldmann, H. (2017). Crimean-Congo hemorrhagic fever virus: A review. *Antiviral Research*, 141, 16–26. https://doi.org/10.1016/j.antiviral.2017.02.010.

3 Infectious Hemorrhagic Fevers
Rift Valley Fever Virus (RVF)

Sudeepto Biswas, Senjuti Bhattacharjee, Anirbandeep Bose, and Sukanta Roy

3.1 INTRODUCTION

Hemorrhagic fevers (HFs) are a group of viral infections characterized by fever, bleeding tendencies, and high mortality rates. They primarily affect the vascular system, causing increased permeability, which can lead to severe complications such as shock and multi-organ failure (Bray, 2005; Peters, 2002). Viral hemorrhagic fevers (VHFs) are caused by various RNA viruses from different families, including Filoviridae, Arenaviridae, Bunyaviridae, and Flaviviridae (WHO, 2022; Nichol et al., 2003). Notable VHFs include Ebola, dengue, Lassa fever, and Crimean-Congo hemorrhagic fever, all of which pose serious public health challenges, especially in resource-limited settings (CDC, 2023; Goeijenbier et al., 2014). Hemorrhagic fevers have left a lasting impact on human health throughout history. One of the earliest recorded hemorrhagic fevers is dengue, with descriptions dating back to the 18th century (Gubler, 2006). Ebola, a more recent viral threat, was first identified in 1976 in Zaire (now the Democratic Republic of Congo) during a severe outbreak (Dupuy et al., 2023). The spread of VHFs has been exacerbated by global factors such as travel, urbanization, and environmental changes (Morens & Fauci, 2013). Key milestones include the discovery of Lassa fever in 1969 and the emergence of the Marburg virus in 1967, both of which have resulted in high mortality rates (Fauci, 2005; Gear et al., 1975)). Additionally, dengue fever continues to be a leading cause of morbidity in tropical regions, with outbreaks occurring globally (Weaver & Vasilakis, 2009; Gubler, 2006.). Emerging and re-emerging hemorrhagic fevers are significant threats due to their potential for large outbreaks with high fatality rates. Emerging pathogens include viruses such as the Rift Valley fever virus (RVFV), which has spread due to environmental changes, including climate shifts and animal migrations (Hewson, 2024). Re-emerging fevers like dengue have resurged as a result of expanding mosquito vectors, particularly *Aedes aegypti*, driven by global warming (Weaver & Vasilakis, 2009; Messina et al., 2015). Rift Valley fever (RVF), a zoonotic disease that affects both humans and livestock, is of particular concern due to its economic and health impacts. In severe cases, RVF can cause hemorrhagic symptoms, making it a critical disease for global health surveillance (Mahmoud & Ali, 2021). This chapter aims to provide a detailed overview of hemorrhagic fevers, focusing on their causative pathogens, epidemiology, and public health implications. It explores the transmission dynamics of VHFs and the potential for large-scale outbreaks. The chapter is structured to offer foundational knowledge necessary for understanding the public health challenges posed by hemorrhagic fevers and the measures required to mitigate their spread (Daszak et al., 2000). This information will serve as a basis for the more detailed analyses presented in subsequent chapters, which will cover specific VHFs, prevention strategies, and the global health response to these emerging threats. The concepts discussed here are essential for understanding the broader epidemiological patterns and disease control strategies. By framing the discussion around both emerging and re-emerging hemorrhagic fevers, this chapter sets the stage for a comprehensive exploration of the public health response to VHFs.

DOI: 10.1201/9781032721811-3

3.2 DEFINITIONS AND TERMINOLOGIES ASSOCIATED WITH RIFT VALLEY FEVER DISEASE AND RIFT VALLEY FEVER VIRUS

Rift Valley fever (RVF) is a zoonotic disease caused by the Rift Valley fever virus (RVFV), a mosquito-borne virus classified in the Bunyaviridae family, specifically the *Phlebovirus* genus (Bird & McElroy, 2016). The disease primarily affects livestock, including cattle, sheep, and goats, but it can also infect humans, where it presents as a febrile illness with varying degrees of severity. Key terms such as *Arbovirus*, viruses transmitted by arthropods like mosquitoes, are used to describe RVFV's mode of transmission (Pepin et al., 2010). Additional terminology relevant to RVF includes the following:

1. *Enzootic cycle:* Refers to the natural transmission of the virus within animal populations, without triggering large outbreaks.
2. *Epizootic cycle:* Describes sudden, widespread outbreaks of RVFV in animal populations, with potential spillover to humans (Bird & McElroy, 2016).
3. *Viremia:* The presence of the virus in the bloodstream, which allows for further transmission via vectors like mosquitoes (Ikegami & Makino, 2011).

3.3 HISTORY OF RIFT VALLEY FEVER DISEASE AND RIFT VALLEY FEVER VIRUS INFECTION

RVF was first identified in 1931 during an outbreak in sheep in the Rift Valley of Kenya (Daubney et al., 1931). Early investigations revealed that the disease was caused by a virus transmitted by mosquitoes, particularly *Aedes* species. Since then, RVF has remained endemic in many parts of Africa, but significant outbreaks have been recorded, especially during periods of heavy rainfall and flooding that increase mosquito populations (Ikegami & Makino, 2011). Major outbreaks outside of Africa include the 2000 outbreak in Saudi Arabia and Yemen, which marked the first time RVFV was identified outside the African continent (Shoemaker et al., 2002). The disease is now considered an emerging threat to Europe and Asia, especially due to climate change and increased global travel (LaBeaud, 2008).

3.4 TAXONOMIC CLASSIFICATION AND NOMENCLATURE OF RIFT VALLEY FEVER VIRUS

RVFV belongs to the family Phenuiviridae, genus *Phlebovirus*, within the order Bunyavirales (Pepin et al., 2010). This RNA virus has a tripartite genome that consists of the following:

- *L segment:* Encodes the RNA-dependent RNA polymerase
- *M segment:* Encodes the glycoproteins (Gn and Gc) that are essential for viral entry into host cells
- *S segment:* Encodes the nucleocapsid protein and non-structural protein (NSs) that plays a role in immune evasion (Ikegami & Makino, 2011)

The virus is enveloped, which allows it to evade immune detection and establish infection in host cells. RVFV shares genetic similarities with other *Phlebovirus* species, yet its pathogenicity, especially in humans, distinguishes it as a major zoonotic threat (Bird & McElroy, 2016).

3.5 TRANSMISSION PROCESS OF RIFT VALLEY FEVER VIRUS

RVFV is predominantly transmitted through mosquito bites, with species such as *Aedes* and *Culex* serving as primary vectors (Pepin et al., 2010). The virus can be transmitted transovarially, where mosquitoes pass it from one generation to the next. Flooding events lead to increased mosquito breeding, which often precedes large RVF outbreaks (LaBeaud, 2008). The following are other modes of transmission:

- Direct contact: Humans can contract the virus by handling tissues, blood, or other fluids from infected animals. This is particularly common during slaughtering or veterinary practices (Ikegami & Makino, 2011).
- Aerosol transmission: The virus can be aerosolized, posing a risk to laboratory workers and individuals in contact with infected animal products (Bird & McElroy, 2016).
- Vertical transmission: The virus can be transmitted from infected mothers to offspring in both animals and humans (LaBeaud, 2008).
- Human outbreaks typically follow periods of heavy rainfall when mosquitoes breed in large numbers. Preventive measures focus on vector control, vaccination of livestock, and personal protective equipment for individuals working with animals (Bird & McElroy, 2016).

3.6 SYMPTOMS ASSOCIATED WITH RIFT VALLEY FEVER VIRUS INFECTION: CLASSIFICATION BASED ON DISEASE SEVERITY

The clinical presentation of RVF in humans varies widely, from mild febrile illness to life-threatening hemorrhagic fever and encephalitis. Symptoms can be classified into two main categories:

- *Mild or self-limiting illness:* The majority of infected individuals experience mild, flu-like symptoms, including fever, muscle aches, joint pain, and headache (Pepin et al., 2010). These symptoms generally resolve within a week without lasting effects.
- *Severe forms of RVF:*

  Ocular disease: Approximately 1–10% of cases result in retinal damage, which may cause vision impairment or blindness (Al-Hazmi et al., 2005).

  Meningoencephalitis: Neurological symptoms such as confusion, hallucinations, and coma may occur in a small percentage of cases. These patients often require intensive care and may suffer from long-term neurological deficits (Ikegami & Makino, 2011).

  Hemorrhagic fever: This severe manifestation is rare but highly fatal. It presents with bleeding disorders, liver damage, and high mortality rates (Bird & McElroy, 2016). Hemorrhagic fever typically occurs within two to four days of symptom onset and requires immediate medical intervention. Though no specific antiviral treatment exists, supportive care, such as managing fluid balance and bleeding, is crucial for patients with severe RVF. In addition, vector control measures and livestock vaccination are key strategies for preventing the spread of RVFV (Pepin et al., 2010).

3.7 BASIC STRUCTURE OF RIFT VALLEY FEVER VIRUS

Rift Valley fever virus (RVFV) is an enveloped virus belonging to the Phenuiviridae family within the order Bunyavirales. It has a spherical shape, ranging from 80 nm to 120 nm in diameter, with an outer lipid bilayer derived from the host cell membrane. This bilayer is embedded with glycoprotein spikes, Gn and Gc, which play a key role in host cell attachment and entry (Ikegami & Makino, 2011). The matrix protein located beneath the envelope gives structural integrity to the virus. The core of the virus contains the ribonucleoprotein (RNP) complex, made up of the viral RNA genome, the nucleoprotein (N), and the RNA-dependent RNA polymerase (L). This structure is crucial for RNA replication and protects the viral genome (Bird & Nichol, 2012). RVFV has a tripartite genome, which consists of three single-stranded RNA segments: small (S), medium (M), and large (L) (Pepin et al., 2010). This segmented nature enables the virus to evolve rapidly through genetic reassortment, which may enhance its virulence and adaptability to new hosts (Gaudreault et al., 2019).

3.8 GENOMIC ORGANIZATION OF RIFT VALLEY FEVER VIRUS

The RVFV genome is divided into three segments:

3.8.1 Small (S) Segment

The S segment is approximately 1.7 kb in length and exhibits an ambisense coding strategy, where genes are transcribed in both directions. It encodes the nucleoprotein (N), responsible for encapsidating the viral RNA, and the non-structural NSs protein, which plays a vital role in viral pathogenesis. The NSs protein is a key virulence factor that suppresses the host immune response by inhibiting interferon signaling (Bouloy & Weber, 2010).

3.8.2 Medium (M) Segment

The M segment, which is approximately 3.8 kb, encodes the viral glycoproteins Gn and Gc. These proteins are crucial for the virus to attach to and enter host cells (Elliott, 2014). Additionally, the M segment produces the NSm protein, which inhibits apoptosis in infected cells, enhancing viral survival (Won et al., 2006).

3.8.3 Large (L) Segment

The L segment, roughly 6.4 kb long, encodes the viral RNA-dependent RNA polymerase (L protein), which is essential for both transcription and replication of the viral genome. This enzyme replicates the viral RNA and is crucial for the production of new viral particles (Figure 3.1) (Ikegami & Makino, 2011).

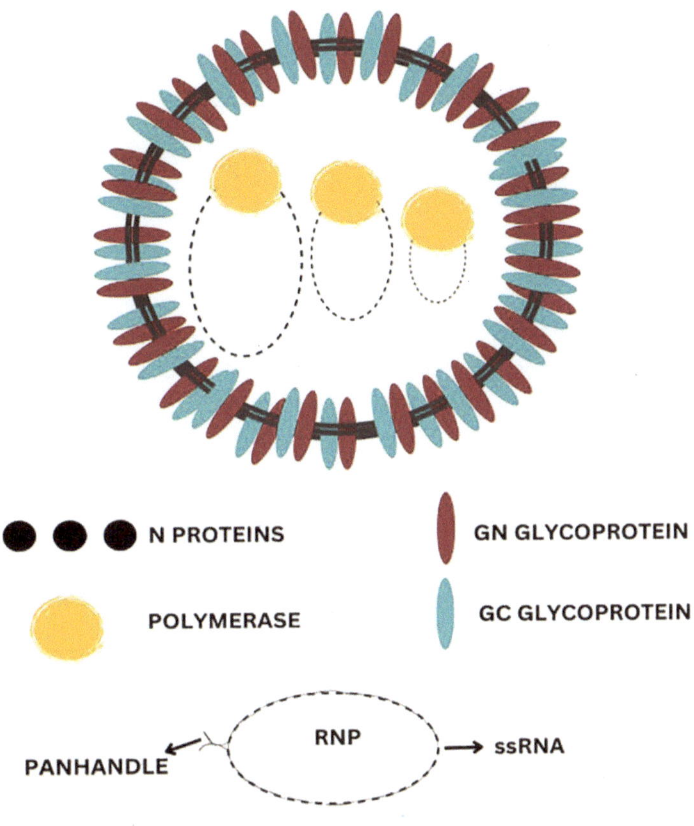

FIGURE 3.1 Genomic Organization of Rift Valley fever virus.

This segmented genome not only allows for efficient replication but also contributes to the virus' ability to reassort genes, potentially creating novel viral strains with increased transmission or pathogenicity (Gaudreault et al., 2019).

3.9 LIFE CYCLE OF RIFT VALLEY FEVER VIRUS

The life cycle of RVFV follows a typical pattern for arboviruses, beginning when mosquitoes infected with RVFV transmit the virus to a vertebrate host via a bite. Alternatively, humans can contract the virus through direct contact with blood or tissues of infected animals (Pepin et al., 2010). The major steps of the life cycle are outlined in the following sections.

3.9.1 Attachment and Entry

Viral entry begins with the attachment of the viral glycoproteins Gn and Gc to specific receptors on the host cell surface, followed by endocytosis. The acidic environment within the endosome triggers fusion between the viral envelope and the endosomal membrane, allowing the release of the viral nucleocapsid into the cytoplasm (Ikegami & Makino, 2011).

3.9.2 Uncoating and Replication

Once inside the host cell cytoplasm, the viral genome is released, and the RNA-dependent RNA polymerase (encoded by the L segment) begins transcription of viral mRNAs. The viral polymerase transcribes the negative-sense RNA into positive-sense mRNA, which serves as a template for protein synthesis and genome replication (Bird & Nichol, 2012). The replication takes place in virus-induced cytoplasmic compartments, where both positive-sense and negative-sense RNA strands are synthesized.

3.9.3 Translation and Protein Synthesis

Host ribosomes translate the viral mRNA into viral proteins, including structural proteins (N, Gn, Gc) and non-structural proteins (NSs, NSm). The glycoproteins (Gn and Gc) are synthesized in the rough endoplasmic reticulum and are transported to the Golgi apparatus, where they are integrated into the viral envelope (Bouloy & Weber, 2010).

3.9.4 Assembly and Maturation

Newly synthesized RNA segments are packaged with nucleoproteins to form new ribonucleoprotein complexes. These complexes, along with the glycoproteins, assemble into mature virions, which are transported to the host cell membrane via the secretory pathway (Pepin et al., 2010).

3.9.5 Release

Virions are released from the host cell through a process of budding, where they acquire a portion of the host cell membrane enriched with viral glycoproteins. These newly formed virions can then infect additional host cells or be transmitted to other organisms through mosquitoes, perpetuating the infection cycle (Bird & Nichol, 2012).

Throughout this process, RVFV employs several strategies to evade the host immune response. For example, the NSs protein inhibits the host interferon response, allowing the virus to replicate unchecked in the host (Bouloy & Weber, 2010). The NSm protein further supports viral persistence by inhibiting apoptosis of infected cells, ensuring sustained viral replication (Won et al., 2006).

3.10 MOLECULAR PATHOPHYSIOLOGY AND CLINICAL PRESENTATION

Rift Valley fever virus (RVFV), a member of the Phenuiviridae family within the order Bunyavirales, is a mosquito-borne pathogen that causes severe disease in humans and livestock. The molecular pathophysiology of RVFV infection is driven by viral replication strategies, immune evasion mechanisms, and tissue tropism (Figure 3.2). The virus' ability to subvert host defenses and promote widespread inflammation underpins the development of clinical disease.

3.10.1 Molecular Mechanisms Underlying Pathophysiology of Rift Valley Fever Virus Infection

3.10.1.1 Viral Entry and Replication

RVFV primarily infects hepatocytes, endothelial cells, macrophages, and dendritic cells, leading to systemic viral dissemination. The virus enters cells via receptor-mediated endocytosis, where the Gn and Gc glycoproteins mediate the fusion of the viral envelope with the host cell (Bird & Nichol, 2012; Craven et al., 2009). Upon entry, the viral genome is released into the cytoplasm, where viral RNA-dependent RNA polymerase (L protein) begins the transcription and replication of viral RNA. The segmented genome of RVFV allows for rapid viral replication and the production of viral proteins necessary for infection (Ikegami & Makino, 2011).

3.10.1.2 NSs Protein: Immune Suppression and Interferon Inhibition

One of the key pathogenic mechanisms of RVFV is the suppression of the host's antiviral immune response. The RVFV non-structural protein (NSs), encoded by the small (S) genome segment, is the primary virulence factor. It suppresses the host's interferon (IFN) response by inhibiting the transcription of interferon-stimulated genes (ISGs) through degradation of the transcription factor TFIIH (Kalveram et al., 2011). This impairs the production of type I IFNs, allowing the virus to evade immune detection and replicate unchecked within host cells (Ikegami & Makino, 2011; Habjan et al., 2009). The NSs protein also interferes with host gene transcription by forming

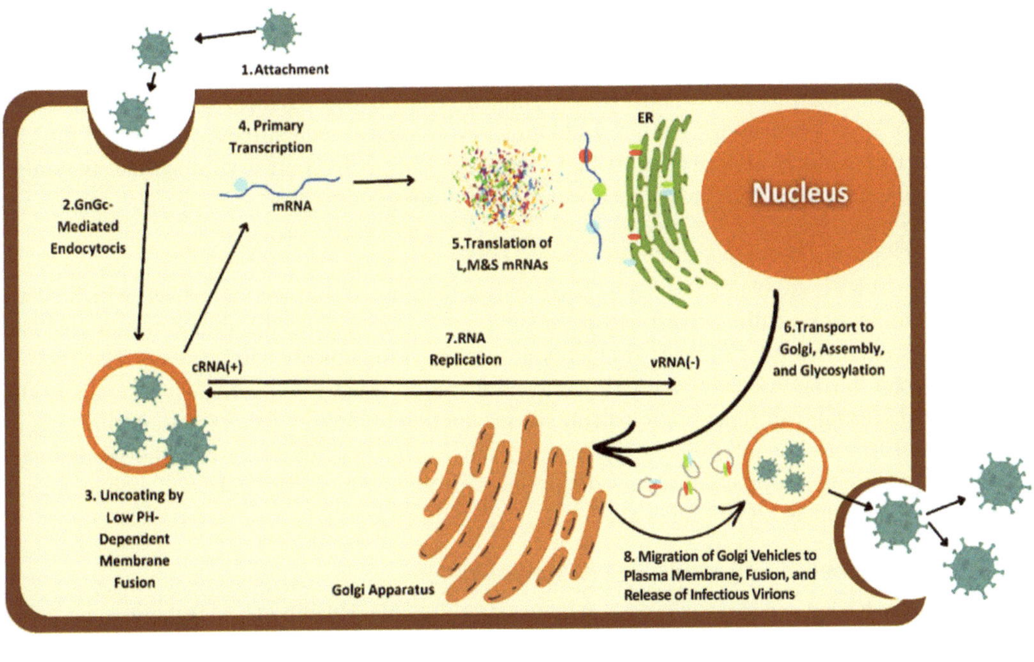

FIGURE 3.2 Pathophysiology of RVF virus.

filamentous structures that sequester transcription factors, further contributing to immune suppression (Bouloy & Weber, 2010).

3.10.1.3 NSm Protein: Inhibition of Apoptosis

Another non-structural protein encoded by the medium (M) segment, NSm, plays a crucial role in blocking host cell apoptosis. By inhibiting programmed cell death, NSm promotes viral persistence within infected cells, allowing prolonged viral replication (Won et al., 2006; Billecocq et al., 2004). This anti-apoptotic effect is critical for viral survival, particularly in macrophages and hepatocytes, which are primary targets during RVFV infection (Won et al., 2006).

3.10.1.4 Tissue Tropism and Systemic Inflammation

RVFV exhibits a strong tropism for the liver, where it induces hepatic necrosis and contributes to systemic inflammation. The virus replicates in hepatocytes, causing direct cellular damage and triggering the release of pro-inflammatory cytokines. This cytokine storm, mediated by infected macrophages and dendritic cells, results in systemic inflammation that contributes to the severe outcomes observed in RVF, such as hemorrhagic fever and multi-organ failure (Pepin et al., 2010; Baudin et al., 2016). The liver pathology is characterized by massive hepatocyte necrosis and hemorrhage, which are driven by both viral replication and immune-mediated cytotoxicity.

3.10.2 Molecular Mechanisms Underlying the Clinical Manifestation of Rift Valley Fever Virus Infection

The clinical manifestations of Rift Valley fever (RVF) range from mild febrile illness to severe hemorrhagic fever, encephalitis, and retinitis. The underlying molecular mechanisms responsible for these outcomes involve both direct viral effects and the host's immune response. Several key processes contribute to the clinical picture of RVF infection.

3.10.2.1 Acute Febrile Illness

In the majority of cases, RVFV infection results in a self-limiting febrile illness, characterized by fever, myalgia, and malaise (Ikegami & Makino, 2011; McElroy & Nichol, 2012). These symptoms are largely attributable to the host's innate immune response to viral infection. During the early stages of infection, dendritic cells and macrophages detect viral replication via pattern recognition receptors (PRRs) such as RIG-I and MDA5 (Habjan et al., 2009). This recognition leads to the production of type I interferons and pro-inflammatory cytokines, which induce fever and flu-like symptoms (Kalveram et al., 2011). Although these responses help control viral spread, they also contribute to the systemic symptoms observed in RVFV infection.

3.10.2.2 Hepatic Damage and Hemorrhagic Fever

Severe cases of RVF are characterized by hepatic dysfunction and hemorrhagic fever, which occur due to a combination of direct viral cytopathic effects and dysregulated immune responses. The virus causes extensive liver damage by inducing hepatocyte apoptosis and necrosis. The liver is a central target during infection, and the massive replication of RVFV in hepatocytes leads to the disruption of normal liver function, impairing the production of clotting factors and causing coagulopathy (McElroy & Nichol, 2012). This coagulopathy, coupled with direct endothelial damage, contributes to the development of hemorrhagic manifestations, including epistaxis, hematemesis, and melena (Pepin et al., 2010; Liu et al., 2017).

3.10.2.3 Encephalitis and Neurological Manifestations

In some cases, RVFV invades the central nervous system (CNS), leading to encephalitis and other neurological complications. The mechanism by which RVFV crosses the blood–brain barrier (BBB) remains unclear, but it is believed that infected macrophages and endothelial cells play a role in facilitating viral entry into the brain. Once inside the CNS, RVFV causes neuronal damage and inflammation, leading to symptoms such as confusion, seizures, and coma (Ikegami & Makino, 2011). The virus' ability to suppress the interferon response in the CNS may exacerbate the severity of neurological disease (Kalveram et al., 2011; Liu et al., 2017).

3.10.2.4 Retinitis and Ocular Complications

RVFV infection can also result in ocular disease, particularly retinitis, which can lead to visual impairment or blindness (Al-Hazmi et al., 2005). The virus infects retinal cells, causing inflammation and necrosis of the retina. The pathogenesis of RVFV-induced retinitis is thought to involve both direct viral damage to retinal tissues and immune-mediated injury caused by an overactive inflammatory response (Pepin et al., 2010).

3.11 INFECTION PREVENTION AND CONTROL: CHALLENGES AND FUTURE PROSPECTS

3.11.1 Vector Control Difficulties

The primary vector for RVFV transmission is the *Aedes* mosquito, which acts as both a reservoir and a vector. Other mosquito species, such as *Culex*, also play a significant role in the virus' spread. The complexity of mosquito control presents a major challenge for RVFV prevention. *Aedes* mosquitoes are particularly difficult to control due to their ability to lay desiccation-resistant eggs, which can remain dormant during dry seasons and hatch when conditions become favorable (Linthicum et al., 2016). This makes it difficult to predict and prevent outbreaks based solely on weather patterns, as mosquito populations can explode following seasonal rains. Traditional mosquito control strategies such as insecticide spraying and larviciding are often inadequate in rural areas with limited access to infrastructure and public health resources (Pepin et al., 2010).

3.11.2 Environmental and Climatic Factors

RVF outbreaks are closely tied to environmental and climatic conditions, especially periods of heavy rainfall and flooding, which create ideal breeding grounds for mosquito vectors (Linthicum et al., 2016; Anyamba et al., 2019). Predicting outbreaks based on climatic models has improved, but these models are not foolproof. Variability in rainfall patterns due to climate change may further complicate prevention efforts, as it becomes more difficult to anticipate periods of heightened risk (Bathiany et al., 2018; Baudin et al., 2016). The virus can also persist in the environment within mosquito eggs, posing a long-term threat that can re-emerge years after an initial outbreak Linthicum et al., 2016).

3.11.3 Limitations in Livestock Management

Livestock plays a critical role in the epidemiology of RVFV, acting as an amplifying host during outbreaks. Infected animals often show no clinical signs initially, but they can shed the virus in body fluids, making livestock-to-human transmission a significant risk for agricultural workers. Effective control of RVFV requires the implementation of biosecurity measures in livestock farms, such as restricting animal movement, vaccination, and surveillance (Bird & Nichol, 2012). However, in many endemic regions, such measures are difficult to enforce due to lack of resources, political instability, and the nomadic or transhumant nature of livestock farming. Consequently, infected animals often travel long distances, spreading the virus to new areas (Pepin et al., 2010).

3.11.4 Human Behavior and Cultural Factors

In addition to mosquito bites, humans can contract RVFV through direct contact with the blood, tissues, or body fluids of infected animals. This route of transmission is particularly common during slaughtering practices or handling of contaminated meat. Cultural practices and local traditions regarding livestock farming, slaughtering, and consumption of meat often make it difficult to implement stringent IPC measures. Education and awareness campaigns are essential to inform communities about the risks associated with handling livestock during outbreaks, but these efforts are often underfunded or poorly executed in high-risk areas (WHO, 2020).

3.11.5 Vaccine and Therapeutic Development Challenges

Although several RVFV vaccines exist for livestock, no licensed human vaccine is currently available. The development of an effective human vaccine is complicated by the need for a balance between safety and immunogenicity, particularly in populations living in endemic areas where mass vaccination would be ideal (Faburay et al., 2017). In addition, RVFV can reassort genetically with other viruses, potentially leading to the emergence of new strains, which complicates vaccine design and efficacy. Furthermore, the absence of antiviral treatments for RVFV poses a significant challenge for managing human cases, especially those presenting with severe complications such as hemorrhagic fever or encephalitis.

3.11.6 Limitations in Public Health Infrastructure

The effectiveness of RVFV prevention strategies is often hindered by limited public health infrastructure in affected regions. Many RVF-endemic countries face challenges such as inadequate disease surveillance systems, insufficient funding for mosquito control programs, and limited access to healthcare services, particularly in rural areas (Linthicum et al., 2016). These gaps make it difficult to implement timely and coordinated responses to RVF outbreaks, contributing to the virus' rapid spread (Baudin et al., 2016).

3.12 FUTURE PROSPECTS FOR RIFT VALLEY FEVER VIRUS CONTROL

3.12.1 Improved Vector Control Strategies

Advances in vector control technologies offer promising avenues for reducing RVFV transmission. Genetic modification of mosquito populations, such as the use of sterile insect techniques or gene-drive systems, could potentially reduce mosquito populations or prevent them from transmitting the virus (Alphey et al., 2013). Additionally, new insecticides that target mosquito larvae or disrupt mosquito breeding cycles may offer more sustainable and effective control strategies. Integrated vector management (IVM) programs that combine environmental, biological, and chemical control methods should be prioritized, particularly in regions with complex ecologies.

3.12.2 Development of Vaccines

The development of a safe and effective vaccine for humans represents one of the most promising prospects for controlling RVFV. Several candidates are currently in clinical trials, including live-attenuated and recombinant vaccines (Faburay et al., 2017). Advances in vaccine technology, such as mRNA-based vaccines, have shown success in combating other viral diseases like COVID-19 and could potentially be adapted for RVFV. The widespread deployment of an effective RVFV vaccine would significantly reduce the incidence of both human and livestock infections and help prevent future outbreaks.

3.12.3 Improved Surveillance and Early Warning Systems

Advancements in surveillance technologies, including satellite-based weather prediction systems, have improved the ability to predict RVF outbreaks (Anyamba et al., 2019). These tools can help public health authorities anticipate high-risk periods based on rainfall and vector activity, allowing for preemptive vector control and vaccination campaigns (Baudin et al., 2016). Enhanced disease surveillance, including improved reporting systems and diagnostic capabilities, will also be crucial in identifying and responding to RVF outbreaks in a timely manner.

3.12.4 Educational Campaigns and Community Engagement

Future efforts to control RVFV will require greater emphasis on education and community engagement. Public awareness campaigns that target high-risk populations, such as agricultural workers and rural communities, are essential for reducing human-to-human transmission and encouraging the adoption of safer livestock-handling practices. Involving local leaders and stakeholders in the design and implementation of these campaigns can improve their effectiveness and ensure that IPC measures are culturally appropriate (WHO, 2020).

3.12.5 International Collaboration and Funding

Controlling RVFV on a global scale will require international cooperation and sustained funding. RVFV poses a potential threat to non-endemic regions due to the global movement of people and livestock. Strengthening global surveillance networks, sharing data across borders, and coordinating research efforts will be essential to mitigate the impact of future RVFV outbreaks. Organizations such as the World Health Organization (WHO) and the Food and Agriculture Organization (FAO) will play critical roles in fostering collaboration and ensuring that resources are allocated to high-risk areas.

3.13 CONCLUSION

The prevention and control of RVFV present significant challenges, particularly in resource-limited settings. Vector control, vaccine development, and public health interventions face numerous obstacles, including environmental unpredictability, limited infrastructure, and cultural practices. However, advancements in technology, enhanced surveillance systems, and international collaboration offer promising avenues for future control efforts. By addressing these challenges and investing in long-term prevention strategies, it is possible to reduce the impact of RVFV and prevent large-scale outbreaks.

REFERENCES

Al-Hazmi, M., Ayoola, E. A., Abdurahman, M., Banzal, S., Ashraf, J., El-Bushra, A., & Gadour, M. (2005). Ocular complications of Rift Valley fever outbreak in Saudi Arabia. *Ophthalmology*, 112(2), 313–318. https://doi.org/10.1016/j.ophtha.2004.06.032.

Alphey, L., Benedict, M., Bellini, R., Clark, G. G., Dame, D. A., Service, M. W., & Dobson, S. L. (2013). Sterile-insect methods for control of mosquito-borne diseases: An analysis. *Vector-Borne and Zoonotic Diseases*, 10(3), 295–311. https://doi.org/10.1089/vbz.2009.0014.

Anyamba, A., Linthicum, K. J., Small, J. L., Collins, K. M., Tucker, C. J., Pak, E. W., & Britch, S. C. (2019). Climate teleconnections and recent patterns of human and animal disease outbreaks. *PLOS Neglected Tropical Diseases*, 13(7), e0007215. https://doi.org/10.1371/journal.pntd.0007215.

Bathiany, S., Dakos, V., Scheffer, M., & Lenton, T. M. (2018). Climate models predict increasing temperature variability in poor countries. *Sci Adv*, 4(5), eaar5809. https://doi.org/10.1126/sciadv.aar5809.

Baudin, M., Jumaa, A. M., Jomma, H. J. E., Karsany, M. S., Bucht, G., Naslund, J., & Evander, M. (2016). Association of Rift Valley fever virus infection with miscarriage in Sudanese women: A cross-sectional study. *Lancet Global Health*, 4(11), e864–e871. https://doi.org/10.1016/S2214-109X(16)30176-0.

Billecocq, A., Gauliard, N., Le May, N., Elliott, R. M., & Bouloy, M. (2004). RNA polymerase I-mediated expression of viral RNA for the rescue of infectious virulent and avirulent Rift Valley fever viruses. *Journal of Virology*, 78(16), 8464–8471. https://doi.org/10.1128/JVI.78.16.8464-8471.2004.

Bird, B. H., & McElroy, A. K. (2016). Rift Valley fever virus: Unanswered questions. *Antiviral Research*, 132, 44–52. https://doi.org/10.1016/j.antiviral.2016.05.010.

Bird, B. H., & Nichol, S. T. (2012). Breaking the chain: Rift Valley fever virus control via livestock vaccination. *Current Opinion in Virology*, 2(3), 315–323. https://doi.org/10.1016/j.coviro.2012.02.012.

Bouloy, M., & Weber, F. (2010). Molecular biology of Rift Valley fever virus. *Open Virology Journal*, 4, 8–14. https://doi.org/10.2174/1874357901004010008.

Bray, M. (2005). Pathogenesis of viral hemorrhagic fever. *Current Opinion in Immunology*, 17(4), 399–403. https://doi.org/10.1016/j.coi.2005.07.001.

Centers for Disease Control and Prevention (CDC). (2023). *Viral hemorrhagic Fevers (VHFs)*. https://www.cdc.gov.

Craven, R. B., Eliason, D. A., Francy, D. B., Reiter, P., Campos, E. G., Jakob, W. L., & Moore, C. G. (2009). Aedes aegypti mosquitoes collected in South America: Susceptibility to Rift Valley fever virus. *Emerging Infectious Diseases*, 8(2), 492–497. https://doi.org/10.3201/eid0802.020401.

Daszak, P., Cunningham, A. A., & Hyatt, A. D. (2000). Emerging infectious diseases of wildlife – threats to biodiversity and human health. *Science*, 287(5452), 443–449. https://doi.org/10.1126/science.287.5452.443. Erratum in: *Science*, 287(5459), 1756.

Daubney, R., Hudson, J. R., & Garnham, P. C. (1931). Enzootic hepatitis or Rift Valley fever: An undescribed virus disease of sheep, cattle, and man from East Africa. *Journal of Pathology and Bacteriology*, 34(4), 545–579.

Dupuy, L. C., Spiropoulou, C. F., Towner, J. S., Spengler, J. R., Sullivan, N. J., & Montgomery, J. M. (2023). Filoviruses: Scientific gaps and prototype pathogen recommendation. *J Infect Dis*, 228(Suppl 6), S446–S459. https://doi.org/10.1093/infdis/jiad362.

Elliott, R. M. (2014). Orthobunyaviruses: Recent genetic and structural insights. *Nature Reviews Microbiology*, 12(10), 673–685. https://doi.org/10.1038/nrmicro3332.

Faburay, B., Wilson, W. C., Gaudreault, N. N., Davis, A. S., & Richt, J. A. (2017). Rift Valley fever: Molecular and cellular pathogenesis. *Pathogens*, 6(4), 33. https://doi.org/10.3390/pathogens6040033.

Fauci, A. S. (2005). Emerging and reemerging infectious diseases: The perpetual challenge. *Acad Med*, 80(12), 1079–1085. https://doi.org/10.1097/00001888-200512000-00002.

Gaudreault, N. N., Indran, S. V., Balaraman, V., Wilson, W. C., & Richt, J. A. (2019). Molecular aspects of Rift Valley fever virus and the emergence of reassortants. *Viruses*, 11(8), 765. https://doi.org/10.3390/v11080765.

Gear, J. H. S., Cassel, G. A., & Gear, A. J. (1975). The outbreak of Marburg virus disease in southern Africa. *South African Medical Journal*, 49, 767–781.

Goeijenbier, M., van Kampen, J. J., Reusken, C. B., Koopmans, M. P., & van Gorp, E. C. (2014). Viral infections and mechanisms of thrombosis and bleeding. *Journal of Medical Virology*, 86(9), 1502–1516. https://doi.org/10.1002/jmv.23961.

Gubler, D. J. (2006). Dengue/dengue hemorrhagic fever: History and current status. *Novartis Foundation Symposium*, 277, 3–22. https://doi.org/10.1002/0470035792.ch1.

Habjan, M., Penski, N., Wagner, V., Spiegel, M., Overby, A. K., Kochs, G., & Weber, F. (2009). NSs protein of Rift Valley fever virus induces the specific degradation of the double-stranded RNA-dependent protein kinase. *Journal of Virology*, 83(9), 4365–4375. https://doi.org/10.1128/JVI.02248-08.

Hewson, R. (2024). Understanding viral haemorrhagic fevers: Virus diversity, vector ecology, and public health strategies. *Pathogens*, 13(10), 909. https://doi.org/10.3390/pathogens13100909.

Ikegami, T., & Makino, S. (2011). The pathogenesis of Rift Valley fever. *Viruses*, 3(5), 493–519. https://doi.org/10.3390/v3050493.

Kalveram, B., Ikegami, T., & Ikegami, S. (2011). Molecular mechanisms of Rift Valley fever virus infection. *Journal of Virology*, 85(24), 13718–13728. https://doi.org/10.1128/JVI.05618-11.

LaBeaud, A. D. (2008). Rift Valley fever virus infection in Africa and the Arabian Peninsula: A review of the history, distribution, and burden of disease. *American Journal of Tropical Medicine and Hygiene*, 78(2), 169–178.

Linthicum, K. J., Britch, S. C., & Anyamba, A. (2016). Rift Valley fever: An emerging mosquito-borne disease. *Annual Review of Entomology*, 61, 395–415. https://doi.org/10.1146/annurev-ento-010715-023819.

Liu, L., Celma, C. C. P., Roy, P., & Duprex, W. P. (2017). Rift Valley fever virus NSm protein inhibits apoptosis. *Virology Journal*, 14(1), 129. https://doi.org/10.1186/s12985-017-0784-3.

Mahmoud, H. Y. A. H., & Ali, A. O. (2021) Epidemiology and serological detection of Rift Valley Fever disease in farm animals in southern Egypt. *Onderstepoort J Vet Res*, 88(1), e1–e5. https://doi.org/10.4102/ojvr.v88i1.1877.

McElroy, A. K., & Nichol, S. T. (2012). Rift Valley fever virus infection of mice. *Virology*, 424(1), 93–100. https://doi.org/10.1016/j.virol.2011.12.019.

Messina, J. P., Brady, O. J., Golding, N., Kraemer, M. U., Wint, G. R., Ray, S. E., & Hay, S. I. (2015). The current and future global distribution and population at risk of dengue. *Nature Microbiology*, 1(5), 15029. https://doi.org/10.1038/nmicrobiol.2015.29.

Morens, D. M., & Fauci, A. S. (2013). Emerging infectious diseases: Threats to human health and global stability. *PLoS Pathogens*, 9(7), e1003467. https://doi.org/10.1371/journal.ppat.1003467.

Nichol, S. T., Beaty, B. J., Elliott, R. M., Goldbach, R., Plyusnin, A., & Schmaljohn, C. S. (2003). Family Bunyaviridae. In C. M. Fauquet, M. A. Mayo, J. Maniloff, U. Desselberger, & L. A. Ball (Eds.), *Virus Taxonomy* (pp. 599–621). Academic Press.

Paweska, J. (2015). Rift valley fever. *Revue scientifique et technique (International Office of Epizootics)*, 34, 375–389. https://doi.org/10.1016/B978-0-12-416975-3.00006-6.

Pepin, M., Bouloy, M., Bird, B. H., Kemp, A., & Paweska, J. (2010). Rift Valley fever virus (Bunyaviridae: Phlebovirus): An update on pathogenesis, molecular epidemiology, vectors, diagnostics and prevention. *Veterinary Research*, 41(6), 61. https://doi.org/10.1051/vetres/2010033.

Peters, C. J. (2002). Human infection with Rift Valley fever virus. *Infectious Disease Clinics of North America*, 16(1), 29–51. https://doi.org/10.1016/S0891-5520(01)00004-8.

Shoemaker, T., Boulianne, C., Vincent, M. J., Pezzanite, L., Al-Qahtani, M. M., Al-Mazrou, Y., & Rollin, P. E. (2002). Genetic analysis of viruses associated with emergence of Rift Valley fever in Saudi Arabia and Yemen, 2000–01. *Emerging Infectious Diseases*, 8(12), 1415–1420. https://doi.org/10.3201/eid0812.020269.

Weaver, S. C., & Vasilakis, N. (2009) Molecular evolution of dengue viruses: Contributions of phylogenetics to understanding the history and epidemiology of the preeminent arboviral disease. *Infect Genet Evol.*, 9(4), 523–540. https://doi.org/10.1016/j.meegid.2009.02.003.

Won, S., Ikegami, T., Peters, C. J., & Makino, S. (2006). NSm protein of Rift Valley fever virus suppresses virus-induced apoptosis. *Journal of Virology*, 81(24), 13335–13345. https://doi.org/10.1128/JVI.01165-07.

World Health Organization (WHO). (2020). *Rift Valley Fever*. https://www.who.int.

World Health Organization (WHO). (2022). *Viral Haemorrhagic Fevers*. https://www.who.int.

4 Re-Emerging Zoonotic Viruses

Lassa Fever Viruses and Nipah Viruses

Abhishek Jana, Dibya Das, Himangshu Sekhar Maji, Pintu Kumar De, Sudipta Saha, Anupa Ruwali, Easha Biswas, and Debjani Sarkar

4.1 INTRODUCTION

The Greek term "zoon," which denoted animal, and "nosos," which denoted illness, are the sources of the term "zoonoses." The World Health Organization (WHO) explains zoonosis as any infection or illness that can be naturally transmitted from animals to humans and vertebrate animals. The environment, animals, and people all play significant roles in the development and spread of various infectious diseases (Thompson & Kutz, 2019). Most of the infectious diseases that impact humans have animal origins; approximately 60% of newly discovered infections in humans are zoonotic, with over 70% of these pathogens originating from wildlife animals (Rahman et al., 2020). Recent decades have witnessed the emergence of diseases in humans directly linked to animal products. Zoonoses represent a serious threat to human health, potentially even fatal, and thus constitute a major public health concern. Worldwide, they are responsible for an estimated 2.7 million human fatalities and 2.4 billion cases of disease annually (Grace et al., 2012). These diseases have particularly adverse effects on low- and middle-class livestock workers in affected nations. Zoonoses are categorized into various groups based on their etiology: parasitic (e.g., trichinosis, trematodosis, and malaria), viral (e.g., rabies, AIDS, Ebola, Nipah, and Lassa), and bacterial (e.g., anthrax, salmonellosis, and tuberculosis) (Burrell et al., 2016). Two RNA-based viral zoonotic viruses, Lassa fever virus and Nipah virus, have caused various outbreaks worldwide. Both are classified as level IV pathogens and listed as epidemic-causing viruses by the WHO.

4.2 LASSA FEVER VIRUS: THE FUNDAMENTALS

The notable incident occurred in 1969, thrusting Lassa fever (LF) into the spotlight. A small town named Lassa, situated in the northeastern part of Nigeria, reported the deaths of two nurses in a missionary hospital. The first nurse became infected with a hemorrhagic fever, and the second nurse, attending to her, also contracted the disease. Both nurses succumbed within a few days, shedding light on the existence of LFV. Surprisingly, around 20 years earlier, in the early 1950s, similar symptoms were observed in Sierra Leone. Unfortunately, due to negligence, this incident was not reported in the 1950s. Following the 1969 incident, several outbreaks with similar clinical symptoms were reported in various towns and villages of Nigeria. Reports of Lassa fever outbreaks began emerging from Nigeria, Liberia, Senegal, Guinea, Mali, and Sierra Leone after 1970. Epidemics of this disease began to be documented in West African and Central African countries. International travel and the posting of UN peacekeeping forces in LF endemic zones spread the disease to different portion of the world (McCormick et al., 1987a). The first LF case in Europe was reported in the year 2000, from a patient traveling in an epidemic area in Ghana. The natural reservoir host of this virus is an African rodent of the genus *M. natalensis*. Around 50–100% of

DOI: 10.1201/9781032721811-4

rodents caught in endemic regions are infected with the Lassa virus (McCormick et al., 1987a). Urine and contaminated food of Mastomys are likely the main sources of LFV transmission to humans. Year by year, the tally of annually infected persons in West Africa is increasing, from thousands of cases annually in old reports to 300,000–500,000 cases annually in present reports. Among all those cases, only 0–15% of people in Guinea, 10–16% in Sierra Leone, and 14.3% in Liberia require hospitalization (Nwafor et al., 2020; Frame, 1989; McCormick & Fisher-Hoch, 2002). The fatality rate among hospitalized patients ranges from 9.3% to 18% in epidemic situations. Annually, 5,000 people die due to LFV in the present situation, compared to 1,000 deaths per year previously. In 1979, an antiviral drug, ribavirin, began to be used as a treatment medication against Lassa hemorrhagic fever (Bausch & Mills, 2014). Since then, ribavirin has been continuously used for the treatment of LF and remains the first-choice drug. This treatment has saved thousands of lives of patients around the world infected by the Lassa virus.

4.3 HISTORY OF LASSA FEVER VIRUS AND LASSA FEVER

Lassa fever became an epidemic in West African nations stretching from Cameroon to Senegal. Since 1969, several outbreaks have been reported worldwide, with most occurring in West Africa. Notable outbreaks in Nigeria include those in 1970 in the Plateau State capital, Jos; 1974–1977 in Zonkwua; 1985 in Owerri and Abo Mbaise, Imo State; 1990–1992 and 2012 in Ekpoma, Delta State; 1992 and 2012 in Lafiya, Nasarawa State; 2005, 2007, 2009, 2011, and 2012 in Abakaliki, Ebonyi State; 2012 in Lagos; 2012 in Ondo; and 2012 in Port Harcourt, Rivers State (Bausch & Mills, 2014). As per the report of WHO, in 2018, an epidemic of LF occurred in Nigeria, where approximately1,495 people were infected, resulting in 119 deaths across 19 states of Nigeria (Tomori et al., 1988). Population data shows the presence of LFV antibodies in Nigeria (21%) (McCormick et al., 1987b), Sierra Leone (8–52%) (Lukashevich et al., 1993), and Guinea (4–55%) (Richmond & Baglole, 2003). The prevalence of LF is high in the West African region due to poor healthcare systems, unstable governments, and civil unrest. The number of infections increases during the dry season, which may be related to the virus' better stability in lower humidity.

4.4 TAXONOMIC CLASSIFICATION AND NOMENCLATURE

Lassa hemorrhagic fever (LHF) known as Lassa fever, is a zoonotic disease caused by a highly contagious Level IV pathogen, the Lassa fever virus (LFV). LFV belongs to the viral family Arenaviridae, which is a negative-stranded RNA virus family (Günther & Lenz, 2004). This Arenaviridae family can be further separated into two clusters, one of which is the Old World complex, to which LFV belongs (Nwafor et al., 2020).

4.5 TRANSMISSION PROCESS OF LASSA FEVER VIRUS

The natural vector of this virus is the African native multimammate rat (*M. natalensis*) (Ter Meulen et al., 2001). This rodent is asymptomatic and spreads throughout sub-Saharan Africa, predominantly in rural areas close to human civilization. Transmission from these rats occurs through various means, including their excreted products, infected leftover food, and aerosols produced from rat urine. Additionally, consumption of rat meat as a delicacy in the West African region provides another route of transmission of LFV. Human-to-human transmission of LFV occurs through exposure to infectious blood and bodily fluids from oral and mucosal membranes. Nosocomial transmission significantly increases the infection rate due to poor medical infrastructure and faulty nursing practices in this economically backward region. Transmission from West Africa to other parts of the world occurs through the movement of infected persons. Foreign medical aid workers and travelers returning from epidemic areas predominantly carry this disease to other parts of the world. Also, foreign military personnel and UN peacekeepers posted in those regions of civil unrest carry LFV back home.

4.6 LFV: MORPHOLOGY, GENOMIC ORGANIZATION, AND LIFE CYCLE

4.6.1 Basic Structure of LFV (Figure 4.1)

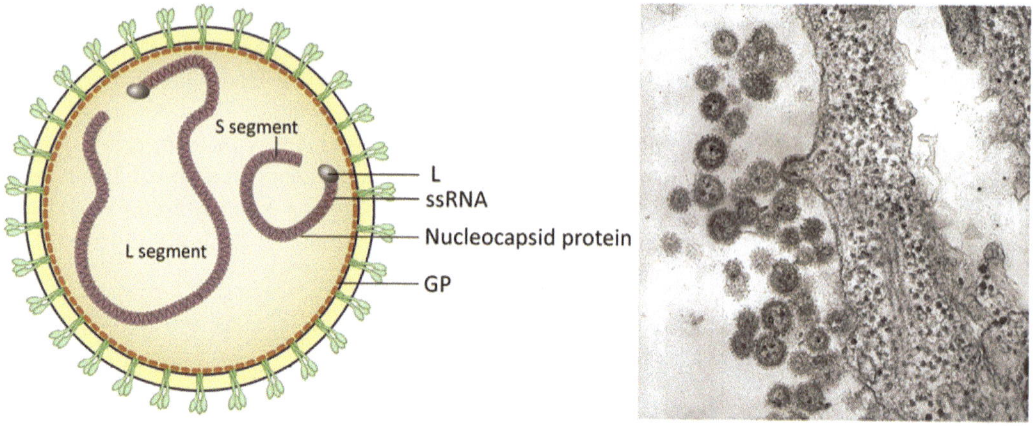

FIGURE 4.1 The morphology of a virion (*left*). The Machupo virus electron micrograph (*right*). The Machupo virus, which belongs to the Arenavirus family, was discovered in isolation in Bolivia's Beni Province in 1963.

Source: **Reprinted from Payne. S, Family Arenaviridae, *Viruses*, 215–218, 2017, with permission from Elsevier.**

4.6.2 Genomic Organization of LFV (Figure 4.2)

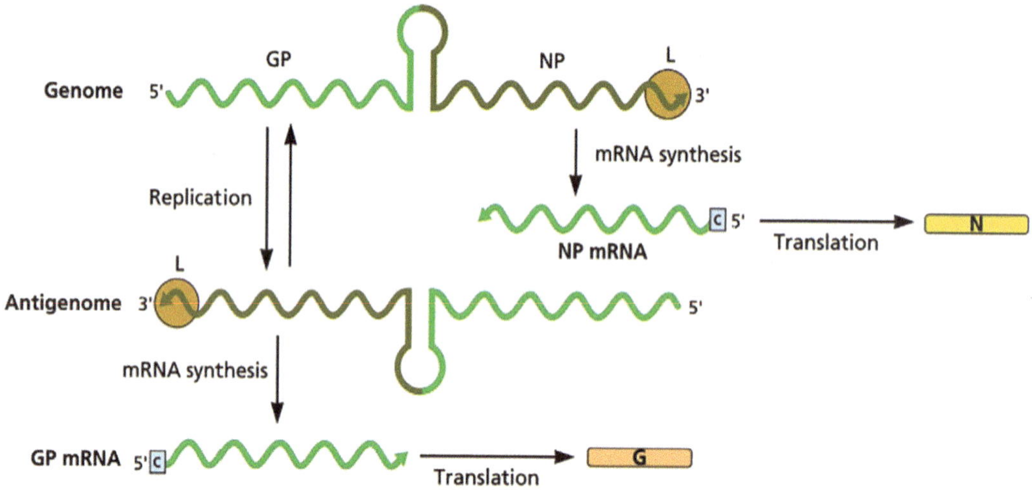

FIGURE 4.2 Arenavirus genome organization and replication. Two-number of ambisense RNA segments, large segments denoted as (L; 7.2 kb) and short segments denoted as (S; 3.5 kb), make up the viral genome. An accessory protein (Z) and RNA-dependent RNA polymerase (L) involved in budding particle assembly, and genome packaging are encoded by the L segment. The nucleocapsid in conjunction with the viral RNAs formed by the S segment encodes a histone-like nucleocapsid protein (NP), as well as the surface glycoprotein (GP) that is attached to the viral receptor and promotes the target-cell identification and entrance. Just as the S segment's gene expression is displayed for simplicity, the L segment goes through the identical procedure. The L protein present in the virus, which enters the cell along with the infecting particle, is attached to the 3′-end of the RNA (represented as an orange ball) upon entry of the viral RNA into the cytoplasm of the host cell. In the synthesis of the (+) strand NP mRNA, this result is subsequently translated. The stem-loop structure of the virus may temporarily prevent the copying of the remaining genome. GP mRNA production is subsequently made possible by the replication of the genome of RNA into a full-length antigenome which is complementary. Temporarily, the regulation of viral gene expression is the outcome of this gene expression mechanism, and many viruses with DNA genomes share this characteristic in their reproductive cycles.

Source: **Reprinted from Burrell. Christopher J, Arenaviruses, *Fenner and White's Medical Virology*, 425–436, 2017, with permission from Elsevier.**

4.7 LASSA FEVER: PATHOPHYSIOLOGY AND CLINICAL PRESENTATION

4.7.1 Pathophysiology of LFV Infection

The entry and attachment of the Lassa virus to the host cell involve several mechanisms. The viral envelope contains glycoproteins, specifically GP1, GP2, two sites on GP1, and six sites on GP2, which facilitate viral attachment to the host cell (Yun & Walker, 2012; Pennington, 2021). GP1 and GP2 are responsible for binding to the cell surface receptors (Murphy & Ly, 2021). When the Lassa virus enters the body, it primarily targets the human or animal respiratory system, causing symptoms such as fever, cough, and difficulty breathing. The virus then disseminates to various tissues and organs, including the liver and spleen (Schlie et al., 2010).

4.7.1.1 Entry and Attachment to Host Cell

Lassa virus moves into the body through respiratory droplets or contacts with contaminated surfaces (Lopez & Arias, 2010). The virus binds to specific receptors on the cells of hosts, such as endothelial cells, and its entry is motivated by the macrophage facility. Following attachment, the virus is internalized by endocytosis, a process where engulfed bodies are transported into the host cell through the formation of vesicles. Once inside the host cell, the virus enters the cytoplasm through the margins of the viral envelope with the endosomal membrane (Hensley et al., 2011).

4.7.1.2 Transcription and Translation

Once in the cytoplasm, the Lassa virus RNA genome is transcribed into mRNA. The viral RNA is once inside the cell, the Lassa virus hijacks the cellular machinery and replicates its genetic material, producing new copies of the virus. The viral RNA utilizes host cell enzymes to synthesize viral proteins, which assemble into new viral particles. This process continues, leading to an exponential increase in viral progeny (Hensley et al., 2011).

4.7.1.3 Host Cell Damage

Lassa virus primarily targets endothelial and epithelial cells, causing widespread inflammation and tissue damage (Schlie et al., 2010). The virus replicates primarily in the cytoplasm, leading to the production and release of more viral particles. The infected cells also undergo lysis, releasing the virus into the surrounding tissues (Pennington & Lee, 2022). In Lassa virus infection, the liver is one of the primary targets. The liver cells are replicated by the virus and release virions into the bloodstream, causing liver damage. The spleen becomes inflamed, leading to enlargement and hemorrhagic complications. The spleen plays a vital role in the filtration of blood and the removal of damaged or infected cells. However, Lassa virus infection can impair its normal function, leading to bleeding disorders and an increased risk of infection by other microorganisms (Talukdar et al., 2023).

4.7.1.4 Hemorrhagic Complications

One of the hallmark features of Lassa fever is hemorrhage, particularly thrombocytopenia, which is a low platelet count in the blood. Thrombocytopenia leads to increased bleeding and clotting complications, which can contribute to organ damage. The virus can also damage the vascular system, leading to hemorrhage (Murphy & Ly, 2022).

4.7.1.5 Immune Response

The host's immune response to the Lassa virus also contributes to the pathogenesis. The virus elicits both humoral and cellular immune responses, including the production of neutralizing antibodies and the activation of immune cells. However, the severity of the disease is often influenced by the balance between the immune response and viral replication. This response involves both innate and adaptive immune responses (Russier et al., 2012). The innate immune system serves as the first line of defense against pathogens. Upon infection, cells release cytokines and chemokines, leading to the recruitment of immune cells. Natural killer (NK) cells, macrophages, and dendritic cells play a role in the initial recognition and destruction of infected cells (Russier et al., 2012; Torriani et al., 2017). The adaptive immune response involves the production of antibodies by B cells and the activation of T cells. Antigen-presenting cells (APCs) present viral antigens to T cells, leading to their activation and differentiation into specific subtypes of T cells. T helper cells (Th1 and Th2) secrete cytokines that stimulate the production of antiviral antibodies and recruit more immune cells (Torriani et al., 2017).

The pathogenesis of Lassa fever involves several stages, including viral entry, replication, immune response, and organ involvement. The virus primarily infects the respiratory system, disseminating to various tissues and organs. The immune response plays a crucial role in limiting viral replication and modulating the disease course, but can also lead to complications. The liver, spleen, and blood vessels are the primary targets of Lassa virus, leading to liver dysfunction, splenic involvement, and hemorrhagic complications. Understanding the pathogenesis is essential for developing effective control measures and improving patient care.

4.7.2 Clinical Manifestation of LFV Infection

4.7.2.1 Symptomatology

Apart from the humans, no clinical evidence has been found that other animals can show symptoms in normal conditions. However, some laboratory experiments on guinea pigs indicated fatal disease conditions after controlled inoculation. The host rodent *M. natalensis* does not show clinical symptoms of this disease (Brammer, 2012) The LF symptoms are indistinguishable from many febrile illnesses like dengue, malaria and hemorrhagic fever Ebola. It is very difficult to diagnose LF due to the variable in its presentation. The main clinical response came in patients are high fever (≥38°C) and not responding to antibiotic and antimalarial drugs. Other associated symptoms of LF are sore throat, vomiting, pharyngitis, retrosternal pain, joint pain, and diarrhea. Apart from this, complications like hearing difficulty, mucosal bleeding, and pleural and pericardial effusion can be observed. In savior clinical conditions, symptoms like pneumonia, hemorrhage, myocarditis, hepatitis, and hepatic necrosis happen. Twenty percent of people among hospitalized patients pass away after cardiovascular collapse. In the case of pregnant women, the mortality rates are high, and fatal loss is common (Brammer, 2012).

4.7.2.2 Diagnosis

At the time of acute symptomatic phase, viral nucleic acid can be detected in blood by doing RT-PCR. The enzyme-linked immunosorbent assay (ELISA) can also be performed to check the presence of antibodies against the virus in patients, where 67% are with IgM and 53% with IgG (Ibekwe, 2012). In fatal cases, viral antigens are detected by the liver tissue by using the instrument immunofluorescence, ELISA, and RT-PCR.

4.8 NIPAH VIRUS DISEASE: THE FUNDAMENTALS

NiV was founded in 1998 during its first outburst in Sungai Nipah village in Malaysia (Luby et al., 2006). At that time humans got infected with this virus due to the consumption of NiV-infected pigs. Since then, NiV spread over the South and Southeast Asian region and re-emerges in various locations. After the first outbreak, in 2001 another outbreak happened in Bangladesh due to the utilization of raw date palm as food which was infected with bat saliva or excreted products. Also, in the same year, similar incidents are reported in Siliguri, the district of West Bengal, India (Arunkumar et al., 2019). In 2007, another outbreak was reported in the Nadia district of West Bengal, India (Arunkumar et al., 2019). The most recent case in 2018 was also from the Kozhikode district of Kerala, India (Yob et al., 2001). This virus now spreads around Malaysia, India, Bangladesh, Singapore, and in most Southeast Asian countries. The natural reservoir of this NiV is giant fruit bats, five species of *Pteropus* spp. are currently known. Partially eaten fruit of these bats contains NiV also from the intermediate hosts like pigs can spread this virus (Nor et al., 2000). More than 250 people got infected by NiV during the first outbreak (Luby et al., 2006). In comparison to the LFV, infection rate of NiV is low. However, the mortality rate of NiV is very high, the recent outbreak in Kerala has a mortality rate of 91% (Yob et al., 2001). The antiviral drug Ribavirin works as a supportive treatment against NiV (Luby et al., 2006). WHO puts NiV in the list of pathogens which can cause outbreaks and needs urgent research and development action.

4.9 HISTORY OF NIV DISEASE AND NIV INFECTION

When the first outbreak happened in Malaysia and Singapore during 1998–1999, it was initially diagnosed as Japanese encephalitis (JE). But when JE treatments did not work on those patients, an investigation was done. Then it was found as Nipah virus which was closely related to HeV. In 2001, the first outbreak of NiV happened in Bangladesh; during that winter, 20 districts faced this outbreak. From 2001 to 2015, a total of 17 outbreaks were reported in Bangladesh [31]. In India, the outbreak happened in 2001 and 2007, and the most recent one in 2018 in Kerala (Banerjee et al., 2019; Ibekwe, 2012; Ching et al., 2015), with an 82% fatality rate. In 2014, In the Philippines NiV outbreak happened (Looi & Chua, 2007).

4.10 TAXONOMIC CLASSIFICATION AND NOMENCLATURE OF NIV

The Nipah virus (NiV) is a bat-borne highly infectious level IV pathogen that causes lethal encephalitis in humans. NiV comes from the negative single-stranded RNA viral family Paramyxoviridae (Shariff, 2019). NiV is closely related to another zoonotic virus Hendra virus (HeV), which was discovered in 1994 in Australia (Banerjee et al., 2019).

4.11 TRANSMISSION PROCESS OF NIV

The natural source of NiV is Pteropid bats (*Pteropus* spp.), which are found around Southeast Asia. Partially eaten fruits and contaminated fruit by excretory products of Ni- infected bats spread the virus (Luby, 2013). It can be transmitted to humans by consuming those fruits, and direct contact with body fluid or aerosols. Also, intermediate transmission of virus happens through pigs, like pig consumes contaminated food and human consumes that pig meat (Hayman & Johnson, 2014). Even those pigs' urine, respiratory secretion, and saliva contain NiV (Ching et al., 2015). Apart from all this, inter-person human to human transmission is also reported, and due to this nosocomial spread happens a lot.

4.12 NIV: MORPHOLOGY, GENOMIC ORGANIZATION, AND LIFE CYCLE

4.12.1 Basic Structure of Nipah Virus (Figure 4.3)

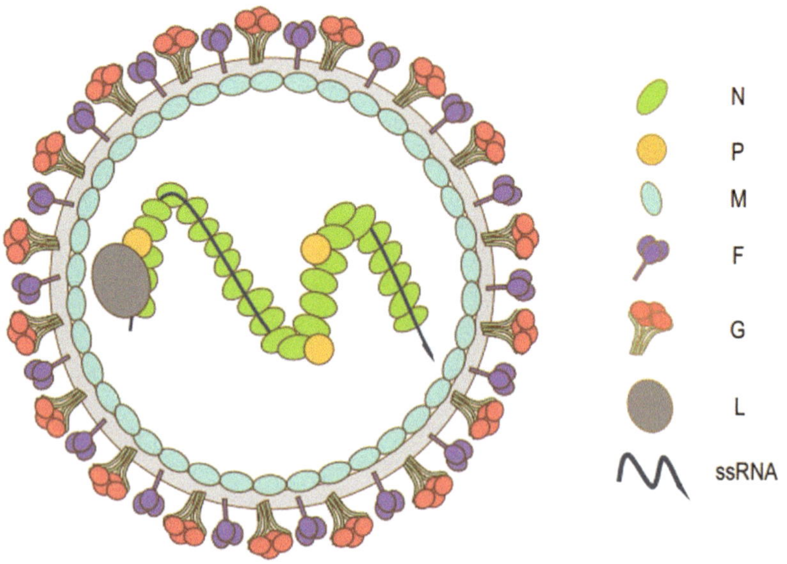

FIGURE 4.3 Schematic representation of the NiV viral structure.

Source: Reprinted from Bangyao. S, Phylogeography, Transmission and Viral Proteins of Nipah Virus, *Virological Sinica*, 215–218, 2017, with permission from Springer.

4.12.2 Genomic Organization of Nipah Virus (Figure 4.4)

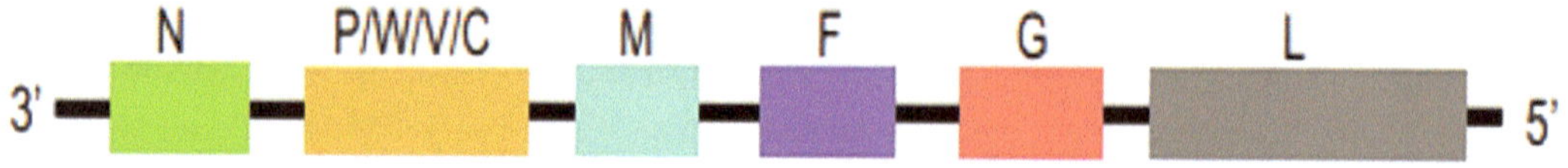

FIGURE 4.4 Genome organization of NiV.

Source: Reprinted from Bangyao. S, Phylogeography, Transmission and Viral Proteins of Nipah Virus, *Virological Sinica*, 215–218, 2017, with free-to-use permission from Springer.

4.13 MOLECULAR PATHOPHYSIOLOGY AND CLINICAL PRESENTATION

4.13.1 Pathophysiology of NiV

Normally, the NiV is transmitted through respiratory droplets from an infected person when they cough or sneeze. In addition to direct contact with contaminated surfaces, it can also be contaminated through close contact with saliva, the secretion of respiratory, or other body fluids (Hassan et al., 2018). The NiV can survive on inanimate surfaces for several days (from a few hours to up to two days), and is highly dependent on pH, temperature, and desiccation, making it more likely to spread through the environment (Lamp et al., 2013).

4.13.1.1 Entry and Replication of NiV

Once inhaled by an infected individual, the Nipah virus spreads through the respiratory tract by the epithelial cells of the nasal cavity, throat, and lungs. When viral surface glycoprotein G binds to Ephrin-B2 or Ephrin-B3 receptors on the cell surface, then the cellular infections begin. In the host cell membrane, the fusion protein, F, facilitates pH-independent fusion between the viral envelope and the host cell membrane. Following initial replication in the airways, NiV spreads throughout the body. As the infection progresses into the viremic phase, endothelial cells in many organs are targeted by NiV (Prescott et al., 2012).

4.13.1.2 Inflammatory Response

In response to Nipah virus infection, the body's immune system mounts an inflammatory response. This includes the production of cytokines and chemokines, which can attract immune cells to the site of infection. However, the exaggerated immune response can actually contribute to the severity of disease, leading to cytokine storm, inflammation, and damage to various organs (Singh et al., 2019).

4.13.1.3 Neurological Involvement

Nipah virus infection can also affect the central nervous system, leading to neurological complications. As a result of NiV infection, several neurological symptoms develop, such as encephalopathy, cerebral atrophy, changes in behavior, ocular motility palsies, cervical dystonia, weakness, and facial paralysis over time (Baseler et al., 2016).

4.13.1.4 Respiratory Distress

One of the hallmark features of Nipah virus infection is the respiratory distress it causes. The virus can damage the respiratory epithelium, leading to inflammation, fluid buildup in the lungs, and impaired lung function. This can result in difficulty in breathing, chest pain, and hypoxia (low oxygen levels in the blood).

4.13.1.5 Organ Dysfunction and Multi-Organ Failure

Nipah virus infection can lead to organ dysfunction and multi-organ failure. The virus can invade multiple organs, including the brain, kidneys, heart, and lungs. This can cause severe complications, such as encephalitis, acute respiratory distress syndrome (ARDS), kidney failure, and cardiac arrhythmias.

4.13.2 Clinical Manifestation of NiV Infection

4.13.2.1 Symptomatology

The major clinical symptoms include fever with encephalitis along with respiratory problems. Apart from this dizziness, vomiting, seizures, headache, loose stools, and myalgia are also being reported during various outbreaks (Luby et al., 2006). Clinical data from Malaysia also indicate deamination of consciousness and dysfunction of the brainstem among 55% of patients. Due to this brainstem dysfunction in some critical patients' hypotonia, tachycardia, areflexia, and hypertension are common (Daniels et al., 2001). Also, vasomotor changes, myoclonic jerks, and abnormal doll's eye reflexes happen due to brain-stem dysfunction. In Malaysia, 8% of patients are found to be asymptomatic, but in India and Bangladesh, the asymptomatic patients are not recorded (Luby et al., 2006).

4.13.2.2 Diagnosis

Preliminary diagnosis can show hematological abnormalities like thrombocytopenia and leukopenia (Luby et al., 2006). Elevation of liver enzymes and lymphocytic pleocytosis along with the raised proteins, any other viral meningitis in cerebrospinal fluid (CSF) can be observed. RT-PCR can be done by using nasal swabs, cerebrospinal fluid, urine, and blood of an infected person (Lam, 2003) Detection of serum or CSF IgM and IgG antibodies can detect infected persons. Apart from them ELISA, Serum Neutralization Test and Immunohistochemistry can be performed (Banerjee et al., 2019).

4.14 INFECTION PREVENTION AND CONTROL: CHALLENGES AND FUTURE PROSPECTS

4.14.1 Prevention and Control of Lassa Fever Virus (LFV)

First, the rodent population should be controlled as they are the natural host for LFV. The sanitation system and the medical system should be overhauled in the West African region as they are the main reasons for making LF an epidemic. The diagnosis procedure should be improved to date due to the false diagnosis of patients with LF getting treated with antimalarial drugs. Nosocomial transmission can be controlled by continuously monitoring those who have direct contact with Lassa fever patients. Previously, efforts have been made to make the Lassa vaccine. Inactivated vaccines have been tested which provide a good humoral response but are unable to provide complete solutions. Then another attempt has been done to make a recombinant vaccine which was highly successful but due to its side effects, it was not used in clinical practice (Nwafor et al., 2020). The most effective drug for LF is ribavirin, which is a guanosine analog antiviral drug. Some of the clinical data shows ribavirin is twofold more effective when used intravenously than orally. Intravenous administration of a 2-g loading dose, followed by 1 g every six hours for four days and then 0.5 g every eight hours for six more days is the most effective way to provide medication (McCormick & Fisher-Hoch, 2002). It can reduce the death of critical patients by 90%. Favipiravir. an alternative to Ribavirin, is now used which shows positive results; in a few conditions, Ribavirin and Favipiravir are even given in combination (Bausch & Mills, 2014).

4.14.2 Probable Bioweapon Application of Lassa Fever Virus (LFV)

Meyer et al. have indicated how infectious diseases can be used as a bioweapon by making the artificial product of contagious microorganisms (Meyer & Morse, 2008). Ibekwe et al. also show concern over the use of LFV as a bioweapon due to its aerosol transmission pattern (Ibekwe, 2012). Some evidence is also available on how the United States and the Soviet Union have experimented on LFV in their bioweapon program (Borio et al., 2002).

4.14.3 Prevention and Control of NiV Infection

Preventive measures can be taken, such as visually inspecting fruits before consuming them, and keeping sanitation standards high in pig farms as they are the intermediate source of contamination. Proper surveillance and improving medical facilities can reduce the risk of nosocomial transmission. Many vaccination approaches against NiV have been proposed, and some of them have been evaluated in animal models using the G-glycoprotein of NiV (Bossart et al., 2009). The ephrin-B2 and ephrin-B3 receptors targeted the binding region of the hepatitis virus G envelope glycoprotein, the experimental monoclonal antibody m102.4 is a strong cross-reactive neutralizing antibody that works effectively against NiV (Aljofan et al., 2009). The antiviral drug ribavirin has also proven clinically successful in treating NiV-infected patients (Dawes et al., 2018). Similar to LFV, Favipiravir also works against NiV infection due to its RNA polymerase inhibitory mechanism (Ruo et al., 1991).

4.14.4 Probable Bioweapon Application of NiV Infection

According to Lam et al., NiV possesses numerous physical characteristics that could make it a viable bioterrorism agent (Prasad et al., 2020). The Nipah Virus (NiV) is a serious public health concern due to its high case fatality rates and potential for use as a biological weapon, as stated by Prasad et al. (Pallister et al., 2013). NiV was found to cause brain damage, and although this was a naturally occurring phenomenon, many experts suspected it might be a biological weapon.

4.15 CONCLUSION

The discussion above on Lassa fever virus (LFV) and Nipah virus (NiV) highlights the critical importance of understanding and addressing zoonotic diseases at both regional and global levels. Zoonoses, diseases transmitted between animals and humans, pose significant threats to public health, as evidenced by the devastating impact of LFV and NiV outbreaks. LFV, originating from the *M. natalensis* rodent in West Africa, has manifested as a persistent public health challenge, leading to recurrent epidemics with increasing annual infection rates. NiV, on the other hand, emerges from giant fruit bats in Southeast Asia. It presents a similar narrative of zoonotic spillover and subsequent human infections. The consumption of NiV-infected fruits and the intermediate transmission through pigs have contributed to outbreaks with alarmingly high mortality rates. The virus' ability to spread from rodents to humans through urine and contaminated food, coupled with human-to-human transmission, highlights the complex interplay between ecological factors, human behavior, and healthcare infrastructure. Likewise, the respiratory distress, neurological complications, and multi-organ failure associated with NiV infection underline the diverse clinical manifestations of zoonotic viruses. Moreover, the mention of potential bioweapon applications raises concerns about the intentional misuse of these infectious agents. Historical experimentation by nations such as the United States and the Soviet Union underscores the dual-use nature of research on these pathogens. Vigilance and international cooperation are paramount to prevent the deliberate weaponization of such viruses and to ensure the responsible conduct of scientific investigations. Preventive measures for both LFV and NiV involve controlling the natural reservoirs, improving sanitation practices, and enhancing medical facilities, especially in regions with a high prevalence of these diseases. Research into vaccines and antiviral drugs, such as ribavirin and favipiravir, demonstrate ongoing efforts to develop effective tools for treatment and prevention. A comprehensive, interdisciplinary approach is essential in confronting the challenges posed by LFV, NiV, and other zoonotic threats. Collaboration between nations, researchers, and healthcare providers is crucial to understanding the complex dynamics of these viruses, developing effective interventions, and safeguarding global public health. The lessons learned from LFV and NiV emphasize the need for sustained investment in research, surveillance, and international cooperation to mitigate the impact of emerging zoonotic diseases on humanity.

REFERENCES

Aljofan, M., Saubern, S., Meyer, A. G., Marsh, G., Meers, J., & Mungall, B. A. (2009). Characteristics of Nipah virus and Hendra virus replication in different cell lines and their suitability for antiviral screening. *Virus Research*, 142(1–2), 92–99.

Arunkumar, G., Chandni, R., Mourya, D. T., Singh, S. K., Sadanandan, R., Sudan, P., & Bhargava, B. (2019). Outbreak investigation of Nipah virus disease in Kerala, India, 2018. *The Journal of Infectious Diseases*, 219(12), 1867–1878.

Banerjee, S., et al. (2019). Nipah virus disease: A rare and intractable disease. *Intractable & Rare Diseases Research*, 8(1), 1–8.

Baseler, L., et al. (2016). Identifying early target cells of Nipah virus infection in Syrian hamsters. *PLoS Neglected Tropical Diseases*, 10(11), e0005120.

Bausch, D. G., & Mills, J. N. (2014). Arenaviruses: Lassa fever, lujo hemorrhagic fever, lymphocytic choriomeningitis, and the South American hemorrhagic fevers. In *Viral Infections of Humans: Epidemiology and Control* (pp. 147–171). Boston, MA: Springer.

Borio, L., et al. (2002). Hemorrhagic fever viruses as biological weapons: Medical and public health management. *Jama*, 287(18), 2391–2405.

Bossart, K. N., et al. (2009). A neutralizing human monoclonal antibody protects against lethal disease in a new ferret model of acute nipah virus infection. *PLoS Pathogens*, 5(10), e1000642.

Brammer, D. W. (2012). *Zoonoses and Occupational Health. The Laboratory Rabbit, Guinea Pig, Hamster, and Other Rodents* (pp. 141–153).

Burrell, C. J., Howard, C. R., & Murphy, F. A. (2016). *Fenner and White's Medical Virology*. Academic Press.

Ching, P. K. G., et al. (2015). Outbreak of henipavirus infection, Philippines, 2014. *Emerging Infectious Diseases*, 21(2), 328.

Daniels, P., Ksiazek, T., & Eaton, B. T. (2001). Laboratory diagnosis of Nipahand Hendra virus infections. *Microbes and Infection*, 3(4), 289–295.

Dawes, B. E., et al. (2018). Favipiravir (T-705) protects against Nipah virus infection in the hamster model. *Scientific Reports*, 8(1), 7604.

Frame, J. D. (1989). Clinical features of Lassa fever in Liberia. *Reviews of Infectious Diseases*, 11(Suppl. 4), S783–S789.

Grace, D., et al. (2012). Mapping of poverty and likely zoonoses hotspots. Chomel, B. B. Zoonoses. In *Encyclopedia of Microbiology* (3rd ed., pp. 820–829). Davis, CA, USA: Elsevier Inc., University of California (Original work published 2009).

Günther, S., & Lenz, O. (2004). Lassa virus. *Critical Reviews in Clinical Laboratory Sciences*, 41(4), 339–390.

Hassan, M. Z., et al. (2018). Nipah virus contamination of hospital surfaces during outbreaks, Bangladesh, 2013–2014. *Emerging Infectious Diseases*, 24(1), 15.

Hayman, D. T., & Johnson, N. (2014). Nipah virus: A virus with multiple pathways of emergence. In *The Role of Animals in Emerging Viral Diseases* (pp. 293–315). Academic Press.

Hensley, L. E., Smith, M. A., Geisbert, J. B., Fritz, E. A., Daddario-DiCaprio, K. M., Larsen, T., & Geisbert, T. W. (2011). Pathogenesis of Lassa fever in cynomolgus macaques. *Virology Journal*, 8(1), 1–15.

Ibekwe, T. (2012). Letter to the editors-Lassa fever: The challenges of curtailing a deadly disease. *Pan African Medical Journal*, 11(1).

Lam, S. K. (2003). Nipah virus—a potential agent of bioterrorism? *Antiviral Research*, 57(1–2), 113–119.

Lamp, B., Dietzel, E., Kolesnikova, L., Sauerhering, L., Erbar, S., Weingartl, H., & Maisner, A. (2013). Nipah virus entry and egress from polarized epithelial cells. *Journal of Virology*, 87(6), 3143–3154.

Looi, L. M., & Chua, K. B. (2007). Lessons from the Nipah virus outbreak in Malaysia. *Malaysian Journal of Pathology*, 29(2), 63–67.

Lopez, S., & Arias, C. (2010). How viruses hijack endocytic machinery. *Nature Education*, 3(9), 16.

Loureiro, M. E., D'Antuono, A., & López, N. (2019). Virus–host interactions involved in Lassa virus entry and genome replication. *Pathogens*, 8(1), 17.

Luby, S. P. (2013). The pandemic potential of Nipah virus. *Antiviral Research*, 100(1), 38–43.

Luby, S. P., et al. (2006). Foodborne transmission of Nipah virus, Bangladesh. *Emerging Infectious Diseases*, 12(12), 1888.

Lukashevich, I. S., Clegg, J. C. S., & Sidibe, K. (1993). Lassa virus activity in Guinea: Distribution of human antiviral antibody defined using enzyme-linked immunosorbent assay with recombinant antigen. *Journal of Medical Virology*, 40(3), 210–217.

McCormick, J. B., & Fisher-Hoch, S. P. (2002). Lassa fever. In *Arenaviruses I: The Epidemiology, Molecular and Cell Biology of Arenaviruses* (pp. 75–109).

McCormick, J. B., et al. (1987a). A case-control study of the clinical diagnosis and course of Lassa fever. *Journal of Infectious Diseases*, 155(3), 445–455.

McCormick, J. B., Webb, P. A., Krebs, J. W., Johnson, K. M., & Smith, E. S. (1987b). A prospective study of the epidemiology and ecology of Lassa fever. *Journal of Infectious Diseases*, 155(3), 437–444.

Meyer, R. F., & Morse, S. A. (2008). Viruses and bioterrorism. *Encyclopedia of Virology*, 406.

Murphy, H. L., & Ly, H. (2021). Pathogenicity and virulence mechanisms of Lassa virus and its animal modeling, diagnostic, prophylactic, and therapeutic developments. *Virulence*, 12(1), 2989–3014.

Murphy, H. L., & Ly, H. (2022). Understanding immune responses to Lassa virus infection and to its candidate vaccines. *Vaccines*, 10(10), 1668.

Nor, M., Gan, C. H., & Ong, B. L. (2000). Nipah virus infection of pigs in peninsular Malaysia. *Revue scientifique et technique (International Office of Epizootics)*, 19(1), 160–165.

Nwafor, I. E., et al. (2020). Prevalence and outcome of Lassa fever among hospitalized patients in Ebonyi State, Nigeria, 2018–2019. *Virus Research*, 285, 198000.

Pallister, J. A., Klein, R., Arkinstall, R., Haining, J., Long, F., White, J. R., & Middleton, D. (2013). Vaccination of ferrets with a recombinant G glycoprotein subunit vaccine provides protection against Nipah virus disease for over 12 months. *Virology Journal*, 10, 1–7.

Pennington, H. N., & Lee, J. (2022). Lassa virus glycoprotein complex review: Insights into its unique fusion machinery. *Bioscience Reports*, 42(2), BSR20211930.

Prasad, A. N., et al. (2020). A lethal aerosol exposure model of Nipah virus strain Bangladesh in African green monkeys. *The Journal of Infectious Diseases*, 221(Suppl. 4), S431–S435.

Prescott, J., de Wit, E., Feldmann, H., & Munster, V. J. (2012). The immune response to Nipah virus infection. *Archives of Virology*, 157, 1635–1641.

Rahman, M. T., Sobur, M. A., Islam, M. S., Ievy, S., Hossain, M. J., El Zowalaty, M. E., & Ashour, H. M. (2020). Zoonotic diseases: Etiology, impact, and control. *Microorganisms*, 8(9), 1405.

Richmond, J. K., & Baglole, D. J. (2003). Lassa fever: Epidemiology, clinical features, and social consequences. *BMJ*, 327(7426), 1271–1275.

Ruo, S. L., Mitchell, S. W., Kiley, M. P., Roumillat, L. F., Fisher-Hoch, S. P., & McCormick, J. B. (1991). Antigenic relatedness between arenaviruses defined at the epitope level by monoclonal antibodies. *Journal of General Virology*, 72(3), 549–555.

Russier, M., Pannetier, D., & Baize, S. (2012). Immune responses and Lassa virus infection. *Viruses*, 4(11), 2766–2785.

Schlie, K., Maisa, A., Freiberg, F., Groseth, A., Strecker, T., & Garten, W. (2010). Viral protein determinants of Lassa virus entry and release from polarized epithelial cells. *Journal of Virology*, 84(7), 3178–3188.

Shariff, M. (2019). Nipah virus infection: A review. *Epidemiology & Infection*, 147, e95.

Singh, R. K., et al. (2019). Nipah virus: Epidemiology, pathology, immunobiology and advances in diagnosis, vaccine designing and control strategies–a comprehensive review. *Veterinary Quarterly*, 39(1), 26–55.

Talukdar, P., Dutta, D., Ghosh, E., Bose, I., & Bhattacharjee, S. (2023). Molecular pathogenesis of Nipah virus. *Applied Biochemistry and Biotechnology*, 195(4), 2451–2462.

Ter Meulen, J., Lenz, O., Koivogui, L., Magassouba, N. F., Kaushik, S. K., Lewis, R., & Aldis, W. (2001). Lassa fever in Sierra Leone: UN peacekeepers are at risk. *Tropical Medicine & International Health*, 6(1), 83–84.

Thompson, A., & Kutz, S. (2019). Introduction to the special issue on 'emerging zoonoses and wildlife'. *International Journal for Parasitology: Parasites and Wildlife*, 9, 322.

Tomori, O., Fabiyi, A., Sorungbe, A., Smith, A., & McCormick, J. B. (1988). Viral hemorrhagic fever antibodies in Nigerian populations. *The American Journal of Tropical Medicine and Hygiene*, 38(2), 407–410.

Torriani, G., Galan-Navarro, C., & Kunz, S. (2017). Lassa virus cell entry reveals new aspects of virus-host cell interaction. *Journal of Virology*, 91(4), 10–1128.

Yob, J. M., et al. (2001). Nipah virus infection in bats (order Chiroptera) in peninsular Malaysia. *Emerging Infectious Diseases*, 7(3), 439.

Yun, N. E., & Walker, D. H. (2012). Pathogenesis of Lassa fever. *Viruses*, 4(10), 2031–2048.

5 Filoviridae
Ebola Virus and Marburg Virus

Senjuti Bhattacharjee, Srijan Goswami, Sangita Agarwal, Anirbandeep Bose, and Sukanta Roy

5.1 EBOLA VIRUS: INTRODUCTION

The Filoviridae family encompasses a group of viruses that are notorious for causing severe hemorrhagic fevers in humans and non-human primates. These viruses are filamentous, enveloped, and possess a non-segmented, negative-sense RNA genome. Filoviridae consists of three genera: *Ebolavirus*, *Marburgvirus*, and *Cuevavirus*. Members of this family are known for their high case fatality rates, rapid transmission, and potential for causing large-scale outbreaks, making them a significant public health concern (Kuhn et al., 2019). The history of Filoviridae infections is marked by several notable outbreaks that have had a profound impact on global health. The first recorded outbreak of Marburg virus occurred in 1967 in Marburg, Germany, and Belgrade, Yugoslavia, associated with laboratory workers exposed to infected African green monkeys imported from Uganda (Martini et al., 1968). The first known outbreak of Ebola virus was documented in 1976 in two simultaneous outbreaks in Sudan and the Democratic Republic of Congo (formerly Zaire), leading to high mortality rates and raising global awareness about the potential threat posed by these viruses (Johnson et al., 1977). Over the following decades, sporadic outbreaks continued to occur, particularly in sub-Saharan Africa. The most devastating of these was the West African Ebola outbreak between 2014 and 2016, which resulted in over28,000 cases and more than 11,000 deaths, highlighting the virus' potential to spread rapidly in regions with weak healthcare infrastructures (CDC, 2016). Filoviridae viruses have demonstrated a capacity for both emergence and re-emergence, often linked to ecological disturbances and zoonotic transmission. *Ebolaviruses* and *Marburgviruses* are of particular concern due to their high transmissibility and lethality. *Ebolaviruses* include several species, such as *Zaire ebolavirus*, which has been responsible for the majority of human outbreaks. In contrast, *Marburgvirus* has caused smaller, but equally concerning outbreaks, with case fatality rates sometimes exceeding 80% (WHO, 2021). The emergence of new species and the re-emergence of known pathogens in previously unaffected areas underscore the unpredictable nature of Filoviridae infections. Factors such as deforestation, climate change, and increased human–wildlife interaction are believed to play significant roles in the spillover of these viruses from animal reservoirs, primarily fruit bats, to humans (Biedenkopf et al., 2024; Leroy et al., 2009).

This chapter aims to provide a foundational understanding of the Filoviridae family, highlighting their epidemiological significance, historical context, and the challenges they pose to global health. Given the ongoing threat of emerging infectious diseases, it is crucial to have a comprehensive grasp of the biology, transmission dynamics, and preventive strategies associated with Filoviridae infections. The information presented here serves as a prelude to the more detailed discussions in subsequent chapters, which will delve into the molecular biology, pathogenesis, clinical management, and public health implications of Filoviridae. Understanding these fundamental concepts will not only facilitate a deeper appreciation of the complexities involved in managing these infections but also underscore the importance of a coordinated global response to future outbreaks. The current chapter sets the stage for exploring the multifaceted nature of Filoviridae infections. By providing a concise yet thorough overview of the family, it prepares readers to engage with more complex

 DOI: 10.1201/9781032721811-5

topics such as vaccine development, therapeutic interventions, and the socioeconomic impacts of these diseases. As such, this chapter is essential for building a comprehensive understanding of Filoviridae and its implications for public health, ultimately aiding in the development of effective strategies to combat these formidable pathogens.

5.2 EBOLA VIRUS DISEASE: THE FUNDAMENTALS

5.2.1 Definition

Ebola virus disease (EVD), formerly known as Ebola hemorrhagic fever, is a severe, often fatal illness in humans and non-human primates caused by the *Ebolavirus* genus. This genus belongs to the Filoviridae family, which also includes the *Marburgvirus* and *Cuevavirus* genera. The term "*ebolavirus*" refers to any virus species within the genus, including *Zaire ebolavirus*, *Sudan ebolavirus*, *Bundibugyo ebolavirus*, *Tai Forest ebolavirus*, and *Reston ebolavirus*. Among these, *Zaire ebolavirus* is the most virulent and responsible for the majority of EVD outbreaks in Africa (Kuhn et al., 2019). EVD is characterized by its rapid onset and progression, involving symptoms such as fever, severe headache, muscle pain, and ultimately leading to multi-organ dysfunction and hemorrhaging. The disease is transmitted through direct contact with infected bodily fluids, making it highly contagious during the symptomatic and postmortem stages. Understanding the specific terminologies associated with EVD, such as "case fatality rate," "index case," and "zoonotic transmission," is essential for grasping the dynamics of EVD outbreaks and control measures (World Health Organization, 2021).

5.2.2 History of Ebola Virus Disease and Ebola Virus Infection

The history of Ebola virus is relatively recent compared to other viral pathogens, with the first known outbreaks occurring in 1976. The simultaneous emergence of two outbreaks in Nzara, Sudan, and Yambuku, Democratic Republic of Congo (formerly Zaire), marked the initial recognition of the virus, which was subsequently named after the Ebola River near the site of the Congo outbreak (Johnson et al., 1977). These early outbreaks were characterized by high mortality rates, with case fatality rates of 53% in Sudan and 88% in Congo, respectively (Centers for Disease Control and Prevention, 2019).

Subsequent decades saw sporadic outbreaks in sub-Saharan Africa, with varying degrees of severity. Notably, the 1995 outbreak in Kikwit, Congo, and the 2000 outbreak in Gulu, Uganda, further highlighted the virus' devastating impact on affected communities (Feldmann & Geisbert, 2011). However, the 2014–2016 West African Ebola outbreak, the largest and most complex to date, brought unprecedented global attention to EVD. This outbreak, affecting Guinea, Liberia, and Sierra Leone, resulted in over 28,000 cases and more than 11,000 deaths, prompting an international public health emergency response (Centers for Disease Control and Prevention, 2016) (Figure 5.1 and Table 5.1).

5.2.3 Taxonomic Classification

The genus *Ebolavirus* is classified within the order Mononegavirales, family Filoviridae. The genus is divided into six recognized species, each associated with different geographical regions and varying pathogenicity in humans:

- *Zaire ebolavirus (EBOV):* This species is the most pathogenic, with case fatality rates often exceeding 80%. It is responsible for the majority of EVD outbreaks, including the devastating 2014–2016 West African outbreak (WHO, 2021).
- *Sudan ebolavirus (SUDV):* Historically, SUDV has caused several outbreaks in Sudan and Uganda with a lower case fatality rate, typically between 40% and 60%.

FIGURE 5.1 Map of countries which have reported EVD cases, up to January 2023, including year of reporting and *Ebolavirus* species (Emerging Infections: How and Why They Arise, 2023).

Source: **Available under Open Government Licence v3.0.**

TABLE 5.1
EVD Cases that Occurred during the West Africa Outbreak, 2014–2016, Caused by *Zaire ebolavirus* (Emerging Infections: How and Why They Arise, 2023)

Country	Cases	Deaths	Case Fatality Rate
Guinea, Liberia, and Sierra Leone	28,610	11,308*	40%
Mali	8	6	75%
Nigeria	20	8	40%
Senegal	1	0	0%
Italy	1	0	0%
Spain	1	0	0%
UK	1	0	0%
USA	4	1	25%
Total	**28,646**	**11,323***	**40%**

**Note:* The number of deaths reported during this outbreak was likely to have been underestimated.
Source: Available under Open Government Licence v3.0.

- *Bundibugyo ebolavirus (BDBV):* First identified during an outbreak in Uganda in 2007, BDBV has caused fewer cases and has a case fatality rate of approximately 25%.
- *Tai Forest ebolavirus (TAFV):* The least understood of the *ebolaviruses*, TAFV was discovered in Côte d'Ivoire in 1994 and has only been confirmed in a single non-fatal human case.
- *Reston ebolavirus (RESTV):* Unlike other species, RESTV primarily affects non-human primates and pigs and is not known to cause disease in humans.
- *Bombali ebolavirus (BOMV):* Discovered in bats in Sierra Leone in 2018, BOMV's pathogenic potential in humans remains unknown (Amarasinghe et al., 2019).

5.2.4 Transmission of the Disease

Ebola virus transmission occurs primarily through direct contact with the blood, secretions, organs, or other bodily fluids of infected individuals, or through contact with surfaces and materials (e.g., bedding, clothing) contaminated with these fluids. The virus can also be transmitted through handling or consumption of infected animals, such as fruit bats or non-human primates, which are considered the natural reservoir hosts of the virus (Leroy et al., 2009).

Human-to-human transmission is most intense in healthcare settings or during traditional burial practices that involve close contact with the body of the deceased. This transmission dynamic makes EVD a significant challenge in regions with inadequate healthcare infrastructure and cultural practices that involve extensive contact with the deceased (CDC, 2019).

5.2.5 Symptoms Associated with Ebola Infection

The clinical presentation of EVD varies but typically progresses through several stages:

1. *Early stage (days 1–3):* Initial symptoms are non-specific and include sudden onset of fever, severe headache, muscle pain, and fatigue. These symptoms are often mistaken for more common tropical diseases such as malaria or typhoid fever (WHO, 2021).
2. *Acute stage (days 4–7):* As the disease progresses, gastrointestinal symptoms such as severe diarrhea, vomiting, and abdominal pain become prominent. This stage is often associated with severe dehydration, electrolyte imbalances, and systemic inflammatory responses (Feldmann & Geisbert, 2011).
3. *Severe stage (days 7–10):* In severe cases, patients may develop multi-organ failure, including hepatic, renal, and central nervous system involvement. Hemorrhagic manifestations, such as bleeding from the mucous membranes, gastrointestinal tract, and puncture sites, are common but not present in all cases. The likelihood of severe hemorrhage is influenced by factors such as viral load, host immunity, and supportive care availability (Feldmann & Geisbert, 2011).
4. *Recovery or death (after day 10):* Patients who survive beyond the acute phase may begin to show signs of recovery, although convalescence can be prolonged and fraught with complications such as severe fatigue, joint pain, and psychological disturbances. The case fatality rate varies widely depending on the virus species and the quality of supportive care available (WHO, 2021).

5.3 *EBOLAVIRUS*: MORPHOLOGY, GENOMIC ORGANIZATION, AND LIFE CYCLE

5.3.1 Basic Structure of *Ebolavirus*

The *Ebolavirus* is an enveloped, filamentous virus that exhibits a characteristic elongated shape, typically measuring 800–1,000 nm in length and approximately 80 nm in diameter (Figures 5.2 and 5.3). It belongs to the family Filoviridae and is known for its helical symmetry. This unique filamentous morphology, resembling a shepherd's crook, differentiates it from other viral pathogens (Kuhn et al., 2019). The viral particle is primarily composed of a lipid envelope derived from the host cell membrane. This envelope is embedded with glycoproteins (GP) that protrude from the surface, forming spikes approximately 7–10 nm long. These glycoproteins are crucial for mediating the virus' attachment and entry into host cells (Lee et al., 2008). Beneath the lipid envelope lies the viral matrix, a protein layer composed mainly of the viral matrix protein VP40. This layer provides structural integrity to the virion and plays a pivotal role in virion assembly and budding (Bornholdt & Prasad, 2008). Encased within this structure is the nucleocapsid, which consists of the viral RNA genome tightly associated with the nucleoprotein (NP) and other proteins such as VP24, VP30, and VP35. These proteins are essential for maintaining the stability of the viral genome and are involved in the replication and transcription processes within the host cell (Muhlberger, 2007).

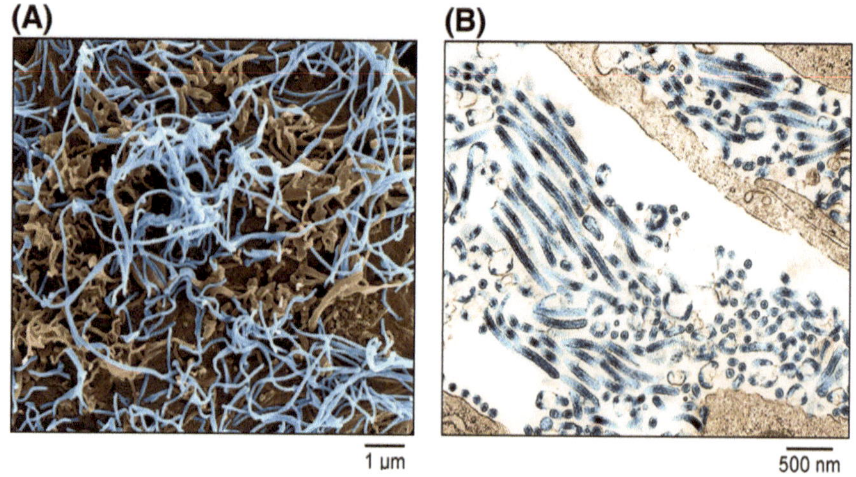

FIGURE 5.2 *Ebolavirus.* (A) Scanning electron micrograph of *Ebolavirus* particles (blue) budding from an infected grivet (*Chlorocebus aethiops*) (Linnaeus, 158) Vero E6 cells. (B) Transmission electron micrograph of *Ebolavirus* particles (blue) found both as extracellular particles and budding particles from infected Vero E6 cells.

Source: **Courtesy of John G. Bernbaum and Jiro Wada, NIH/NIAID/DCR/IRF-Frederick, Fort Detrick, MD, USA. International Committee on Taxonomy of Viruses, Licensed under Creative Commons Attribution—ShareAlike 4.0 International License.**

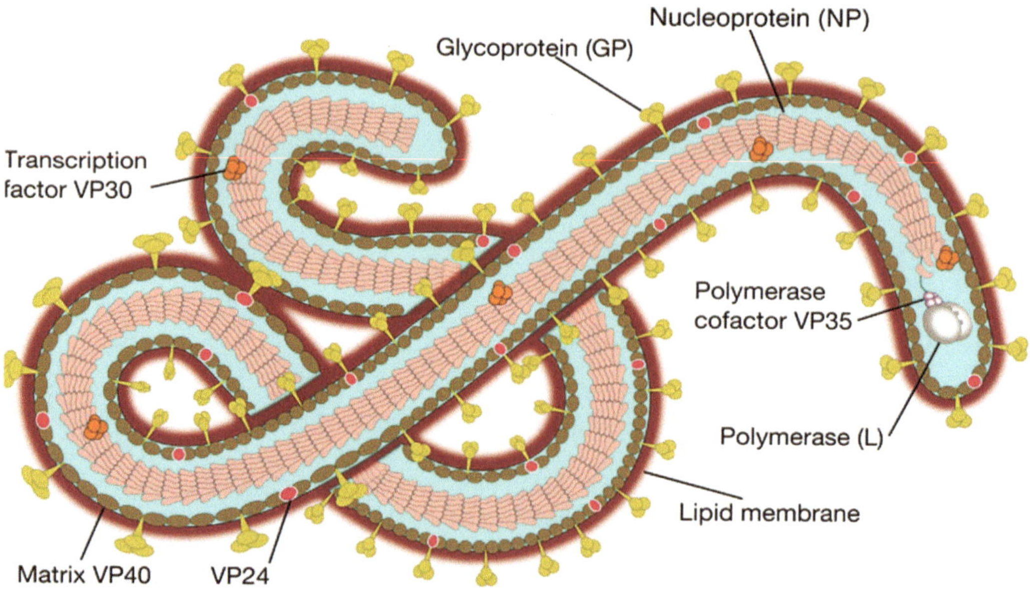

FIGURE 5.3 *Ebolavirus* molecular structure: the Ebola genome is composed of three leaders, nucleoprotein (NP), virion protein 35 (VP35), VP40, glycoprotein (GP), VP30, VP24, polymerase (L) protein, and five trailers.

Source: **Adapted from SIB SWISS Institute of Bioinformatics, 2014; Nguyen et al., 2015). Creative Commons Attribution License (CC BY) (Falasca et al., 2015).**

Understanding the intricate architecture of *Ebolavirus* is foundational for comprehending its pathogenicity and interaction with host cells.

5.3.2 Genomic Organization of Ebolavirus

The Ebolavirus genome is a single-stranded, negative-sense RNA approximately 19 kb in length, encoding seven genes that produce the following viral proteins: nucleoprotein (NP), viral proteins VP35, VP40, GP, VP30, VP24, and the RNA-dependent RNA polymerase (L) (Sanchez et al., 1996). These genes are arranged sequentially along the genome, flanked by highly conserved untranslated regions at both the 3′- and 5′-ends, which are critical for replication and transcription (Muhlberger, 2007). The general organization of these genes is as follows (Figure 5.4):

- *Nucleoprotein (NP):* Encoded by the first gene, NP encapsulates the viral RNA genome, forming the ribonucleoprotein complex. It is essential for the replication and packaging of the viral genome.
- *VP35:* This multifunctional protein acts as a cofactor for the viral polymerase and also inhibits the host's innate immune response by antagonizing interferon production.
- *VP40:* Serving as the primary matrix protein, VP40 regulates virion assembly and release. It is crucial for budding, as it interacts with the host cell membrane and drives the formation of new viral particles.
- *Glycoprotein (GP):* The sole surface protein of the virus, GP mediates viral entry into host cells by binding to cellular receptors and facilitating fusion with the host cell membrane. The GP gene also undergoes transcriptional editing to produce two different proteins: the full-length transmembrane GP and a secreted glycoprotein (sGP) that may play a role in immune evasion.

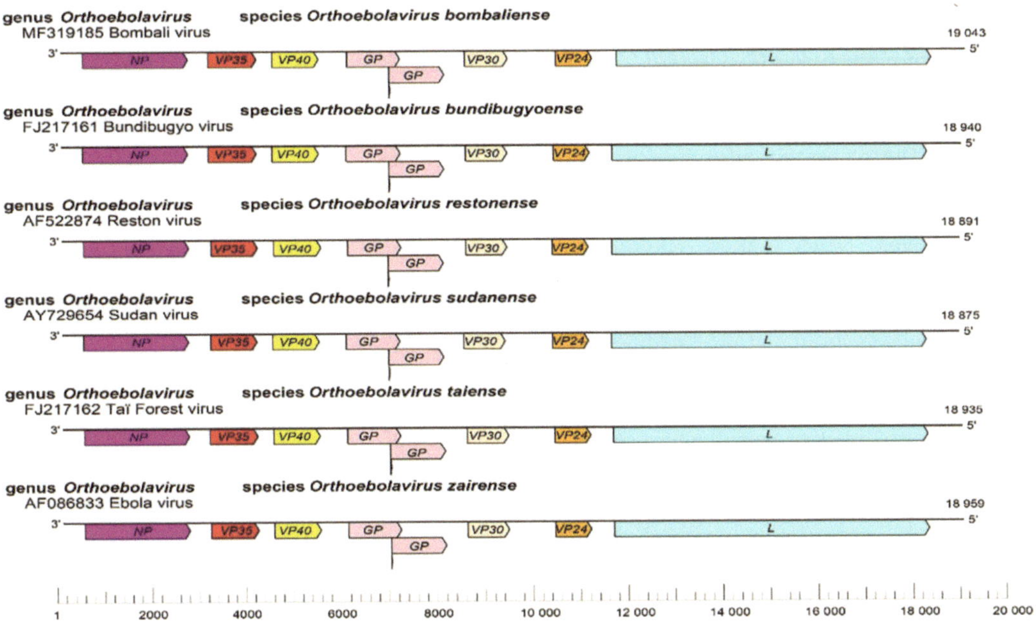

FIGURE 5.4 General schematic and comparative representation of Ebolavirus genome organization (types include EBOV, SUDV, BDBV, TAFV, RESTV, and BOMV).

Source: International Committee on Taxonomy of Viruses,

- *VP30:* This minor nucleoprotein acts as a transcriptional activator. It is essential for initiating transcription by interacting with the RNA-dependent RNA polymerase.
- *VP24:* Another multifunctional protein, VP24, contributes to the inhibition of the host immune response by interfering with interferon signaling pathways.
- *L (RNA-dependent RNA polymerase):* The largest and most complex protein, the L protein is responsible for viral RNA synthesis and processing. It carries out the transcription and replication of the viral genome in concert with VP35 (Feldmann et al., 1993).

The tight regulation and coordinated expression of these genes are essential for the efficient replication and pathogenicity of Ebolavirus. Any disruption in this genomic organization can lead to attenuation of the virus, making it a key focus for antiviral research and vaccine development.

5.3.3 Life Cycle of *Ebolavirus*

The life cycle of *Ebolavirus* can be divided into several key stages: attachment and entry, replication and transcription, assembly, and budding. Understanding this life cycle is crucial for identifying potential therapeutic targets to disrupt the viral replication process (Muhlberger, 2007).

- *Attachment and entry: Ebolavirus* initiates infection by attaching to host cells through its glycoprotein (GP). GP binds to cellular receptors such as C-type lectins (e.g., DC-SIGN), folate receptor-α, and the recently identified receptor, Niemann-Pick C1 (NPC1) (Carette et al., 2011). After attachment, the virus enters the host cell through macropinocytosis, a process in which the virus is engulfed in a vesicle and transported into the cell. Once inside the cell, the viral particle is trafficked to endosomes, where the acidic environment triggers the cleavage of GP by host cell proteases such as cathepsin B and L. This cleavage is essential for the subsequent fusion of the viral and endosomal membranes, a process mediated by the fusion loop of GP, releasing the viral ribonucleoprotein complex into the cytoplasm (Sullivan et al., 2005).
- *Replication and transcription:* The released ribonucleoprotein complex serves as the template for transcription and replication. The L protein, in association with VP35, transcribes the viral genes into positive-sense mRNAs. These mRNAs are then translated by host ribosomes into viral proteins. For replication, the L protein synthesizes a full-length positive-sense RNA, which is subsequently used as a template to produce new negative-sense genomic RNA (Muhlberger, 2007).
- *Assembly:* Newly synthesized viral proteins and genomic RNA are transported to the host cell membrane, where they assemble into new viral particles. VP40, the matrix protein, plays a central role in this process by forming a scaffold that coordinates the assembly of the nucleocapsid and the budding of the virion from the host cell surface (Bornholdt & Prasad, 2008).
- *Budding:* The final stage of the viral life cycle is budding, where the assembled viral particles are released from the host cell. VP40 interacts with the host cell's endosomal sorting complexes required for transport (ESCRT) machinery to facilitate the scission of the viral particle from the plasma membrane. This process results in the release of mature virions into the extracellular space, where they can go on to infect new cells and propagate the infection (Bavari et al., 2002).

5.4 *EBOLAVIRUS*: MOLECULAR PATHOPHYSIOLOGY AND CLINICAL PRESENTATION

The pathophysiology of Ebola virus infection is characterized by a complex interplay between viral replication, immune evasion, and host inflammatory responses. Understanding these molecular mechanisms is crucial for developing effective therapeutic and preventive strategies.

5.4.1 Viral Entry and Immune Evasion

Ebolavirus initiates infection by targeting specific host cells, including dendritic cells, macrophages, and monocytes, which are pivotal components of the immune system. The virus utilizes its glycoprotein (GP) to bind to cellular receptors such as C-type lectins (e.g., DC-SIGN), T-cell immunoglobulin and mucin domain 1 (TIM-1), and the Niemann-Pick C1 (NPC1) protein within endosomes (Carette et al., 2011). This binding facilitates viral entry through macropinocytosis, a process where the virus is engulfed by the host cell membrane, forming an endocytic vesicle. Once inside the host cell, *Ebolavirus* deploys several strategies to evade the immune response. One such strategy involves the secretion of a soluble form of its glycoprotein, known as sGP, which acts as a decoy to neutralize anti-GP antibodies, thereby impeding the host's antibody-mediated immune response (Falzarano et al., 2006). Additionally, the viral protein VP35 plays a critical role in antagonizing the host's innate immune response by inhibiting the production of interferons (IFNs) through the sequestration of double-stranded RNA, which prevents the activation of key immune signaling pathways (Hartman et al., 2008).

5.4.2 Viral Replication and Systemic Inflammation

Following entry, *Ebolavirus* replicates rapidly within the cytoplasm of infected cells. The virus employs its RNA-dependent RNA polymerase (L protein) in conjunction with the VP35 cofactor to transcribe and replicate its genome. This robust viral replication leads to high viral loads and widespread dissemination throughout the host's body (Sanchez et al., 2004a). Infected macrophages and dendritic cells release large amounts of pro-inflammatory cytokines and chemokines, including tumor necrosis factor-alpha (TNF-α), interleukin-6 (IL-6), and interleukin-8 (IL-8). This cytokine storm contributes to the dysregulation of the vascular endothelium, leading to increased vascular permeability and tissue damage (Hensley et al., 2002). Furthermore, the activation of these inflammatory pathways causes apoptosis of lymphocytes and hepatocytes, exacerbating immunosuppression and hepatic dysfunction, respectively.

5.4.3 Coagulation and Hemorrhage

One of the hallmark features of Ebola virus infection is its impact on the coagulation system, leading to disseminated intravascular coagulation (DIC). Ebola virus infection induces the expression of tissue factor (TF) on the surface of infected monocytes and macrophages, which triggers the extrinsic coagulation pathway (Geisbert et al., 2003). This results in the formation of microthrombi, consumption of clotting factors, and ultimately, widespread hemorrhaging. The dysregulation of coagulation, coupled with endothelial damage and platelet dysfunction, underlies the severe bleeding tendencies observed in Ebola virus disease.

The clinical manifestations of Ebola virus infection are diverse, ranging from mild febrile illness to severe hemorrhagic fever with multi-organ failure. These manifestations are a direct consequence of the molecular interactions between the virus and the host.

5.4.4 Early Phase of Infection

Non-specific symptoms: The initial phase of Ebola virus infection, typically occurring within 2–21 days of exposure, is characterized by non-specific flu-like symptoms such as fever, fatigue, headache, and myalgia (World Health Organization, 2021). These symptoms arise as a result of the systemic inflammatory response triggered by the rapid replication of the virus in primary target cells. The release of cytokines such as IL-6 and TNF-α contributes to fever and malaise, while viral replication in muscle tissue and hepatocytes may cause myalgia and abdominal pain.

5.4.5 Progression to Severe Disease: Hemorrhagic and Gastrointestinal Symptoms

As the infection progresses, *Ebolavirus* spreads to various organs, including the liver, spleen, and gastrointestinal tract. The involvement of these organs leads to more severe symptoms, including vomiting, diarrhea, abdominal pain, and in some cases, hemorrhagic manifestations such as petechiae, ecchymoses, and bleeding from mucosal surfaces (Feldmann & Geisbert, 2011). The extensive liver damage caused by viral replication and immune-mediated injury results in impaired synthesis of clotting factors, contributing to coagulopathy and bleeding.

In the gastrointestinal tract, *Ebolavirus* targets epithelial cells and disrupts the integrity of the mucosal barrier, leading to severe diarrhea and fluid loss. This not only exacerbates dehydration and electrolyte imbalances but also facilitates the dissemination of the virus within the host and the environment, increasing the risk of transmission.

5.4.6 Multi-Organ Failure and Shock

In severe cases, the disease may progress to multi-organ failure and shock. The systemic inflammatory response, coupled with the direct cytopathic effects of the virus on endothelial cells, leads to profound vascular leakage, hypovolemia, and shock (Geisbert et al., 2003). The failure of multiple organ systems, including the kidneys, liver, and lungs, is a common cause of death in Ebola virus infection.

5.4.7 Immune Dysregulation and Late-Stage Complications

Ebola virus infection is marked by a paradoxical immune response characterized by early immunosuppression followed by hyperactivation of the immune system. The initial inhibition of interferon responses and lymphocyte apoptosis result in a compromised immune system, allowing unchecked viral replication (Sullivan et al., 2003). In the later stages, the excessive activation of immune cells leads to a hyperinflammatory state, contributing to tissue damage and organ dysfunction. Survivors of Ebola virus infection often experience long-term sequelae, including arthritis, uveitis, and chronic fatigue, collectively known as "post-Ebola syndrome" (Mattia et al., 2016). These complications may be attributable to the persistence of viral antigens in immune-privileged sites or ongoing inflammatory responses.

The molecular mechanisms underlying the pathophysiology and clinical manifestations of Ebola virus infection reveal the complex interplay between the virus and host immune responses. From immune evasion strategies that enable unchecked viral replication to the dysregulation of coagulation and immune responses that lead to severe hemorrhagic disease, Ebola virus poses significant challenges for treatment and prevention. Understanding these mechanisms is crucial for developing effective therapeutic interventions and improving outcomes for patients affected by this devastating disease.

5.5 *MARBURGVIRUS*: INTRODUCTION

Marburgvirus is a genus of viruses within the Filoviridae family, known for causing severe viral hemorrhagic fever in humans and non-human primates. The term *Marburgvirus* refers to the genus, which includes two distinct species: *Marburg marburgvirus* and *Ravn marburgvirus*. The disease caused by these viruses is referred to as Marburg virus disease (MVD). Similar to Ebola virus, Marburg virus is associated with high fatality rates and severe clinical presentations, including hemorrhagic symptoms, multi-organ dysfunction, and high morbidity (World Health Organization [WHO], 2021). MVD is characterized by a sudden onset of fever, severe headache, and muscle pain, followed by the rapid progression to gastrointestinal symptoms, hemorrhaging, and, in severe cases, shock and multi-organ failure. The high pathogenicity and lack of specific treatment options make MVD a significant public health concern, particularly in regions where outbreaks have occurred.

5.5.1 History of Marburg Virus Disease

Marburg virus was first identified in 1967 during simultaneous outbreaks in Marburg and Frankfurt, Germany, and Belgrade, Serbia. These initial cases were linked to laboratory workers exposed to infected African green monkeys (*Cercopithecus aethiops*) imported from Uganda (Martini et al., 1968). This outbreak resulted in 31 cases, with seven fatalities, marking the first known instance of Marburg virus disease. Subsequent outbreaks have primarily occurred in Africa, with significant incidents reported in Angola (2004–2005), the Democratic Republic of Congo (1998–2000), and Uganda (2012). These outbreaks have been associated with high case fatality rates, often exceeding 80% (Bausch et al., 2006). The Angola outbreak remains the largest and deadliest to date, with 374 reported cases and 329 deaths, illustrating the virus' potential for widespread and devastating impact (Figure 5.5).

5.5.2 Taxonomic Classification of *Marburgvirus*

The Filoviridae family, to which *Marburgvirus* belongs, is divided into several genera, including *Ebolavirus*, *Marburgvirus*, and *Cuevavirus*. Within the genus *Marburgvirus*, there are two recognized species: *Marburg marburgvirus* and *Ravn marburgvirus* (Kuhn et al., 2019). These species are further classified into distinct viral strains based on genetic variation. *Marburg marburgvirus* was the first species to be identified and includes the *Marburgvirus* (MARV) strain, while *Ravn marburgvirus* was identified from a fatal case in Kenya in 1987. Despite genetic differences, both species exhibit similar clinical presentations and pathogenicity in humans.

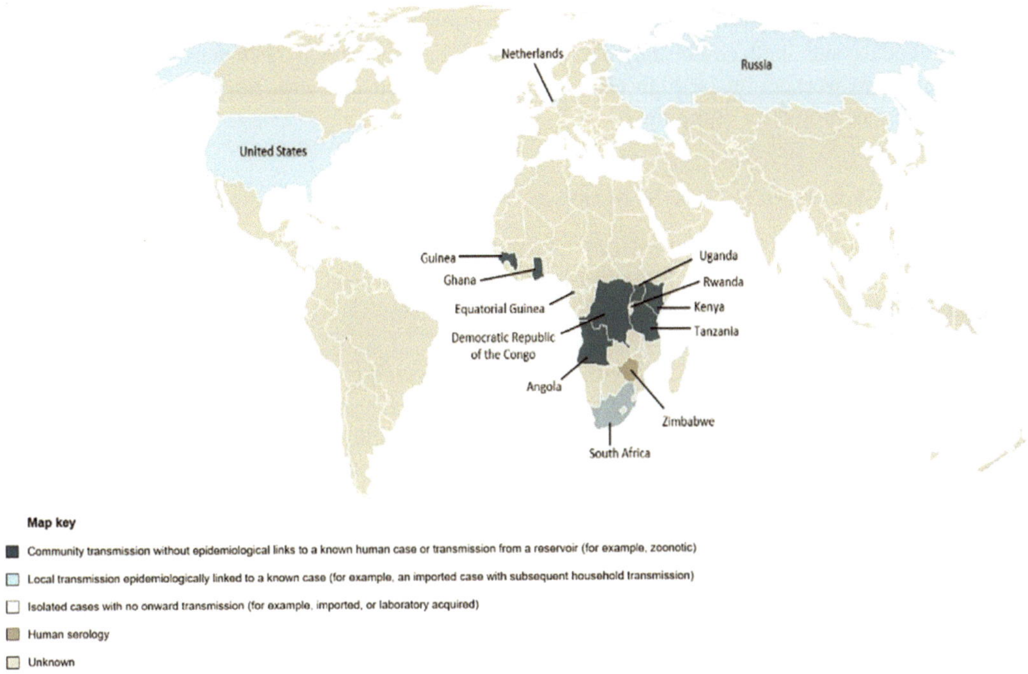

FIGURE 5.5 Map of countries which have reported human MVD cases, up to August 2023 (Marburg Virus Disease: Origins, Reservoirs, Transmission and Guidelines, 2024).

Source: **Available under Open Government Licence v3.0.**

The taxonomic classification of *Marburgvirus* is as follows:

Order: Mononegavirales
Family: Filoviridae
Genus: *Marburgvirus*
Species: *Marburg marburgvirus* and *Ravn marburgvirus*

The nomenclature reflects the geographic and clinical history of the virus, with *Marburg marburgvirus* named after the site of the first outbreak, and *Ravn marburgvirus* named after the initial patient in Kenya (Kuhn et al., 2019).

5.5.3 Transmission Process of *Marburgvirus*

Marburgvirus is transmitted to humans through contact with the bodily fluids or tissues of infected animals, particularly fruit bats of the Pteropodidae family, which are considered the natural reservoir hosts of the virus (Kumar et al., 2023; Towner et al., 2009). Transmission can also occur through direct contact with blood, secretions, organs, or other bodily fluids of infected persons or contaminated surfaces and materials.

- *Zoonotic transmission:* The primary mode of transmission to humans is believed to be through exposure to infected bats, through direct contact with either bat excreta or aerosols in bat-inhabited caves and mines. Human-to-human transmission occurs via contact with blood, bodily fluids, or contaminated surfaces, often during the care of infected individuals or through unsafe burial practices (WHO, 2021).
- *Nosocomial transmission:* Nosocomial transmission, or the spread of the virus within healthcare settings, has been a significant factor in several outbreaks. This mode of transmission typically results from inadequate infection control practices, including the reuse of needles and inadequate sterilization of medical equipment (Borchert et al., 2007). Healthcare workers are at high risk, particularly in settings with limited resources and insufficient protective equipment.

5.5.4 Symptoms Associated with Marburg Virus Infection

The clinical presentation of Marburg virus disease can range from mild to severe, with progression often rapid and potentially fatal. Symptoms typically appear 2–21 days after exposure, with the course of the disease divided into distinct phases.

- *Early phase (days 1–5):* The early phase is characterized by non-specific symptoms such as high fever, severe headache, chills, and myalgia. These symptoms are often mistaken for more common febrile illnesses, such as malaria or typhoid fever. Gastrointestinal symptoms, including nausea, vomiting, and diarrhea, typically begin within a few days of onset, contributing to severe dehydration (CDC, 2022).
- *Severe phase (days 6–10):* As the disease progresses, patients may develop severe hemorrhagic manifestations, including petechiae, ecchymoses, and bleeding from mucosal surfaces such as the gums and gastrointestinal tract. Multi-organ dysfunction, liver failure, and shock are common in severe cases, leading to a high case fatality rate (Martini & Schmidt, 1968). Neurological symptoms, such as confusion, irritability, and seizures, may also occur, reflecting the impact of the virus on the central nervous system.
- *Recovery and complications:* Patients who survive the acute phase of the disease may experience prolonged convalescence, with persistent fatigue, muscle pain, and psychological symptoms. Long-term complications, such as uveitis and arthralgia, have been reported in survivors (Borchert et al., 2007).

5.6 BASIC STRUCTURE OF *MARBURGVIRUS*

Marburgvirus is a member of the Filoviridae family, characterized by its filamentous, pleomorphic structure. The virus particles, or virions, exhibit a characteristic filamentous shape, often appearing as elongated threads, with a diameter of approximately 80 nm and varying lengths that can reach up to 14,000 nm (Kuhn et al., 2019). The virion's structure is helical and encapsulates a single-stranded, negative-sense RNA genome. The morphology of *Marburgvirus* is similar to that of its close relative, *Ebolavirus*, but there are subtle differences in the surface glycoprotein structures and the organization of the viral matrix (Figure 5.6). The virion is composed of several key structural components:

- *Nucleocapsid:* The nucleocapsid consists of the viral RNA genome encapsulated by the nucleoprotein (NP), which protects the viral RNA and is crucial for the replication process. It also includes other structural proteins like VP30 and VP35, which play roles in viral transcription and replication (Muhlberger, 2007).

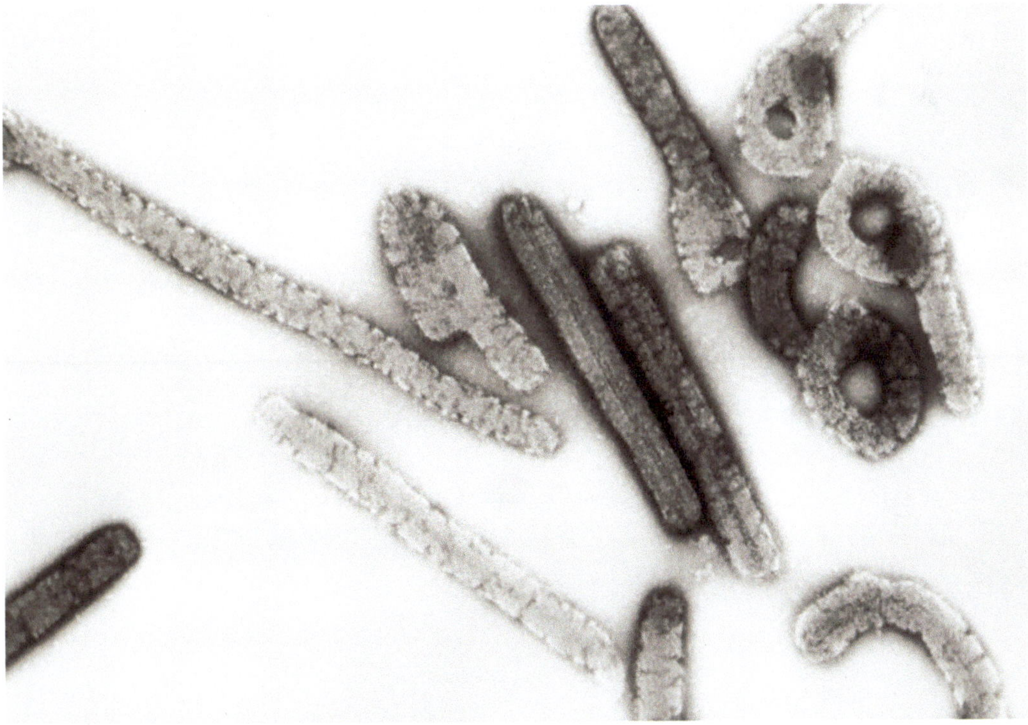

FIGURE 5.6 This negative stained transmission electron micrograph (TEM) depicts a number of filamentous Marburg virions, which had been cultured on Vero cell cultures, and purified on sucrose, rate-zonal gradients. Note the virus' morphologic appearance with its characteristic "Shepherd's Crook" shape; magnified approximately 100,000×. Marburg hemorrhagic fever is a rare, severe type of hemorrhagic fever which affects both humans and non-human primates. Caused by a genetically unique zoonotic (that is, animal-borne) RNA virus of the Filovirus family, its recognition led to the creation of this virus family. The four species of *Ebolavirus* are the only other known members of the Filovirus family. *Marburgvirus* was first recognized in 1967, when outbreaks of hemorrhagic fever occurred simultaneously in laboratories in Marburg and Frankfurt, Germany, and in Belgrade, Yugoslavia (now Serbia).

***Source:* Dr. Erskine Palmer, Russell Regnery, Ph.D., 1981. Centers for Disease Control and Prevention's Public Health Image Library—PHIL #275—. Centers of Disease Control and Prevention—United States Department of Health and Human Services (Public Domain).**

- *Matrix layer:* Surrounding the nucleocapsid is the matrix layer, which is formed by the viral proteins VP40 and VP24. VP40, the most abundant protein, is essential for virion assembly and release from the host cell. VP24 is involved in viral particle formation and the inhibition of host immune responses (Watanabe et al., 2007).
- *Envelope and glycoproteins:* The outer layer of the virion is a lipid bilayer derived from the host cell membrane, embedded with the viral glycoprotein (GP). The GP spikes protrude from the surface and are responsible for binding to the host cell receptors, facilitating viral entry. The GP is also a significant factor in the virus' pathogenicity, contributing to immune evasion and cellular damage (Sanchez et al., 2004a).

This structural complexity allows *Marburgvirus* to effectively invade host cells and evade immune responses, contributing to its high virulence and mortality rates.

5.7 GENOMIC ORGANIZATION OF *MARBURGVIRUS*

The genome of *Marburgvirus* is a single-stranded, negative-sense RNA approximately 19,000 nucleotides in length. It encodes seven genes arranged in the following order: NP, VP35, VP40, GP, VP30, VP24, and L (Sanchez et al., 2004b). Each gene is separated by conserved non-coding intergenic regions, which play a crucial role in transcriptional regulation and gene expression (Figure 5.7).

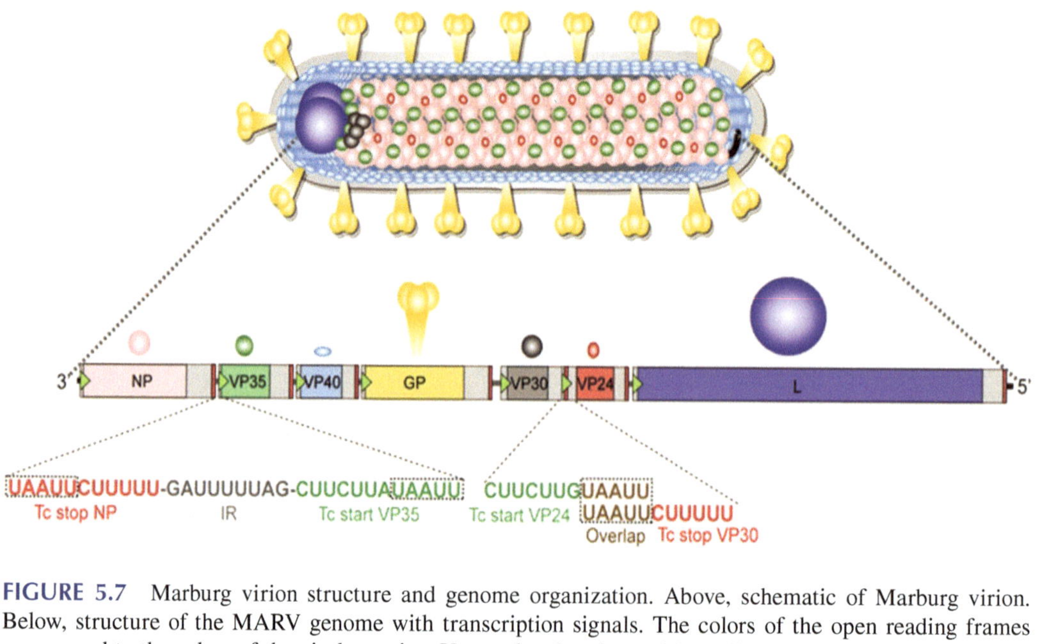

FIGURE 5.7 Marburg virion structure and genome organization. Above, schematic of Marburg virion. Below, structure of the MARV genome with transcription signals. The colors of the open reading frames correspond to the colors of the viral proteins. Untranslated regions of the different genes are shown as light gray boxes, intergenic regions (IR) are shown as dark gray lines, and the leader and trailer of the genome are colored in black. Transcription start signals (Tc start) are represented by green triangles, while transcription stop signals (Tc stop) are shown as red bars. The sequence of two gene borders (NP/VP35 and VP30/VP24) is shown in 3'–5' orientation, as it occurs in the negative sense RNA genome (MARV Musoke, GenBank accession number: NC_001608). The gene border between VP30 and VP24 contains overlapping transcription signals, with the start signal of VP24 upstream of the stop signal of VP30 (Brauburger et al., 2012).

Source: **Licensed under Creative Commons Attribution 3.0 Unported.**

- *NP (nucleoprotein):* Encodes the nucleoprotein that binds to the viral RNA, forming the nucleocapsid and ensuring the stability of the viral genome (Muhlberger, 2007).
- *VP35:* Functions as a cofactor in viral RNA synthesis and acts as an antagonist to host innate immune responses by inhibiting the production of interferons (Leung et al., 2010b).
- *VP40:* The primary matrix protein involved in the assembly and budding of new virions from the host cell membrane. It plays a pivotal role in virion morphogenesis and release (Watanabe et al., 2007).
- *GP (glycoprotein):* Responsible for mediating viral attachment and entry into host cells. The GP gene undergoes transcriptional editing, resulting in two forms of glycoprotein: the transmembrane GP1,2 and the secreted soluble glycoprotein (sGP). These glycoproteins are crucial for the virus' interaction with host cells and immune evasion (Feldmann et al., 1994).
- *VP30:* An RNA-binding protein that acts as a transcriptional activator. It is essential for the initiation of viral transcription and the regulation of RNA synthesis (Martínez et al., 2007).
- *VP24:* Plays a role in inhibiting the host immune response by blocking interferon signaling pathways. This helps the virus evade the host's antiviral defense mechanisms (Watanabe et al., 2007).
- *L (large protein):* The RNA-dependent RNA polymerase responsible for the replication and transcription of the viral genome. It works in conjunction with other viral proteins to replicate the RNA and produce messenger RNAs (mRNAs) for protein synthesis (Feldmann et al., 1994).

This genomic organization allows *Marburgvirus* to efficiently control its replication cycle and subvert host cellular processes to its advantage.

5.8 LIFE CYCLE OF *MARBURGVIRUS*

The life cycle of *Marburgvirus* involves a complex series of steps that allow the virus to enter host cells, replicate, assemble new virions, and release them to infect other cells. The process can be divided into several key stages:

- *Attachment and entry:* The life cycle begins with the attachment of the viral glycoprotein (GP) to host cell receptors. *Marburgvirus* predominantly targets cells expressing the NPC1 receptor, which is crucial for viral entry. The GP mediates fusion between the viral envelope and the host cell membrane, allowing the nucleocapsid to enter the cytoplasm (Sanchez et al., 2004b).
- *Transcription and replication:* Once inside the host cell, the nucleocapsid is released, and the RNA-dependent RNA polymerase (L protein) initiates transcription of the viral RNA genome. This process results in the production of viral mRNAs, which are translated into viral proteins using the host cell's ribosomes. Concurrently, the L protein replicates the viral RNA genome to generate new nucleocapsids (Muhlberger, 2007).
- *Assembly:* The newly synthesized viral proteins and RNA genomes are transported to specific sites within the host cell, where they assemble into new nucleocapsids. VP40, the matrix protein, plays a critical role in coordinating the assembly process and driving the formation of new virions (Watanabe et al., 2007).
- *Budding and release:* The assembled nucleocapsids are transported to the host cell membrane, where they bud off from the cell surface, acquiring a lipid envelope in the process. This budding is facilitated by the interaction between VP40 and the host cell's membrane. The newly formed virions are then released into the extracellular environment, where they can infect adjacent cells (Feldmann et al., 1994).

- *Immune evasion and spread:* Throughout its life cycle, *Marburgvirus* employs various strategies to evade the host immune system. VP35 inhibits interferon production, while VP24 blocks the interferon signaling pathway, allowing the virus to replicate without being detected and destroyed by the host's immune defenses (Leung et al., 2010a).

This life cycle not only enables efficient viral replication and spread within the host but also contributes to the high pathogenicity of *Marburgvirus*, resulting in severe disease outcomes.

5.8 MOLECULAR PATHOPHYSIOLOGY AND CLINICAL PRESENTATION OF MARBURG VIRUS INFECTION

- *Viral entry and initial replication:* Marburg virus enters host cells through macropinocytosis, a process facilitated by the viral glycoprotein (GP). The GP interacts with the Niemann-Pick C1 (NPC1) receptor on the surface of host cells, particularly those of macrophages, dendritic cells, and endothelial cells (Kuhn et al., 2019). Upon entry, the virus rapidly replicates in the cytoplasm, using the host's cellular machinery. The replication occurs in specialized cytoplasmic structures termed viral inclusions, where newly synthesized nucleocapsids are assembled (Kühne et al., 2013).
- *Evasion of host immune responses:* Marburg virus employs multiple strategies to evade the host's immune responses. The VP35 protein, an interferon antagonist, inhibits the host's innate immune response by blocking the detection of viral RNA. This inhibition prevents the activation of interferon regulatory factors (IRFs) and, subsequently, the production of type I interferons (Leung et al., 2010b). Furthermore, the VP24 protein interferes with the Jak-STAT signaling pathway, which is crucial for the antiviral state induced by interferons. This suppression allows the virus to replicate unchecked in the early stages of infection (Valmas et al., 2010).
- *Induction of apoptosis and cytopathic effects:* The replication of Marburg virus in host cells leads to significant cytopathic effects. The GP of Marburg virus is highly cytotoxic, inducing apoptosis in infected cells. This apoptosis is partly mediated through the activation of pro-apoptotic pathways, such as the mitochondrial pathway, involving caspases and the release of cytochrome c. The extensive cell death contributes to the disruption of tissue architecture and organ function, particularly in the liver, spleen, and lymph nodes (Ravichandran & Nalini, 2011).
- *Systemic inflammation and immune dysregulation:* As the infection progresses, the massive replication of the virus and the release of viral particles lead to a hyperinflammatory state known as a cytokine storm. Elevated levels of pro-inflammatory cytokines, such as TNF-α, IL-6, and IL-1β, are observed, which further exacerbates tissue damage and promotes vascular leakage (Sanchez et al., 2004b). This cytokine dysregulation is a hallmark of severe Marburg virus disease and contributes significantly to the pathogenesis of the infection.
- *Vascular dysfunction and hemorrhage:* One of the most severe manifestations of Marburg virus infection is the breakdown of the vascular endothelial barrier, leading to hemorrhage. The GP of Marburg virus directly damages endothelial cells, causing them to become leaky and lose their integrity. The matrix protein VP40 also plays a role in vascular permeability by modulating cellular junctions. This endothelial dysfunction, combined with coagulation abnormalities such as disseminated intravascular coagulation (DIC), leads to widespread hemorrhage and shock, which are often fatal (Mühlberger et al., 2012).
- *Molecular mechanisms:* The clinical presentation of Marburg virus infection is the result of complex molecular interactions between the virus and host, leading to a spectrum of symptoms that reflect the severity of the disease. Understanding these mechanisms is essential for accurate diagnosis and management.

- *Initial symptoms and viral spread:* The incubation period for Marburg virus ranges from 2 to 21 days. Early symptoms include fever, chills, headache, and myalgia, which are nonspecific and resemble other febrile illnesses. These symptoms correspond to the initial replication of the virus in lymphoid tissues and its dissemination to other organs (Brauburger et al., 2012). As the virus spreads to the liver, spleen, and adrenal glands, it causes damage to these organs, leading to more severe symptoms.
- *Gastrointestinal and hepatic manifestations:* Gastrointestinal symptoms such as nausea, vomiting, abdominal pain, and diarrhea are common as the infection progresses. These symptoms are linked to the viral replication in the gastrointestinal tract and liver. Liver involvement is particularly significant, as the virus induces hepatocellular necrosis and impairs the liver's ability to produce clotting factors, contributing to coagulopathy and hemorrhagic symptoms (Feldmann & Geisbert, 2011).
- *Neurological and hemorrhagic complications:* Severe cases of Marburg virus infection can lead to neurological symptoms such as confusion, delirium, and coma, which are associated with direct viral invasion of the central nervous system and secondary effects of systemic inflammation. Hemorrhagic manifestations, including petechiae, ecchymoses, and mucosal bleeding, are indicative of advanced disease and are associated with severe endothelial damage and coagulopathy (Ravichandran & Nalini, 2011).
- *Multi-organ failure and shock:* The combination of vascular leakage, immune dysregulation, and direct tissue destruction leads to multi-organ failure. The most commonly affected organs include the liver, kidneys, and lungs. Hepatic dysfunction, acute kidney injury, and respiratory failure are common in the terminal stages of the disease. The massive fluid loss due to vascular leakage and bleeding results in hypovolemic shock, which is often the immediate cause of death in Marburg virus infection (Sanchez et al., 2004b).
- *Recovery and long-term sequelae:* Survivors of Marburg virus infection may experience prolonged recovery periods with lingering symptoms such as fatigue, joint pain, and vision problems. These post-recovery symptoms are likely due to the long-term effects of immune activation and residual organ damage (Mühlberger et al., 2012). Understanding the pathophysiological mechanisms underlying these sequelae is crucial for providing adequate care to survivors.

5.9 INFECTION PREVENTION AND CONTROL: *EBOLAVIRUSES* AND *MARBURGVIRUSES*

Ebolavirus and *Marburgvirus* are among the most lethal pathogens known to humans, causing severe viral hemorrhagic fevers with high mortality rates. These diseases pose significant challenges to public health systems, particularly in regions where resources are limited. The complex nature of these viruses, combined with their potential for rapid transmission and high case fatality rates, necessitates robust infection prevention and control (IPC) strategies. This chapter explores the unique challenges associated with the IPC of Ebola virus and Marburg virus diseases and discusses future prospects for improving control measures (Valmas et al., 2010).

5.9.1 Challenges in Infection Prevention and Control

- *High transmission potential and case-fatality rates:* Both Ebola virus and Marburg virus can be transmitted through direct contact with the blood, secretions, organs, or other bodily fluids of infected people or animals. Healthcare workers and family members are at high risk of infection, particularly in the absence of appropriate personal protective equipment (PPE) and strict IPC protocols (World Health Organization [WHO], 2021). The case-fatality rates for Ebola virus can range from 25% to 90%, while for Marburg virus, they range from 23% to 88%, depending on the outbreak and the quality of care provided (Centers for Disease Control and Prevention [CDC], 2022).

- *Healthcare system constraints:* In many of the regions where Ebola virus and Marburg virus outbreaks occur, healthcare systems are already overburdened. Inadequate infrastructure, limited access to essential medical supplies, and a shortage of trained healthcare workers contribute to the difficulty of implementing effective IPC measures (Fischer et al., 2020). These constraints are exacerbated during outbreaks, as the demand for resources like PPE, isolation units, and trained personnel rapidly outstrips supply.
- *Cultural practices and misconceptions:* Cultural practices, such as traditional burial rites, which often involve direct contact with the deceased, can significantly increase the risk of transmission. In addition, misconceptions and fear surrounding these diseases can lead to mistrust of healthcare workers and avoidance of medical treatment, further complicating IPC efforts (Richards et al., 2019). Community engagement and culturally sensitive education are crucial components of effective IPC strategies in affected regions.
- *Delay in diagnosis and case detection:* Early symptoms of Ebola virus and Marburg virus infections, such as fever, fatigue, and muscle pain, are non-specific and can be mistaken for other diseases like malaria or typhoid fever, leading to delays in diagnosis. Rapid and accurate diagnostic tools are essential for identifying cases early and implementing isolation and treatment protocols to prevent further transmission (Jacob et al., 2020). However, in many resource-limited settings, access to diagnostic facilities is limited, and delays in diagnosis can result in uncontrolled spread.

5.10 FUTURE PROSPECTS IN INFECTION PREVENTION AND CONTROL

- *Improved diagnostic and surveillance tools:* Advancements in diagnostic technology, including rapid antigen and nucleic acid tests, offer the potential for earlier and more accurate detection of Ebola virus and Marburg virus infections. Implementing these tools in outbreak-prone regions can improve case detection and help control the spread of these diseases (Loy et al., 2021). In addition, integrating these diagnostics into surveillance systems can enhance real-time monitoring and early warning capabilities, facilitating more rapid responses to emerging outbreaks.
- *Development and deployment of vaccines:* Recent advances in vaccine development have provided new tools for the prevention of Ebola virus disease. The rVSV-ZEBOV vaccine, for example, has shown efficacy in preventing Ebola virus disease and has been used in ring vaccination strategies during outbreaks (Henao-Restrepo et al., 2017). Similar efforts are underway for the development of vaccines against Marburg virus. Widespread deployment of these vaccines in at-risk populations and healthcare workers could significantly reduce the impact of future outbreaks.
- *Strengthening health systems and infrastructure:* Building robust healthcare infrastructure is essential for effective IPC. This includes not only physical infrastructure, such as isolation units and laboratories but also human resources, such as training healthcare workers in IPC protocols and emergency response (Gostin et al., 2019). Strengthening health systems in outbreak-prone regions will improve the capacity to respond to Ebola virus and Marburg virus outbreaks and other infectious disease threats.
- *Community engagement and risk communication:* Effective IPC requires the cooperation of the community. Public health initiatives that engage community leaders and use culturally appropriate communication strategies can help build trust and encourage behaviors that reduce the risk of transmission, such as safe burial practices and early seeking of medical care (Gillespie et al., 2016). Continuous community education and involvement are key to sustaining IPC efforts beyond the acute phase of an outbreak.
- *Global collaboration and funding:* Addressing the challenges of Ebola virus and Marburg virus IPC requires coordinated global efforts. International organizations, governments, and NGOs must collaborate to provide funding, technical support, and

resources to affected regions. Initiatives such as the Global Health Security Agenda and the WHO's International Health Regulations (IHR) provide frameworks for international cooperation and capacity-building in IPC (Gostin et al., 2019).

REFERENCES

Amarasinghe, G. K., Crameri, G., Bente, D. A., Rodriguez, L. L., & Bowen, R. A. (2019). Ebolaviruses: New insights into pathology and pathogenesis. *Current Opinion in Virology*, 34, 1–8. https://doi.org/10.1016/j.coviro.2018.11.002.

Bausch, D. G., Nichol, S. T., Muyembe-Tamfum, J. J., Borchert, M., Rollin, P. E., Sleurs, H., Campbell, P., Tshioko, F. K., Roth, C., Colebunders, R., Pirard, P., & Sprecher, A. (2006). Marburg hemorrhagic fever associated with multiple genetic lineages of virus. *The New England Journal of Medicine*, 355(9), 909–919. https://doi.org/10.1056/NEJMoa051465.

Bavari, S., Bosio, C. M., Wiegand, E., Ruthel, G., Will, A. B., Geisbert, T. W., Hevey, M., Schmaljohn, C., & Schmaljohn, A. (2002). Lipid raft microdomains: A gateway for compartmentalized trafficking of Ebola and Marburg viruses. *The Journal of Experimental Medicine*, 195(5), 593–602. https://doi.org/10.1084/jem.20012027.

Biedenkopf, N., Bukreyev, A., Chandran, K., Di Paola, N., Formenty, P. B. H., Griffiths, A., Hume, A. J., Mühlberger, E., Netesov, S. V., Palacios, G., Pawęska, J. T., Smither, S., Takada, A., Wahl, V., & Kuhn, J. H. (2024). ICTV virus taxonomy profile: *Filoviridae* 2024. *Journal of General Virology*, 105, 001955.

Borchert, M., Mutyaba, I., Van Kerkhove, M. D., Lutwama, J., Luwaga, H., Bisoborwa, G., Turyagaruka, J., Pirard, P., Ndayimirije, N., Roddy, P., Van Der Stuyft, P., & Colebunders, R. (2007). Ebola hemorrhagic fever outbreak in Masindi District, Uganda: Outbreak description and lessons learned. *BMC Infectious Diseases*, 7(1), 1–12. https://doi.org/10.1186/1471-2334-7-11.

Bornholdt, Z. A., & Prasad, B. V. (2008). X-ray structure of Ebola VP35 oligomerization domain and its implications for filoviral assembly. *Journal of Virology*, 82(7), 4030–4037. https://doi.org/10.1128/JVI.02677-07.

Brauburger, K., Hume, A. J., Mühlberger, E., & Olejnik, J. (2012). Forty-five years of Marburg virus research. *Viruses*, 4(10), 1878–1927. https://doi.org/10.3390/v4101878.

Carette, J. E., Raaben, M., Wong, A. C., Herbert, A. S., Obernosterer, G., Mulherkar, N., Kuehne, A. I., Kranzusch, P. J., Griffin, A. M., Ruthel, G., Dal Cin, P., Dye, J. M., Whelan, S. P., Chandran, K., Brummelkamp, T. R., & Basler, C. F. (2011). Ebola virus entry requires the cholesterol transporter Niemann-Pick C1. *Nature*, 477(7364), 340–343. https://doi.org/10.1038/nature10348.

Centers for Disease Control and Prevention (CDC). (2016). *2014–2016 Ebola Outbreak in West Africa*. https://www.cdc.gov/vhf/ebola/history/2014-2016-outbreak/index.html.

Centers for Disease Control and Prevention (CDC). (2019). *Healthcare-Associated Infections*. https://www.cdc.gov/hai/index.html.

Centers for Disease Control and Prevention (CDC). (2022). *Marburg Virus Disease (MVD)*. https://www.cdc.gov/vhf/marburg/index.html.

Falasca, L., et al. (2015). Molecular mechanisms of Ebola virus pathogenesis: Focus on cell death. *Cell Death & Differentiation*, 22(8), 1250–1259. https://doi.org/10.1038/cdd.2015.67.

Falzarano, D., Feldmann, H., & Jones, S. M. (2006). Immunotherapeutic approaches to Ebola virus infection. *Immunotherapy*, 12(2), 223–240. https://doi.org/10.1016/j.tim.2005.12.005.

Feldmann, H., & Geisbert, T. W. (2011). Ebola haemorrhagic fever. *The Lancet*, 377(9768), 849–862. https://doi.org/10.1016/S0140-6736(10)60667-8.

Feldmann, H., Kiley, M. P., & Mahy, B. W. J. (1994). Molecular biology and evolution of filoviruses. *Archives of Virology*, 9, 81–100. https://doi.org/10.1007/978-3-7091-9283-6_8.

Feldmann, H., Volchkov, V. E., Volchkova, V. A., Stroher, U., & Klenk, H. D. (1993). Biosynthesis and role of filoviral glycoproteins. *Journal of General Virology*, 74(12), 2819–2824. https://doi.org/10.1099/0022-1317-74-12-2819.

Fischer, W. A., Hynes, N. A., & Perl, T. M. (2020). Protecting health care workers from Ebola: Personal protective equipment is critical but is not enough. *Annals of Internal Medicine*, 161(10), 753–754. https://doi.org/10.7326/M14-1953.

Frost, I., Van Boeckel, T. P., Pires, J., Craig, J., Global, G. E., & Laxminarayan, R. (2019). Global geographic trends in antimicrobial resistance: The role of international travel. *Journal of Travel Medicine*, 26(7), taz036. https://doi.org/10.1093/jtm/taz036.

Gates, B. (2018). Innovation for pandemics. *New England Journal of Medicine*, 378(22), 2057–2060. https://doi.org/10.1056/NEJMp1806283.

Geisbert, T. W., Hensley, L. E., Gibb, T. R., Steele, K. E., Jaax, N. K., & Jahrling, P. B. (2003). Apoptosis induced in vitro and in vivo during infection by Ebola and Marburg viruses. *Laboratory Investigation*, 83(12), 1837–1850. https://doi.org/10.1097/01.lab.0000101479.63939.7c.

Gillespie, A. M., Obregon, R., El Asawi, R., Richey, C., Manoncourt, E., Joshi, K., & Quereshi, S. (2016). Social mobilization and community engagement central to the Ebola response in West Africa: Lessons for future public health emergencies. *Global Health: Science and Practice*, 4(4), 626–646. https://doi.org/10.9745/GHSP-D-16-00226.

Gostin, L. O., et al. (2019). Toward a common secure future: Four global commissions in the wake of Ebola. *PLOS Medicine*, 13(5), e1002042. https://doi.org/10.1371/journal.pmed.1002042.

Hartman, A. L., Towner, J. S., Nichol, S. T., & Comer, J. A. (2008). Interferon antagonism of the ebolaviruses. *Journal of Virology*, 82(4), 1890–1900. https://doi.org/10.1128/JVI.01347-07.

Henao-Restrepo, A. M., Camacho, A., Longini, I. M., Watson, C. H., Edmunds, W. J., Egger, M., & Kieny, M. P. (2017). Efficacy and effectiveness of an rVSV-vectored vaccine in preventing Ebola virus disease: Final results from the Guinea ring vaccination, open-label, cluster-randomised trial (Ebola Ça Suffit!). *The Lancet*, 389(10068), 505–518. https://doi.org/10.1016/S0140-6736(16)32621-6.

Hensley, L. E., Young, H. A., Jahrling, P. B., & Geisbert, T. W. (2002). Proinflammatory response during Ebola virus infection of primate models: Possible involvement of the tumor necrosis factor receptor superfamily. *Immunology and Cell Biology*, 80(4), 399–410. https://doi.org/10.1046/j.1440-1711.2002.01090.x.

Jacob, S. T., Crozier, I., Fischer, W. A., Hewlett, A., Kraft, C. S., Vega, M. A., & Williams, D. (2020). Ebola virus disease. *Nature Reviews Disease Primers*, 6(1), 1–31. https://doi.org/10.1038/s41572-020-0147-3.

Johnson, K. M., Lange, J. V., Webb, P. A., & Murphy, F. A. (1977). Isolation and partial characterization of a new virus causing acute hemorrhagic fever in Zaire. *The Lancet*, 309(8011), 569–571. https://doi.org/10.1016/S0140-6736(77)92000-1.

Kuhn, J. H., Amarasinghe, G. K., Basler, C. F., Bavari, S., Bukreyev, A., Jahrling, P. B., & Feldmann, H. (2019). ICTV virus taxonomy profile: Filoviridae. *Journal of General Virology*, 100(5), 911–912. https://doi.org/10.1099/jgv.0.001252.

Kühne, A., Becker, S., & Ebihara, H. (2013). Virus-host cell interplay during Marburg virus infection. *Current Opinion in Virology*, 3(1), 78–84. https://doi.org/10.1016/j.coviro.2012.11.005.

Kumar, A., Gangwar, R., Zargar, A., Kumar, H., Kumar, R., & Sharma, A. (2023). An outbreak of Marburg virus disease: World Health Organization. *Pharmaspire*, 15(1), 28–32. https://doi.org/10.56933/Pharmaspire.2023.15105.

Lee, J. E., Fusco, M. L., Hessell, A. J., Oswald, W. B., Burton, D. R., & Saphire, E. O. (2008). Structure of the Ebola virus glycoprotein bound to an antibody from a human survivor. *Nature*, 454(7201), 177–182. https://doi.org/10.1038/nature07082.

Leroy, E. M., Epelboin, A., Mondonge, V., Pourrut, X., Gonzalez, J. P., Muyembe-Tamfum, J. J., & Formenty, P. (2009). Human Ebola outbreak resulting from direct exposure to fruit bats in Luebo, Democratic Republic of Congo, 2007. *Vector-Borne and Zoonotic Diseases*, 9(6), 723–728. https://doi.org/10.1089/vbz.2008.0167.

Leung, D. W., et al. (2010a). Structural and functional characterization of Reston Ebola virus VP35 interferon inhibitory domain. *Journal of Molecular Biology*, 399(3), 347–357. https://doi.org/10.1016/j.jmb.2010.04.034.

Leung, D. W., et al. (2010b). Structural basis for dsRNA recognition and interferon antagonism by Ebola VP35. *Nature Structural & Molecular Biology*, 17(2), 165–172. https://doi.org/10.1038/nsmb.1765.

Loy, J. D., Paul, P., & Mores, C. N. (2021). Rapid diagnostics for viral hemorrhagic fevers: Current capabilities and future directions. *Viruses*, 13(5), 733. https://doi.org/10.3390/v13050733.

Martínez, M. J., Biedenkopf, N., Volchkova, V., Hartlieb, B., Becker, S., & Feldmann, H. (2007). Role of Ebola virus VP30 in transcription reinitiation. *Journal of Virology*, 81(20), 10812–10822. https://doi.org/10.1128/JVI.00829-07.

Martini, G. A., Knauff, H. G., Schmidt, H. A., Mayer, G., & Baltzer, G. (1968). On the etiology of an unknown human infection originating from monkeys. *The Lancet*, 291(7536), 513–514. https://doi.org/10.1016/S0140-6736(68)91518-9.

Martini, G. A., & Schmidt, H. A. (1968). Clinical course and prognosis of Marburg virus disease. *Die Medizinische Welt*, 19(12), 487–490.

Mattia, J. G., Vandy, M. J., Chang, J. C., Platt, D. E., Dierberg, K., & Squire, J. S. (2016). Early clinical sequelae of Ebola virus disease in Sierra Leone: A cross-sectional study. *The Lancet Infectious Diseases*, 16(3), 331–338. https://doi.org/10.1016/S1473-3099(15)00489-2.

Muhlberger, E. (2007). Filovirus replication and transcription. *Future Virology*, 2(2), 205–215. https://doi.org/10.2217/17460794.2.2.205.

Mühlberger, E., Sánchez, A., Randolf, A., & Kiley, M. P. (2012). The nucleotide sequence of the Marburg virus genome: Conservation of the gene order and genome organization in the *Filoviridae*. *Virology*, 233(1), 16–27. https://doi.org/10.1006/viro.1997.8592.

Nguyen, V. K., Binder, S. C., Boianelli, A., Meyer-Hermann, M., & Hernandez-Vargas, E. A. (2015). Ebola virus infection modeling and identifiability problems. *Frontiers in Microbiology,* 6. https://doi.org/10.3389/fmicb.2015.00257.

Ravichandran, R., & Nalini, R. (2011). Mechanisms of Marburg and Ebola virus mediated pathogenesis. *Archives of Virology*, 156(10), 1711–1723. https://doi.org/10.1007/s00705-011-1058-0.

Richards, P., Amara, J., Ferme, M. C., Kamara, P., Mokuwa, E., Sheriff, A. I., & Suluku, R. (2019). Village responses to Ebola virus disease in rural central Sierra Leone: An anthropological perspective. *Journal of African Cultural Studies*, 31(3), 363–378. https://doi.org/10.1080/13696815.2018.1542629.

Sanchez, A., Geisbert, T. W., & Feldmann, H. (2004a). Filoviridae: Marburg and Ebola viruses. In D. M. Knipe, & P. M. Howley (Eds.), *Fields Virology* (5th ed., pp. 1409–1448). Lippincott Williams & Wilkins.

Sanchez, A., Kiley, M. P., & Mahy, B. W. J. (2004b). The filoviruses. In D. M. Knipe, & P. M. Howley (Eds.), *Fields Virology* (pp. 1279–1304). Lippincott Williams & Wilkins.

Sanchez, A., Trappier, S. G., Mahy, B. W., Peters, C. J., & Nichol, S. T. (1996). The virion glycoproteins of Ebola viruses are encoded in two reading frames and are expressed through transcriptional editing. *Proceedings of the National Academy of Sciences*, 93(8), 3602–3607. https://doi.org/10.1073/pnas.93.8.3602.

Sullivan, N. J., Sanchez, A., Rollin, P. E., Yang, Z. Y., & Nabel, G. J. (2003). Development of a preventive vaccine for Ebola virus infection in primates. *Nature*, 424(6949), 681–684. https://doi.org/10.1038/nature01832.

Sullivan, N. J., Yang, Z. Y., & Nabel, G. J. (2005). Ebola virus pathogenesis: Implications for vaccines and therapies. *Journal of Virology*, 77(18), 9733–9737. https://doi.org/10.1128/JVI.77.18.9733-9737.2003.

Towner, J. S., et al. (2009). Isolation of genetically diverse Marburg viruses from Egyptian fruit bats. *PLoS Pathogens*, 5(7), e1000536. https://doi.org/10.1371/journal.ppat.1000536.

Valmas, C., & Basler, C. F. (2010). Marburg virus VP40 antagonizes interferon signaling in a species-specific manner. *Journal of Virology*, 84(17), 8819–8828. https://doi.org/10.1128/JVI.00794-10.

Valmas, C., Grosch, M. N., Schümann, M., Olejnik, J., Martinez, O., Best, S. M., et al. (2010) Marburg virus evades interferon responses by a mechanism distinct from ebola virus. *PLoS Pathog*, 6(1), e1000721. https://doi.org/10.1371/journal.ppat.1000721.

Watanabe, S., Noda, T., & Kawaoka, Y. (2007). Functional mapping of the nucleoprotein of Ebola virus. *Journal of Virology*, 81(8), 4166–4177. https://doi.org/10.1128/JVI.02570-06.

World Health Organization (WHO). (2021). *Marburg Virus Disease*. https://www.who.int/news-room/fact-sheets/detail/marburg-virus-disease.

6 Coronaviruses
MERS-CoV, SARS-CoV, and SARS-CoV-2

Srijan Goswami, Manash Mukherjee, and Rajesh Das

6.1 INTRODUCTION TO CORONAVIRUSES

The coronavirus family, Coronaviridae, consists of enveloped, positive-sense, single-stranded RNA viruses known for their crown-like appearance under electron microscopy due to the club-shaped glycoprotein spikes on their surfaces. This family is taxonomically divided into four genera: *Alphacoronavirus*, *Betacoronavirus*, *Gammacoronavirus*, and *Deltacoronavirus*, with the first two primarily infecting mammals, including humans, and the latter two generally targeting birds and other animals. Coronaviruses have the largest RNA genomes among viruses, measuring about 26–32 kb, which encode multiple structural and non-structural proteins essential for viral replication and host interaction. Human coronaviruses have been associated with diseases ranging from mild respiratory infections, such as the common cold, to severe conditions like SARS (severe acute respiratory syndrome), MERS (Middle East respiratory syndrome), and COVID-19 (caused by SARS-CoV-2). These viruses are zoonotic in origin, with bats often serving as reservoirs. Transmission to humans typically occurs via intermediate hosts such as civets, camels, or pangolins. Their ability to mutate and adapt facilitates cross-species transmission and makes them a significant global health concern. Recent studies highlight their genomic diversity, which supports their capacity to infect a broad range of hosts and evade immune defenses. This genetic variability underscores the challenges in managing their associated diseases and developing effective vaccines and therapeutics (Brant et al., 2021; Goswami et al., 2022).

The history and discovery of coronaviruses began with their identification as causative agents of diseases in animals in the 1930s. In 1931, infectious bronchitis virus (IBV), an avian pathogen, was the first recognized coronavirus, later classified as a gammacoronavirus. This marked the foundation for understanding coronaviruses as filterable agents causing respiratory diseases. By the mid-20th century, murine hepatitis virus (MHV) and transmissible gastroenteritis virus (TGEV) in swine were identified, further illustrating the diverse host range of these viruses. In 1965, David Tyrrell and Mark Bynoe isolated a human coronavirus, B184 , from patients with the common cold. They were the first to describe these particles morphologically distinct under an electron microscope. In 1967, the term "coronavirus" was coined by Tyrrell and June Almeida, inspired by the crown-like appearance of the glycoprotein spikes. These findings laid the groundwork for the identification of additional human coronaviruses, including HCoV-229E and HCoV-OC43, which primarily cause mild respiratory illnesses. The 21st century has seen the emergence of more pathogenic human coronaviruses, such as SARS-CoV, MERS-CoV, and SARS-CoV-2, highlighting their zoonotic origins and potential for pandemic spread. Advances in molecular virology and genomic sequencing have since provided deeper insights into their evolutionary biology and epidemiology (Mahase et al., 2020; Attia et al., 2021; Lv et al., 2024). Coronaviruses (CoVs) are enveloped viruses with a positive-sense single-stranded RNA genome, classified within the family Coronaviridae. Their genome, ranging from 26 kb to 32 kb, is among the largest of all RNA viruses. The genome's organization reflects a conserved structure across CoVs, with a 5′-untranslated region (UTR), a replication–transcription complex (ORF1a/b), structural proteins (spike [S],

DOI: 10.1201/9781032721811-6

membrane [M], envelope [E], and nucleocapsid [N]), and accessory proteins that modulate host interactions and pathogenesis. A 3′-UTR supports RNA stability and replication. The S protein, critical for host cell entry, facilitates viral attachment to receptors like ACE2 in coronaviruses (SARS-CoV-2), enabling membrane fusion. This protein's receptor binding domain (RBD) shows significant variation among CoVs, influencing host specificity and infectivity. The N protein packages the RNA genome into a helical nucleocapsid, while the M and E proteins contribute to virion assembly and budding. Accessory proteins, although not necessary for replication, play roles in immune evasion and pathogenicity. Structurally, CoVs exhibit lipid envelopes derived from host membranes, with spikes projecting outward, giving the characteristic crown-like appearance. The genomic flexibility, driven by high mutation rates and recombination events, underpins their zoonotic spillover potential and adaptation to new hosts (Yang et al., 2020; Brant et al., 2021; Goswami et al., 2022).

6.2 MIDDLE EAST RESPIRATORY SYNDROME CORONAVIRUS (MERS-CoV)

Middle East respiratory syndrome coronavirus (MERS-CoV) is a zoonotic virus first identified in Saudi Arabia in 2012. It is part of the Coronaviridae family, known for causing severe respiratory and systemic infections in humans. MERS-CoV primarily infects humans through contact with infected dromedary camels, either directly or indirectly. Human-to-human transmission occurs predominantly in healthcare settings, highlighting the critical need for infection control measures. The symptoms of MERS range from mild respiratory distress to severe pneumonia, often accompanied by fever and gastrointestinal manifestations. Patients with chronic health conditions or weakened immune systems are at higher risk of severe disease or death, with an estimated case fatality rate of 35% based on reported cases, though mild infections may be underreported. Geographically, the majority of MERS cases have occurred in the Arabian Peninsula, with notable outbreaks in South Korea and Europe resulting from imported cases. Genetic studies indicate that MERS-CoV undergoes minimal mutation, maintaining its zoonotic reservoir in camels. Despite intensive research, effective vaccines or antiviral treatments remain unavailable. As of recent reports, surveillance and preventive strategies focus on managing zoonotic risks and early detection of human cases to mitigate outbreaks globally (MERS-CoV Worldwide Overview, 2024; World Health Organization: WHO, 2022; Zhang et al., 2021).

The Middle East respiratory syndrome coronavirus (MERS-CoV) was first identified in 2012 in Saudi Arabia, marking the beginning of a global effort to understand and combat this novel pathogen (Figure 6.1). It is part of the *Betacoronavirus* genus and is believed to have originated from bats, with camels identified as intermediate hosts transmitting the virus to humans (Zaki et al., 2012; de Groot et al., 2013). Early studies suggested that dromedary camels, which have close contact with humans, play a significant role in the zoonotic transmission of MERS-CoV (Memish et al., 2014). MERS-CoV infections are primarily transmitted through direct or indirect contact with camels or their products, such as camel milk, or through human-to-human transmission, particularly in healthcare settings (Assiri et al., 2013; Omrani et al., 2015). Research also points to seasonal fluctuations in camel infections, which may contribute to periodic outbreaks (Baharoon & Memish, 2019). Further studies on the animal-to-human transmission dynamics are ongoing to fully elucidate how the virus is spread in different settings (Hemida et al., 2017). Middle East respiratory syndrome coronavirus (MERS-CoV) first emerged in 2012 and has since been a focal point of global health monitoring due to its high fatality rate and sporadic outbreaks. The virus was initially identified in Saudi Arabia, where it spread to other countries, with a total of 2,609 confirmed cases reported globally as of January 2024. The majority of these cases occurred in the Middle East, with Saudi Arabia being the most affected country (European Centre for Disease Prevention and Control [ECDC], 2024). MERS-CoV has a case fatality rate (CFR) of approximately 36%, with the highest mortality observed among older individuals, particularly those over

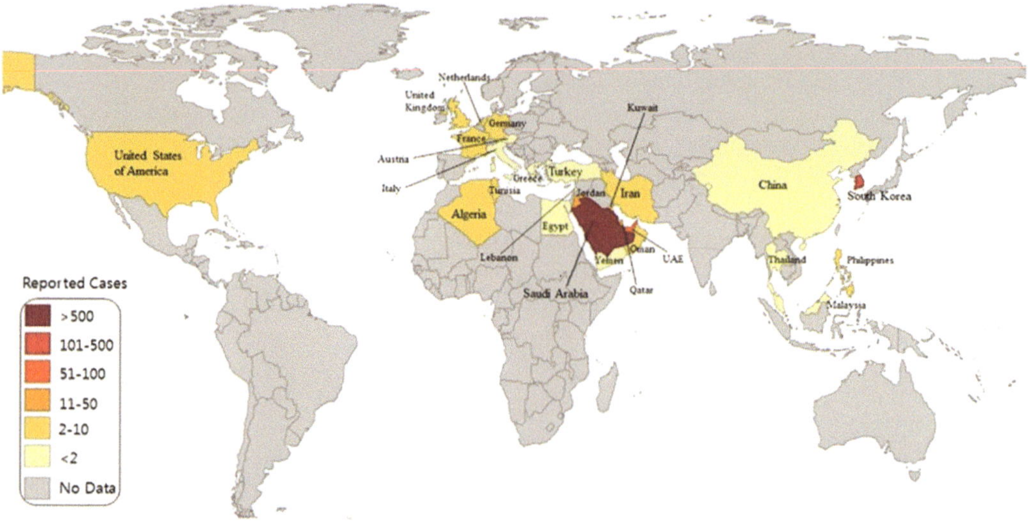

FIGURE 6.1 Global distribution map of Middle East respiratory syndrome coronavirus (MERS-CoV). Individuals in 26 countries have been infected by MERS-CoV. The infographic was generated based on MERS-CoV updates released on July 7, 2015, by World Health Organization (WHO; http://www.who.int/csr/disease/coronavirus_infections/risk-assessment-7july2015/en/).

Source: **Durai et al., 2015. Creative Commons Attribution 4.0 International License.**

70 years old (European Centre for Disease Prevention and Control [ECDC], 2024). Outbreaks of MERS-CoV have primarily been linked to nosocomial (hospital-acquired) infections, though community transmission has also been reported. Healthcare workers are at heightened risk, with numerous infections reported in this group, underscoring the importance of infection control measures in clinical settings (European Centre for Disease Prevention and Control [ECDC], 2024; Csr, n.d.). As the virus predominantly circulates in camels, zoonotic transmission plays a critical role in the epidemiology of MERS-CoV, with limited person-to-person spread typically occurring in close-contact settings such as healthcare environments (Cui et al., 2019; Csr, n.d.). The continuing global surveillance of MERS-CoV remains essential to prevent future outbreaks and mitigate its impact.

The clinical manifestations of Middle East respiratory syndrome coronavirus (MERS-CoV) are diverse and often severe. Common symptoms include fever, cough, and shortness of breath, with some patients experiencing gastrointestinal issues such as diarrhea and nausea (Zaki et al., 2012; Azhar et al. 2017). Severe cases may progress to acute respiratory distress syndrome (ARDS), which is a hallmark of MERS infection (Zaki et al., 2012). The clinical course is frequently complicated by pneumonia, and in some cases, renal failure. While mild cases have been documented, the majority of patients present with severe illness, particularly those with underlying health conditions such as diabetes, hypertension, and chronic renal disease (Zumla et al., 2015). Fatalities are common, with case fatality rates as high as 36%, reflecting the severity of the disease (Azhar et al., 2017). In addition to respiratory symptoms, MERS-CoV can also result in neurological symptoms, including confusion, encephalitis, and even seizures in some patients, although these manifestations are less common (Alraddadi et al., 2016). Notably, a substantial proportion of cases involve multi-organ failure, which significantly contributes to the high mortality rate (Cevik et al., 2021). The severity and progression of MERS-CoV infection are influenced by factors such as age, comorbidities, and the time to intervention.

6.3 SEVERE ACUTE RESPIRATORY SYNDROME CORONAVIRUS (SARS-CoV)

The emergence and outbreaks of severe acute respiratory syndrome coronavirus (SARS-CoV) have been significant milestones in the history of global health. The virus was first identified in 2002 in Guangdong Province, China, after an outbreak of severe pneumonia-like illness began to spread across Asia. SARS-CoV caused a global health crisis in 2003, infecting over 8,000 individuals in more than 26 countries, with a fatality rate of approximately 10% (World Health Organization: WHO, 2019a). The virus was linked to a zoonotic transmission from animals, particularly civet cats, to humans, highlighting the dangers of wildlife as intermediaries in the spread of new diseases. While the SARS outbreak was eventually contained, the economic and social impact was immense, particularly in affected countries such as China, Hong Kong, and Toronto, Canada (Zhu et al., 2023; Gu et al., 2007). The SARS-CoV epidemic is often cited as an early warning for future coronavirus outbreaks. Lessons learned from the rapid global response to SARS, such as the establishment of better surveillance systems and international collaboration, were critical during the subsequent emergence of MERS-CoV in 2012 and the COVID-19 pandemic in 2019 (Liu et al., 2007). Understanding SARS-CoV's origins and transmission dynamics has been crucial in shaping strategies for preventing similar outbreaks in the future (World Health Organization: WHO, 2019a). The clinical features of severe acute respiratory syndrome coronavirus (SARS-CoV) are varied, with most patients presenting with fever, chills, and respiratory symptoms such as cough and shortness of breath. These initial symptoms often progress to severe pneumonia, and some cases can develop into acute respiratory distress syndrome (ARDS) or multi-organ failure (World Health Organization: WHO, 2019a). Gastrointestinal symptoms, including diarrhea, were also reported, though less commonly than respiratory issues (World Health Organization: WHO, 2019a). The clinical course of the disease can be severe, particularly in individuals with underlying health conditions, leading to a higher risk of complications and mortality. SARS-CoV infection is often complicated by secondary bacterial infections, which exacerbate disease progression (Peiris et al., 2003). The mortality rate for SARS-CoV was significant, with an overall fatality rate of approximately 10%, although this varied across different regions and patient groups (Zaki et al., 2012). The risk of death was notably higher among older adults and individuals with preexisting comorbidities, such as diabetes, heart disease, and immunosuppression (Azhar et al., 2017). Despite the high mortality rate, the global outbreak of SARS was effectively contained through rapid intervention and stringent public health measures, but the long-term health impacts on survivors, including chronic fatigue and lung dysfunction, were notable (Zumla et al., 2015).

6.4 SEVERE ACUTE RESPIRATORY SYNDROME CORONAVIRUS 2 (SARS-CoV-2)

The emergence of severe acute respiratory syndrome coronavirus 2 (SARS-CoV-2) marked the beginning of the COVID-19 pandemic in December 2019, following its identification in Wuhan, China. Initially reported as a cluster of pneumonia cases of unknown origin, the virus quickly spread globally, reaching pandemic status by March 2020. SARS-CoV-2 is believed to have zoonotic origins, with the virus likely crossing over from animals to humans, similar to its predecessors, SARS-CoV and MERS-CoV (Hu et al., 2021). Its rapid transmission was fueled by global mobility, with high-density travel networks playing a key role in the worldwide spread of the virus (Zhou et al., 2020). Within a matter of months, the virus had spread to nearly every country, leading to millions of infections and deaths, and overwhelming healthcare systems. The global spread was also compounded by the emergence of several variants, including the highly transmissible Delta and Omicron variants, which contributed to waves of infections even as vaccination efforts were underway. The pandemic caused significant social, economic, and public health disruptions, underscoring the need for coordinated international responses, including travel restrictions, lockdowns, and widespread testing (Mefsin et al., 2022). Continued genomic surveillance remains crucial to understanding and mitigating the spread of new variants, which could potentially evade immunity or vaccines (Aleem et al., 2024).

The transmission dynamics of SARS-CoV-2 have been a critical area of study during the pandemic, reflecting how the virus spreads among populations and influencing control measures. The primary mode of transmission is via respiratory droplets, which can be inhaled or deposited on mucous membranes in the nose, mouth, or eyes (Cevik et al., 2020). Close contact particularly within households and at gatherings has been identified as a significant risk factor, with secondary attack rates ranging from 4% to 35%, depending on factors such as environment and contact patterns (Cevik et al., 2020; Ma et al., 2024). Aerosol transmission also occurs in enclosed spaces with poor ventilation, facilitating spread beyond the 2-meter range commonly advised for physical distancing (Cevik et al., 2020). Furthermore, asymptomatic and pre-symptomatic individuals contribute notably to the transmission dynamics, as they can spread the virus before showing symptoms (Gowler et al., 2022; Su & Kao, 2020; Zsichla & Müller, 2023). The viral load in the upper respiratory tract peaks in the first week after infection, which corresponds to the highest potential for transmission, decreasing as the infection progresses (Gowler et al., 2022; Ma et al., 2024). Understanding these dynamics is essential for developing effective public health strategies and intervention measures (Figure 6.2).

The clinical symptoms and severity spectrum of SARS-CoV-2 infection range from mild to severe, with varying degrees of respiratory, systemic, and post-infection effects. Acute symptoms commonly include fever, cough, fatigue, shortness of breath, and loss of taste or smell. However, the clinical presentation can also involve gastrointestinal symptoms such as diarrhea and nausea. Severe cases often lead to pneumonia, acute respiratory distress syndrome (ARDS), and multi-organ failure, particularly in individuals with comorbidities like diabetes, hypertension, and cardiovascular diseases. The mortality rate for SARS-CoV-2 varies globally, with older adults and those with preexisting health conditions being at greater risk of severe disease outcomes (Zsichla & Müller, 2023; Su et al., 2020; Rodriguez-Morales et al., 2023). Moreover, some patients experience persistent symptoms after the acute phase of infection, referred to as "long COVID." These symptoms can include fatigue, cognitive impairments, and joint pain, which can last for months after

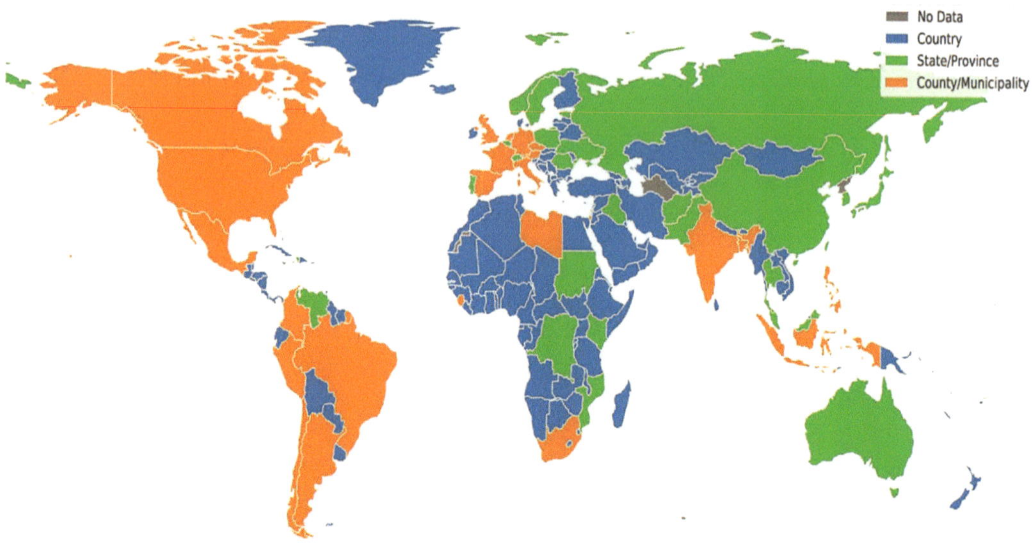

FIGURE 6.2 Visualization of the countries for which we have country-, state-, and county-level epidemiological data. We focus here on data sources that report confirmed cases. We have coverage of 232 different countries or independent territories, from 1,118 different states within these countries (of which, 56 are US states/territories), and 18,478 counties (including all 3,228 US counties).

Source: **Wahltinez et al., 2022. Creative Commons Attribution 4.0 International License.**

recovery. Recent studies have shown that up to 39% of individuals report symptoms lasting longer than 28 days, with a notable portion of cases involving long-term pulmonary, neurological, and psychological issues (Zsichla & Müller, 2023; Su et al., 2020; Rodriguez-Morales et al., 2023). These findings highlight the diverse and evolving nature of SARS-CoV-2, underscoring the complexity of the clinical course and the need for ongoing management of post-infection sequelae.

6.5 COMPARATIVE ANALYSIS OF MERS-CoV, SARS-CoV, AND SARS-CoV-2

6.5.1 Genetic and Structural Differences

The genetic variations between MERS-CoV, SARS-CoV, and SARS-CoV-2 are central to their distinct pathogenicity and transmission dynamics. All three viruses belong to the *Betacoronavirus* genus, yet they exhibit notable genetic differences that influence their ability to infect humans and their virulence profiles (Goyal et al., 2022; Rabaan et al., 2020). SARS-CoV, first identified in 2003, has a genome length of approximately 29.7 kb and is closely related to coronaviruses found in civets and bats. MERS-CoV, which emerged in 2012, shares similarities with SARS-CoV but is genetically distinct. It also has a genome of around 30.1 kb and primarily uses the DPP4 receptor for cell entry, differing from SARS-CoV's ACE2 receptor. These differences affect the host range and pathogenesis of the viruses. SARS-CoV-2, responsible for the COVID-19 pandemic, was identified in December 2019 and shares 79% genetic similarity with SARS-CoV but with a number of key mutations in its spike protein. These mutations enhance its ability to bind to the ACE2 receptor, which contributes to its efficient human-to-human transmission (Goyal et al., 2022; Rabaan et al., 2020; Zhou et al., 2021). Genomic analyses reveal that SARS-CoV-2 has undergone significant mutations since its emergence, influencing both its transmissibility and the effectiveness of vaccines. Comparative studies of the viral genomes highlight the role of specific genes in immune response modulation and viral replication, contributing to the severity and spread of each virus (Goyal et al., 2022; Rabaan et al., 2020; Zhou et al., 2021).

The structural differences between MERS-CoV, SARS-CoV, and SARS-CoV-2 highlight significant variations in their spike proteins and receptor binding mechanisms, despite all being members of the *Betacoronavirus* genus (Baharoon & Memish, 2019; Gu & Korteweg, 2007; Mahase, 2020).

1. *Spike protein variations:* All three coronaviruses utilize spike proteins (S) to mediate entry into host cells via specific receptor binding. However, while SARS-CoV and SARS-CoV-2 bind to the *ACE2* receptor on human cells, MERS-CoV binds to *Dipeptidyl peptidase 4 (DPP4)* (Goyal et al., 2022; Zhou et al., 2021). This receptor specificity contributes to different tissue tropisms and clinical manifestations of the diseases caused by these viruses (Belouzard et al., 2009).
2. *SARS-CoV-2's unique features:* SARS-CoV-2 differs from both SARS-CoV and MERS-CoV in the presence of a furin cleavage site in its spike protein, which is believed to enhance its infectivity and ability to enter human cells more efficiently (Zhou et al., 2021). This feature is unique to SARS-CoV-2 and differentiates it from SARS-CoV, even though both share high sequence similarity in their spike proteins (Goyal et al., 2022; Zhou et al., 2021; Belouzard et al., 2009).
3. *Genomic and structural similarities:* While these viruses belong to different lineages within the Betacoronavirus subfamily, they all share similar genomic architectures, with structural proteins encoding the spike (S), membrane (M), nucleocapsid (N), and envelope (E) proteins. However, genetic analysis reveals that SARS-CoV-2 and SARS-CoV are more genetically related, with a sequence identity of 79.5%, compared to only 40% between MERS-CoV and SARS-CoV-2 (Belouzard et al., 2009; Goyal et al., 2022; Zhou et al., 2021).

These structural differences play critical roles in determining the viruses' transmission, pathogenicity, and clinical outcomes.

6.5.2 Pathogenicity and Virulence

The pathogenicity and virulence of MERS-CoV, SARS-CoV, and SARS-CoV-2 differ based on several factors, including their ability to bind to host receptors, induce immune responses, and cause organ damage. All three coronaviruses share a similar structural mechanism, using spike (S) proteins to enter host cells via binding to ACE2 receptors. However, SARS-CoV-2 demonstrates a higher affinity for ACE2 compared to SARS-CoV and MERS-CoV, potentially leading to more widespread tissue infection and greater transmission potential (Zhang et al., 2020). This increased binding affinity may contribute to the higher transmissibility of SARS-CoV-2, despite its lower fatality rate compared to MERS-CoV, which has a much higher mortality rate of around 34% (Rabaan et al., 2020). While SARS-CoV-2's lower lethality can be attributed to a more robust immune response and fewer severe organ impacts in many patients, MERS-CoV's virulence is linked to its rapid progression to severe respiratory distress and organ failure (Pustake et al., 2022). Notably, SARS-CoV-2's induction of cytokine storms in severe cases has been shown to contribute to multi-organ failure, similar to SARS-CoV, but more frequently and with a broader spectrum of symptoms (Alfarouk et al., 2021). Thus, these coronaviruses exhibit distinct pathogenic features, which are crucial for determining public health strategies and treatment approaches.

6.5.3 Epidemiological Trends

The epidemiological trends of MERS-CoV, SARS-CoV, and SARS-CoV-2 have varied significantly, reflecting differences in transmission patterns, geography, and public health responses. MERS-CoV, which emerged in 2012, was primarily confined to the Arabian Peninsula, with Saudi Arabia reporting the highest number of cases. The disease spread to other countries primarily through healthcare-associated transmissions and international travelers. Notably, the virus was mainly transmitted from camels to humans, with sporadic human-to-human transmission in healthcare settings (Alfarouk et al., 2021; Pustake et al., 2022; Zhang et al., 2020; Zaki et al., 2012). In contrast, SARS-CoV, which caused the 2002–2003 outbreak, spread more widely across the globe, with major outbreaks in Hong Kong, mainland China, and other parts of Asia, as well as North America and Europe. The rapid human-to-human transmission and high mortality rates led to significant global containment efforts, and the outbreak was eventually brought under control in 2003 (Pustake et al., 2022; Zhang et al., 2020; Cheng et al., 2007). SARS-CoV-2, responsible for the COVID-19 pandemic, demonstrated unparalleled global spread. From its initial detection in Wuhan, China, in December 2019, the virus has spread worldwide, causing millions of cases. Unlike MERS-CoV and SARS-CoV, SARS-CoV-2 showed a higher asymptomatic carrier rate and a higher level of community transmission, contributing to the pandemic's widespread impact (Zhang et al., 2020; Alfarouk et al., 2021; Pustake et al., 2022). These differing epidemiological patterns underscore the challenges faced by public health authorities in managing these coronaviruses and highlight the role of global surveillance and swift response systems in controlling outbreaks (Table 6.1).

6.6 FUTURE DIRECTIONS AND CHALLENGES: THE CONCLUSION

6.6.1 Emerging Variants and Mutation Risks

The emergence of variants and mutations in coronaviruses, particularly in SARS-CoV-2, has raised significant concerns regarding public health, immunity, and disease management. SARS-CoV-2 exhibits rapid mutation rates due to its RNA genome, with an average of one to two changes per month, leading to the emergence of several variants of concern, including Alpha, Beta, Delta, and Omicron (Day et al., 2020). These mutations, particularly in the spike (S) protein, which facilitates viral entry into host cells, can enhance transmissibility, evade neutralizing antibodies, and impact vaccine efficacy (Amanat & Krammer, 2020). Similar mutations have been observed in MERS-CoV

TABLE 6.1
Comparative Analysis of MERS-CoV, SARS-CoV, and SARS-CoV-2 on the Basis of Genetic Differences, Structural Differences, Pathogenicity, and Virulence and Epidemiological Trends

Parameter	MERS-CoV	SARS-CoV	SARS-CoV-2
Genetic differences	Belongs to the *Betacoronavirus* genus; shares ~50% genome similarity with SARS-CoV and ~80% with SARS-CoV-2 (Zaki et al., 2012)	Belongs to the *Betacoronavirus* genus; shares ~80% genome similarity with SARS-CoV-2 and ~50% with MERS-CoV (Ksiazek et al., 2003)	Shares ~80% genome similarity with SARS-CoV and ~50% with MERS-CoV; has a unique furin cleavage site enhancing infectivity (Zhou et al., 2020)
Structural differences	Spike protein binds to *DPP4* (dipeptidyl peptidase-4) receptor (Raj et al., 2013)	Spike protein binds to *ACE2* (angiotensin-converting enzyme 2) receptor (Li et al., 2003)	Spike protein binds to *ACE2* receptor with higher affinity due to a unique receptor binding domain (RBD) (Shang et al., 2020)
Pathogenicity and virulence	Causes *severe lower respiratory tract infections,* with high case fatality rate (CFR) of ~34.4% (WHO, 2019b)	Causes *acute respiratory distress syndrome (ARDS)*; CFR ~9.6% (WHO, 2019b)	Causes *broad-spectrum disease* (asymptomatic to ARDS); CFR ~1–3% globally, though higher in vulnerable populations (WHO, 2019b)
Epidemiological trends	Primarily localized in the Middle East; outbreaks linked to camels as reservoirs; limited human-to-human transmission (Zumla et al., 2015)	Originated in China; human-to-human transmission mainly via respiratory droplets; extinguished by 2004 (WHO, 2019b)	Pandemic spread worldwide since late 2019; highly transmissible due to asymptomatic spread and airborne transmission (WHO, 2019b)

and SARS-CoV, though SARS-CoV-2's mutations are more frequent and diverse, contributing to its global spread. The high mutation potential of these viruses, coupled with interspecies transmission risks, continues to pose challenges in controlling outbreaks and adapting vaccines and therapeutics (Goswami et al., 2022).

6.6.2 Advances in Vaccine Technology

Advances in vaccine technology have been critical in the fight against coronaviruses, particularly MERS-CoV, SARS-CoV, and SARS-CoV-2. Vaccine development for MERS-CoV has primarily focused on the spike protein, which is essential for the virus to enter human cells. Early trials demonstrated that spike protein-based vaccines were effective in eliciting immune responses, although no licensed MERS vaccine is yet available. SARS-CoV, responsible for the 2003 outbreak, also spurred vaccine development, with early research focusing on the spike protein and receptor binding domain (RBD). However, the lack of a sustained outbreak limited progress. In contrast, the global spread of SARS-CoV-2 accelerated vaccine research, leading to the rapid development of mRNA vaccines, such as the Pfizer-BioNTech and Moderna vaccines, which target the spike protein of the virus. These vaccines have shown high efficacy in preventing symptomatic COVID-19 infection and severe outcomes. In addition to mRNA vaccines, viral vector-based vaccines, protein subunit vaccines, and inactivated vaccines have been developed and distributed worldwide. The success of these vaccines in controlling the COVID-19 pandemic has also led to new approaches for future coronavirus vaccine development, including the exploration of universal coronavirus vaccines that could address a broader range of coronavirus strains (Al Fayez et al., 2023; Peng et al., 2021; Goswami et al., 2022).

6.6.3 Global Strategies for Pandemic Preparedness

Global strategies for pandemic preparedness, particularly for outbreaks such as MERS-CoV, SARS-CoV, and SARS-CoV-2, require coordinated, multifaceted approaches that address health, governance, and economic dimensions. Effective responses are heavily reliant on early detection, robust surveillance systems, and rapid containment measures. A critical aspect of preparedness is the establishment of strong global health governance structures, such as the proposed Global Health Threats Council, which would provide consistent political attention and enhance accountability across nations (Jung et al., 2021). Countries that employed aggressive containment strategies, such as New Zealand and South Korea, succeeded in minimizing transmission by taking swift actions based on scientific data. Additionally, international collaboration, such as the Access to COVID-19 Tools Accelerator (ACT-A), has been pivotal in ensuring equitable vaccine distribution and addressing disparities in health resources (Pandemic Prevention, Preparedness and Response Accord, 2023). Furthermore, enhancing the authority and independence of the World Health Organization (WHO) is essential for quicker response times and more effective pandemic management (Jung et al., 2021). The need for new international frameworks, such as a pandemic treaty, has been emphasized to ensure that legal frameworks can swiftly respond to future health threats (Jung et al., 2021; Pandemic Prevention, Preparedness and Response Accord, 2023; WHO, 2019b).

REFERENCES

Al Fayez, N., Nassar, M. S., Alshehri, A. A., Alnefaie, M. K., Almughem, F. A., Alshehri, B. Y., Alawad, A. O., & Tawfik, E. A. (2023, July 18). Recent advancement in mRNA vaccine development and applications. *Pharmaceutics*, 15(7), 1972. https://doi.org/10.3390/pharmaceutics15071972. PMID: 37514158; PMCID: PMC10384963.

Aleem, A., Akbar Samad, A. B., & Vaqar, S. (2024, January). Emerging variants of SARS-CoV-2 and novel therapeutics against coronavirus (COVID-19). In *StatPearls [Internet]*. Treasure Island (FL): StatPearls Publishing. Updated 2023 May 8. https://www.ncbi.nlm.nih.gov/books/NBK570580/.

Alfarouk, K. O., AlHoufie, S. T. S., Ahmed, S. B. M., Shabana, M., Ahmed, A., Alqahtani, S. S., Alqahtani, A. S., Alqahtani, A. M., Ramadan, A. M., Ahmed, M. E., Ali, H. S., Bashir, A., Devesa, J., Cardone, R. A., Ibrahim, M. E., Schwartz, L., & Reshkin, S. J. (2021). Pathogenesis and management of COVID–19. *Journal of Xenobiotics*, 11(2), 77–93. https://doi.org/10.3390/jox11020006.

Alraddadi, B. M., et al. (2016). Risk factors for primary Middle East respiratory syndrome Coronavirus illness in humans, Saudi Arabia, 2014. *Emerging Infectious Diseases*, 22(1), 49–55. https://doi.org/10.3201/eid2201.151340.

Amanat, F., & Krammer, F. (2020). SARS-CoV-2 vaccines: Status report. *Immunity*, 52(4), 583–589. https://doi.org/10.1016/j.immuni.2020.03.007.

Assiri, A., Al-Tawfiq, J. A., Al-Rabeeah, A. A., Al-Rabiah, F. A., Al-Hajjar, S., Al-Barrak, A., Flemban, H., Al-Nassir, W. N., Balkhy, H. H., Al-Hakeem, R. F., Makhdoom, H. Q., Zumla, A. I., & Memish, Z. A. (2013). Epidemiological, demographic, and clinical characteristics of 47 cases of Middle East respiratory syndrome coronavirus disease from Saudi Arabia: A descriptive study. *The Lancet. Infectious Diseases*, 13(9), 752–761. https://doi.org/10.1016/S1473-3099(13)70204-4.

Attia, Y. A., et al. (2021). COVID-19: Pathogenesis, advances in treatment and vaccine development and environmental impact—an updated review. *Environmental Science and Pollution Research*, 28, 22241–22264. https://doi.org/10.1007/s11356-021-13018-1.

Azhar, E. I., Lanini, S., Ippolito, G., & Zumla, A. (2017). The Middle East respiratory syndrome coronavirus—a continuing risk to global health security. *Advances in Experimental Medicine and Biology*, 972, 49–60. https://doi.org/10.1007/5584_2016_133.

Baharoon, S., & Memish, Z. A. (2019, November–December). MERS-CoV as an emerging respiratory illness: A review of prevention methods. *Travel Medicine and Infectious Disease*, 32, 101520. https://doi.org/10.1016/j.tmaid.2019.101520. Epub 2019 Nov 12. PMID: 31730910; PMCID: PMC7110694.

Belouzard, S., Chu, V. C., & Whittaker, G. R. (2009). Activation of the SARS coronavirus spike protein via sequential proteolytic cleavage at two distinct sites. *Proceedings of the National Academy of Sciences of the United States of America*, 106(14), 5871–5876. https://doi.org/10.1073/pnas.0809524106.

Brant, A. C., et al. (2021). SARS-CoV-2: From its discovery to genome structure, transcription, and replication. *Cell & Bioscience*, 11, 136. https://doi.org/10.1186/s13578-021-00643-z.

Cevik, M., Kuppalli, K., Kindrachuk, J., & Peiris, M. (2020). Virology, transmission, and pathogenesis of SARS-CoV-2. *BMJ (Clinical Research Edition)*, 371, m3862. https://doi.org/10.1136/bmj.m3862.

Cevik, M., Tate, M., Lloyd, O., Maraolo, A. E., Schafers, J., & Ho, A. (2021, January). SARS-CoV-2, SARS-CoV, and MERS-CoV viral load dynamics, duration of viral shedding, and infectiousness: A systematic review and meta-analysis. *Lancet Microbe*, 2(1), e13–e22. https://doi.org/10.1016/S2666-5247(20)30172-5. Epub 2020 Nov 19. PMID: 33521734; PMCID: PMC7837230.

Cheng, V. C., Lau, S. K., Woo, P. C., & Yuen, K. Y. (2007). Severe acute respiratory syndrome coronavirus as an agent of emerging and reemerging infection. *Clinical Microbiology Reviews*, 20(4), 660–694. https://doi.org/10.1128/CMR.00023-07.

Csr. (n.d.). *MERS Outbreaks*. World Health Organization—Regional Office for the Eastern Mediterranean. https://www.emro.who.int/health-topics/mers-cov/mers-outbreaks.html.

Cui, J., Li, F., & Shi, Z. L. (2019). Origin and evolution of pathogenic coronaviruses. *Nature Reviews. Microbiology*, 17(3), 181–192. https://doi.org/10.1038/s41579-018-0118-9.

Day, T., Gandon, S., Lion, S., & Otto, S. P. (2020, August 3). On the evolutionary epidemiology of SARS-CoV-2. *Current Biology*, 30(15), R849–R857. https://doi.org/10.1016/j.cub.2020.06.031. Epub 2020 Jun 11. PMID: 32750338; PMCID: PMC7287426.

de Groot, R. J., Baker, S. C., Baric, R. S., Brown, C. S., Drosten, C., Enjuanes, L., Fouchier, R. A., Galiano, M., Gorbalenya, A. E., Memish, Z. A., Perlman, S., Poon, L. L., Snijder, E. J., Stephens, G. M., Woo, P. C., Zaki, A. M., Zambon, M., & Ziebuhr, J. (2013). Middle East respiratory syndrome coronavirus (MERS-CoV): Announcement of the Coronavirus study group. *Journal of Virology*, 87(14), 7790–7792. https://doi.org/10.1128/JVI.01244-13.

Durai, P., et al. (2015). Middle East respiratory syndrome coronavirus: Transmission, virology and therapeutic targeting to aid in outbreak control. *Experimental & Molecular Medicine*, 47, e181. https://doi.org/10.1038/emm.2015.76.

European Centre for Disease Prevention and Control (ECDC). (2024). *MERS-CoV Worldwide Overview*. https://www.ecdc.europa.eu/en/mers.

Goswami, S., & Dey, C. (Eds.). (2022). *COVID-19 and SARS-CoV-2: The Science and Clinical Application of Conventional and Complementary Treatments (1st ed.)*. CRC Press. https://doi.org/10.1201/9781003178514

Gowler, C. D., Paul, P., & Reddy, S. C. (2022). Viral shedding kinetics and transmission of emerging SARS-CoV-2 variants-critical components of study design. *JAMA Network Open*, 5(5), e2213614. https://doi.org/10.1001/jamanetworkopen.2022.13614.

Goyal, R., Gautam, R. K., Chopra, H., Dubey, A. K., Singla, R. K., Rayan, R. A., Kamal, M. A. (2022, September 29). Comparative highlights on MERS-CoV, SARS-CoV-1, SARS-CoV-2, and NEO-CoV. EXCLI Journal, 21, 1245–1272. https://doi.org/10.17179/excli2022-5355. PMID: 36483910; PMCID: PMC9727256.

Gu, J., & Korteweg, C. (2007, April). Pathology and pathogenesis of severe acute respiratory syndrome. *The American Journal of Pathology*, 170(4), 1136–1147. https://doi.org/10.2353/ajpath.2007.061088. PMID: 17392154; PMCID: PMC1829448.

Hemida, M. G., Elmoslemany, A., Al-Hizab, F., Alnaeem, A., Almathen, F., Faye, B., Chu, D. K., Perera, R. A., & Peiris, M. (2017). Dromedary Camels and the transmission of Middle East respiratory syndrome Coronavirus (MERS-CoV). *Transboundary and Emerging Diseases*, 64(2), 344–353. https://doi.org/10.1111/tbed.12401.

Hu, B., Guo, H., Zhou, P., & Shi, Z. L. (2021). Characteristics of SARS-CoV-2 and COVID–19. *Nature Reviews. Microbiology*, 19(3), 141–154. https://doi.org/10.1038/s41579-020-00459-7.

Jung, A., et al. (2021). National responses to covid-19: Drivers, complexities, and uncertainties in the first year of the pandemic. *BMJ*, 375, e068954. https://doi.org/10.1136/bmj-2021-068954.

Ksiazek, T. G., et al. (2003). A novel coronavirus associated with severe acute respiratory syndrome. *The New England Journal of Medicine*, 348(20), 1953–1966. https://doi.org/10.1056/NEJMoa030781.

Li, W., Moore, M. J., Vasilieva, N., Sui, J., Wong, S. K., Berne, M. A., Somasundaran, M., Sullivan, J. L., Luzuriaga, K., Greenough, T. C., Choe, H., & Farzan, M. (2003). Angiotensin-converting enzyme 2 is a functional receptor for the SARS coronavirus. *Nature*, 426(6965), 450–454. https://doi.org/10.1038/nature02145.

Liu, M., Yang, Y., Gu, C., Yue, Y., Wu, K. K., Wu, J., & Zhu, Y. (2007). Spike protein of SARS-CoV stimulates cyclooxygenase-2 expression via both calcium-dependent and calcium-independent protein kinase C pathways. *FASEB Journal: Official publication of the Federation of American Societies for Experimental Biology*, 21(7), 1586–1596. https://doi.org/10.1096/fj.06-6589com (Retraction published FASEB J. 2024 Nov 30;38(22):e70142. doi: 10.1096/fj.FSB2-2024-m10191109).

Lv, J.-X., Liu, X., Pei, Y.-Y., Song, Z.-G., Chen, X., Hu, S.-J., She, J.-L., Liu, Y., Chen, Y.-M., Zhang, Y.-Z. (2024). Evolutionary trajectory of diverse SARS-CoV-2 variants at the beginning of COVID-19 outbreak. *Virus Evolution*, 10(1), veae020. https://doi.org/10.1093/ve/veae020.

Ma, T., Chen, C., Wang, J., Wang, H., Zhao, Y., Zhu, Y., Yan, Z., Ding, S., & Ding, J. (2024). Parametric analysis of the transmission dynamics during indigenous aggregated outbreaks caused by five SARS-CoV-2 strains in Nanjing, China. *Frontiers in Public Health*, 12, 1358577. https://doi.org/10.3389/fpubh.2024.1358577.

Mahase E. (2020). Coronavirus covid-19 has killed more people than SARS and MERS combined, despite lower case fatality rate. *BMJ* (Clinical research ed.), 368, m641. https://doi.org/10.1136/bmj.m641

Mahase, E. (2020). Covid-19: Coronavirus was first described in *The BMJ* in 1965. *BMJ (Clinical Research Edition)*, 369, m1547. https://doi.org/10.1136/bmj.m1547.

Mefsin, Y. M., et al. (2022, January–March). Epidemiology of Infections with SARS-CoV-2 Omicron BA.2 variant, Hong Kong. *Emerging Infectious Diseases*, 28(9), 1856–1858. https://doi.org/10.3201/eid2809.220613.

Memish, Z. A., Cotten, M., Meyer, B., Watson, S. J., Alsahafi, A. J., Al Rabeeah, A. A., Corman, V. M., Sieberg, A., Makhdoom, H. Q., Assiri, A., Al Masri, M., Aldabbagh, S., Bosch, B. J., Beer, M., Müller, M. A., Kellam, P., & Drosten, C. (2014, June). Human infection with MERS coronavirus after exposure to infected camels, Saudi Arabia, 2013. *Emerging Infectious Diseases*, 20(6), 1012–1015. https://doi.org/10.3201/eid2006.140402. PMID: 24857749; PMCID: PMC4036761.

MERS-CoV Worldwide Overview. (2024, November 11). European Centre for Disease Prevention and Control. https://www.ecdc.europa.eu/en/middle-east-respiratory-syndrome-coronavirus-mers-cov-situation-update.

Omrani, A. S., Al-Tawfiq, J. A., & Memish, Z. A. (2015). Middle East respiratory syndrome coronavirus (MERS-CoV): Animal to human interaction. *Pathogens and Global Health*, 109(8), 354–362. https://doi.org/10.1080/20477724.2015.1122852.

Pandemic Prevention, Preparedness and Response Accord. (2023, June 17). https://www.who.int/news-room/questions-and-answers/item/pandemic-prevention-preparedness-and-response-accord.

Peiris, J. S., Chu, C. M., Cheng, V. C., Chan, K. S., Hung, I. F., Poon, L. L., Law, K. I., Tang, B. S., Hon, T. Y., Chan, C. S., Chan, K. H., Ng, J. S., Zheng, B. J., Ng, W. L., Lai, R. W., Guan, Y., Yuen, K. Y., & HKU/UCH SARS Study Group. (2003). Clinical progression and viral load in a community outbreak of coronavirus-associated SARS pneumonia: A prospective study. *Lancet (London, England)*, 361(9371), 1767–1772. https://doi.org/10.1016/s0140-6736(03)13412-5.

Peng, X. L., Cheng, J. S., Gong, H. L., Yuan, M. D., Zhao, X. H., Li, Z., & Wei, D. X. (2021). Advances in the design and development of SARS-CoV-2 vaccines. *Military Medical Research*, 8(1), 67. https://doi.org/10.1186/s40779-021-00360-1.

Pustake, M., Tambolkar, I., Giri, P., & Gandhi, C. (2022, January). SARS, MERS and CoVID-19: An overview and comparison of clinical, laboratory and radiological features. *Journal of Family Medicine and Primary Care*, 11(1), 10–17. https://doi.org/10.4103/jfmpc.jfmpc_839_21. Epub 2022 Jan 31. PMID: 35309670; PMCID: PMC8930171.

Rabaan, A. A., Al-Ahmed, S. H., Haque, S., Sah, R., Tiwari, R., Malik, Y. S., Dhama, K., Yatoo, M. I., Bonilla-Aldana, D. K., & Rodriguez-Morales, A. J. (2020). SARS-CoV-2, SARS-CoV, and MERS-COV: A comparative overview. *Le infezioni in medicina*, *28*(2), 174–184.

Raj, V. S., Mou, H., Smits, S. L., Dekkers, D. H., Müller, M. A., Dijkman, R., Muth, D., Demmers, J. A., Zaki, A., Fouchier, R. A., Thiel, V., Drosten, C., Rottier, P. J., Osterhaus, A. D., Bosch, B. J., & Haagmans, B. L. (2013). Dipeptidyl peptidase 4 is a functional receptor for the emerging human coronavirus-EMC. *Nature*, 495(7440), 251–254. https://doi.org/10.1038/nature12005.

Rodriguez-Morales, A. J., Lopez-Echeverri, M. C., Perez-Raga, M. F., Quintero-Romero, V., Valencia-Gallego, V., Galindo-Herrera, N., López-Alzate, S., Sánchez-Vinasco, J. D., Gutiérrez-Vargas, J. J., Mayta-Tristan, P., Husni, R., Moghnieh, R., Stephan, J., Faour, W., Tawil, S., Barakat, H., Chaaban, T., Megarbane, A., Rizk, Y., Sakr, R., Escalera-Antezana, J. P., Alvarado-Arnez, L. E., Bonilla-Aldana, D. K., Camacho-Moreno, G., Mendoza, H., Rodriguez-Sabogal, I. A., Millán-Oñate, J., Lopardo, G., Barbosa, A. N., Cimerman, S., Chaves, T. D. S. S., Orduna, T., Lloveras, S., Rodriguez-Morales, A. G., Thormann, M., Zambrano, P. G., Perez, C., Sandoval, N., Zambrano, L., Alvarez-Moreno, C. A., Chacon-Cruz, E., Villamil-Gomez, W. E., Benites-Zapata, V., Savio-Larriera, E., Cardona-Ospina, J. A., Risquez, A., Forero-Peña, D. A., Henao-Martínez, A. F., Sah, R., Barboza, J. J., León-Figueroa, D. A., Acosta-España, J. D., Carrero-Gonzalez, C. M., Al-Tawfiq, J. A., Rabaan, A. A., Leblebicioglu, H., Gonzales-Zamora, J. A., & Ulloa-Gutiérrez, R. (2023, July-August). The global challenges of the long COVID-19 in adults and children. *Travel Medicine and Infectious Disease*, 54, 102606. https://doi.org/10.1016/j.tmaid.2023.102606. Epub 2023 Jun 8. PMID: 37295581; PMCID: PMC10247301.

Shang, J., Wan, Y., Luo, C., Ye, G., Geng, Q., Auerbach, A., & Li, F. (2020). Cell entry mechanisms of SARS-CoV-2. *Proceedings of the National Academy of Sciences of the United States of America*, 117(21), 11727–11734. https://doi.org/10.1073/pnas.2003138117.

Su, T. H., & Kao, J. H. (2020, June). The clinical manifestations and management of COVID-19-related liver injury. *Journal of the Formosan Medical Association*, 119(6), 1016–1018. https://doi.org/10.1016/j.jfma.2020.04.020. Epub 2020 Apr 24. PMID: 32345544; PMCID: PMC7180368.

Su, Y., Chen, D., Yuan, D., Lausted, C., Choi, J., Dai, C. L., Voillet, V., Duvvuri, V. R., Scherler, K., Troisch, P., Baloni, P., Qin, G., Smith, B., Kornilov, S. A., Rostomily, C., Xu, A., Li, J., Dong, S., Rothchild, A., Zhou, J., . . . Heath, J. R. (2020). Multi-omics resolves a sharp disease-state shift between mild and moderate COVID-19. *Cell*, 183(6), 1479–1495.e20. https://doi.org/10.1016/j.cell.2020.10.037.

Wahltinez, O., et al. (2022). COVID-19 open-data a global-scale spatially granular meta-dataset for *coronavirus* disease. *Scientific Data*, 9, 162. https://doi.org/10.1038/s41597-022-01263-z.

World Health Organization: WHO. (2019a, November 1). *Severe Acute respiratory Syndrome (SARS)*. https://www.who.int/health-topics/severe-acute-respiratory-syndrome#tab=tab_1.

World Health Organization: WHO. (2019b, November 8). *Middle East Respiratory Syndrome Coronavirus (MERS-CoV)*. https://www.who.int/health-topics/middle-east-respiratory-syndrome-coronavirus-mers#tab=tab_1.

World Health Organization: WHO. (2022, August 5). *Middle East Respiratory Syndrome Coronavirus (MERS-CoV)*. https://www.who.int/news-room/fact-sheets/detail/middle-east-respiratory-syndrome-coronavirus-(mers-cov).

Yang, Y., et al. (2020). SARS-CoV-2: Characteristics and current advances in research. *Virology Journal*, 17, 117. https://doi.org/10.1186/s12985-020-01369-z

Zaki, A. M., van Boheemen, S., Bestebroer, T. M., Osterhaus, A. D., & Fouchier, R. A. (2012). Isolation of a novel coronavirus from a man with pneumonia in Saudi Arabia. *The New England Journal of Medicine*, 367(19), 1814–1820. https://doi.org/10.1056/NEJMoa1211721.

Zhang, A. R., et al. (2021). Epidemiology and evolution of Middle East respiratory syndrome coronavirus, 2012–2020. *Infectious Diseases of Poverty*, 10, 66. https://doi.org/10.1186/s40249-021-00853-0.

Zhang, L., Jackson, C. B., Mou, H., Ojha, A., Peng, H., Quinlan, B. D., Rangarajan, E. S., Pan, A., Vanderheiden, A., Suthar, M. S., Li, W., Izard, T., Rader, C., Farzan, M., & Choe, H. (2020). SARS-CoV-2 spike-protein D614G mutation increases virion spike density and infectivity. *Nature Communications*, 11(1), 6013. https://doi.org/10.1038/s41467-020-19808-4.

Zhou, H., Yang, J., Zhou, C., Chen, B., Fang, H., Chen, S., Zhang, X., Wang, L., & Zhang, L. (2021, December 7). A review of SARS-CoV2: Compared with SARS-CoV and MERS-CoV. *Frontiers in Medicine (Lausanne)*, 8, 628370. https://doi.org/10.3389/fmed.2021.628370. PMID: 34950674; PMCID: PMC8688360.

Zhou, P., et al. (2020). A pneumonia outbreak associated with a new coronavirus of probable bat origin. *Nature*, 579(7798), 270–273. https://doi.org/10.1038/s41586-020-2012-7.

Zhu, Y., Sharma, L., & Chang, D. (2023, August 14). Pathophysiology and clinical management of coronavirus disease (COVID-19): A mini-review. *Frontiers in Immunology*, 14, 1116131. https://doi.org/10.3389/fimmu.2023.1116131. PMID: 37646038; PMCID: PMC10461092.

Zsichla, L., & Müller, V. (2023, January 7). Risk factors of severe COVID-19: A review of host, viral and environmental factors. *Viruses*, 15(1), 175. https://doi.org/10.3390/v15010175. PMID: 36680215; PMCID: PMC9863423.

Zumla, A., Hui, D. S., & Perlman, S. (2015). Middle East respiratory syndrome. *Lancet (London, England)*, 386(9997), 995–1007. https://doi.org/10.1016/S0140-6736(15)60454-8.

7 Mosquito-Borne Viral Infections

Chikungunya, Dengue, and Zika Virus

Pallavi Tiwari, Mamta Tiwari, Theeshan Bahorun, Apoorv Mishra, and Nisha Sharma

7.1 CHIKUNGUNYA (CHIK): THE FUNDAMENTALS

7.1.1 Definitions and Terminologies Associated with Chikungunya and CHIKV

Over the past 15 years, CHIK, a tropical disease spread by mosquitoes, has become a serious threat to public health (Ganesan et al., 2017)—a developing arbovirus affecting humans, producing chikungunya fever, is known as CHIKV. As an Alphavirus, the CHIKV, belongs to the Togaviridae family (Kariuki Njenga et al., 2008). The term "chikungunya" is derived from the Kimakonde root verb "kungunyala" (Weinbren et al., 1958).

7.1.2 History of Chikungunya and CHIKV Infection

In 1823, the first human case of a disease resembling CHIK were reported in Zanzibar, Africa. Two years later, in 1827 and 1828, there were similar cases reported. Since then, no other reports of a comparable illness have been found until 1952 when multiple persons in Tanzania experienced rheumatic fever episodes; it was during this period that CHIKV was initially isolated, recognized, and classified as an arbovirus (Constant et al., 2021). The first substantial urban outbreaks of chikungunya fever (CHIKF) were described in the early 1960s in Bangkok and from 1963 to 1973 in India. There were occasional small outbreaks during the next 30 years; however, no notable outbreaks until 2004, when a major epidemic started on Kenya's coast (Nimmannitya et al., 1969). Millions of people became ill in the nations surrounding the Indian Ocean as a result of the 2004 re-emergence of CHIKV in Kenya, which then moved eastward. During the four years that followed this outbreak, CHIKV extended to multiple Indian Ocean islands, India, and portions of Southeast Asia (Staples et al., 2009). The last significant outbreak began in Kenya in 2004 and moved to La Reunion in 2005 via nearby islands (Pena-Garcia et al., 2017). The disease impacted more than one-third of people living on the island during this outbreak, which resulted in 284 fatalities with around 40,000 cases every week (Josseran et al., 2006). Additionally, Aedes albopictus, a newly discovered species, also contributed in the spread. The discovery of the first outbreak in Italy in 2007 was indicative of this outbreak because this mosquito has a remarkable adaptability to temperate regions (Thiboutot et al., 2010). CHIKF was not known to have entered the Americas until 2013, when 658 cases of CHIKV were confirmed with infection rate of 1.76% during the first outbreak in Saint Martin. However, it is likely that CHIKV had already entered the continent at that time (Morens & Fauci, 2014). Since then, around 45 American nations and territories reported local transmission cases, leading to the confirmation of almost three million cases in the Americas (Wahid et al., 2017). Since its discovery, several genotypes of CHIKV have been

 DOI: 10.1201/9781032721811-7

identified, including Asian, East Indian (IOL), West African (WA), and East/Central/South African (ECSA) (Nunes et al., 2015).

7.1.3 Taxonomic Classification and Nomenclature of CHIKV

Alphaviruses are a family of viruses that affect animals worldwide. Their geographic distribution allows them to be divided into two groups: Old and New World viruses. The New World alphaviruses are widespread throughout the Americas that cause encephalitis in both humans and horses (equine). The Old World alphaviruses that cause fever, rash, and arthritis in humans include CHIKV and other viruses present in Asia, Europe, Australia, and some regions of Africa along with the viruses from the Mayaro, Barmah Forest, Ross River, and O'nyong'nyong. The Semliki Forest Virus Complex includes the alphavirus CHIKV (King et al., 2012).

Realm: Riboviria	Class: Alsuviricetes
Kingdom: Orthornavirae	Order: Martellivirales
Phyllum: Kitrinoviricota	Family: Togaviridae
Genus: Alphavirus	Species: Chikungunya virus

7.1.4 Transmission Process of CHIKV

The transmission of CHIKV occurs through two different cycles, Sylvan and Urban. The term "urban cycle" describes the process by which a disease spreads from people to mosquitoes and then back to people, whereas sylvatic involves transfer of disease from animals to mosquitos and subsequently to humans (Figure 7.1) (Ganesan et al., 2017).

In urban cycle, non-human primates and forest-dwelling Aedes mosquitoes sustain a sylvatic transmission cycle of CHIKV, which causes occasional human instances and minor breakouts (Petersen et al., 2016). The most important mosquito vectors in the urban cycle, Aegypti and Albopictus, also contribute to the other vector, which is more important for public health and involves the recurring spread of CHIKV from infected to uninfected humans (Wahid et al., 2017). Apart from the traditional urban and sylvatic transmission cycles, vertical transmission of CHIKV infection can also transpire during pregnancy and blood transfusion. All phases of pregnancy were found to have vertical transmission. The most significant indicators of severity are meningoencephalitis and/or myocarditis; however,

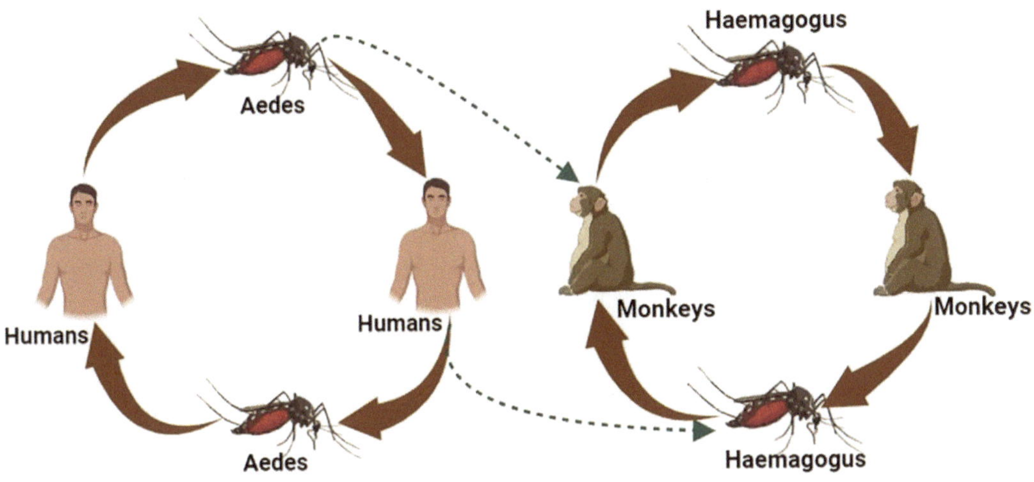

FIGURE 7.1 Transmission cycle of CHIKV (created with BioRender).

the consequences of contracting CHIKV in newborns can range greatly from asymptomatic to severe (Cardona-Correa et al., 2017). People with CHIKV infection may have the ability to transmit the virus through organ or blood transfusions, especially in times of epidemic (Appassakij et al., 2020).

7.1.5 Symptoms Associated with CHIKV Infection, Classification Based on Disease Severity

The usual incubation period for CHIKF is 3–7 days, with a range of 2–12 days. Not everyone who contracts the virus experiences symptoms. Asymptomatic infections are present in 3–25% in people with CHIKV antibodies (Staples et al., 2009). Fever (typically 38.9°C) is the first and most severe symptom of CHIKV infection. The fevers can be biphasic in character and usually last for a few days to two weeks (Silva et al., 2018). The early onset of fever, polyarthralgia, and maculopapular rash are the hallmarks of CHIKF. In 5–15% of instances, asymptomatic infection occurs during the two to four days incubation period, the remaining individuals typically experience the progression of acute to chronic phase symptoms (Bautista-Reyes et al., 2017).

7.1.5.1 Acute Phase

The acute phase is marked by an abrupt onset of fever (≥39°C) and lasts for three to ten days. Over the next three days, more symptoms appear. In 87–98% of the instances, joint pain is recorded (Gerardin et al., 2013). The patient presents with normal radiological findings and moderately high levels of erythrocyte sedimentation rate and C-reactive protein, two physiological markers of inflammation (Galan-Huerta et al., 2015). Pain in several joints is typically bilateral (symmetrical), primarily affecting the peripheral joints (knees, ankles, hands, and wrists (Gerardin et al., 2013). In 46–72% of the cases, there is myalgia, which affects the calves, thighs, and arms (Thiberville et al., 2013). CHIKF is the name given to this clinical condition, limiting ability of an individual to engage in regular physical activities (Simon et al., 2011). The petechial or maculopapular type of exanthema, which causes pruritus and mainly affects the face, trunk, and limbs, is seen in 40–50% of cases. The duration of acute symptoms is often less than two weeks, arthralgia can persist for more than weeks. This particular symptom can help differentiate DENGV infection from CHIKV infection (Schilte et al., 2013). According to laboratory results, patients in the acute stage of the illness show considerable thrombocytopenia (100,000–150,000 cells/mm3) and lymphopenia (500–1,000 cells/mm3) (Chow et al., 2011). Neonates with CHIKF may also experience podalicedoema, epithelial vesicular lesions that finally get dry and peel off, convulsions, and peripheral cyanosis. Typically, this illness in youngsters is regarded as benign (Valamparampil et al., 2009).

7.1.5.2 Chronic Phase

Subsequent research produced enough proof to show that infection with this virus might cause chronic rheumatic disorders following the several outbreaks and epidemics of CHIK in various parts of Africa, Asia, and Europe. In children it is less prevalent, but can persist in adults for months or even years following the infection (Thiberville et al., 2013). Among the musculoskeletal conditions, worsening of preexisting rheumatic disease is the main (Burt et al., 2012). Neuropathy, neurosensory deficiency, cubita tunnel syndrome, gastrointestinal diseases, exanthema, pruritus, bursitis, and synovitis are other less prevalent symptoms (Gasque et al., 2015). Since viral RNA has been found in perivascular macrophages, it is possible that the development of arthralgia is the result of ongoing joint inflammation in response to viral antigens (Thiberville et al., 2013; Kam et al., 2012).

7.2 CHIKV: MORPHOLOGY, GENOMIC ORGANIZATION, AND LIFE CYCLE

7.2.1 Basic Structure and Genomic Organization of CHIKV

Alphaviruses are composed of an icosahedral capsid that houses a single positive-sense strand RNA. A single genome RNA molecule, the capsid (C), a lipid bilayer produced by the host, and repeated units of the E1 and E2 transmembrane glycoproteins make up the 70 nm-diameter virion

(Tang et al., 2011). A capsid protein and two envelope glycoproteins (C, E1, E2) make up the three structural proteins and four non-structural proteins (NSP1–4) encoded with the positive polarity RNA genome. The 5′-end of the genome encodes non-structural proteins (nsPs), while the 3′-end encodes structural proteins. The genome is roughly 12 kb long. A subgenomic RNA is translated into structural proteins, whereas genomic RNA is used to translate the nsPs (Abu Bakar & Ng, 2018). CHIK genotypes are found in four lineages: East-Central South Africa, West Africa, Asian, and Indian Ocean based on geographic distribution (Kindhauser et al., 2016).

7.2.2 Life Cycle of CHIKV

The CHIKV has three distinct stages in its life cycle:

7.2.2.1 The CHIKV Infection at Its Initial Stage

The CHIKV can enter the human body through the salivary glands of mosquitoes. The virus enters the host's bloodstream by female Aedes mosquito bite carrying the infection. The virus attaches itself to receptive cells quickly after entering the bloodstream, particularly those found in the mouth, throat, and nose.

7.2.2.2 Cellular and Infectious Stage

After passing through the cytoplasm of cells, the CHIKV eventually reaches the nucleus. When the CHIKV enters the nucleus, it releases genetic material and replicates its genome there. Following its passage through the cellular stage, the virus invades the tissues in its vicinity and starts to infect further cells. It spreads throughout the body and bloodstream, leading to the primary infection. After an infected mosquito bite, symptoms take 2–12 days to manifest.

7.2.2.3 The Mosquito Stage

A mosquito transmits the virus to humans by bites, and the virus replicates in the insect's fat bodies, ovaries, midgut, and brain tissues. This happens when the virus bites the human suffering from the infection. The virus multiplies while spreading to the salivary glands of mosquito and with subsequent biting to human by infected mosquito, the virus transfers to other individuals producing a new host (Auta et al., 2020).

7.3 CHIKUNGUNYA: MOLECULAR PATHOPHYSIOLOGY AND CLINICAL PRESENTATION

7.3.1 Molecular Mechanisms Underlying Pathophysiology of CHIKV Infection

Although the pathophysiology of infection with CHIKV in humans is still not well known, after being injected intradermally by infected mosquitoes, CHIKV penetrates the capillaries beneath the skin directly and spreads to the vulnerable cells in the skin, such as endothelial, fibroblast, and macrophage cells, where it replicates only partially (Sourisseau et al., 2007). Locally generated viruses travel to proceed to the secondary lymphoid organs., where they release viruses and infect migrant cells into the lymphatic system, eventually reaching the bloodstream (Caglioti et al., 2013). The virus can enter the bloodstream and go to the liver, muscles, joints, and brain, among other organs (Couderc & Lecuit, 2009). A notable infiltration of mononuclear cells is linked to the infection in these organs. Pain is linked to viral replication and mononuclear cell infiltration in the muscles and joints (Dupuis-Maguiraga et al., 2012). Strong system responses are triggered by CHIKV infection, primarily including the production of antiviral IFN-α along with several, and growth factors, chemokines, and cytokines that promote inflammation. After that, in the early stages of the illness, CD8+ T-cell activation and proliferation trigger the adaptive immune response. Later phases of the acute phase are characterized by the synthesis of anti-inflammatory proteins, IL-1RA and IL-2RA in addition to a normal transition to the CD4+ T-cell response (Wauquier et al., 2011). A CHIKV infection triggers a strong inflammatory response that may be regulated by the synthesis of MIP-1, IP-10, IL-16, IL-17, and monocyte chemo attractant protein 1 (MCP-1). Proinflammatory MIF, MIP-1β, SDF-1α, IL-6, and IL-8 production are indicative of the end of the acute phase. Activated macrophages that are prone to CHIKV infection generate CCL5, MCP-1, IP-10, MIP-1, and IL-8 (Sourisseau et al., 2007). Strong cellular immune response is also induced by CHIKV infection. Elevated plasma concentrations of cytokines that support adaptive immunity, such as IFN-γ, IL-4, IL-7, and IL-12p40, indicate the potential involvement of cellular responses. Natural killer cells offer a crucial act in the elimination of infected cells and the onset of CHIKV arthralgia (Abu Bakar & Ng, 2018). IgG antibodies are found within the first week of infection, suggesting that CHIKV-infected people have high levels of antibody responses and quick seroconversion (Kam et al., 2012). After the disease starts, specific IgM lasts for three to four months, but specific IgG lasts for more than six months (Aoyama et al., 2010).

7.3.2 Molecular Mechanisms Underlying the Clinical Manifestation of CHIKV Infection

Humans are known to contract CHIKV by a specific mechanism, and some cell types are more vulnerable to infection than others. Monocyte-derived macrophages, primary fibroblasts, and human epithelial and endothelial cells are prone to viral proliferation and infection. On the other hand, there were no indications of CHIKV replication in lymphoid and monocytoid cells, primary lymphocytes and monocytes, or monocyte-derived dendritic cells (Sourisseau et al., 2007). The immune system of the host reacts to the virus after its initial replication cycle, but the virus subsequently spreads to lymph nodes and eventually other tissues through the circulatory system. The disease enters its viremic phase where it replicates in various tissues, and this is also the time when mosquitoes can spread the infection through blood (Kam et al., 2009).

After infection, CHIKV activates non-hematopoietic cells, enhancing the synthesis of interferon (IFN) type I, important for the body to eliminate CHIKV (Schilte et al., 2012). In individuals with acute cutaneous rashes, CD8+ lymphocytes are present, whereas in patients with chronic synovial effusions, CD4+ T cells predominate (Pardigon, 2009). Experiments conducted by Wauguier et al. provided support for this theory, showing that CD8+ T-cell activation is found early in the infection, followed by CD4+ T-cell death produced by CHIKV, and that later stages of the acute stage showed a larger CD4+ T-cell count (Wauquier et al., 2011). Bone marrow stromal antigen 2, or BST-2, has the potential to modulate host inflammatory responses caused by CHIKV and save lymphatic tissues (Mahauad et al., 2014). The effects of chronic CHIKV may be exacerbated by findings that inflammation resulting from CHIKV infection affects osteoblast and osteoclast growth and function. Numerous cytokines, including TNF-α, IL-6, and IL-1, have also been linked to infection, osteoclastogenesis, and osteoclast activity (Chen et al., 2015). It is well-known that the brain has an inherent immune system of its own. Neurological symptoms are rare but can be rather serious, especially in newborns. The brain defends against CHIKV infection in a number of ways. In mouse models, it was shown that microglial cells were more resistant to infection, whereas astrocytes and oligodendrocytes were more vulnerable (Das et al., 2015). CHIKV triggered apoptosis in astrocytes by means of the external caspase-9 pathway. Microglial cells release pro-apoptotic chemicals such as TNF-α, IFN-α, and IL-12 in response to CHIKV (Van Duijl-Richter et al., 2015).

7.4 MOSQUITO-BORNE INFECTIONS: DENGUE VIRUS

7.4.1 Dengue: The Fundamentals

7.4.1.1 Definitions and Terminologies Associated with Dengue and DENV

Dengue is a virus that is transmitted by mosquitoes and is the primary cause of viral diseases transmitted by arthropods worldwide. Dengue fever is an arboviral disease transmitted by mosquitoes that has experienced fast dissemination throughout several locations in recent years. Dengue fever manifests as pyrexia, cephalalgia, epistaxis, arthromyalgia, emesis, cutaneous eruption, enteritis, hematemesis, cognitive impairment, convulsions, and pruritus (Qureshi, 2016). It is also known as dandy fever or seven-day fever, as symptoms typically persist for about a week. Some individuals who have previously been infected with one strain of the dengue virus experience significant capillary permeability and bleeding when they are subsequently infected with a different strain of the virus. The medical condition is referred to as dengue hemorrhagic fever. The main carriers of the disease are female mosquitoes of the species A. aegypti and A. albopictus (Schaefer et al., 2022) The DENV belongs to the Flaviviridae family, which comprises over 70 significant viruses that cause diseases in humans. These viruses predominantly affect intertropical regions, where approximately 3.9 billion people reside (Zerfu et al., 2023).

7.4.1.2 History of Dengue and DENV Infection

The earliest documented instance of a likely case of dengue fever can be found in a Chinese medical encyclopedia from the Jin period (AD 265–420). The encyclopedia mentions a condition called "water poison" that is linked to winged insects (Gubler, 1998). The initial documented epidemic of dengue came up in 1779 in Jakarta, Indonesia, and Cairo. The inaugural documented occurrence of a dengue outbreak in North America occurred in Philadelphia in 1780. From 2010 to 2016, the WHO documented the mean number of globally reported dengue cases, which were either suspected or confirmed (Beaumier et al., 2014). Across North and South America in 2010, a total of 1.6 million occurrences of dengue occurred, among which 49,000 were classified as serious. In 2016, the United States saw its most significant dengue epidemic, with a staggering 2.38 million reported cases. Brazil was the primary contributor to this epidemic, accounting for 1.5 million infections. The number of recorded dengue infections in the United States had a significant surge in 2019, surpassing three million cases (Jain & Jain, 2014). Dengue outbreaks have proven a persistent problem in East, West, and South Africa since the early 1800s. From 1980 to 2000, many epidemics of DENV serotypes 1, 2, and 3 occurred in various countries in East and West Africa. Several big outbreaks in several African nations have recorded a significant number of dengue cases (Sang, 2007). These include Seychelles from 1977 to 1979, Réunion Island from 1977 to 1978, Djibouti and Comoros from 1992 to 1993, and Cape Verde in 2009 (Guerra Nunes et al., 2019).

The Indian subcontinent has witnessed multiple epidemics of dengue, including all serotypes except DENV-5, owing to its favorable habitat. In 1996, there were a total of 16,000 reported cases and 545 deaths due to the outbreak that occurred in India. Since 2010, the prevalence of dengue has risen to around 15 cases per million individuals per year (Mustafa et al., 2015). India annually registers more than 100,000 illnesses and 200–400 fatalities. In 2017, there was a notable outbreak of dengue fever (DENGF), which resulted in 188,401 cases of illness and325 fatalities. The initial clinical presentation of a dengue-like disease was documented in Madras in 1780 (Gupta et al., 2012). Nevertheless, the confirmation of DENGF having been brought about by a virus wasn't confirmed till 1946 (Mutheneni et al., 2017). Subsequently, there were no significant outbreaks of dengue until 1963. In the period from 2019 to 2020, all serotypes of DENV were found to be present in the northern region of West Bengal (Gupta & Ballani, 2014).

7.4.1.3 Taxonomic Classification and Nomenclature of DENV

The DENGV is categorized within the genus Flavivirus and the family Flaviviridae which also encompasses other notable viruses, including West Nile virus and YFV. The taxonomy categorization of DENGV is as follows:

Kingdom: Animalia	Phylum: Chordata
Class: Mammalia	Order: Primates
Family: Flaviviridae	Genus: Flavivirus

DENGF, an infection imitating the flu, and DHF are the two serious kinds of virus triggered by the four serotypes of DENV (DEN-1, DEN-2, DEN-3, and DEN-4) (Kularatne & Dalugama, 2022).

7.4.1.4 Transmission Process of DENV

The transmission of the DENV takes place by mosquito bite, namely, A. aegypti and A. albopictus, specifically A. aegypti that often inhabit regions with latitudes ranging from 35°N to 35°S and altitudes below 1,000 m (3,300 feet) (WHO, 2020). They commonly engage in biting activity during daylight hours, namely, in the early morning and evening. Additional Aedes species that carry the disease are also present. A female mosquito gets infected with the virus while feeding on the blood of an infected host during the 2–12-day period when the person has fever and the virus in their bloodstream. The virus infects the cells lining the mosquito's gut. Approximately eight to ten days later, the virus disseminates to various tissues, such as the salivary glands of the mosquito, and is then discharged into its saliva (Gubler, 2010). Dengue can also be spread through contaminated blood products and by organ transplantation (Wilder-Smith et al., 2009).

7.4.1.5 Symptoms Associated with DENV Infection and Classification Based on Disease Severity

The symptoms of DENGF appear after four to ten days of bite by an infected mosquito, and can be misinterpreted as other diseases, such as influenza. DENGF is distinguished by a severe fever of 104°F (40°C) and can be associated with manifestations such as headache and pain, vomiting, discomfort behind the eyes, and rash. The majority of persons experience a full recovery within one week (Zhang et al., 2014). Nevertheless, in specific instances, manifestations can deteriorate and reach a life-threatening level, resulting in illnesses referred to as severe dengue, DHF, or dengue shock syndrome. Some people may be asymptomatic when infected with dengue. Severe dengue is distinguished by vascular damage and leakage, as well as a reduction in the quantity of platelets (cells responsible for blood clotting) in the bloodstream. These consequences may include shock, internal bleeding, organ dysfunction, and even fatal outcomes (Yeh et al., 2017). Indicators of severe DENGF, a potentially lethal illness, can manifest quickly. Indications of these symptoms encompass intense abdominal discomfort, continuous emesis, hemorrhaging from the gingiva or nasal passages, and the existence of blood in urine, feces, or vomitus. These symptoms usually appear within one to two days after the fever has decreased. This explanation offers a scholarly viewpoint on the manifestations and possible ramifications of DENGF (Htun et al., 2021). The WHO categorizes DENGF into two types: uncomplicated and severe. Substantial cases are associated with significant hemorrhage, organ impairment, or substantial plasma escape, while the remainder are deemed uncomplicated (Ajlan et al., 2019). The mild symptoms of DENGF include vomiting, rash, headache, eye pain, muscular aching, body pain, and leukopenia (Ganeshkumar et al., 2018). Common symptoms of DENGF in the moderate stage include abdominal pain, vomiting, fluid accumulation, and mucosal bleeding. The occurrence of an elevated hematocrit level together with abrupt decline in platelet count, accompanied by severe symptoms, is observed in cases of DENGF. Severe organ involvement is indicated by elevated levels of aspartate aminotransferase or alanine aminotransferase (Akbar et al., 2012).

7.5 DENV: MORPHOLOGY, GENOMIC ORGANIZATION, AND LIFE CYCLE

7.5.1 Basic Structure of DENV

DENGV, family Flaviviridae, is a pathogenic arthropod-borne flavivirus (arbovirus), characterized by RNA molecule with a single strand and positive sense. The Flavivirus genus is characterized by its tiny size, approximately 50 nm, a single RNA chain with three structural proteins capsid, precursor membrane, envelope, and seven proteins that are not structural (NS1, N2A, N2B, N3, N4A, N4B, and N5) (Bhatnagar et al., 2021). Its nucleocapsid has structural proteins M and E making genome of the virus. A non-structural NS1 protein, interacts with the immune system of the host and triggers responses from T cells. Viral propagation and antiviral response are related to non-structural proteins. DENV-1 to DENV-4 are four distinct serotypes. Currently, a total of 47 strains of DENV have been officially recognized and documented (Gowri Sankar et al., 2021).

7.5.2 Genomic Organization of DENV

The genome of the DENGV is a single-strand +ve sense RNA that is 11 kb long. The genome that the virus-encoded protein uses to translate the single predictable polyprotein results in seven non-structural proteins and three structural proteins, which are described as follows: Gene 1 non-structural (NS1), Genes known as non-structural 2A (NS2A) and 2B (NS2b), NS3 or non-structural Gene 3, Gene 4A that is not structural (NS4A), Gene 4b that is non-structural, (NS4b), the structural proteins, such as membrane protein (MP), envelope protein (EP), and capsid protein (CP), as well as non-structural Gene 5 (NS5) (El Sahili & Lescar, 2017). On the 3'-end of the genome, there are also some non-coding regions (NTR). The non-structural protein is essential for virus entrance, and the structural protein is dynamic in the structure of viral proteins (Lan et al., 2021).

The DENGV possesses an icosahedral structure, featuring a covering composed of a protein layer surrounding its outer core. The viral genome of the C-protein is located within the icosahedral core, which has a diameter of 40–50 nm. The central part is encircled by two widely recognized proteins, namely, membrane (M) and envelope proteins (E), which collectively form the lipid envelope. In addition, non-structural proteins perform a pivotal task in regulating DENV replication by modulating the immune system, inducing vascular permeability, facilitating vRNA synthesis, and cleaving the dengue polyprotein (Li & Kang, 2022).

7.5.3 Life Cycle of DENV

The DENGV acquires entry into host cells by receptor-mediated endocytosis, a mechanism in which the envelope protein (E) of the virus binds to certain cellular receptors. In order to enter a

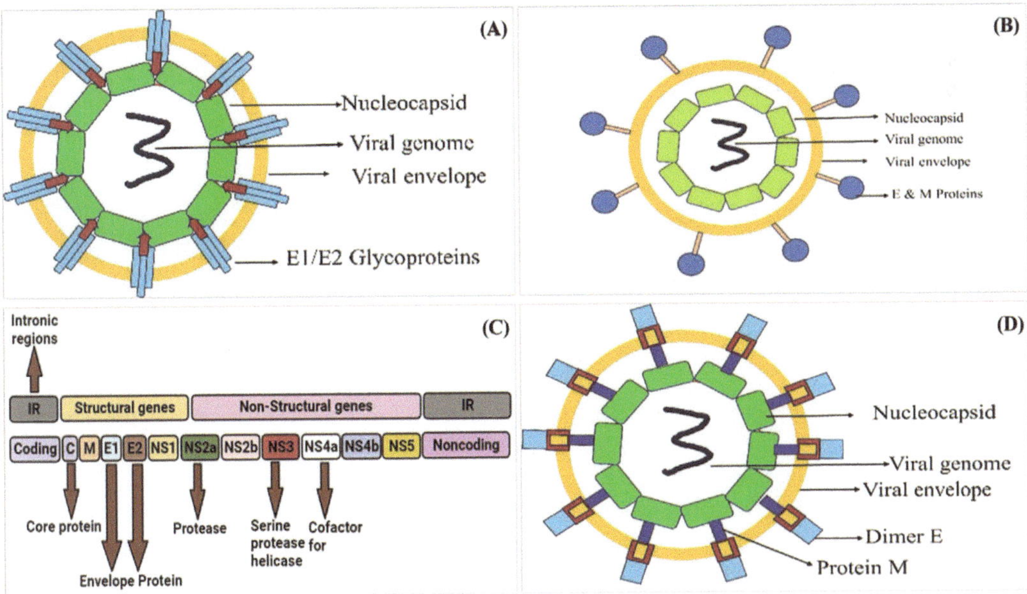

FIGURE 7.2 (A) Basic structure of chikungunya virus. (B) Basic structure of dengue virus. (C) Genomic structure of dengue virus. (D) Basic structure of Zika virus.

cell, DENGV must either identify a widely present molecule on the cell surface or employ several receptors. The E protein's interaction with glycosaminoglycans, specifically heparan sulfate, on the cell surface, along with various other surface proteins, is believed to be involved in the process of internalization (Miller et al., 2008). Clathrin-mediated endocytosis lets DENGV into mosquito and human cells (Cruz-Oliveira et al., 2015). DENGV localizes to the clathrin-coated pit on the cell surface. The dynamin protein facilitates the transformation of a clathrin-coated pit into a vesicle containing virus and transmits to early endosomes (Khan et al., 2008). After endosome internalization and acidification, viral and vesicular membranes fuse, releasing the nucleocapsid by changing the envelope protein (E) shape (Chen et al., 2018). The recently generated genomes possess two possible courses of action: they can either be employed in the synthesis of supplementary viral proteins or they can combine with the capsid to generate novel viral particles. The capsid is dispersed across the endoplasmic reticulum (ER) membranes, the nucleus, and the lipid droplet surface (Bhalerao et al., 2023). After nucleocapsid formation, it enters the lumen of the endoplasmic reticulum (ER) and picks up the viral E and prM proteins in a lipid bilayer. The secretory pathway is then followed by these recently created particles. Mature viral particles are produced in the trans-Golgi network through a sequence of events that include the rearrangement and homodimerization of E, which are triggered by furin-mediated proteolysis of prM. Partially developed, immature, and mature viruses are all present in the excreted viruses (Yu et al., 2008).

7.6 DENGUE: MOLECULAR PATHOPHYSIOLOGY AND CLINICAL PRESENTATION

7.6.1 Molecular Mechanisms Underlying Pathophysiology of DENV Infection

A nucleotide sequence homology of 65–70% is shared by the four dengue virus serotypes (DENGV1–4), suggesting a tight relationship between them. A primary infection is an early infection caused by a particular serotype. The majority of initial infections either are asymptomatic or cause a low-grade fever. They can, however, occasionally result in hemorrhagic fever, especially in infants born to mothers who are immune to DENGV. A secondary dengue infection is a recurrent infection caused by another serotype. Serious clinical consequences like dengue shock syndrome (DSS) or DHF can result from this kind of illness (Usuda et al., 2023). Numerous viral and host variables are combined to cause dengue and its pathogenesis. The non-structural protein 1 (NS1) viral antigen, DENGV genome variants, subgenomic RNA, memory cross-reactive T cells, antibody-dependent enhancement, anti-DENGV NS1 antibodies, and autoimmunity are a few of these. The combined influence of these variables is principally responsible for the acute human manifestations of dengue (Teo et al., 2021).

7.6.1.1 Molecular Mechanisms Underlying the Clinical Manifestation of DENGV Infection

The course of clinical manifestation typically consists of three stages: febrile, critical, and recovery. Patients with DHF experience all three stages, while those with dengue fever do not undergo the critical phase.

- Febrile phase: Symptoms of this phase, which lasts from three to seven days, include anorexia, arthralgia, myalgia, elevated fever, and headache (Blackhurst et al., 2023). Occasionally, upper respiratory tract and gastrointestinal issues may arise. Patients may exhibit a sickly appearance, generalized skin flushing, blanching with pressure, morbilliform erythematous eruptions, and pallid regions. Manifestations of cutaneous hemorrhage, including petechiae, purpura, and ecchymosis, may manifest as the febrile episode draws to a close. Additionally, moderate hepatomegaly or tenderness in the right hypochondrium may be present (Trivedi & Chakravarty, 2022). Starting on the second day of fever, the results of the complete blood count show a decrease in white blood cells, a decrease in platelets, and an increasing concentration of red blood cells. Frequently, hepatic transaminases, specifically alanine transaminase (ALT) and aspartate transaminase (AST), exhibit increased levels. The temperature pattern may exhibit a biphasic nature (Yong et al., 2022).

- Critical phase: Some individuals enter the critical phase, which involves systemic vascular leak and temporary defervescence. This phase increases plasma concentration as hematocrit rises. Vascular leak into peritoneal spaces can be diagnosed early by abdominal ultrasound, which shows gall bladder wall edema and pericholecystic fluid buildup. The critical phase begins with warning indications. Plasma leaks' physiological compensatory processes narrow the pulse pressure, causing shock and multi-organ failure if not diagnosed or controlled (Srikiatkhachorn et al., 2017).
- Recovery phase: As the plasma leakage terminates during the recovery phase, the patient's condition significantly improves. A steady decrease in hematocrit, accompanied by a rise in platelets and white blood cells, and recuperation from other symptoms characterize this phase. Certain individuals may encounter fatigue, as well as a "recovery rash" and bradycardia, during this phase (Kularatne & Dalugama, 2022).

7.7 MOSQUITO-BORNE INFECTIONS: ZIKA FEVER

7.7.1 Zika Fever and Virus: The Fundamentals

7.7.1.1 Definitions and Terminologies Associated with Zika Fever (ZIKF) and Zika Virus (ZIKV)

ZIKV is a re-emerging arbovirus, meaning it is spread by arthropods such as mosquitoes, specifically the Aedes mosquito, primarily by A. aegypti causing ZIKF. The ZIKV is an RNA virus with a single strand. It is classified under the family Flaviviridae and the genus Flavivirus. It is further divided into two phylogenetic types: Asian and African. Most individuals get a moderate and self-limiting infection when exposed to the ZIKV (Gharbaran & Somenarain, 2019). The ZIKV was first identified in rhesus monkeys in the Zika forest of Uganda during a study on yellow fever (Campos et al., 2015).

7.7.1.2 History of Zika Fever and Zika Virus Infection

The first human outbreak was documented in 2007 in Yap located in the Pacific area with subsequent outbreak in October 2013, which was five years after the occurrence of that initial epidemic in French Polynesia (Gourinat et al., 2015). A 1947 scientific endeavor effectively extracted a new virus from the blood of a sentry rhesus macaque located in Uganda's Zika forest, as part of an initiative to combat yellow fever (Petersen & Epstein, 2014). The ZIKV remained largely inactive for close to seven decades before it abruptly proliferated across the Americas, subsequent to its emergence in the Pacific Islands and Brazil (Chang et al., 2016). In November 2015, Brazil reported three fatalities resulting from Zika infection in two adults and a newborn infant. By the start of 2016, a total of 3,893 cases of microcephaly, including 49 fatalities, were recorded in Brazil (Morens & Fauci, 2014).

By December 2015, the Pan American Health Organization confirmed the presence of ZIKV in nine of its member states (PAHO, 2015). ZIKV infections were recorded in tourists who have returned to Germany, Japan, and Australia from different destinations, such as Thailand, French Polynesia, and Indonesia. Furthermore, there exist reports indicating possible cases of ZIKV transmission via sexual intercourse, and it has been hypothesized that ZIKV could potentially be transmitted from mothers to their infants during pregnancy (Summers et al., 2015). In May 2017, three instances of ZIKV infection were officially recorded in Gujarat, India. By the end of 2018, Rajasthan had registered a minimum of 159 cases, whereas Madhya Pradesh had documented 127 incidents. The first case of ZIKV in Kerala, India, was diagnosed in July 2021. Subsequently, 19 persons displaying symptoms were examined, and 13 of them tested positive for Zika, indicating that the virus had been circulating in Kerala from at least May 2021 (Bhardwaj et al., 2017). As of August 6, 2021, Kerala had recorded a total of 65 reported cases. Moreover, on October 22, 2021, a member of the Indian Air Force contracted the virus, representing the most recent occurrence in the state. The transmission of ZIKV has been confirmed in 55 countries and territories worldwide (Rolph & Mahalingam, 2019). However, between 2015 and 2016, native transmission of the virus was reported in 41 of these countries, with indirect evidence of virus circulation in 6 additional countries (Gourinat et al., 2015).

7.7.1.3 Taxonomic Classification and Nomenclature of Zika Virus

Realm: Riboviria	Kingdom: Orthornavirae
Phylum: Kitrinoviricota	Class: Flasuviricetes
Order: Amarillovirales	Family: Flaviviridae
Genus: Flavivirus	Species: Zika virus

The ZIKV, belonging to the Flaviviridae family, is a notable pathogen linked to several infectious illnesses. The main carriers of this virus are Aedes mosquitoes, notably A. aegypti and A. albopictus essential in transmitting ZIKV to humans via their bites (Candelo et al., 2021). Within the Flaviviridae family, ZIKV is classified under the Flavivirus genus, which also includes other notable viruses such as dengue, yellow fever, Japanese encephalitis, and West Nile viruses (Masmejan et al., 2020).

7.7.1.4 Transmission Process of Zika Virus

The ZIKV, similar to other arboviruses such as DENV and YFV, exhibits a complex transmission dynamic involving direct transmission between humans and Aedes mosquitoes, facilitating the rapid spread of the virus during epidemics. There are two distinct transmission cycles for ZIKV: an urban cycle, where the virus is transmitted between humans and domestic Aedes mosquitoes, and a sylvatic cycle, where ZIKV circulates between non-human primates and Aedes mosquitoes (Miner & Diamond, 2017). The primary vector for ZIKV transmission is A. aegypti, which acquires the virus after feeding on ZIKV-infected individuals, leading to the amplification and spread of the virus in both urban and sylvatic environments (Simmonds et al., 2017). In addition to the commonly recognized mode of transmission through mosquito bites, the ZIKV can also be propagated through alternative means, including sexual contact, mother-to-child transmission, and blood transfusion. The majority of ZIKV transmission occurs through sexual contact, with viral RNA predominantly detected in male semen. Pregnant women who contract ZIKV are at risk of transmitting the virus to their fetus, leading to fetal abnormalities such as developmental restrictions, microcephaly, and miscarriage (Victora et al., 2016). Most ZIKV-infected patients are asymptomatic, but about 3% of blood donors test positive for ZIKV. The virus can survive in whole blood for up to two months, facilitating its spread to blood donors (Magnus et al., 2018). Due to the diverse and complex patterns of ZIKV transmission, formulating strategies to combat ZIKV infection is challenging (Hirsch et al., 2018).

7.7.1.5 Symptoms Associated with Zika Virus Infection: Classification Based on Severity

ZIKV infection is characterized by a range of clinical manifestations, which can vary in severity among individuals. Mild symptoms may include a fever, skin rashes, joint discomfort, and muscle spasms. In more severe cases, ZIKV infection can lead to conjunctivitis and the development of mouth sores (stomatitis). Severe manifestations, although rare, can occur, such as Guillain–Barré syndrome, a neurological disorder that can last for a prolonged period (Krauer et al., 2017).

7.8 ZIKA VIRUS: STRUCTURE, GENOMIC ORGANIZATION, AND LIFE CYCLE

7.8.1 Structure and Genomic Organization of Zika Virus

The nucleocapsid protein of ZIKV exhibits an icosahedral structure with a diameter of approximately 50–60 nm, as depicted in Figure 7.2 (Dasti, 2016). ZIKV genome is a single-stranded RNA molecule and is 10,794 kb long, consisting of two non-coding regions (NCRs), namely, the 39 and 59 NCR, and a lengthy open reading frame (ORF). ORF encodes a polyprotein that consists of 59 amino acids. The protein undergoes proteolytic breakdown to generate the capsid (C), envelope (E), precursor of membrane (prM), and seven non-structural proteins. The main protein E on the surface of the virus, plays a crucial role in life cycle of virus, such as facilitating the fusing of membranes and mediating the binding process (Kaur et al., 2023). The 3' NCR of the ZIKV genome consists of 428 nucleotides and exhibits 27 distinct folding patterns. These patterns potentially have roles in many processes such as cyclization, translation, interaction with cellular proteins, RNA packaging, interaction with viral factors, and genome stabilization (Hu et al., 2023).

The architecture of flaviviruses, including the ZIKV, vary depending on the amino acids that surround a glycosylation site in the viral envelope. The envelope is made up of two unique proteins, each having 180 copies. ZIKV is characterized by a glycosylation site on its viral surface, which sets it apart from other flaviviruses. At this location, a carbohydrate molecule binds several sugars to the surface of the viral protein. The adjacent residues and the glycosylation site on ZIKV likely have a pivotal function in the virus' adherence to human cells (Figure 7.3) (Lindenbach & Rice, 2003).

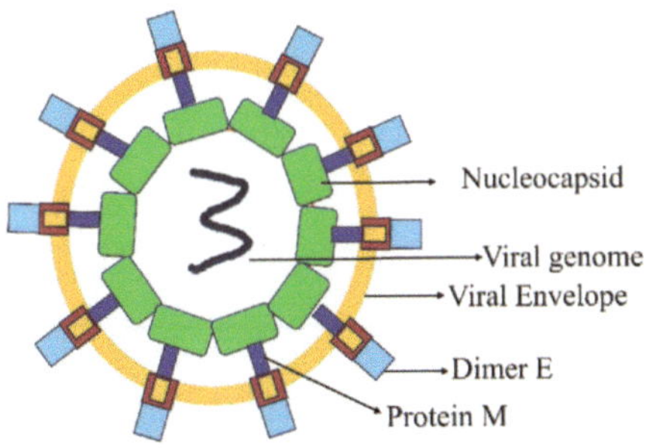

FIGURE 7.3 Structure of Zika virus.

7.8.2 Life Cycle of Zika Virus

The ZIKV, like other well-known flaviviruses, follows an intricate life cycle that can be broadly categorized into stages of binding, entrance, translation, replication, assembly, and release. At first, the virus adheres to the surface of the host cell by means of interactions among viral surface glycoproteins and cell surface receptors. Later, the virus infiltrates the cell via receptor-mediated endocytosis and is then taken up into clathrin-coated pits. Upon entering the cell, the viral RNA is liberated into the cytoplasm by the merging of the viral and host membranes (Hamel et al., 2015). The genomic RNA with a positive sense is translated into a single polyprotein. This polyprotein is then prepared by both cellular and viral proteases, both during and after translation. Consequently, three structural proteins and seven non-structural proteins are formed. Genomic replication takes place within vesicle packets, which aid in the development of the viral replication complex. Virus aggregation happens on the exterior of the endoplasmic reticulum. The freshly produced particles then move via the host's secretory system, specifically the trans-Golgi network, whence the virion matures. Ultimately, the fully developed viral particle is discharged through exocytosis (Roby et al., 2015).

7.9 ZIKA FEVER: MOLECULAR PATHOPHYSIOLOGY AND CLINICAL PRESENTATION

7.9.1 Molecular Mechanisms Underlying Pathophysiology of Zika Virus Infection

Once inside the host, the ZIKV infest and continue in many host cells like skin cells, blood cells, placental cells, retinal cells, etc. ZIKV demonstrates a predilection for neural stem cells and progenitor cells. The virus gains entry into host cells by binding to Axl receptors found on the host cell membrane. Within the host cells, the ZIKV has four stages: (1) translation of viral proteins, (2) replication of ZIKV RNA, (3) accumulation of viral particles in the endoplasmic reticulum, and (4) release of the virion from the host cell (Graciaa et al., 2018). The ZIKV encodes three structural proteins: capsid (C), envelope (E), and prM, as well as seven non-structural proteins. The prM and E proteins, which are part of the structural components, have a vital function in facilitating the virus' attachment to the neural cell membrane of the host. The C protein, in conjunction with the non-structural proteins, is present in multiple organelles inside the neuron. The mentioned components are the nucleus and its nucleoli, the Golgi apparatus, and lipid droplets found in the cytoplasm. The existence of these proteins can induce a cessation in the cell cycle, initiate apoptosis, and eventually culminate in cellular demise (Lee & Shin, 2019).

ZIKV has been discovered to cause DNA damage in the cells of the host. This is accomplished by creating double-strand breaks in the DNA and causing the host cell to pause in its S phase, so limiting its replication. The biological outcomes of ZIKV are believed to cause the cessation of growth in neural cells in the cortex of developing fetuses, resulting in undeveloped brains (Serman & Gack, 2019). Furthermore, ZIKV is speculated to bypass human cellular defense mechanisms by using interferon antagonism, a process facilitated by its non-structural proteins. It can also reduce the synthesis of interferons that normally hinder ZIKV replication in human cells. Several more evasion tactics utilized by ZIKV have been discovered, primarily through the inhibition of innate cellular immune pathways (Hammack et al., 2019).

7.9.2 Molecular Mechanisms Underlying the Clinical Manifestation of Zika Virus Infection

Most individuals infected with the ZIKV either show no symptoms or exhibit mild clinical signs, especially during the period from the 1960s to the 1980s. However, if the infection occurs during pregnancy, it can result in congenital microcephaly, other brain anomalies, Guillain–Barré syndrome, stillbirths, and miscarriages (Kakooza-Mwesige et al., 2019). Serological studies have associated ZIKV infection with neurological disorders, such as elevated intraocular pressure that can lead to damage to the optic nerve and a potential link with glaucoma (Song et al., 2017). While the majority of ZIKV patients are asymptomatic, about 20–25% of infected individuals develop symptoms like skin rashes, headaches, fever, joint pains, and conjunctivitis, typically with an incubation period of approximately a week (Adebayo et al., 2017). Some patients might also experience symptoms like vomiting, diarrhea, redness in the eyes, weakness, edema, abdominal pain, loss of appetite, and hematospermia. ZIKV primarily targets immature brain cells, including forebrain-specific human neural progenitor cells and embryonic fetal cells, leading to cell necrosis and disruption of the cell cycle (Tang et al., 2016).

7.10 ARBOVIRUS TRANSMISSION IN A CHANGING WORLD

Arboviruses possess a distinctive characteristic within the realm of animal viruses, since they depend on arthropods that feed on blood (known as vectors) to transmit the virus to vertebrates. This mechanism of transmission is generally referred to as biological transmission.

7.10.1 Interactions with the Abiotic Environment

Climate conditions, including temperature, humidity, and rainfall, have a significant contribution to the facilitation of vector-borne disease transmission. Temperature fluctuation impacts mosquito physiology, life history, and arbovirus fitness, including immunity, development, survival, reproduction, bite rates, vector competence, and extrinsic incubation durations (Robert et al., 2020). It is anticipated that in 2018, 55% of world population were city dwellers; by 2050, that number will rise to 68%. Urbanization is characterized by either population density or the expansion of artificial geographic space, and increase the prevalence and quantity of Aedes mosquitoes because of factors like availability of suitable breeding locations, accelerated larval development, etc. The rapid urbanization, resulting in concentration of populations in unplanned urban, areas favors the proliferation of the Aedes mosquito (Kolimenakis et al., 2021). This species has a reproductive behavior characterized by the deposition of its eggs in water-storage containers and garbage. It frequently engages in many blood feasts throughout each reproductive cycle, serving the dual purpose of egg generation and nutritional sustenance (Weaver & Barrett, 2004).

7.10.2 Interactions with the Biotic Environment

Anthropological Aedes aegypti follows humans and was observed breeding in Australian gold mine shafts. Mosquitoes possess the ability to disseminate, with human beings, to many nations The primary vectors of these viruses are A. albopictus and A. aegypti mosquitoes. The combination of climate change and the globalization of trade and tourism has facilitated their global expansion, leading to the transmission ZIKV and DENV as well as the appearance of new outbreaks (Silva et al., 2020). There exists a multitude of mosquito species that engage in hematophagy, consuming the blood of diverse hosts within the vertebrate category, encompassing mammals, birds, reptiles, amphibians, and fish species. During the hematophagous phase, the majority of mosquito species engage in the process of depositing their saliva at the site of the bite, resulting in the manifestation of an irritating rash in the vicinity of the individual's skin (Islam et al., 2020).

7.11 INFECTION PREVENTION AND CONTROL: CHALLENGES AND FUTURE PROSPECTS

One of the issues faced is resistance exhibited by mosquitoes of the Aedes genus toward several pesticides, including DDT (dichlorodiphenyltrichloroethane), organophosphate adulticides, pyrethroids, and carbamates (Kamgang et al., 2017). Another notable challenge relates to the manufacturing of vaccines. Despite significant investments in vaccine development for arboviral illnesses, the majority of viruses, currently lack a licensed vaccine for efficient prevention (Girard et al., 2020). The implementation of a singular, efficacious approach to mitigating flavivirus infection within regions impacted by this disease, such as tropical countries, poses significant challenges. Nevertheless, the effective reduction of infection risk can be achieved through a comprehensive study of the biological behavior and feeding habits of the vector. The A. aegypti female mosquitoes have a preference for indoor feeding, displaying their highest levels of biting activity roughly 2–3 hours after daybreak and 2–3 hours before to sunset. Preventive measures encompass the use of protective attire, application of mosquito repellent, and refraining from the accumulation of uncontaminated water within residential premises, plants, and coolers, since these actions serve to impede the proliferation of mosquitoes. The timely identification, intervention, and implementation of preventive strategies can effectively mitigate the transmission of the DENV, CHIKV, and ZIKV (Soni et al., 2023). The management of arboviruses presents significant challenges, including the absence of efficacious vaccines, limited availability of antiviral medications, pesticide resistance in vector species Aedes, etc. (Batool et al., 2018). In this scenario, the prioritization and demand of public health necessitate the discovery of novel methodologies for prompt identification and managing the spread of arboviruses. With the advent of nanotechnology, technology of the future, the future prospects to combat these seems to be possible (Suganya et al., 2017; Tajudeen et al., 2022). The prompt and timely identification of viral infection is of utmost importance to confirm its presence, facilitate subsequent clinical intervention, and implement essential measures for the surveillance and safeguarding of public health (Rashid & Yusof, 2018). Technological development has potential applications in vector control, including the advancement of pharmaceuticals with enhanced efficacy, reduced toxicity, and prolonged release. Additionally, nanotechnology can be utilized in the formulation of novel repellents, derived from either natural or synthetic compounds. Nanotechnology can also be employed for vector control through their larvicidal, insecticidal, or repellant properties (Magro et al., 2019). Furthermore, it holds a promise in the process of creating biosensors that can identify and diagnose viral infections spread by mosquitoes (Singhal et al., 2018). Virus isolation and molecular approaches are considered to yield the most precise test results compared to other existing methods for diagnosing DENV, CHIKV, and ZIKV (Soni et al., 2023).

7.12 CONCLUSION

Arboviruses provide significant challenges to global public health systems, resulting in the manifestation of diseases and fatalities in both human and animal populations. Currently, there exists no targeted therapeutic intervention, and the number of authorized vaccinations available to combat these viral pathogens is limited. The allocation of research priority should be directed toward the investigation of surveillance systems and the comprehension of the elements that contribute to transmission cycles, vectorial competency, and host infection factors. Additionally, emphasis should be made on the advancement of antiviral compounds or potential vaccines. In order to effectively manage future arboviral outbreaks, it is crucial to ascertain and comprehend the virus genomes, antigenic characteristics, patterns of virulence, interactions between vectors, and mechanisms of maintenance. The existing requirements in the field of public health encompass enhanced dissemination of information regarding vector-borne illness to both the general populace and medical practitioners, assurance of vector control programs, maintaining reliable monitoring systems by assigning appropriately skilled staff, and having quick diagnostic skills.

REFERENCES

Abu Bakar, F., & Ng, L. F. P. (2018). Nonstructural proteins of alphavirus-potential targets for drug development. *Viruses*, 10(2), 71. https://doi.org/10.3390/v10020071.

Adebayo, G., Neumark, Y., Gesser-Edelsburg, A., Ahmad, W. A., & Levine, H. (2017). Zika pandemic online trends, incidence and health risk communication: A time trend study. *BMJ Global Health*, 2(3).

Ajlan, B. A., Alafif, M. M., Alawi, M. M., Akbar, N. A., Aldigs, E. K., & Madani, T. A. (2019). Assessment of the new World Health Organization's dengue classification for predicting severity of illness and level of healthcare required. *PLoS Neglected Tropical Diseases*, 13(8), e0007144.

Akbar, N. A., Allende, I., Balmaseda, A., Coelho, I. C., Da Cunha, R. V., Datta, B., & Wills, B. (2012). Regarding "dengue—how best to classify it". *Clinical Infectious Diseases*, 54(12), 1820–1821.

Aoyama, I., Uno, K., Yumisashi, T., Takasaki, T., Lim, C. K., Kurane, I., Kase, T., & Takahashi, K. (2010). A case of chikungunya fever imported from India to Japan, follow-up of specific IgM and IgG antibodies over a 6-month period. *Japanese Journal of Infectious Diseases*, 63(1), 65–66.

Appassakij, H., Silpapojakul, K., Promwong, C., & Rujirojindakul, P. (2020). The potential impact of Chikungunya virus outbreaks on blood transfusion. *Transfusion Medicine Reviews*, 34(1), 23–28. https://doi.org/10.1016/j.tmrv.2019.06.002.

Auta, S. A., Kumurya, A. S., & Hauwa, A. (2020). Chikungunya virus infection: A review. *Bayero Journal of Medical Laboratory Science*, 5(2), 207–216.

Batool, K., et al. (2018). Transcriptomic analysis of aedes aegypti in response to mosquitocidal Bacillus thuringiensis LLP29 toxin. *Scientific Reports*, 8, 12650. https://doi.org/10.1038/s41598-018-30741-x.

Bautista-Reyes, E., Nunez-Avellaneda, D., Alonso-Palomares, L. A., & Salazar, M. I. (2017). Chikungunya: Molecular aspects, clinical outcomes and pathogenesis. *Revista de investigacion clinica; organo del Hospital de Enfermedades de la Nutricion*, 69(6), 299–307. https://doi.org/10.24875/RIC.17002029.

Beaumier, C., Garcia, M. N., & Murray, K. O. (2014). The history of dengue in the United States and its recent emergence. *Current Tropical Medicine Reports*, 1, 32–35.

Bhalerao, U., Preethi, L., Bhukya, P. L., & Mhaske, S. T. (2023). Dengue haemorrhagic fever: A resurgent arbovirosis in humans. In *Emerging Human Viral Diseases, Volume I: Respiratory and Haemorrhagic Fever* (pp. 315–343). Singapore: Springer Nature.

Bhardwaj, S., Gokhale, M. D., & Mourya, D. T. (2017). Zika virus: Current concerns in India. *The Indian Journal of Medical Research*, 146(5), 572.

Bhatnagar, P., Sreekanth, G. P., Murali-Krishna, K., Chandele, A., & Sitaraman, R. (2021). Dengue virus non-structural protein 5 as a versatile, multi-functional effector in host–pathogen interactions. *Frontiers in Cellular and Infection Microbiology*, 11, 574067.

Blackhurst, B. M., & Funk, K. E. (2023). Molecular and cellular mechanisms underlying neurologic manifestations of mosquito-borne flavivirus infections. *Viruses*, 15(11), 2200.

Burt, F. J., Rolph, M. S., Rulli, N. E., Mahalingam, S., & Heise, M. T. (2012). Chikungunya: A re-emerging virus. *Lancet (London, England)*, 379(9816), 662–671. https://doi.org/10.1016/S0140-6736(11)60281-X.

Caglioti, C., Lalle, E., Castilletti, C., Carletti, F., Capobianchi, M. R., & Bordi, L. (2013). Chikungunya virus infection: An overview. *The New Microbiologica*, 36(3), 211–227.

Campos, G. S., Bandeira, A. C., & Sardi, S. I. (2015). Zika virus outbreak, Bahia, Brazil. *Emerging Infectious Diseases*, 21, 1885–1886.

Candelo, E., Sanz, A. M., Ramirez-Montaño, D., Diaz-Ordoñez, L., Granados, A. M., Rosso, F., & Pachajoa, H. (2021). A possible association between Zika virus infection and CDK5RAP2 mutation. *Frontiers in Genetics*, 12, 530028.

Cardona-Correa, S. E., Castaño-Jaramillo, L. M., & Quevedo-Vélez, A. (2017). Reporte de caso de transmisión vertical de la infección por virus chikungunya [Vertical transmission of chikungunya virus infection. Case report]. *Revista chilena de pediatria*, 88(2), 285–288. https://doi.org/10.4067/S0370-41062017000200015.

Chang, C., Ortiz, K., Ansari, A., & Gershwin, M. E. (2016). The Zika outbreak of the 21st century. *Journal of Autoimmunity*, 68, 1–13. https://doi.org/10.1016/j.jaut.2016.02.006.

Chen, H. R., Lai, Y. C., & Yeh, T. M. (2018). Dengue virus non-structural protein 1: A pathogenic factor, therapeutic target, and vaccine candidate. *Journal of Biomedical Science*, 25, 1–11.

Chen, W., Foo, S. S., Sims, N. A., Herrero, L. J., Walsh, N. C., & Mahalingam, S. (2015). Arthritogenic alphaviruses: New insights into arthritis and bone pathology. *Trends in Microbiology*, 23(1), 35–43. https://doi.org/10.1016/j.tim.2014.09.005.

Chow, A., Her, Z., Ong, E. K., Chen, J. M., Dimatatac, F., Kwek, D. J., Barkham, T., Yang, H., Rénia, L., Leo, Y. S., & Ng, L. F. (2011). Persistent arthralgia induced by Chikungunya virus infection is associated with interleukin-6 and granulocyte macrophage colony-stimulating factor. *The Journal of Infectious Diseases*, 203(2), 149–157. https://doi.org/10.1093/infdis/jiq042.

Constant, L. E. C., Rajsfus, B. F., Carneiro, P. H., Sisnande, T., Mohana-Borges, R., & Allonso, D. (2021). Overview on Chikungunya virus infection: From epidemiology to state-of-the-art experimental models. *Frontiers in Microbiology*, 12, 744164. https://doi.org/10.3389/fmicb.2021.744164.

Couderc, T., & Lecuit, M. (2009). Focus on Chikungunya pathophysiology in human and animal models. *Microbes and Infection*, 11(14–15), 1197–1205. https://doi.org/10.1016/j.micinf.2009.09.002.

Cruz-Oliveira, C., Freire, J. M., Conceição, T. M., Higa, L. M., Castanho, M. A., & Da Poian, A. T. (2015). Receptors and routes of dengue virus entry into the host cells. *FEMS Microbiology Reviews*, 39(2), 155–170.

Das, T., Hoarau, J. J., Bandjee, M. C. J., Maquart, M., & Gasque, P. (2015). Multifaceted innate immune responses engaged by astrocytes, microglia and resident dendritic cells against Chikungunya neuroinfection. *The Journal of General Virology*, 96(Pt 2), 294–310. https://doi.org/10.1099/vir.0.071175-0.

Dasti, J. I. (2016). Zika virus infections: An overview of current scenario. *Asian Pacific Journal of Tropical Medicine*, 9(7), 621–625.

Dupuis-Maguiraga, L., Noret, M., Brun, S., Le Grand, R., Gras, G., & Roques, P. (2012). Chikungunya disease: Infection-associated markers from the acute to the chronic phase of arbovirus-induced arthralgia. *PLoS Neglected Tropical Diseases*, 6(3), e1446. https://doi.org/10.1371/journal.pntd.0001446.

El Sahili, A., & Lescar, J. (2017). Dengue virus non-structural protein 5. *Viruses*, 9(4), 91.

Galan-Huerta, K. A., Rivas-Estilla, A. M., & Fernández-Salas, I. (2015). Chikungunya virus: A general overview. *Medicina Universitaria*, 17(68), 175–183. http://dx.doi.org/10.1016/j.rmu.2015.06.001.

Ganesan, V. K., Duan, B., & Reid, S. P. (2017). Chikungunya virus: Pathophysiology, mechanism, and modeling. *Viruses*, 9(12), 368. https://doi.org/10.3390/v9120368.

Ganeshkumar, P., et al. (2018). Dengue infection in India: A systematic review and meta-analysis. *PLoS Neglected Tropical Diseases*, 12(7), e0006618.

Gasque, P., Couderc, T., Lecuit, M., Roques, P., & Ng, L. F. (2015). Chikungunya virus pathogenesis and immunity. *Vector Borne and Zoonotic Diseases (Larchmont, N.Y.)*, 15(4), 241–249. https://doi.org/10.1089/vbz.2014.1710.

Gerardin, P., Fianu, A., Michault, A., Mussard, C., Boussaid, K., Rollot, O., Grivard, P., Kassab, S., Bouquillard, E., Borgherini, G., Gaüzère, B. A., Malvy, D., Bréart, G., & Favier, F. (2013). Predictors of Chikungunya rheumatism: A prognostic survey ancillary to the TELECHIK cohort study. *Arthritis Research & Therapy*, 15(1), R9. https://doi.org/10.1186/ar4137.

Gharbaran, R., & Somenarain, L. (2019). Putative cellular and molecular roles of Zika virus in fetal and pediatric neuropathologies. *Pediatric and Developmental Pathology*, 22(1), 5–21. [PubMed] [Reference list].

Girard, M., Nelson, C. B., Picot, V., & Gubler, D. J. (2020). Arboviruses: A global public health threat. *Vaccine*, 38, 3989–3994.

Gourinat, A. C., O' Connor, O., Calvez, E., Goarant, C., & Dupont-Rouzeyrol, M. (2015). Detection of Zika virus in urine. *Emerging Infectious Diseases*, 21(1), 84–86.

Gowri Sankar, S., Mowna Sundari, T., & Alwin Prem Anand, A. (2021). Emergence of dengue 4 as dominant serotype during 2017 outbreak in South India and associated cytokine expression profile. *Frontiers in Cellular and Infection Microbiology*, 11, 681937.

Graciaa, D. S., Collins, M. H., & Wu, H. M. (2018). Zika in 2018: Advising travelers amid changing incidence. *Annals of Internal Medicine*, 169(5), 337–338.

Gubler, D. J. (1998). Dengue and dengue hemorrhagic fever. *Clinical Microbiology Reviews*, 11(3), 480–496. [PMC free article] [PubMed] [Google Scholar] [Ref list].

Gubler, D. J. (2010). Dengue viruses. In B. W. Mahy, & M. H. Regenmortel (Eds.), *Desk Encyclopedia of Human and Medical Virology* (pp. 372–381). Boston, MA: Academic Press. [Google Scholar] [Ref list].

Guerra Nunes, P. C., Daumas, R. P., Sanchez-Arcila, J. C., Ribeiro Nogueira, R. M., Pereira Horta, M. A., & dos Santos, F. B. (2019). 30 years of fatal dengue cases in Brazil. *BMC Public Health*, 19.

Gupta, E., & Ballani, N. (2014). Current perspectives on the spread of dengue in India. *Infection and Drug Resistance*, 337–342.

Gupta, N., Srivastava, S., Jain, A., & Chaturvedi, U. C. (2012). Dengue in India. *The Indian Journal of Medical Research*, 136(3), 373–390.

Hamel, R., Dejarnac, O., Wichit, S., Ekchariyawat, P., Neyret, A., Luplertlop, N., Perera-Lecoin, M., Surasombatpattana, P., Talignani, L., Thomas, F., Cao-Lormeau, V. M., Choumet, V., Briant, L., Desprès, P., Amara, A., Yssel, H., & Missé, D. (2015). Biology of Zika virus infection in human skin cells. *Journal of Virology*, 89(17), 8880–8896.

Hammack, C., Ogden, S. C., Madden Jr, J. C., Medina, A., Xu, C., Phillips, E., & Tang, H. (2019). Zika virus infection induces DNA damage response in human neural progenitors that enhances viral replication. *Journal of Virology*, 93(20), 10–1128.

Hirsch, A. J., Roberts, V. H., Grigsby, P. L., Haese, N., Schabel, M. C., Wang, X., & Streblow, D. N. (2018). Zika virus infection in pregnant rhesus macaques causes placental dysfunction and immunopathology. *Nature Communications*, 9(1), 263.

Htun, T. P., Xiong, Z., & Pang, J. (2021). Clinical signs and symptoms associated with WHO severe dengue classification: A systematic review and meta-analysis. *Emerging Microbes & Infections*, 10(1), 1116–1128.

Hu, D., Zou, H., Chen, W., Li, Y., Luo, Z., Wang, X., & Li, G. (2023). ZDHHC11 suppresses Zika virus infections by palmitoylating the envelope protein. *Viruses*, 15(1), 144.

Islam, M. A., Haque, E., Sharif, M., Islam, S., & Amin, M. R. 2020. Vector borne viral diseases: An emerging threat and their control strategies in Bangladesh perspectives. *International Research Journal of Medicine and Medical Sciences*, 8(1), 8–17.

Jain, B., & Jain, A. (2014). Dengue: A water-related mosquito-borne disease. *Water and Health*, 1–15.

Josseran, L., Paquet, C., Zehgnoun, A., Caillere, N., Le Tertre, A., Solet, J. L., & Ledrans, M. (2006). Chikungunya disease outbreak, Reunion Island. *Emerging Infectious Diseases*, 12(12), 1994–1995. https://doi.org/10.3201/eid1212.060710.

Kakooza-Mwesige, A., Tshala-Katumbay, D., & Juliano, S. L. (2019). Viral infections of the central nervous system in Africa. *Brain Research Bulletin*, 145, 2–17.

Kam, Y. W., Lum, F. M., Teo, T. H., Lee, W. W., Simarmata, D., Harjanto, S., Chua, C. L., Chan, Y. F., Wee, J. K., Chow, A., Lin, R. T., Leo, Y. S., Le Grand, R., Sam, I. C., Tong, J. C., Roques, P., Wiesmüller, K. H., Rénia, L., Rötzschke, O., & Ng, L. F. (2012). Early neutralizing IgG response to Chikungunya virus in infected patients targets a dominant linear epitope on the E2 glycoprotein. *EMBO Molecular Medicine*, 4(4), 330–343. https://doi.org/10.1002/emmm.201200213.

Kam, Y. W., et al. (2009). Immuno-biology of Chikungunya and implications for disease intervention. *Microbes and Infection*, 11(14–15), 1186–1196. https://doi.org/10.1016/j.micinf.2009.09.003.

Kamgang, B., Yougang, A. P., Tchoupo, M., Riveron, J. M., & Wondji, C. (2017). Temporal distribution and insecticide resistance profile of two major arbovirus vectors Aedes aegypti and Aedes albopictus in Yaoundé, the capital city of Cameroon. *Parasites Vectors*, 10, 469.

Kariuki Njenga, M., Nderitu, L., Ledermann, J. P., Ndirangu, A., Logue, C. H., Kelly, C. H. L., Sang, R., Sergon, K., Breiman, R., & Powers, A. M. (2008). Tracking epidemic Chikungunya virus into the Indian Ocean from East Africa. *The Journal of General Virology*, 89(Pt 11), 2754–2760. https://doi.org/10.1099/vir.0.2008/005413-0.

Kaur, G., Pant, P., Bhagat, R., & Seth, P. (2023). Zika virus E protein modulates functions of human brain microvascular endothelial cells and astrocytes: Implications on blood-brain barrier properties. *Frontiers in Cellular Neuroscience*, 17.

Khan, A. M., et al. (2008). Conservation and variability of dengue virus proteins: Implications for vaccine design. *PLoS Neglected Tropical Diseases*, 2(8), e272.

Kindhauser, M. K., Allen, T., Frank, V., Santhana, R. S., & Dye, C. (2016). Zika: The origin and spread of a mosquito-borne virus. *Bulletin of the World Health Organization*, 94(9), 675–686C. https://doi.org/10.2471/BLT.16.171082.

King, A. M. Q., Adams, M. J., Carstens, E. B., & Lefkowitz, E. J. (2012). *Virus Taxonomy: Classification and Nomenclature of Viruses: Ninth Report of the International Committee on Taxonomy of Viruses*. San Diego: Elsevier Academic Press.

Kolimenakis, A., Heinz, S., Wilson, M. L., Winkler, V., Yakob, L., Michaelakis, A., Papachristos, D., Richardson, C., & Horstick, O. (2021). The role of urbanisation in the spread of Aedes mosquitoes and the diseases they transmit: A systematic review. *PLoS Neglected Tropical Diseases*, 15(9), e0009631. https://doi.org/10.1371/journal.pntd.0009631.

Krauer, F., Riesen, M., Reveiz, L., Oladapo, O. T., Martínez-Vega, R., Porgo, T. V., & WHO Zika Causality Working Group. (2017). Zika virus infection as a cause of congenital brain abnormalities and Guillain–Barré syndrome: Systematic review. *PLoS Medicine*, 14(1), e1002203.

Kularatne, S. A., & Dalugama, C. (2022). Dengue infection: Global importance, immunopathology and management. *Clinical Medicine*, 22(1), 9.

Lan, Q., et al. (2021). Molecular characterization of structural protein genes of dengue virus serotype 1 epidemic in Yunnan, Southwest China, in 2018. *Archives of Virology*, 166, 863–870.

Lee, J. K., & Shin, O. S. (2019). Advances in Zika virus-host cell interaction: Current knowledge and future perspectives. *International Journal of Molecular Sciences*, 20(5), 1101.

Li, Q., & Kang, C. (2022). Structures and dynamics of dengue virus nonstructural membrane proteins. *Membranes*, 12(2), 231.

Lindenbach, B. D., & Rice, C. M. (2003). Molecular biology of flaviviruses. *Advances in Virus Research*, 59, 23–61.

Magnus, M. M., Espósito, D. L. A., Costa, V. A. D., Melo, P. S. D., Costa-Lima, C., Fonseca, B. A. L. D., & Addas-Carvalho, M. (2018). Risk of Zika virus transmission by blood donations in Brazil. *Hematology, Transfusion and Cell Therapy*, 40, 250–254.

Magro, M., et al. (2019). Self-assembly of chlorin-e6 on γ-Fe2O3 nanoparticles: Application for larvicidal activity against Aedes aegypti. *Journal of Photochemistry and Photobiology*, 194, 21–31. https://doi.org/10.1016/j.jphotobiol.2019.03.004.

Mahauad, F. W. D., Jones, P. H., & Okeoma, C. M. (2014). Critical role for bone marrow stromal antigen 2 in acute Chikungunya virus infection. *Journal of General Virology*, 95(Pt 11), 2450–2461. https://doi.org/10.1099/vir.0.068643-0.

Masmejan, S., Musso, D., Vouga, M., Pomar, L., Dashraath, P., Stojanov, M., & Baud, D. (2020). Zika virus. *Pathogens*, 9(11), 898.

Miller, J. L., deWet, B. J. M., Martinez-Pomares, L., Radcliffe, C. M., Dwek, R. A., Rudd, P. M., & Gordon, S. (2008). The mannose receptor mediates dengue virus infection of macrophages. *PLoS Pathogens*, 4(2), e17.

Miner, J. J., & Diamond, M. S. (2017). Zika virus pathogenesis and tissue tropism. *Cell Host & Microbe*, 21(2), 134–142.

Morens, D. M., & Fauci, A. S. (2014). Chikungunya at the door-déjà vu all over again? *The New England Journal of Medicine*, 371(10), 885–887. https://doi.org/10.1056/NEJMp1408509.

Mustafa, M. S., Rasotgi, V., Jain, S., & Gupta, V. J. (2015). Discovery of fifth serotype of dengue virus (DENV-5): A new public health dilemma in dengue control. *Medical Journal Armed Forces India*, 71(1), 67–70.

Mutheneni, S. R., Morse, A. P., Caminade, C., & Upadhyayula, S. M. (2017). Dengue burden in India: Recent trends and importance of climatic parameters. *Emerging Microbes & Infections*, 6(1), 1–10.

Nimmannitya, S., Halstead, S. B., Cohen, S. N., & Margiotta, M. R. (1969). Dengue and chikungunya virus infection in man in Thailand, 1962–1964. I. Observations on hospitalized patients with hemorrhagic

fever. *The American Journal of Tropical Medicine and Hygiene*, 18(6), 954–971. https://doi.org/10.4269/ajtmh.1969.18.954.
Nunes, M. R., et al. (2015). Emergence and potential for spread of Chikungunya virus in Brazil. *BMC Medicine*, 13, 102. https://doi.org/10.1186/s12916-015-0348-x.
Pan American Health Organization. (2015). Neurological Syndrome, Congenital Malformations and Zika Virus Infection. Implications for Public Health in the America.
Pardigon, N. (2009). The biology of chikungunya: A brief review of what we still do not know. *Pathologie. Biologie*, 57(2), 127–132. https://doi.org/10.1016/j.patbio.2008.02.016.
Pena-Garcia, V. H., McCracken, M. K., & Christofferson, R. C. (2017). Examining the potential for South American arboviruses to spread beyond the new world. *Current Clinical Microbiology Reports*, 4, 208–217. https://doi.org/10.1007/s40588-017-0076-4.
Petersen, L. R., & Epstein, J. S. (2014). Chikungunya virus: New risk to transfusion safety in the Americas. *Transfusion*, 54(8), 1911–1915. https://doi.org/10.1111/trf.12790.
Petersen, L. R., Jamieson, D. J., Powers, A. M., & Honein, M. A. (2016). Zika virus. *The New England Journal of Medicine*, 374(16), 1552–1563. https://doi.org/10.1056/NEJMra1602113.
Qureshi, A. I. (2016). *Chapter 4—Ebola Virus Disease Epidemic in Light of Other Epidemics, Ebola Virus Disease* (pp. 39–65, A. I. Qureshi, Ed.). Academic Press. https://doi.org/10.1016/B978-0-12-804230-4.00004-2.
Rashid, J. I. A., & Yusof, N. A. (2018). Laboratory diagnosis and potential application of nucleic acid biosensor approach for early detection of dengue virus infections. *Biosciences Biotechnology Research Asia*, 15, 245–255. https://doi.org/10.13005/bbra/2628.
Robert, M. A., Stewart-Ibarra, A. M., & Estallo, E. L. (2020). Climate change and viral emergence: Evidence from Aedes-borne arboviruses. *Current Opinion in Virology*, 40, 41–47. https://doi.org/10.1016/j.coviro.2020.05.001.
Roby, J. A., Setoh, Y. X., Hall, R. A., & Khromykh, A. A. (2015). Post-translational regulation and modifications of flavivirus structural proteins. *The Journal of General Virology*, 96(Pt 7), 1551–1569.
Rolph, M. S., & Mahalingam, S. (2019). Zika's passage to India. *The Lancet Infectious Diseases*, 19(5), 469–470.
Sang, R. C. (2007). Dengue in Africa. In *Report of the Scientific Working Group Meeting on Dengue* (pp. 50–52). Geneva: WHO Special Programme for Research and Training in Tropical Diseases.
Schaefer, T. J., Panda, P. K., & Wolford, R. W. (2024, January). Dengue fever. In *StatPearls* [Internet]. Treasure Island, FL: StatPearls Publishing. Updated 2022 Nov 14. https://www.ncbi.nlm.nih.gov/books/NBK430732/.
Schilte, C., Buckwalter, M. R., Laird, M. E., Diamond, M. S., Schwartz, O., & Albert, M. L. (2012). Cutting edge: Independent roles for IRF-3 and IRF-7 in hematopoietic and nonhematopoietic cells during host response to Chikungunya infection. *Journal of Immunology (Baltimore, MD: 1950)*, 188(7), 2967–2971. https://doi.org/10.4049/jimmunol.1103185.
Schilte, C., Staikowsky, F., Couderc, T., Madec, Y., Carpentier, F., Kassab, S., Albert, M. L., Lecuit, M., & Michault, A. (2013). Chikungunya virus-associated long-term arthralgia: A 36-month prospective longitudinal study. *PLoS Neglected Tropical Diseases*, 7(3), e2137. https://doi.org/10.1371/journal.pntd.0002137.
Serman, T. M., & Gack, M. U. (2019). Evasion of innate and intrinsic antiviral pathways by the Zika virus. *Viruses*, 11(10), 970.
Silva Jr, J. V. J., Ludwig-Begall, L. F., Oliveira-Filho, E. F., Oliveira, R. A. S., Duraes-Carvalho, R., Lopes, T. R. R., Silva, D. E. A., & Gil, L. H. V. G. (2018). A scoping review of Chikungunya virus infection: Epidemiology, clinical characteristics, viral co-circulation complications, and control. *Acta Tropica*, 188, 213–224. https://doi.org/10.1016/j.actatropica.2018.09.003.
Silva, N. M., Santos, N. C., & Martins, I. C. (2020). Dengue and Zika viruses: Epidemiological history, potential therapies, and promising vaccines. *Tropical Medicine and Infectious Disease*, 5(4), 150. https://doi.org/10.3390/tropicalmed5040150.
Simmonds, P., et al. (2017). ICTV virus taxonomy profile: Flaviviridae. *Journal of General Virology*, 98(1), 2–3.
Simon, F., Javelle, E., Oliver, M., Leparc-Goffart, I., & Marimoutou, C. (2011). Chikungunya virus infection. *Current Infectious Disease Reports*, 13(3), 218–228. https://doi.org/10.1007/s11908-011-0180-1.
Singhal, C., Dubey, A., Mathur, A., Pundir, C. S., & Narang, J. (2018). Paper based DNA biosensor for detection of chikungunya virus using gold shells coated magnetic nanocubes. *Process Biochemistry*, 74, 35–42. https://doi.org/10.1016/j.procbio.2018.08.020.
Song, B. H., Yun, S. I., Woolley, M., & Lee, Y. M. (2017). Zika virus: History, epidemiology, transmission, and clinical presentation. *Journal of Neuroimmunology*, 308, 50–64.

Soni, S., Gill, V. J. S., Anusheel, Singh, J., Chhabra, J., Gill, G. J. S., & Bakshi, R. (2023). Dengue, Chikungunya, and Zika: The causes and threats of emerging and re-emerging arboviral diseases. *Cureus*, 15(7), e41717. https://doi.org/10.7759/cureus.41717.

Sourisseau, M., Schilte, C., Casartelli, N., Trouillet, C., Guivel-Benhassine, F., Rudnicka, D., Sol-Foulon, N., Le Roux, K., Prevost, M. C., Fsihi, H., Frenkiel, M. P., Blanchet, F., Afonso, P. V., Ceccaldi, P. E., Ozden, S., Gessain, A., Schuffenecker, I., Verhasselt, B., Zamborlini, A., Saïb, A., & Schwartz, O. (2007). Characterization of reemerging chikungunya virus. *PLoS Pathogens*, 3(6), e89. https://doi.org/10.1371/journal.ppat.0030089.

Srikiatkhachorn, A., Mathew, A., & Rothman, A. L. (2017). Immune-mediated cytokine storm and its role in severe dengue. In *Seminars in Immunopathology* (Vol. 39, pp. 563–574). Berlin, Heidelberg: Springer.

Staples, J. E., Breiman, R. F., & Powers, A. M. (2009). Chikungunya fever: An epidemiological review of a re-emerging infectious disease. *Clinical Infectious Diseases: An Official Publication of the Infectious Diseases Society of America*, 49(6), 942–948. https://doi.org/10.1086/605496.

Suganya, P., et al. (2017). Biopolymer zein-coated gold nanoparticles: Synthesis, antibacterial potential, toxicity and histopathological effects against the Zika virus vector Aedes aegypti. *Journal of Photochemistry and Photobiology*, 173, 404–411. https://doi.org/10.1016/j.jphotobiol.2017.06.004.

Summers, D. J., Acosta, R. W., & Acosta, A. M. (2015). Zika virus in an American recreational traveler. *Journal of Travel Medicine*, 22(5), 338–340.

Tajudeen, Y. A., Oladipo, H. J., Oladunjoye, I. O., Yusuf, R. O., Sodiq, H., Omotosho, A. O., Adesuyi, D. S., Yusuff, S. I., & El-Sherbini, M. S. (2022). Emerging arboviruses of public health concern in Africa: Priorities for future research and control strategies. *Challenges*, 13, 60. https://doi.org/10.3390/challe13020060.

Tang, H., Hammack, C., Ogden, S. C., Wen, Z., Qian, X., Li, Y., Yao, B., Shin, J., Zhang, F., Lee, E. M., Christian, K. M., Didier, R. A., Jin, P., Song, H., & Ming, G. (2016). Zika virus infects human cortical neural progenitors and attenuates their growth. *Cell Stem Cell*, 18(5), 587–590. https://doi.org/10.1016/j.stem.2016.02.016.

Tang, J., Jose, J., Chipman, P., Zhang, W., Kuhn, R. J., & Baker, T. S. (2011). Molecular links between the E2 envelope glycoprotein and nucleocapsid core in Sindbis virus. *Journal of Molecular Biology*, 414(3), 442–459. https://doi.org/10.1016/j.jmb.2011.09.045.

Teo, A., Chua, C. L. L., Chia, P. Y., & Yeo, T. W. (2021). Insights into potential causes of vascular hyperpermeability in dengue. *PLoS Pathogens*, 17(12), e1010065.

Thiberville, S. D., Moyen, N., Dupuis-Maguiraga, L., Nougairede, A., Gould, E. A., Roques, P., & de Lamballerie, X. (2013). Chikungunya fever: Epidemiology, clinical syndrome, pathogenesis and therapy. *Antiviral Research*, 99(3), 345–370. https://doi.org/10.1016/j.antiviral.2013.06.009.

Thiboutot, M. M., Kannan, S., Kawalekar, O. U., Shedlock, D. J., Khan, A. S., Sarangan, G., Srikanth, P., Weiner, D. B., & Muthumani, K. (2010). Chikungunya: A potentially emerging epidemic? *PLoS Neglected Tropical Diseases*, 4(4), e623. https://doi.org/10.1371/journal.pntd.0000623.

Trivedi, S., & Chakravarty, A. (2022). Neurological complications of dengue fever. *Current Neurology and Neuroscience Reports*, 22(8), 515–529.

Usuda, J. N., Plaça, D. R., Fonseca, D. L. M., Marques, A. H., Filgueiras, I. S., Chaves, V. G. B., & Cabral-Marques, O. (2023). Interferome signature dynamics during the anti-dengue immune response: A systems biology characterization. *Frontiers in Immunology*, 14.

Valamparampil, J. J., et al. (2009). Clinical profile of Chikungunya in infants. *Indian Journal of Pediatrics*, 76, 151–155. https://doi.org/10.1007/s12098-009-0045-x.

van Duijl-Richter, M. K., Hoornweg, T. E., Rodenhuis-Zybert, I. A., & Smit, J. M. (2015). Early events in Chikungunya virus infection-from virus cell binding to membrane fusion. *Viruses*, 7(7), 3647–3674. https://doi.org/10.3390/v7072792.

Victora, C. G., Schuler-Faccini, L., Matijasevich, A., Ribeiro, E., Pessoa, A., & Barros, F. C. (2016). Microcephaly in Brazil: How to interpret reported numbers? *The Lancet*, 387(10019), 621–624.

Wahid, B., Ali, A., Rafique, S., & Idrees, M. (2017). Global expansion of chikungunya virus: Mapping the 64-year history. *International Journal of Infectious Diseases: IJID: Official Publication of the International Society for Infectious Diseases*, 58, 69–76. https://doi.org/10.1016/j.ijid.2017.03.006.

Wauquier, N., Becquart, P., Nkoghe, D., Padilla, C., Ndjoyi-Mbiguino, A., & Leroy, E. M. (2011). The acute phase of Chikungunya virus infection in humans is associated with strong innate immunity and T CD8 cell activation. *The Journal of Infectious Diseases*, 204(1), 115–123. https://doi.org/10.1093/infdis/jiq006.

Weaver, S. C., & Barrett, A. (2004). Transmission cycles, host range, evolution and emergence of arboviral disease. *Nature Reviews Microbiology*, 2, 789–801. https://doi.org/10.1038/nrmicro1006.

Weinbren, M. P., Haddow, A. J., & Williams, M. C. (1958). The occurrence of Chikungunya virus in Uganda. I. Isolation from mosquitoes. *Transactions of the Royal Society of Tropical Medicine and Hygiene*, 52(3), 253–257. https://doi.org/10.1016/0035-9203(58)90084-1.

Wilder-Smith, A., Chen, L. H., Massad, E., & Wilson, M. E. (2009). Threat of dengue to blood safety in dengue-endemic countries. *Emerging Infectious Diseases*, 15(1), 8–11. [PMC free article] [PubMed] [Google Scholar] [Ref list].

World Health Organization (WHO). (2020). *Evaluation of Genetically Modified Mosquitoes for the Control of Vector-Borne Diseases* (pp. 1–7). Geneva: World Health Organization.

Yeh, C. Y., Chen, P. L., Chuang, K. T., Shu, Y. C., Chien, Y. W., Perng, G. C., & Ko, N. Y. (2017). Symptoms associated with adverse dengue fever prognoses at the time of reporting in the 2015 dengue outbreak in Taiwan. *PLoS Neglected Tropical Diseases*, 11(12), e0006091.

Yong, Y. K., Wong, W. F., Vignesh, R., Chattopadhyay, I., Velu, V., Tan, H. Y., & Shankar, E. M. (2022). Dengue infection-recent advances in disease pathogenesis in the era of COVID–19. *Frontiers in Immunology*, 13, 889196.

Yu, I. M., Zhang, W., Holdaway, H. A., Li, L., Kostyuchenko, V. A., Chipman, P. R., & Chen, J. (2008). Structure of the immature dengue virus at low pH primes proteolytic maturation. *Science*, 319(5871), 1834–1837.

Zerfu, B., Kassa, T., & Legesse, M. (2023). Epidemiology, biology, pathogenesis, clinical manifestations, and diagnosis of dengue virus infection, and its trend in Ethiopia: A comprehensive literature review. *Tropical Medicine and Health*, 51(1), 11.

Zhang, H., Zhou, Y. P., Peng, H. J., Zhang, X. H., Zhou, F. Y., Liu, Z. H., & Chen, X. G. (2014). *Predictive Symptoms and Signs of Severe Dengue Disease for Patients with Dengue Fever: A Meta-Analysis.* BioMed Research International.

8 Pharmacological Strategies and Investigational Drugs for Emerging and Re-Emerging Diseases of Viral Origin

Srijan Goswami, Manash Mukherjee, and Rajesh Das

8.1 INTRODUCTION

Emerging and re-emerging viral diseases pose significant challenges to global public health due to their capacity to cause widespread morbidity and mortality. Emerging viral diseases are often associated with novel pathogens or those previously unrecognized in humans, such as severe acute respiratory syndrome coronavirus 2 (SARS-CoV-2), which led to the COVID-19 pandemic. Re-emerging diseases, on the other hand, involve pathogens that were once under control but have resurged, such as measles and dengue, often due to factors like vaccine hesitancy, ecological changes, or urbanization. The drivers of these outbreaks are multifaceted, including environmental changes, globalization, and increased human–animal interactions, which facilitate zoonotic spillovers. For instance, the Nipah virus in India and the Marburg virus in Africa have highlighted the role of wildlife reservoirs in sustaining these pathogens. Global health inequities further exacerbate the impact, as regions with limited access to healthcare struggle to contain outbreaks effectively (Hakim et al., 2024; Ha et al., 2023). Genomic surveillance and advanced diagnostics have become critical tools in identifying and managing these diseases. Technologies like next-generation sequencing enable the rapid identification of novel viral strains, aiding in the timely implementation of control measures. However, these efforts must be complemented by robust public health interventions, including vaccination programs, vector control, and public education (Hakim et al., 2024; Ha et al., 2023). Overall, addressing the challenges posed by emerging and re-emerging viral diseases requires a coordinated global approach that combines scientific innovation, healthcare system strengthening, and proactive policy measures. This is crucial to mitigate their far-reaching health, social, and economic impacts.

Addressing viral epidemics presents significant challenges due to the complexity of pathogen dynamics, societal structures, and global interconnectedness. One key issue is the unpredictability of emerging pathogens, driven by environmental changes, such as deforestation, urbanization, and climate change, which increase human–animal interactions and facilitate zoonotic spillovers. Moreover, re-emerging diseases often exploit gaps in surveillance and healthcare systems, especially in resource-limited regions (Branda et al., 2024; Writer, 2024). The rapid development and deployment of diagnostics, therapeutics, and vaccines during the COVID-19 pandemic demonstrate the potential of modern science to combat outbreaks. However, disparities in global healthcare infrastructure and vaccine distribution remain significant barriers. Vulnerable populations are disproportionately affected due to socioeconomic inequalities, leading to higher morbidity and mortality in low-income regions (Branda et al., 2024; Writer, 2024). Furthermore, the reliance on reactive rather than proactive measures often delays containment efforts. Insufficient investment in public health infrastructure, vaccine research, and global surveillance systems exacerbates this issue. Additionally, public mistrust of health authorities, fueled by misinformation and inconsistent

DOI: 10.1201/9781032721811-8

communication, hampers compliance with preventive measures like vaccination and social distancing (Branda et al., 2024; Writer, 2024). Future epidemic responses must emphasize equitable resource allocation, enhanced global coordination, and stronger public health systems. Proactive investments in research, early detection, and preparedness will be crucial in mitigating the devastating impacts of viral epidemics.

8.2 MECHANISM OF VIRAL PATHOGENESIS AND DRUG TARGETS

The mechanism of viral pathogenesis is a complex interplay between the virus' ability to replicate and evade the host immune response, often leading to cellular damage and systemic effects. Viral pathogenesis begins with the attachment of the virus to specific host cell receptors, a process mediated by viral surface proteins. For instance, the SARS-CoV-2 virus uses the spike protein to bind to the ACE2 receptor, facilitating entry via endocytosis or membrane fusion. Once inside the host cell, the virus utilizes its replication machinery to synthesize viral RNA and proteins, often causing cell damage or death. Key targets in the replication cycle include viral proteases, polymerases, and structural proteins, which are crucial for the assembly of new virions (Huang et al., 2024; Bergmann & Silverman, 2020). Understanding these mechanisms has been instrumental in drug development. Antiviral drugs target critical points in the viral life cycle. For example, remdesivir inhibits RNA-dependent RNA polymerase in coronaviruses, halting viral replication. Similarly, protease inhibitors like nirmatrelvir (part of Paxlovid) block viral proteases such as the main protease (Mpro) in SARS-CoV-2, preventing the cleavage of polyproteins necessary for viral maturation. Other investigational drugs focus on host cell pathways exploited by the virus, such as inhibitors of endocytosis or translation mechanisms (Bege & Borbás, 2024). In addition to antiviral drugs, immunomodulatory agents aim to mitigate the severe inflammatory responses, such as the cytokine storm observed in diseases like COVID-19 and influenza. Monoclonal antibodies targeting viral proteins have also emerged as potent therapeutic options, offering both direct antiviral effects and passive immunity. However, the development of drug resistance through mutations, as seen with variants of SARS-CoV-2, remains a significant challenge. This underscores the need for a combination of therapeutic strategies, including direct antivirals and host-targeting agents, to address viral pathogenesis effectively (Bergmann & Silverman, 2020; Feng, 2024). The viral life cycle is a complex series of steps that viruses undergo to infect a host cell, replicate, and spread. It includes key stages: attachment, entry, replication, assembly, and egress, each of which presents unique therapeutic targets.

- *Attachment and entry:* Viruses begin by binding to specific host cell surface receptors. For example, influenza viruses use hemagglutinin to attach to sialic acid residues on host cells, while HIV employs gp120 to interact with CD4 receptors. The entry can occur via membrane fusion, as seen in enveloped viruses like HIV, or endocytosis, utilized by non-enveloped viruses such as adenoviruses. These stages are prime therapeutic targets, with strategies like receptor-blocking agents or fusion inhibitors in development (Jones et al., 2021; Idrees et al., 2024). The initial step in viral infection involves the attachment of viral particles to host cell receptors, followed by entry into the host cell. These processes represent critical therapeutic targets, as blocking them can effectively prevent infection. Numerous pharmacological agents have been developed to disrupt these stages, with some focusing on viral surface proteins, while others target host cell receptors. For instance, the SARS-CoV-2 spike protein has been extensively studied as a primary target for therapeutic interventions. The interaction between the spike protein's receptor binding domain (RBD) and the host cell's angiotensin-converting enzyme 2 (ACE2) receptor is essential for viral entry. Novel spike inhibitors, identified through computational methods and validated in cell-based assays, have shown promising results. For example, compounds such as C8.2 demonstrated significant antiviral efficacy against SARS-CoV-2 and its variants, including Omicron, with low effective concentrations (EC50 values below 10 μM) (Tai et al., 2024). Other viruses, such as respiratory syncytial virus (RSV), are targeted by monoclonal antibodies like palivizumab, which block the viral F protein responsible for attachment. Additionally, compounds like hederasaponin B and other terpenoids exhibit broad-spectrum antiviral activity by interfering with early stages of viral infection, including attachment and entry mechanisms (*Antiviral Potential of Terpenoids Against Major Viral Infections: Recent Advances, Challenges, and Opportunities*, n.d.). Natural products and synthetic derivatives have also gained attention for their potential as entry inhibitors. Glycyrrhizin, derived from licorice root, has shown efficacy against Epstein–Barr virus (EBV) by targeting early replicative steps, further underscoring the versatility of compounds that interfere with attachment (Lin et al., 2008). These strategies highlight the importance of targeting the initial stages of viral infection. By preventing viruses from engaging with host cells, such pharmacological agents offer a potent line of defense, particularly for emerging and re-emerging viral threats.

- *Replication:* Once inside, the virus hijacks the host's cellular machinery to replicate its genome. DNA viruses typically use the host nucleus and enzymes, whereas RNA viruses often replicate in the cytoplasm with their own RNA-dependent RNA polymerases. These enzymes, being error-prone, lead to high mutation rates, enhancing viral adaptability. Retroviruses like HIV also employ reverse transcription, integrating their DNA into the host genome, which poses challenges for treatment (Jones et al., 2021; Idrees et al., 2024). Viral replication is a crucial step in the viral life cycle, providing an attractive target for therapeutic intervention. Numerous pharmacological agents have been developed to inhibit the replication of both RNA and DNA viruses. These agents typically target viral polymerases, which are responsible for synthesizing the viral genome. Viral polymerases, whether RNA-dependent RNA polymerases (RdRp) or DNA-dependent DNA polymerases (DdDp), are highly conserved across different viruses, making them ideal targets for antiviral drugs. For RNA viruses, nucleoside and nucleotide analogs, such as remdesivir, favipiravir, and molnupiravir, are widely used to inhibit RdRp activity. These agents mimic the building blocks of RNA and incorporate into the viral RNA chain, causing premature termination of replication (Palazzotti et al., 2024). Remdesivir, for instance, has shown effectiveness against SARS-CoV-2, inhibiting its replication and reducing viral loads in infected patients. Favipiravir and molnupiravir also target the viral RdRp and are being studied for their potential against other RNA viruses like influenza and Ebola (Palazzotti et al., 2024). For DNA viruses, polymerase inhibitors like acyclovir and sofosbuvir are essential. Acyclovir is a well-established drug used for herpes simplex virus (HSV) infections, targeting HSV's DNA polymerase to prevent viral DNA synthesis. Similarly, sofosbuvir, originally developed for hepatitis C virus (HCV), inhibits the RNA-dependent RNA polymerase of HCV, demonstrating the potential for cross-viral applications (Peng et al., 2022). The development of small-molecule inhibitors has greatly expanded the repertoire of antiviral drugs. These inhibitors are effective because of their high specificity for viral enzymes, minimizing host cell damage while preventing the replication of the virus. However, challenges such as drug resistance and limited spectrum of activity remain. Future research is focused on discovering new polymerase inhibitors and improving the efficacy of existing drugs (Peng et al., 2022; Palazzotti et al., 2024).
- *Assembly and egress:* Newly synthesized viral components are assembled into progeny virions, which then exit the host cell. Enveloped viruses typically bud off, acquiring a lipid layer, whereas non-enveloped viruses cause cell lysis. Targeting egress mechanisms, such as neuraminidase inhibitors for influenza, disrupts the release and spread of the virus (Jones et al., 2021; Idrees et al., 2024). The final stages of the viral life cycle, assembly and egress, involve the formation of new viral particles and their release from host cells. These stages represent critical therapeutic targets for antiviral drug development. Several

approaches focus on inhibiting the proteins and host machinery involved in the assembly and release of virions. In particular, the role of viral matrix proteins, such as the VP40 protein in filoviruses like Ebola, is essential for the assembly and budding processes. VP40 facilitates the recruitment of host proteins and contributes to the formation of new viral particles. Inhibition of VP40 or its interactions with host proteins can prevent the virus from efficiently assembling and exiting the cell (Wijesinghe & Stahelin, 2015; Kolesnikova et al., 2002). Similarly, the ESCRT (endosomal sorting complex required for transport) pathway, which is crucial for viral budding, has become a target for antiviral drugs. Small molecules that disrupt ESCRT machinery have shown promise in inhibiting the egress of several viruses, including HIV and hepatitis C virus (HCV) (Srivastava et al., 2024; Pincetic & Leis, 2009; Votteler & Sundquist, 2013; Hoang et al., 2022). In addition to targeting viral proteins, monoclonal antibodies and host-targeting antiviral agents have been employed. For instance, monoclonal antibodies that block viral entry and egress have been effective against influenza, RSV, and Ebola (Pantaleo et al., 2022; Hoenen et al., 2019). These antibodies can either neutralize viral particles directly or prevent viral proteins from interacting with host cell machinery, impeding the virus' ability to assemble and bud from infected cells. Another promising approach is the use of nucleic acid–based therapies. Antisense oligonucleotides (ASOs) and RNA interference (RNAi) have demonstrated potential in inhibiting viral RNA production and protein expression, including viral components necessary for assembly and egress (Tarn et al., 2021). These therapies can effectively block the translation of viral proteins, thus halting the formation of viral particles. Despite the progress, challenges remain, including the risk of resistance development and issues with the stability and delivery of certain drugs. Further research and clinical trials are necessary to refine these therapeutic strategies and evaluate their efficacy across different viral infections (Tan et al., 2023; Panda et al., 2024).

- *Host manipulation:* Viruses modulate host immune responses and metabolic pathways to enhance replication and evade detection. For instance, hepatitis C suppresses interferon signaling to sustain infection. Understanding these manipulations provides insights into novel therapeutic approaches (Jones et al., 2021; Idrees et al., 2024). Viruses have evolved intricate mechanisms to manipulate host cell processes for successful replication, immune evasion, and survival. These host-cell hijacking strategies involve various cellular components, including RNA-binding proteins (RBPs) and signaling pathways, that viruses exploit to modulate the host's immune response and facilitate their replication. For example, the influenza A virus manipulates host RBPs such as the interferon-induced protein with tetratricopeptide repeats (IFIT) family. These proteins, normally induced by interferons to inhibit viral replication, are counteracted by the viral non-structural protein NS1, which prevents their binding to viral RNA, enhancing viral replication (Song et al., 2024). Another critical aspect of viral manipulation is the modulation of host innate immune responses. Many viruses target RIG-I-like receptors (RLRs) and associated adapter proteins like MAVS to disrupt antiviral signaling. For instance, hepatitis C virus NS3/4A protease cleaves MAVS, thereby inhibiting the antiviral cytokine production and promoting viral persistence (Andigema et al., 2024) Additionally, viruses such as human cytomegalovirus (HCMV) exploit RBPs involved in RNA processing and translation. The viral protein pUL69 interacts with host RBPs, such as poly-(A)-binding protein nuclear 1 (PABPN1), enhancing viral gene expression (Andigema et al., 2024). Pharmacological agents targeting these host–pathogen interactions represent a promising avenue for antiviral therapy. By inhibiting viral manipulation of host RBPs or immune pathways, such agents can bolster the host's immune response and suppress viral replication. For example, compounds that interfere with RBP interactions or RLR signaling could restore antiviral defenses, offering novel therapeutic strategies (Song et al., 2024). This approach contrasts with direct-acting antivirals, focusing on host cell processes, making it a promising tool for combating a wide range of viral infections. However, careful evaluation of potential adverse effects and efficacy is necessary to ensure the success of these therapies.

Targeting these stages through antiviral drugs or therapies aimed at critical viral–host interactions is essential for managing infections and reducing the burden of viral diseases effectively. Such strategies not only hinder replication but also limit the spread of infections.

8.3 CURRENT PHARMACOLOGICAL STRATEGIES

The development of pharmacological strategies for viral diseases has made significant strides in recent years, driven largely by the need to address both ongoing viral threats, like HIV, influenza, and SARS-CoV-2, and emerging viral pathogens. Several antiviral drugs have been developed to specifically target viral replication, viral entry, and host–virus interactions, with some also addressing immune modulation. For instance, the use of protease inhibitors, such as lopinavir and ritonavir, has been central in the treatment of HIV, while neuraminidase inhibitors, like oseltamivir, are a mainstay in influenza management. Additionally, novel direct-acting antivirals (DAAs) targeting specific viral enzymes such as polymerase inhibitors, like remdesivir, have been pivotal in the treatment of SARS-CoV-2 (Muthukutty et al., 2024; Kálai et al., 2022). Another key strategy involves the targeting of viral entry into host cells. Recent advances have led to the development of fusion inhibitors and monoclonal antibodies that block viral attachment to the host cell receptor, effectively preventing infection. For example, the monoclonal antibody casirivimab/imdevimab has been shown to prevent SARS-CoV-2 entry by binding to the spike protein. Meanwhile, broader-spectrum approaches, including the development of antiviral compounds that can address various viral families, are gaining attention. One promising direction is the identification of host-targeted therapies that interfere with host cellular machinery utilized by viruses for replication, thus minimizing the risk of resistance (*Scientists Identify Antivirals That Could Combat Emerging Infectious Diseases*, 2023). While the availability of vaccines remains crucial for viral disease prevention, the role of antiviral drugs continues to grow, especially in the context of new variants and unexpected epidemics. For instance, drugs like molnupiravir and nirmatrelvir/ritonavir for SARS-CoV-2 represent the application of innovative antiviral therapy that also addresses concerns about resistance and viral mutation. Importantly, a growing interest in precision medicine, where treatment is tailored based on individual genetic makeup, holds potential in enhancing the efficacy of antiviral therapies, ensuring optimal outcomes in diverse populations (Scientists Identify Antivirals That Could Combat Emerging Infectious Diseases, 2023).

Nucleoside and non-nucleoside analogs have emerged as pivotal classes of antiviral agents in the treatment of various viral infections. Nucleoside analogs, which are modified versions of natural nucleosides, exert their antiviral effects by interfering with viral genome replication. These agents are typically phosphorylated within the host cell to active triphosphate forms, which then get incorporated into the growing viral nucleic acid chain, causing chain termination or mutagenesis (Kamzeeva et al., 2023; Li et al., 2021; Eyer et al., 2018). Among the key nucleoside analogs, drugs like *remdesivir* (an adenosine analog) have shown efficacy against a broad spectrum of RNA viruses, including Ebola, MERS-CoV, SARS-CoV-2, and more (Li et al., 2021; Eyer et al., 2018). Other examples include *tenofovir* and *lamivudine*, widely used in the treatment of HIV and hepatitis B (Kamzeeva et al., 2023; Eyer et al., 2018). On the other hand, non-nucleoside analogs (NNAs) operate differently by binding directly to viral enzymes, often to the active sites of reverse transcriptase (RT) or proteases, thereby inhibiting their catalytic activities. NNAs offer advantages such as fewer mutations and resistance patterns compared to nucleoside analogs. For instance, efavirenz and nevirapine, non-nucleoside inhibitors of reverse transcriptase, have been used extensively in HIV therapy, offering potent viral suppression with a favorable resistance profile (Kamzeeva et al., 2023; Eyer et al., 2018). The development of these antiviral agents continues to evolve, with research focusing on enhancing their specificity, minimizing toxicity, and combating emerging viral resistance (Li et al., 2021; Eyer et al., 2018). As seen with the use of remdesivir in the COVID-19 pandemic, nucleoside analogs play a critical role in rapidly controlling viral outbreaks, highlighting their potential for future antiviral drug development (Li et al., 2021; Kamzeeva et al., 2023; Eyer et al., 2018). Moreover, the ongoing exploration of novel nucleotide modifications promises to expand their efficacy against drug-resistant and newly emerging viral pathogens (Li et al., 2021; Eyer et al., 2018).

Protease and polymerase inhibitors play a pivotal role in the development of antiviral treatments, particularly for RNA viruses. These classes of drugs have shown significant promise in combating viral replication by targeting enzymes essential for the viral life cycle.

- *Protease inhibitors:* Proteases are enzymes critical for processing viral polyproteins into functional viral components during replication. Targeting viral proteases can effectively block virus maturation, halting its ability to spread. For instance, inhibitors of the SARS-CoV-2 main protease (M^pro) have been particularly successful. Nirmatrelvir, the key component of the oral antiviral Paxlovid, inhibits M^pro and has shown efficacy in reducing the severity of COVID-19 infections. This drug functions by preventing the cleavage of viral polyproteins, which is crucial for the production of infectious viral particles (Huang et al., 2024; Hashemian et al., 2023). Additionally, the repurposing of protease inhibitors originally developed for HIV, such as lopinavir and ritonavir, has been explored in the context of SARS-CoV-2, although clinical results have been mixed (Magro et al., 2021; Pinto Pereira et al., 2024). The success of protease inhibitors in the treatment of COVID-19 underscores their potential for broader applications, particularly in respiratory viruses (Lopinavir/Ritonavir: A Rapid Review of Effectiveness in COVID-19 | the Centre for Evidence-Based Medicine, n.d.)
- *Polymerase inhibitors:* Polymerases are enzymes responsible for viral RNA replication. By inhibiting these enzymes, antiviral drugs can block the replication of the virus. For example, remdesivir, a well-known RNA-dependent RNA polymerase (RdRp) inhibitor, has been authorized for use in treating COVID-19. It works by incorporating itself into the viral RNA chain, causing premature termination of transcription. This halts further replication of the virus (Palazzotti et al., 2024). Other polymerase inhibitors are also being explored for their efficacy against a range of RNA viruses, including influenza and hepatitis C virus (HCV), showing promise in inhibiting viral spread by interfering with the viral genome replication process (Deval et al., 2014).

Together, protease and polymerase inhibitors form an essential strategy in the antiviral arsenal. Ongoing research is focused on refining these inhibitors and exploring new variants that could offer broader-spectrum activity against diverse viral pathogens. However, the rapid mutation rates of viruses pose challenges in developing long-term effective treatments, necessitating continued innovation in drug design and therapeutic strategies (Citarella et al., 2023; Papaneophytou, 2024).

Interferon (IFN)-based therapies play a pivotal role in modulating the host immune response against viral infections. IFNs are cytokines that enhance the antiviral state of cells by inducing the expression of numerous interferon-stimulated genes (ISGs). These therapies are especially critical in managing infections caused by viruses that evade immune detection or suppress innate immunity, such as hepatitis B (HBV), hepatitis C (HCV), and SARS-CoV-2. Interferons, primarily IFN-α and IFN-β, bind to specific receptors on host cells, initiating signaling cascades that lead to the activation of ISGs. These genes encode proteins that inhibit various stages of the viral life cycle, including replication, assembly, and release. IFN-based treatments are particularly beneficial in chronic infections like HBV and HCV. Pegylated IFN-α, used in combination with ribavirin, was once the cornerstone for treating HCV prior to the advent of direct-acting antivirals (DAAs) (Doyle, 2016; Bergman et al., 2011). In the context of SARS-CoV-2, recombinant interferons have been explored as potential therapeutic agents. Studies demonstrated that early administration of IFN-β could mitigate disease severity by enhancing the antiviral response, especially when used in combination with other antiviral agents (Salimi-Jeda et al., 2021). However, their efficacy is highly dependent on timing and disease stage, with late-stage treatment potentially exacerbating inflammation (Kow & Hasan, 2021b). Despite their utility, IFN therapies face challenges, such as dose-dependent side effects, including flu-like symptoms, myelosuppression, and neuropsychiatric effects. Resistance to interferon treatment is another limitation, often resulting from mutations in the virus or host factors that impair ISG induction (Tomasello et al., 2014). Efforts to overcome these limitations include the development of pegylated formulations to improve pharmacokinetics and targeted delivery systems that reduce systemic toxicity. Recent advances have also explored synergistic approaches combining interferons with DAAs, monoclonal antibodies, or small-molecule inhibitors. Such combinations aim to maximize antiviral efficacy while minimizing adverse effects. Moreover, the application of artificial intelligence in drug design is being leveraged to identify novel interferon mimetics and optimize therapeutic regimens (Salimi-Jeda et al., 2021). The ongoing refinement of IFN-based therapies highlights their enduring importance in antiviral pharmacology. Emerging strategies focus on improving specificity, reducing adverse effects, and tailoring treatments to individual patient profiles through precision medicine approaches. In parallel, preclinical and clinical trials are exploring novel combinations and formulations to expand the therapeutic potential of interferons, particularly against drug-resistant and emerging viral pathogens.

Monoclonal antibodies (mAbs) represent a transformative approach in the treatment of viral diseases, offering specificity in neutralizing viral particles or modulating host immune responses. These agents are particularly effective in cases where traditional antivirals fall short, such as drug resistance or when precise targeting is necessary. Monoclonal antibodies work by mimicking the immune system's ability to fight pathogens. Neutralizing antibodies bind directly to viral particles, preventing their attachment to host cells and subsequent entry. Examples include REGN-COV2 (casirivimab and imdevimab) for SARS-CoV-2, which targets the spike protein, reducing viral load and mitigating severe COVID-19 symptoms (Focosi et al., 2022). In addition to direct neutralization, some mAbs modulate immune responses. For instance, tocilizumab, an IL-6 receptor blocker, has been used to mitigate cytokine storms in COVID-19 patients, thereby reducing inflammation and improving outcomes (Strait, 2022; Otsubo & Yasui, 2022). Several mAbs have received regulatory approval for viral infections. Palivizumab, for instance, is used prophylactically to prevent respiratory syncytial virus (RSV) in high-risk infants. The FDA has also granted emergency use authorization to numerous mAbs for COVID-19, including bebtelovimab and sotrovimab. These therapies have shown promise in reducing hospitalization rates and mortality (U.S. Food and Drug Administration, 2024; Otsubo & Yasui, 2022). Emerging mAb therapies are under investigation for a broader range of viral pathogens, such as Ebola, HIV, and hepatitis B. mAbs targeting conserved regions of viral proteins are particularly promising as they may provide cross-strain efficacy, a crucial feature for managing rapidly mutating viruses like influenza (*Antiviral Immunotherapy: Current Advancement and Translational Potential*, n.d.; Otsubo & Yasui, 2022). While mAbs hold tremendous promise, challenges such as high production costs, limited scalability, and the emergence of resistant viral strains remain. Innovations in bioengineering, such as bispecific or trispecific antibodies, aim to address these issues by enhancing efficacy and reducing costs. Additionally, leveraging AI for epitope prediction is expediting the development of next-generation mAbs tailored for specific viral pathogens. The ongoing evolution of monoclonal antibody therapies underscores their potential as cornerstone treatments in managing viral diseases, particularly in pandemics and emerging infections (Otsubo & Yasui, 2022; U.S. Food and Drug Administration, 2024; Strait, 2022; Focosi et al., 2022).

8.4 INVESTIGATIONAL DRUGS AND NOVEL THERAPEUTICS

The development of investigational drugs and novel therapeutics is at the forefront of combating emerging and re-emerging viral diseases. These agents are designed to either directly target viral components or modulate host pathways critical for viral replication. Innovations in this area have been accelerated by global pandemics like COVID-19, highlighting the urgent need for versatile and effective therapeutic options. The major categories of investigational drugs are highlighted as follows:

- *Direct-acting antivirals (DAAs):* DAAs focus on inhibiting specific viral enzymes or proteins. Key examples include the following:
  - *Nucleoside analogs:* Compounds like molnupiravir act by incorporating faulty nucleotides into viral RNA, inducing lethal mutagenesis (Karim et al., 2023). Originally developed for alphavirus infections, molnupiravir gained prominence during the COVID-19 pandemic. Nucleoside analogs represent a cornerstone in the development of antiviral therapeutics due to their ability to interfere with viral replication by mimicking natural nucleosides. These compounds integrate into viral RNA or DNA during replication, leading to chain termination or inhibition of viral polymerases, making them potent agents against a broad range of viruses. One prominent investigational nucleoside analog is obeldesivir (GS-5245), a derivative of remdesivir. Unlike its intravenous predecessor, obeldesivir is designed for oral administration, providing a more convenient option for outpatient settings. This drug inhibits viral RNA polymerase, effectively halting viral replication. Although its phase 3 clinical trials did not meet primary endpoints in reducing symptom duration in standard-risk COVID-19 patients, it showed promise when combined with other antivirals like nirmatrelvir, especially in preclinical studies against SARS-CoV-2 (Buchowicz & Koszytkowska-Stawińska, 2024; Technology Networks, 2024). Other novel nucleoside analogs, such as *molnupiravir*, have also demonstrated significant efficacy. Molnupiravir causes lethal mutagenesis in viral genomes, rendering the virus non-viable. It has been approved for emergency use in treating mild to moderate COVID-19, underscoring its utility in addressing public health emergencies (Buchowicz & Koszytkowska-Stawińska, 2024; Technology Networks, 2024). In addition to SARS-CoV-2, these analogs have shown potential against other RNA viruses like hepatitis C virus (HCV) and Ebola. The flexibility of nucleoside analogs in targeting a wide range of viral polymerases positions them as versatile agents in pandemic preparedness. Current efforts focus on enhancing the bioavailability, specificity, and safety profiles of these drugs to expand their applicability in treating diverse viral infections (Buchowicz & Koszytkowska-Stawińska, 2024; Technology Networks, 2024). Emerging strategies also explore combining nucleoside analogs with other antiviral agents to create synergistic effects, potentially enhancing therapeutic outcomes. As research progresses, investigational nucleoside analogs are expected to play a pivotal role in combating viral diseases and preparing for future pandemics (Buchowicz & Koszytkowska-Stawińska, 2024; Technology Networks, 2024).
  - *Protease inhibitors:* Drugs such as nirmatrelvir, the active component of Paxlovid, block the action of viral proteases necessary for protein cleavage and viral assembly (Technology Networks, 2024). Protease inhibitors represent a vital class of antiviral agents, targeting the protease enzymes essential for viral replication. These enzymes process viral polyproteins into functional units necessary for the assembly of new virions. By inhibiting this step, protease inhibitors effectively halt the life cycle of viruses, making them a cornerstone in antiviral pharmacology. One promising investigational protease inhibitor is PF-07321332, developed by Pfizer. This compound, a SARS-CoV-2-3CL protease inhibitor, is specifically designed for oral administration. It has shown potent antiviral activity in preclinical studies, not only against SARS-CoV-2 but also across other coronaviruses, highlighting its potential as a broad-spectrum therapeutic. PF-07321332 works by binding to the active site of the coronavirus protease enzyme, preventing the cleavage of viral polyproteins and subsequent replication. Clinical trials, including the EPIC program (Paraera and Vale., 2024), are evaluating

its efficacy for post-exposure prophylaxis and early-stage treatment in various patient groups. Co-administration with ritonavir extends its half-life, maintaining therapeutic levels for prolonged periods and optimizing its antiviral effects (Pfizer Starts Global Phase 2/3 EPIC-PEP Study of Novel COVID-19 Oral Antiviral Candidate for Post-Exposure Prophylaxis in Adults | Pfizer, n.d.; Pfizer Initiates Phase 1 Study of Novel Oral Antiviral Therapeutic Agent Against SARS-COV-2 | Pfizer, n.d.). Another investigational agent is PF-07304814 , an intravenous protease inhibitor also from Pfizer. It is being explored for use in hospitalized patients with severe COVID-19. By targeting the same viral protease, this compound offers a treatment option for critically ill patients, complementing oral therapies for early-stage intervention (Pfizer Starts Global Phase 2/3 EPIC-PEP Study of Novel COVID-19 Oral Antiviral Candidate for Post-Exposure Prophylaxis in Adults | Pfizer, n.d.; Pfizer Initiates Phase 1 Study of Novel Oral Antiviral Therapeutic Agent Against SARS-COV-2 | Pfizer, n.d.). Protease inhibitors are not limited to SARS-CoV-2. Compounds like danoprevir, initially developed for hepatitis C, have demonstrated potential in clinical trials for COVID-19. Its mechanism of action and safety profile in other viral diseases provide a strong foundation for its repurposing as a COVID-19 therapeutic (Chen et al., 2020). These investigational drugs underscore the versatility and critical role of protease inhibitors in managing viral infections. Their continued development, particularly for emergent and re-emergent viruses, offers hope for more effective and accessible treatments.

- *Polymerase inhibitors:* Remdesivir and its oral derivative obeldesivir target RNA-dependent RNA polymerase, halting viral genome replication. Viral polymerases, particularly RNA-dependent RNA polymerases (RdRPs), are essential enzymes in the replication of viral genomes. As such, they represent crucial targets for antiviral drug development. Polymerase inhibitors can be broadly categorized into nucleoside analogs, which mimic natural substrates and terminate chain elongation, and non-nucleoside inhibitors, which bind to allosteric sites, disrupting enzyme function. Recent investigational agents include remdesivir and molnupiravir, both targeting RdRPs. Remdesivir, initially developed for Ebola, has been repurposed and approved for COVID-19, demonstrating moderate efficacy in reducing hospital stays. Molnupiravir, an orally available prodrug, induces mutagenesis in viral RNA, effectively halting replication. Its mechanism of action and oral bioavailability make it a promising candidate for broader viral applications (Peng et al., 2022; Low et al., 2022). Additionally, baloxavir marboxil, primarily used against influenza, inhibits the cap-dependent endonuclease activity of the viral polymerase, preventing mRNA synthesis. Its broad-spectrum potential is under investigation for other RNA viruses. Structural studies of RdRPs from SARS-CoV-2 have further elucidated binding sites, enabling the design of highly specific inhibitors with minimal off-target effects (Hillen et al., 2020; Gao et al., 2020). Emerging research also focuses on telomerase reverse transcriptase and its potential role in RNA synthesis. Drugs targeting shared structural motifs between viral and host polymerases offer a novel therapeutic avenue but require caution to avoid host toxicity (Maida et al., 2016). While investigational polymerase inhibitors show promise, challenges such as resistance, optimal dosing strategies, and safety profiles in diverse populations remain. Continued advancements in structural biology and high-throughput screening are essential to expand the repertoire of effective polymerase inhibitors.

• *Host-directed therapies:* These therapies target host factors exploited by viruses:
  - *Camostat mesylate:* A serine protease inhibitor that blocks TMPRSS2, a host enzyme critical for SARS-CoV-2 entry. Camostat mesylate, a serine protease inhibitor, has garnered attention for its potential as a host-directed antiviral therapy, particularly in the context of SARS-CoV-2. This compound inhibits the host protease TMPRSS2, which is crucial for the priming of viral spike proteins, a necessary step for viral entry into host

cells. This mechanism highlights its utility in blocking viral entry pathways, making it a promising candidate for treating viral infections beyond SARS-CoV-2. Camostat mesylate has been traditionally used in Japan for conditions like pancreatitis and esophagitis. Preclinical studies revealed its effectiveness in impeding viral entry by targeting TMPRSS2, which is exploited by coronaviruses and influenza viruses. Early laboratory models and animal studies suggested its efficacy in reducing viral replication and associated pathological outcomes in SARS-CoV-2 infections (Fuentes-Prior, 2021; Khan et al., 2024). Clinical trials have been conducted to evaluate camostat's efficacy against COVID-19. For instance, a meta-analysis encompassing six countries with over 400 participants examined its impact on virologic and clinical outcomes. Despite its theoretical promise, camostat did not show significant improvements in viral load reduction or symptom resolution compared to placebo (Hedlin et al., 2024). Similarly, the CANDLE study, a randomized, placebo-controlled trial, reported limited efficacy in hastening viral clearance in COVID-19 patients, though it highlighted the drug's safety profile and tolerability (Kinoshita et al., 2022). The mixed clinical outcomes underscore the challenges in repurposing drugs like camostat for viral infections. Factors such as optimal dosing, timing of administration, and variability in disease severity influence its effectiveness. While camostat mesylate has not demonstrated strong efficacy in clinical settings for COVID-19, its role in combination therapies and other viral diseases remains an area of active research. In conclusion, camostat mesylate exemplifies the potential of host-directed therapies to interfere with key viral–host interactions. Its relatively established safety profile offers a foundation for further investigation, particularly in combination regimens or for other viral pathogens leveraging similar entry mechanisms.

- *Cyclophilin inhibitors:* Alisporivir disrupts host chaperone proteins required for viral replication, showing promise against multiple RNA viruses (Karim et al., 2023). Cyclophilin inhibitors represent a promising class of host-directed therapies (HDTs) aimed at targeting host cellular machinery utilized by viruses for replication and survival. Cyclophilins, a family of cellular proteins with peptidyl-prolyl isomerase activity, play crucial roles in protein folding, trafficking, and immune regulation. Several viruses, including HCV, HIV, coronaviruses, and flaviviruses, exploit cyclophilins for their replication, making these proteins attractive therapeutic targets. Cyclophilins interact with viral proteins to facilitate viral replication and assembly. Cyclophilin A (CypA), for instance, has been shown to bind viral proteins such as HIV Gag and HCV NS5A, enhancing viral replication. Cyclophilin inhibitors, such as cyclosporine A (CsA) and its derivatives, inhibit this interaction, disrupting viral replication pathways. Alisporivir, a non-immunosuppressive derivative of CsA, has shown promise in inhibiting HCV replication by blocking the interaction between CypA and viral proteins, without significant toxicity or immunosuppressive effects. Recent studies have also explored the role of cyclophilin inhibitors against SARS-CoV-2. By targeting host cyclophilins involved in coronavirus replication complexes, these inhibitors have shown broad-spectrum antiviral potential, particularly in preclinical studies. Moreover, they may provide therapeutic benefits across various viral infections, as cyclophilin-dependent mechanisms are conserved among multiple virus families (Thom & D'Elia, 2024; Pleschka, 2024). Cyclophilin inhibitors have following benefits:
  - *Reduced resistance development:* Targeting host factors rather than viral proteins reduces the likelihood of resistance, a common issue with direct-acting antivirals.
  - *Broad-spectrum efficacy:* Cyclophilin inhibitors can be effective against multiple virus families, making them valuable in pandemic preparedness.
  - *Immune modulation:* Beyond their antiviral effects, these inhibitors may modulate host immune responses, reducing inflammation and tissue damage during infections (Thom & D'Elia, 2024; Pleschka, 2024).

- While promising, cyclophilin inhibitors face challenges such as potential off-target effects and toxicity due to the ubiquitous nature of cyclophilins. Ongoing clinical trials are exploring strategies to enhance selectivity and minimize adverse effects. Moreover, combination therapies with direct-acting antivirals or immune modulators may amplify their therapeutic potential (Thom & D'Elia, 2024; Pleschka, 2024). As our understanding of host–pathogen interactions deepens, cyclophilin inhibitors continue to represent a cutting-edge approach in the fight against viral infections. Their development underscores the potential of HDTs to offer versatile and durable solutions against emerging and re-emerging viral threats.

- *Monoclonal antibodies (mAbs):* mAbs offer precision in targeting viral antigens. Casirivimab and imdevimab, approved during the COVID-19 pandemic, neutralize the SARS-CoV-2 spike protein. Advanced mAb platforms are now being developed for Ebola, HIV, and RSV, focusing on conserved viral epitopes to provide cross-strain coverage (Anderson et al., 2021). Monoclonal antibodies (mAbs) have emerged as a pivotal class of therapeutic agents for combating viral infections, particularly due to their specificity and ability to neutralize pathogens effectively. These biologics are engineered to target specific viral antigens, interrupting critical steps in the viral life cycle or modulating immune responses to reduce disease severity. Recent advances have highlighted mAbs as effective tools against SARS-CoV-2, the virus responsible for COVID-19. Antibodies such as sotrovimab, bebtelovimab, and combinations like casirivimab and imdevimab have demonstrated efficacy in neutralizing viral variants, although challenges such as resistance and evolving viral mutations remain a concern (Copin et al., 2021; Crescioli et al., 2024). Efforts are also underway to enhance the durability and breadth of mAbs by engineering them to target conserved viral epitopes or improve binding affinities. For influenza, investigational mAbs are targeting conserved regions of hemagglutinin, a critical viral surface protein, providing cross-strain protection. mAbs such as VIS410 and MHAA4549A are in advanced clinical trials, showing promise in mitigating severe cases of influenza by blocking viral entry into host cells (Kow et al., 2021a). Emerging platforms also include bispecific mAbs and antibody–drug conjugates, which combine antiviral properties with immune modulation. Bispecific antibodies, such as those targeting HIV reservoirs, are designed to simultaneously bind two different epitopes, enhancing neutralization and reducing viral escape mechanisms (Crescioli et al., 2024). Despite their promise, challenges such as high production costs, logistical hurdles in large-scale deployment, and the emergence of resistant viral strains necessitate further innovation. Research is also focusing on extending the half-life of mAbs and improving their pharmacokinetic profiles to enhance their applicability in diverse clinical settings (Copin et al., 2021). Overall, monoclonal antibodies continue to represent a promising frontier in antiviral therapeutics, with ongoing trials aiming to expand their utility for a broader range of viral pathogens.
- *Broad-spectrum antivirals (BSAs):* BSAs, such as ribavirin, offer activity against multiple viral families by targeting conserved pathways like RNA synthesis. Recent developments, including next-generation nucleotide analogs, are enhancing the efficacy and safety profiles of BSAs (ScienceDaily, 2023). Broad-spectrum antivirals (BSAs) represent a pivotal advancement in the fight against viral infections, offering the potential to target multiple viral families through a single therapeutic agent. Unlike narrow-spectrum antivirals that target specific viral proteins, BSAs act on conserved viral mechanisms or host factors essential for the replication of various viruses. Their development is particularly critical in mitigating the global burden of emerging and re-emerging viral threats, as seen with SARS-CoV-2 and other zoonotic pathogens. BSAs generally operate by either targeting viral components shared across families or host cell factors crucial for viral replication. Examples include nucleoside analogs like ribavirin, which inhibit viral RNA synthesis, and protease inhibitors that block the processing of viral polyproteins. Ribavirin, for

instance, has demonstrated efficacy against diverse viruses, including arenaviruses and flaviviruses, by disrupting RNA-dependent RNA polymerase activity. Similarly, remdesivir, a nucleotide analog initially developed for Ebola virus, proved effective against coronaviruses like SARS-CoV-2, underscoring the versatility of this approach (Market Shaping Accelerator at UChicago, 2024; Karim et al., 2023). Recent investigational efforts have yielded promising candidates, such as CRISPR-based gene-editing technologies and DNA origami virus traps, which offer innovative platforms for broad-spectrum antiviral development. For example, molnupiravir, originally developed for alphaviruses, has shown efficacy against COVID-19, demonstrating its utility as a broad-spectrum option. Such drugs not only provide treatment options for current outbreaks but also enhance preparedness for future pandemics (Market Shaping Accelerator at UChicago, 2024; Karim et al., 2023). The development of BSAs faces significant hurdles, including high R&D costs, unpredictable demand, and market failures that disincentivize investment. Governments and global health agencies are addressing these challenges through innovative funding mechanisms like "pull funding," which incentivizes successful development and ensures stockpile readiness. Regulatory pathways are also evolving to facilitate rapid approval of BSAs, ensuring their availability during outbreaks (Market Shaping Accelerator at UChicago, 2024; Karim et al., 2023). Expanding the pipeline of BSAs requires leveraging advanced technologies such as artificial intelligence for drug discovery and high-throughput screening methods to identify compounds with broad efficacy. Moreover, fostering collaborations between public and private sectors can accelerate progress. Ultimately, BSAs are poised to become a cornerstone of global health strategies, reducing reliance on reactive measures like vaccines and enabling proactive management of viral epidemics (Market Shaping Accelerator at UChicago, 2024; Karim et al., 2023). This strategic focus on BSAs represents a transformative shift in combating viral diseases, addressing the need for rapid, adaptable, and scalable antiviral solutions.

- *Emerging therapies and platforms*
  - *RNA-based therapies*: mRNA technology, popularized by COVID-19 vaccines, is being adapted for antiviral therapeutics. RNA interference (RNAi) and small interfering RNAs (siRNAs) are under investigation for targeting viral RNA directly. Investigational RNA-based therapies are gaining momentum as potent tools in the fight against viral infections. These therapies utilize RNA molecules to target viral genomes, inhibit replication, or modulate immune responses. One promising area is the development of RNA interference (RNAi) technologies, which involve small RNA molecules designed to silence specific viral genes. Recent advances have highlighted RNAi's potential in inhibiting the replication of viruses like hepatitis B and HIV (New Worm Study Paves Way for Better RNA-based Drugs to Treat Human Disease, 2024). Additionally, mRNA-based vaccines and therapies have proven instrumental in combating diseases such as COVID-19. These therapies work by delivering messenger RNA to cells, instructing them to produce viral proteins that stimulate immune responses, providing protection against future infections (Karim et al., 2023). Novel investigational approaches also include the use of self-amplifying RNA vaccines, which amplify their RNA payload within the host, increasing the vaccine's potency with lower doses (New Worm Study Paves Way for Better RNA-based Drugs to Treat Human Disease, 2024). As RNA-based therapies evolve, their application expands beyond prevention, targeting viral replication and improving immune system modulation, offering broad-spectrum antiviral potential. However, challenges remain, such as ensuring the stability and delivery of RNA molecules to target cells effectively while minimizing adverse immune reactions (Karim et al., 2023). These breakthroughs underscore the growing importance of RNA technologies in the development of antiviral strategies and treatments.

- *CRISPR-based antivirals:* The CRISPR/Cas system is being engineered to selectively cleave viral genomes. For instance, CRISPR-Cas13 systems show promise against RNA viruses by degrading their RNA with high specificity (Teng et al., 2021). Investigational CRISPR-based antiviral strategies have gained attention as a novel approach to combating viral infections, focusing on manipulating host cellular mechanisms or directly targeting viral genomes. CRISPR technologies, particularly the CRISPR/Cas9 and newer Cas13 systems, are being evaluated for their ability to precisely disrupt viral RNA or DNA, offering potential treatments for chronic viral diseases like HIV, hepatitis B, and even emerging infections like COVID-19. One promising CRISPR-based antiviral application involves using Cas9 to directly cut viral genomes at specific sequences. For example, in HIV therapy, researchers have used CRISPR to target and edit HIV DNA integrated into the host genome, aiming to remove the virus or render it inactive (Lee, 2019). These strategies are enhanced by the ability of Cas9 to recognize long terminal repeat (LTR) regions in retroviruses like HIV, where latent virus resides. Additionally, a variant of Cas9, known as Cas9d, has been modified for transcriptional activation, helping reactivate latent viruses in a "shock and kill" strategy (Lee, 2019). Further, the CRISPR/Cas13 system, which targets RNA instead of DNA, has shown potential against RNA viruses. This system can target and degrade the RNA of viruses like SARS-CoV-2 (Innovative Antiviral Defense with New CRISPR Tool, 2024). New advancements are exploring how to enhance Cas13's efficiency and delivery, potentially overcoming the challenges posed by the need to target viral genomes without inducing host cell toxicity. Moreover, CRISPR technology could also be employed to trigger host cellular responses that combat viral infections. For instance, inducing the expression of host restriction factors like APOBEC3G, which inhibits viral replication, through CRISPR/Cas9 activation has shown promise for controlling HIV and other retroviruses (Lee, 2019). However, challenges remain, including potential off-target effects, delivery issues, and viral mutations that could lead to resistance. In conclusion, CRISPR-based antiviral therapies represent an exciting frontier in the fight against viral infections, with substantial research ongoing to overcome technical and safety challenges before they can be widely applied in clinical settings.
- *Immunomodulators:* Agents like interferon-based therapies and immune checkpoint inhibitors are being studied to enhance the host's antiviral immune response. Combination strategies incorporating these modulators with antivirals have shown synergistic effects. Investigational immunomodulators are emerging as promising therapeutic agents for the treatment of viral infections, particularly those caused by viruses like SARS-CoV-2, influenza, and hepatitis. These therapies aim to regulate the immune system to either enhance its ability to fight off viral infections or dampen an overactive immune response that can result in severe disease pathology. One of the key immunomodulatory approaches involves the use of monoclonal antibodies, which target specific immune checkpoints or inflammatory cytokines. For instance, monoclonal antibodies against interleukins such as IL-6 (e.g., tocilizumab) have been shown to reduce the inflammatory response in severe COVID-19 patients, potentially preventing cytokine storms (Rojas et al., 2024). Additionally, corticosteroids, such as dexamethasone, are also being used to modulate immune responses, with studies showing efficacy in reducing mortality in severe cases of COVID-19 (Strzelec et al., 2023; Alunno et al., 2021). Furthermore, other strategies include the use of Janus kinase (JAK) inhibitors, which block signaling pathways involved in immune cell activation and inflammation. Drugs like baricitinib have demonstrated effectiveness in modulating immune responses in viral infections, particularly in combination with other antiviral therapies. Research is also exploring the potential of using biologic agents to target specific immune cells, such as T cells or macrophages, to either enhance immune response or reduce hyperinflammation

(Strzelec et al., 2023; Alunno et al., 2021). As we continue to better understand the mechanisms of viral immune evasion, the development of targeted immunomodulatory drugs offers a novel way to tackle complex viral infections, reducing the need for broad-spectrum antiviral medications and offering tailored treatments based on immune system modulation (Ha et al., 2024). The challenge, however, lies in ensuring these therapies can be safely administered without inducing adverse side effects or further immunosuppression in patients. Thus, further research and clinical trials will be crucial to define the full therapeutic potential of these immunomodulators (Strzelec et al., 2023; Alunno et al., 2021). Investigational drugs face significant challenges, including resistance development, off-target effects, and production scalability. However, advances in AI-driven drug discovery, high-throughput screening, and biomanufacturing are addressing these hurdles. The integration of multi-omics data is also facilitating precision medicine approaches, ensuring that therapeutic strategies are tailored to specific viral pathogens and patient populations. The pipeline for investigational drugs and novel therapeutics is robust, with innovations aimed at addressing current and future viral threats. These efforts underscore the importance of a multipronged approach, leveraging direct-acting antivirals, host-directed therapies, and cutting-edge platforms like mRNA and CRISPR, to build a resilient arsenal against viral diseases (Strzelec et al., 2023; Alunno et al., 2021).

8.5 REPURPOSED DRUGS IN VIRAL DISEASE MANAGEMENT

Repurposed drugs have become a critical strategy in the rapid development of therapies for emerging viral diseases. The concept of drug repurposing involves taking existing medications, originally developed for other indications, and testing their efficacy against novel viruses. This strategy is particularly valuable in the context of viral outbreaks, where the urgent need for treatments outpaces the typical timeline for drug development. For instance, during the COVID-19 pandemic, several drugs initially designed for other diseases, such as remdesivir and hydroxychloroquine, were repurposed for SARS-CoV-2 infections. Remdesivir, an antiviral originally developed for Ebola and Marburg viruses, has shown efficacy against a variety of RNA viruses, including SARS-CoV-2, by inhibiting the viral RNA polymerase (Sarwar et al., 2021; Martinez, 2022). Other examples include antiviral agents like molnupiravir and nirmatrelvir, which were repurposed to target the replication of SARS-CoV-2. Molnupiravir, originally developed for influenza, causes an increase in viral mutations that lead to the production of non-infectious virus particles (Sarwar et al., 2021; Martinez, 2022). Similarly, nirmatrelvir, a protease inhibitor, was repurposed from its use in HIV treatment and proved highly effective when combined with ritonavir in reducing COVID-19-related hospitalization and death (Sarwar et al., 2021; Martinez, 2022). These examples highlight the potential of repurposed drugs to quickly address public health threats, though challenges remain in optimizing their use and overcoming issues such as viral resistance and side effects (Sarwar et al., 2021; Martinez, 2022). Repurposing drugs is not without challenges, including limited clinical trial data, drug toxicity concerns, and regulatory hurdles. Nonetheless, this approach accelerates the availability of treatments and serves as a bridge while novel antiviral agents are being developed. Additionally, the identification of new uses for existing drugs continues to offer hope in managing emerging viral threats, making it an essential part of pandemic preparedness (Sarwar et al., 2021; Martinez, 2022).

8.6 ROLE OF ADJUNCTIVE THERAPIES: NUTRACEUTICALS

Nutraceuticals play a significant role in managing viral infections, offering therapeutic support alongside conventional antiviral treatments. These bioactive compounds, often derived from food, have immunomodulatory, antiviral, and anti-inflammatory properties that can complement standard pharmacological therapies. For instance, compounds like polyphenols, flavonoids, and certain vitamins (e.g., vitamin C, vitamin D) have demonstrated potential in enhancing immune responses, reducing viral replication, and alleviating inflammation associated with viral diseases (Puri et al., 2022). Additionally, nutraceuticals may help in improving the body's overall resistance to infections, enhance recovery, and reduce the severity of symptoms, especially in chronic viral conditions like HIV and hepatitis (Puri et al., 2022). Supportive care is essential in viral disease management, focusing on relieving symptoms and supporting the body's natural healing processes. This includes maintaining hydration, managing fever, and providing appropriate nutritional support to ensure the immune system functions optimally. Nutraceuticals such as probiotics and prebiotics are also gaining attention for their ability to balance gut microbiota, which in turn can influence viral infection outcomes. Research has shown that nutraceutical interventions can reduce the duration of symptoms and improve overall outcomes in viral infections, particularly in immunocompromised individuals (Puri et al., 2022).

8.7 ETHICAL AND REGULATORY CONSIDERATIONS IN ANTIVIRAL DRUG DEVELOPMENT

The development of antiviral drugs involves a complex interplay of ethical and regulatory challenges aimed at balancing innovation, safety, equity, and public health. Central to these challenges is the ethical principle of beneficence, which mandates that drug development prioritizes patient safety and efficacy. Rigorous preclinical and clinical testing, conducted under ethical guidelines such as the Declaration of Helsinki, ensures that potential benefits outweigh risks. These processes demand informed consent from participants and strict adherence to research protocols to avoid harm and exploitation (Admin, 2024; Salvatore, 2024). Regulatory oversight by agencies like the FDA and EMA plays a critical role in evaluating the safety, efficacy, and quality of new antiviral therapies. These agencies set standards for trial design, monitoring, and post-market surveillance, ensuring that drugs meet high safety thresholds. However, the rapid pace of technological advancements in biomedical research has outpaced existing regulatory frameworks, necessitating adaptive and innovative regulatory strategies. Harmonizing international regulations is also critical, given the global nature of viral pandemics (Leavitt, 2021; Salvatore, 2024). Access and equity remain pressing ethical concerns. The high costs associated with antiviral drug development often translate to limited availability for resource-poor populations, exacerbating health inequities. Strategies such as tiered pricing and public–private partnerships are essential to bridge these gaps and ensure that life-saving drugs reach underserved populations (Admin, 2024; Salvatore, 2024). Another challenge lies in balancing commercial interests with public health priorities. Pharmaceutical companies must navigate the dual pressures of ensuring profitability and addressing urgent public health needs. Transparent communication and accountability mechanisms are crucial to align corporate practices with ethical standards (Admin, 2024; Salvatore, 2024). In summary, addressing ethical and regulatory considerations in antiviral drug development requires a collaborative approach among stakeholders, including researchers, regulatory agencies, and policymakers. This ensures that the process remains ethical, inclusive, and aligned with the overarching goal of improving global health outcomes.

8.8 CONCLUSION

The development of antiviral therapies represents a critical frontier in combating viral diseases, reflecting advancements in pharmacology, biotechnology, and immunology. This chapter has explored the multifaceted strategies underlying antiviral drug development, including the exploitation of viral life cycle vulnerabilities, host-directed approaches, and the innovative use of investigational agents such as CRISPR-based antivirals, RNA therapies, and broad-spectrum antivirals. The exploration of repurposed drugs and nutraceuticals underscores the importance of adaptive strategies in addressing emergent and re-emergent viral threats. Ethical and regulatory considerations

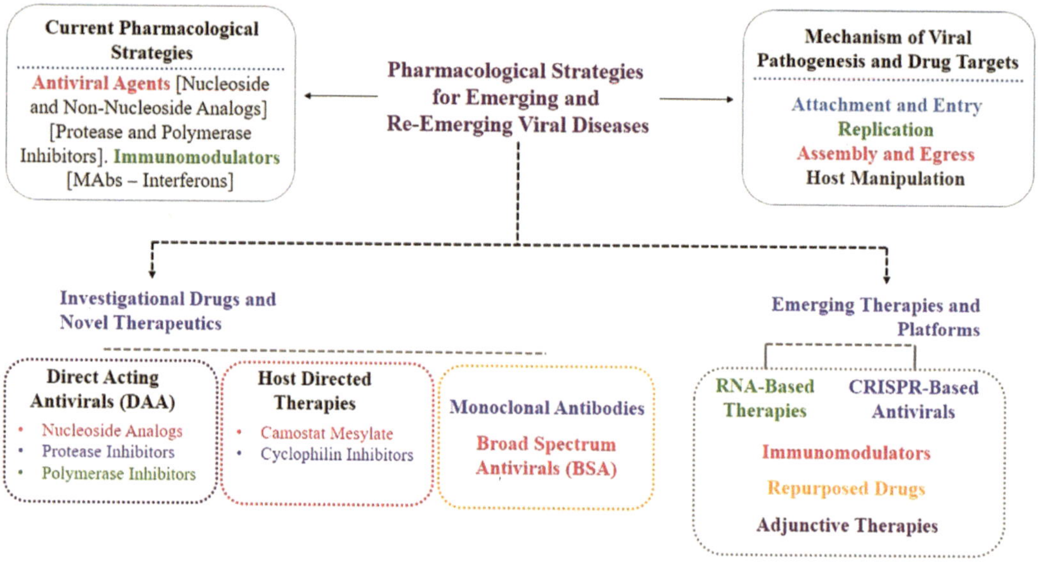

FIGURE 8.1 Conceptual representation of pharmacological strategies for emerging and re-emerging viral diseases.

form the backbone of this endeavor, ensuring the safety, equity, and accessibility of antiviral agents. These principles are particularly crucial in a landscape shaped by the urgency of pandemics and global health crises, demanding a balance between rapid development and rigorous oversight. Collaboration among stakeholders—including researchers, industry leaders, regulatory bodies, and policymakers—is essential to align innovation with ethical imperatives and public health priorities. Moving forward, integrating novel technologies, fostering equitable access, and addressing the ethical complexities of drug development will be pivotal in enhancing the efficacy and reach of antiviral therapies. This collective effort promises to not only mitigate the impact of current viral diseases but also prepare humanity for future outbreaks, safeguarding global health through robust and ethically grounded antiviral strategies. Figure 8.1 provides conceptual representation of the concepts associated with pharmacological strategies for emerging and re-emerging viral diseases.

REFERENCES

Admin. (2024, June 12). Challenges in developing antiviral medications. *Ciron Drugs Blog*. https://www.cironpharma.com/blog/challenges-in-developing-antiviral-medications/.

Alunno, A., et al. (2021). Immunomodulatory therapies for SARS-CoV-2 infection: A systematic literature review to inform EULAR points to consider. *Annals of the Rheumatic Diseases*, 80, 803–815.

Anderson, B., Smith, Z., Edupuganti, S., Yan, X., Masi, C. M., & Wu, H. M. (2021, June 12). Effect of monoclonal antibody treatment on clinical outcomes in ambulatory patients with Coronavirus disease 2019. *Open Forum Infect Dis*, 8(7), ofab315. https://doi.org/10.1093/ofid/ofab315. PMID: 34277887; PMCID: PMC8279096.

Andigema, A. S., Cyrielle, N. N. T., & Chanelle, M. K. (2024). Viral hijacking of host RNA-binding proteins: Implications for viral replication and pathogenesis. *Journal of Clinical and Laboratory Research*, 7(5). https://doi.org/10.31579/2768-0487/136.

Antiviral Immunotherapy: Current Advancement and Translational Potential. (n.d.). Frontiers. https://www.frontiersin.org/research-topics/66736/antiviral-immunotherapy-current-advancement-and-translational-potential.

Antiviral Potential of Terpenoids Against Major Viral Infections: Recent Advances, Challenges, and Opportunities. (n.d.). JABET. https://www.bsmiab.org/jabet/178-1700321148-antiviral-potential-of-terpenoids-against-major-viral-infections-recent-advances-challenges-and-opportunities.

Bege, M., & Borbás, A. (2024). The design, synthesis and mechanism of action of Paxlovid, a protease inhibitor drug combination for the treatment of COVID–19. *Pharmaceutics*, 16(2), 217. https://doi.org/10.3390/pharmaceutics16020217.

Bergman, S. J., Ferguson, M. C., & Santanello, C. (2011, December). Interferons as therapeutic agents for infectious diseases. *Infect Dis Clin North Am*, 25(4), 819–834. https://doi.org/10.1016/j.idc.2011.07.008. PMID: 22054758; PMCID: PMC7134994.

Bergmann, C. C., & Silverman, R. H. (2020). COVID-19: Coronavirus replication, pathogenesis, and therapeutic strategies. *Cleveland Clinic Journal of Medicine*, 87(6), 321–327. https://doi.org/10.3949/ccjm.87a.20047.

Branda, F., & Ciccozzi, M. (2024). Navigating novel viral challenges: Understanding, tracking, and mitigating emerging threats. *Microorganisms*, 12(4), 807. https://doi.org/10.3390/microorganisms12040807.

Buchowicz, W., & Koszytkowska-Stawińska, M. (2024). How much potential do nucleoside analogs offer to combat human Corona viruses? *Organics*, 5(2), 71–110. https://doi.org/10.3390/org5020006.

Chen, H., Zhang, Z., Wang, L., Huang, Z., Gong, F., Li, X., Chen, Y., & Wu, J. J. (2020, November 25). First clinical study using HCV protease inhibitor danoprevir to treat COVID-19 patients. *Medicine (Baltimore)*, 99(48):e23357. https://doi.org/10.1097/MD.0000000000023357. PMID: 33235105; PMCID: PMC7710192.

Citarella, A., Dimasi, A., Moi, D., Passarella, D., Scala, A., Piperno, A., & Micale, N. (2023). Recent advances in SARS-CoV-2 main protease inhibitors: From nirmatrelvir to future perspectives. *Biomolecules*, 13(9), 1339. https://doi.org/10.3390/biom13091339.

Copin, R., Baum, A., Wloga, E., Pascal, K. E., Giordano, S., Fulton, B. O., Zhou, A., Negron, N., Lanza, K., Chan, N., Coppola, A., Chiu, J., Ni, M., Wei, Y., Atwal, G. S., Hernandez, A. R., Saotome, K., Zhou, Y., Franklin, M. C., Hooper, A. T., McCarthy, S., Hamon, S., Hamilton, J. D., Staples, H. M., Alfson, K., Carrion, R. Jr, Ali, S., Norton, T., Somersan-Karakaya, S., Sivapalasingam, S., Herman, G. A., Weinreich, D. M., Lipsich, L., Stahl, N., Murphy, A. J., Yancopoulos, G. D., & Kyratsous, C. A. (2021). The monoclonal antibody combination REGEN-COV protects against SARS-CoV-2 mutational escape in preclinical and human studies. *Cell*, 184(15), 3949–3961.e11. https://doi.org/10.1016/j.cell.2021.06.002. Epub 2021 Jun 5. PMID: 34161776; PMCID: PMC8179113.

Crescioli, S., Kaplon, H., Chenoweth, A., Wang, L., Visweswaraiah, J., & Reichert, J. M. (2024). Antibodies to watch in 2024. *mAbs*, 16(1). https://doi.org/10.1080/19420862.2023.2297450.

Deval, J., Symons, J. A., & Beigelman, L. (2014, December). Inhibition of viral RNA polymerases by nucleoside and nucleotide analogs: Therapeutic applications against positive-strand RNA viruses beyond hepatitis C virus. *Curr Opin Virol*, 9, 1–7. https://doi.org/10.1016/j.coviro.2014.08.004. Epub 2014 Sep 17. PMID: 25194816; PMCID: PMC7102778.

Doyle, T. (2016, April 1). The evolving role of interferons in viral eradication strategies. *J Virus Erad*, 2(2), 121–123. https://doi.org/10.1016/S2055-6640(20)30467-2. PMID: 27482449; PMCID: PMC4965245.

Eyer, L., Nencka, R., de Clercq, E., Seley-Radtke, K., & Růžek, D. (2018). Nucleoside analogs as a rich source of antiviral agents active against arthropod-borne flaviviruses. *Antiviral Chemistry and Chemotherapy*, 26. https://doi.org/10.1177/2040206618761299.

Feng, Y. (2024). Recent advances in the study of zika virus structure, drug targets, and inhibitors. *Frontiers in Pharmacology*, 15. https://doi.org/10.3389/fphar.2024.1418516.

Focosi, D., McConnell, S., Casadevall, A., Cappello, E., Valdiserra, G., & Tuccori, M. (2022, November). Monoclonal antibody therapies against SARS-CoV-2. *Lancet Infect Dis*, 22(11), e311–e326. https://doi.org/10.1016/S1473-3099(22)00311-5. Epub 2022 Jul 5. Erratum in: Lancet Infect Dis. 2022 Sep;22(9):e239. https://doi.org/10.1016/S1473-3099(22)00468-6. PMID: 35803289; PMCID: PMC9255948.

Fuentes-Prior, P. (2021, January-June). Priming of SARS-CoV-2 S protein by several membrane-bound serine proteinases could explain enhanced viral infectivity and systemic COVID-19 infection. *J Biol Chem*, 296, 100135. https://doi.org/10.1074/jbc.REV120.015980. Epub 2020 Dec 6. PMID: 33268377; PMCID: PMC7834812.

Gao, Y., et al. (2020). Structure of the RNA-dependent RNA polymerase from COVID-19 virus. *Science (New York, N.Y.)*, 368(6492), 779–782. https://doi.org/10.1126/science.abb7498.

Ha, D. P., Shin, W.-J., Hernandez, J. C., Neamati, N., Dubeau, L., Machida, K., & Lee, A. S. (2023). GRP78 inhibitor YUM70 suppresses SARS-CoV-2 viral entry, spike protein production and ameliorates lung damage. *Viruses*, 15(5), 1118. https://doi.org/10.3390/v15051118.

Ha, M. K. (2024). How emerging and re-emerging infectious diseases challenge our understanding of viral infections. *Vaccines*, 12(12), 1339. https://doi.org/10.3390/vaccines12121339.

Hakim, M. S., Wang, W., Goeijenbier, M., & Saxena, S. K. (2024). Editorial: Viral emerging and re-emerging diseases: Basic understanding and future intervention strategies. *Frontiers in Microbiology*, 15. https://doi.org/10.3389/fmicb.2024.1395742.

Hashemian, S. M. R., Sheida, A., Taghizadieh, M., Memar, M. Y., Hamblin, M. R., Baghi, H. B., Nahand, J. S., Asemi, Z., & Mirzaei, H. (2023, June). Paxlovid (Nirmatrelvir/Ritonavir): A new approach to Covid-19 therapy? *Biomed Pharmacother*, 162, 114367. https://doi.org/10.1016/j.biopha.2023.114367. Epub 2023 Feb 6. PMID: 37018987; PMCID: PMC9899776.

Hedlin, H., Tobback, E., Lee, J., Wang, Y., Dragoni, I., Anthony, D. C., Dhaliwal, K., Norrie, J., Halford, S., Gotes, J., Moctezuma, M., Olivas-Martinez, A., Khosla, C., Singh, U., Gunst, J. D., Valdez, A., Kershenobich, D., Boutboul, D., Søgaard, O. S., De Scheerder, M.-A., Desai, M., & Parsonnet, J. (2024). *Individual Patient Data Meta-Analysis Evaluating Camostat Mesilate to Treat COVID-19 in Community Settings*. [Preprint] medRxiv 2024.05.15.24307072; https://doi.org/10.1101/2024.05.15.24307072.

Hillen, H. S., et al. (2020). Structure of replicating SARS-CoV-2 polymerase. *Nature*, 584, 154–156. https://doi.org/10.1038/s41586-020-2368-8.

Hoang, P. T., Luong, Q. X. T., Ayun, R. Q., Lee, Y., Vo, T. T. B., Kim, T., & Lee, S. (2022). A novel approach of antiviral drugs targeting viral genomes. *Microorganisms*, 10(8), 1552. https://doi.org/10.3390/microorganisms10081552.

Hoenen, T., Groseth, A., & Feldmann, H. (2019). Therapeutic strategies to target the Ebola virus life cycle. *Nat Rev Microbiol*, 17, 593–606. https://doi.org/10.1038/s41579-019-0233-2.

Huang, J., Ma, Q., Su, Z., & Cheng, X. (2024 October 9). Advancements in the development of anti-SARS-CoV-2 therapeutics. *Int J Mol Sci*, 25(19), 10820. https://doi.org/10.3390/ijms251910820. PMID: 39409149; PMCID: PMC11477007.

Idrees, S., Chen, H., Panth, N., Paudel, K. R., & Hansbro, P. M. (2024). Exploring viral-host protein interactions as antiviral therapies: A computational perspective. *Microorganisms*, 12(3), 630. https://doi.org/10.3390/microorganisms12030630.

Innovative Antiviral Defense with New CRISPR Tool. (2024, April 24). ScienceDaily. https://www.sciencedaily.com/releases/2024/04/240412113401.htm

Jones, J. E., Le Sage, V., & Lakdawala, S. S. (2021, April). Viral and host heterogeneity and their effects on the viral life cycle. *Nat Rev Microbiol*, 19(4), 272–282. https://doi.org/10.1038/s41579-020-00449-9. Epub 2020 Oct 6. PMID: 33024309; PMCID: PMC7537587.

Kálai, T., Pongrácz, J. E., & Mátyus, P. (2022). Recent advances in influenza, HIV and SARS-CoV-2 infection prevention and drug treatment—the need for precision medicine. *Chemistry*, 4(2), 216–258. https://doi.org/10.3390/chemistry4020019.

Kamzeeva, P. N., Aralov, A. V., Alferova, V. A., & Korshun, V. A. (2023). Recent advances in molecular mechanisms of nucleoside antivirals. *Current Issues in Molecular Biology*, 45(8), 6851–6879. https://doi.org/10.3390/cimb45080433.

Karim, M., Lo, C. W., & Einav, S. (2023, June 1). Preparing for the next viral threat with broad-spectrum antivirals. *J Clin Invest*, 133(11), e170236. https://doi.org/10.1172/JCI170236. PMID: 37259914; PMCID: PMC10232003.

Khan, U., et al. (2024). Safety and efficacy of camostat mesylate for Covid-19: A systematic review and meta-analysis of randomized controlled trials. *BMC Infect Dis*, 24, 709. https://doi.org/10.1186/s12879-024-09468-w.

Kinoshita, T., et al. (2022). A multicenter, double-blind, randomized, parallel-group, placebo-controlled study to evaluate the efficacy and safety of camostat mesilate in patients with COVID-19 (CANDLE study). *BMC Med*, 20, 342. https://doi.org/10.1186/s12916-022-02518-7.

Kolesnikova, L., Bugany, H., Klenk, H. D., & Becker, S. (2002, February). VP40, the matrix protein of Marburg virus, is associated with membranes of the late endosomal compartment. *Journal of Virology*, 76(4), 1825–1838. https://doi.org/10.1128/jvi.76.4.1825-1838.2002. PMID: 11799178; PMCID: PMC135914.

Kow, C. S., & Hasan, S. S. (2021a, April). Interferon-based therapies in COVID–19. *Int Immunopharmacol*, 93, 107415. https://doi.org/10.1016/j.intimp.2021.107415. Epub 2021 Jan 28. PMID: 33540249; PMCID: PMC7843103.

Kow, C. S., & Hasan, S. S. (2021b). The risk of mortality in patients with COVID-19 with pre-diagnosis use of NSAIDs: A meta-analysis. *Inflammopharmacol*, 29, 641–644. https://doi.org/10.1007/s10787-021-00810-1.

Leavitt, J. W. (2021). Pandemics and history: Context, context, context. *Am J Public Health*, 111(6), 996–998. https://doi.org/10.2105/AJPH.2021.306270. PMID: 33950732; PMCID: PMC8101584.

Lee, C. (2019). CRISPR/Cas9-based antiviral strategy: Current status and the potential challenge. *Molecules*, 24(7), 1349. https://doi.org/10.3390/molecules24071349.

Leon-Rojas JE, Arias-Erazo F, Jiménez-Arias P, Recalde-Navarrete R, Guevara A, Coloma J, et al. (2024) COVID-19 IgG seropositivity and its determinants in occupational groups of varying infection risks in two Andean cities of Ecuador before mass vaccination. *PLoS ONE*, 19(8): e0309466. https://doi.org/10.1371/journal.pone.0309466

Li, G., Yue, T., Zhang, P., Gu, W., Gao, L.-J., & Tan, L. (2021). Drug discovery of nucleos(t)ide antiviral agents: Dedicated to Prof. Dr. Erik De Clercq on occasion of his 80th birthday. *Molecules*, 26(4), 923. https://doi.org/10.3390/molecules26040923.

Lin, J. C., Cherng, J. M., Hung, M. S., Baltina, L. A., Baltina, L., & Kondratenko, R. (2008) Inhibitory effects of some derivatives of glycyrrhizic acid against Epstein-Barr virus infection: Structure-activity relationships. *Antiviral Res.*, 79(1), 6-11. https://doi.org/10.1016/j.antiviral.2008.01.160. Epub 2008 Mar 31. PMID: 18423902; PMCID: PMC7114209.

Lopinavir/Ritonavir: A Rapid Review of Effectiveness in COVID-19 | The Centre for Evidence-Based Medicine. (n.d.). https://www.cebm.net/covid-19/lopinavir-ritonavir-a-rapid-review-of-the-evidence-for-effectiveness-in-treating-covid/.

Low, Z. Y., Yip, A. J. W., & Lal, S. K. (2022). Repositioning anticancer drugs as novel COVID-19 antivirals: Targeting structural and functional similarities between viral proteins and cancer. *Expert Reviews in Molecular Medicine*, 24, e20. https://doi.org/10.1017/erm.2022.11.

Magro, P., Zanella, I., Pescarolo, M., Castelli, F., & Quiros-Roldan, E. (2021, March). Lopinavir/ritonavir: Repurposing an old drug for HIV infection in COVID-19 treatment. *Biomed J*, 44(1), 43–53. https://doi.org/10.1016/j.bj.2020.11.005. Epub 2020 Nov 10. PMID: 33608241; PMCID: PMC7654234.

Maida, Y., Yasukawa, M., & Masutomi, K. (2016). De novo RNA synthesis by RNA-dependent RNA polymerase activity of telomerase reverse transcriptase. *Molecular and Cellular Biology*, 36(8), 1248–1259. https://doi.org/10.1128/MCB.01021-15.

Market Shaping Accelerator at UChicago. (2024, September 25). *Transforming Pandemic Preparedness with Platform-Based Broad-Spectrum Antivirals—Market Shaping Accelerator at UChicago.* https://marketshaping.uchicago.edu/news/transforming-pandemic-preparedness-with-platform-based-broad-spectrum-antivirals/.

Martinez, M. A. (2022). What should be learned from repurposed antivirals against SARS-COV-2? *Frontiers in Microbiology*, 13. https://doi.org/10.3389/fmicb.2022.843587.

Muthukutty, P., MacDonald, J., & Yoo, S. Y. (2024). Combating emerging respiratory viruses: Lessons and future antiviral strategies. *Vaccines*, 12(11), 1220. https://doi.org/10.3390/vaccines12111220.

New Worm Study Paves Way for Better RNA-Based Drugs to Treat Human Disease. (2024, August 24). ScienceDaily. https://www.sciencedaily.com/releases/2024/08/240820124520.htm.

Otsubo, R., & Yasui, T. (2022, December). Monoclonal antibody therapeutics for infectious diseases: Beyond normal human immunoglobulin. *Pharmacol Ther*, 240, 108233. https://doi.org/10.1016/j.pharmthera.2022.108233. Epub 2022 Jun 20. PMID: 35738431; PMCID: PMC9212443.

Palazzotti, D., Sguilla, M., Manfroni, G., Cecchetti, V., Astolfi, A., & Barreca, M. L. (2024, May 20). Small molecule drugs targeting viral polymerases. *Pharmaceuticals (Basel, Switzerland)*, 17(5), 661. https://doi.org/10.3390/ph17050661. PMID: 38794231; PMCID: PMC11124969.

Panda, K., Alagarasu, K., Tagore, R., Paingankar, M., Kumar, S., Jeengar, M. K., Cherian, S., & Parashar, D. (2024). RNAi-induced gene silencing against Chikungunya and COVID-19: What have we learned so far, and what is the way forward? *Viruses*, 16(9), 1489. https://doi.org/10.3390/v16091489.

Pantaleo, G., Correia, B., Fenwick, C., Joo, V. S., & Perez, L. (2022). Antibodies to combat viral infections: Development strategies and progress. *Nature Reviews. Drug Discovery*, 21(9), 676–696. https://doi.org/10.1038/s41573-022-00495-3.

Papaneophytou, C. (2024). Breaking the chain: Protease inhibitors as game changers in respiratory viruses management. *International Journal of Molecular Sciences*, 25(15), 8105. https://doi.org/10.3390/ijms25158105.

Peng, S., Wang, H., Wang, Z., & Wang, Q. (2022). Progression of antiviral agents targeting viral polymerases. *Molecules (Basel, Switzerland)*, 27(21), 7370. https://doi.org/10.3390/molecules27217370.

Pereira, M., & Vale, N. (2024). Ritonavir's evolving role: A journey from antiretroviral therapy to broader medical applications. *Current Oncology*, 31(10), 6032–6049. https://doi.org/10.3390/curroncol31100450.

Pfizer Initiates Phase 1 Study of Novel Oral Antiviral Therapeutic Agent Against SARS-COV-2 | Pfizer. (n.d.). https://www.pfizer.com/news/press-release/press-release-detail/pfizer-initiates-phase-1-study-novel-oral-antiviral.

Pfizer Starts Global Phase 2/3 EPIC-PEP Study of Novel COVID-19 Oral Antiviral Candidate for Post-Exposure Prophylaxis in Adults | Pfizer. (n.d.). https://www.pfizer.com/news/press-release/press-release-detail/pfizer-starts-global-phase-23-epic-pep-study-novel-covid-19.

Pincetic, A., & Leis, J. (2009). The mechanism of budding of retroviruses from cell membranes. *Advances in Virology*, 2009, 6239691–6239699. https://doi.org/10.1155/2009/623969.

Pinto Pereira, S. M., Nugawela, M. D., Stephenson, T., Foret-Bruno, P., Dalrymple, E., Xu, L., Whittaker, E., Heyman, I., Ford, T., Segal, T., Chalder, T., Ladhani, S. N., Mensah, A. A., McOwat, K., Simmons, R., CLoCk Consortium, & Shafran, R. (2024). Post-Covid-19 condition (Long Covid) in children and young people 12 months after infection or reinfection with the Omicron variant: A prospective observational study. *Sci Rep*, 14(1), 9957. https://doi.org/10.1038/s41598-024-60372-4. PMID: 38693285; PMCID: PMC11063180.

Pleschka, S. (2024). Special issue "host targeted therapeutics against virus infections". *Viruses*, 16(12), 1825. https://doi.org/10.3390/v16121825.

Puri, V., Nagpal, M., Singh, I., Singh, M., Dhingra, G. A., Huanbutta, K., Dheer, D., Sharma, A., & Sangnim, T. (2022). A comprehensive review on nutraceuticals: Therapy support and formulation challenges. *Nutrients*, 14(21), 4637. https://doi.org/10.3390/nu14214637.

Salimi-Jeda, A., Abbassi, S., Mousavizadeh, A., Esghaie, M., Bokharaei-Salim, F., Jeddi, F., Shafaati, M., & Abdoli, A. (2021, December). SARS-CoV-2: Current trends in emerging variants, pathogenesis, immune responses, potential therapeutic, and vaccine development strategies. *Int Immunopharmacol*, 101(Pt A), 108232. https://doi.org/10.1016/j.intimp.2021.108232. Epub 2021 Oct 16. PMID: 34673335; PMCID: PMC8519814.

Salvatore, D. (2024). Ethical and regulatory considerations in the development of advanced biomedical technologies. *J Biomed Syst Emerg Technol*, 11, 202.

Sarwar, S. B., Khondokar, F., Islam, H., Ullah, M., Araf, Y., Sarkar, B., & Rahman, M. (2021). *Assessing Drug Repurposing Option for Emerging Viral Diseases: Concerns, Solutions, and Challenges for Forthcoming Viral Battles*. JABET. https://www.bsmiab.org/jabet/178-1603396401-assessing-drug-repurposing-option-for-emerging-viral-diseases-concerns-solutions-and-challenges-for-forthcoming-viral-battles.

ScientistsIdentify Antivirals that Could Combat Emerging Infectious Diseases. (2023, April 23). ScienceDaily. https://www.sciencedaily.com/releases/2023/04/230428130736.htm.

Song, M. S., Lee, D.-K., Lee, C.-Y., Park, S.-C., & Yang, J. (2024). Host subcellular organelles: Targets of viral manipulation. *International Journal of Molecular Sciences*, 25(3), 1638. https://doi.org/10.3390/ijms25031638.

Srivastava, S., Kumar, S., Ashique, S., Sridhar, S. B., Shareef, J., & Thomas, S. (2024). Novel antiviral approaches for Marburg: A promising therapeutics in the pipeline. *Frontiers in Microbiology*, 15, 1387628. https://doi.org/10.3389/fmicb.2024.1387628.

Strait, J. E. (2022, November 1). *Immune Modulator Drugs for COVID-19 Focus of Major NIH Clinical Trial | WashU Medicine*. WashU Medicine. https://medicine.washu.edu/news/immune-modulator-drugs-for-covid-19-focus-of-major-nih-clinical-trial/.

Strzelec, M., Detka, J., Mieszczak, P., Sobocińska, M. K., & Majka, M. (2023). Immunomodulation—a general review of the current state-of-the-art and new therapeutic strategies for targeting the immune system. *Frontiers in Immunology*, 14. https://doi.org/10.3389/fimmu.2023.1127704.

Tai, L. T., Yeh, C. Y., Chang, Y. J., Liu, J. F., Hsu, K. C., Cheng, J. C., & Lu, C. H. (2024). Discovery of novel spike inhibitors against SARS-CoV-2 infection. *International Journal of Molecular Sciences*, 25(11), 6105. https://doi.org/10.3390/ijms25116105.

Tan, J. S., Jaffar Ali, M. N. B., Gan, B. K., & Tan, W. S. (2023). Next-generation viral nanoparticles for targeted delivery of therapeutics: Fundamentals, methods, biomedical applications, and challenges. *Expert Opinion on Drug Delivery*, 20(7), 955–978. https://doi.org/10.1080/17425247.2023.2228202.

Tarn, W.-Y., Cheng, Y., Ko, S.-H., & Huang, L.-M. (2021). Antisense oligonucleotide-based therapy of viral infections. *Pharmaceutics*, 13(12), 2015. https://doi.org/10.3390/pharmaceutics13122015.

Technology Networks. (2024, June 4). *Investigational Antiviral Progressing Toward Preparing for Future Pandemics*. Drug Discovery From Technology Networks. https://www.technologynetworks.com/drug-discovery/news/investigational-antiviral-progressing-toward-preparing-for-future-pandemics-387385.

Teng, M., Yao, Y., Nair, V., & Luo, J. (2021). Latest advances of virology research using CRISPR/Cas9-based gene-editing technology and its application to vaccine development. *Viruses*, 13(5), 779. https://doi.org/10.3390/v13050779. PMID: 33924851; PMCID: PMC8146441.

Thom, R. E., & D'Elia, R. V. (2024). Future applications of host direct therapies for infectious disease treatment. *Frontiers in Immunology*, 15. https://doi.org/10.3389/fimmu.2024.1436557.

Tomasello, E., Pollet, E., Vu Manh, T. P., Uzé, G., & Dalod, M. (2014, October 30). Harnessing mechanistic knowledge on beneficial versus deleterious IFN-I effects to design innovative immunotherapies targeting cytokine activity to specific cell types. *Front Immunol*, 5, 526. https://doi.org/10.3389/fimmu.2014.00526. PMID: 25400632; PMCID: PMC4214202.

U.S. Food and Drug Administration. FDA Roundup. (2024, March 22). *FDA*. https://www.fda.gov/news-events/press-announcements/fda-roundup-march-22-2024. Epub 2024 Mar 22.

Votteler, J., & Sundquist, W. I. (2013, September 11). Virus budding and the ESCRT pathway. *Cell Host Microbe*, 14(3), 232–241. https://doi.org/10.1016/j.chom.2013.08.012. PMID: 24034610; PMCID: PMC3819203.

Wijesinghe, K. J., & Stahelin, R. V. (2015). Investigation of the lipid binding properties of the Marburg virus matrix protein VP40. *Journal of Virology*, 90(6), 3074–3085. https://doi.org/10.1128/JVI.02607-15.

Writer, S. (2024, November 22). *The Next Pandemic: Not if, But When*. Harvard T. H. Chan School of Public Health. https://hsph.harvard.edu/news/next-pandemic-not-if-but-when/.

9 Treatment Strategies for MERS, SARS, and COVID-19

Kiran Bala, Navneet Kaur, Arushi Dogra, Sakshi Verma, and Sangeeta Jaiswal

9.1 INTRODUCTION

Coronaviruses are members of the enclosed, single-stranded, positive-sense RNA virus family, Coronaviridae. The word "Corona" (meaning "crown") refers to the spike-like glycoproteins that have a morphology resembling a crown and are collectively responsible for a variety of respiratory diseases that affect not only humans but also other animals like bats and camels. SARS-CoV first appeared in November 2003 and caused the short-term respiratory syndrome in Asian countries like China and was lethal than earlier strains. It was a zoonotic virus (animals to humans) cattle, as depicted in Figure 9.1. The latest SARS-CoV-2 (the novel coronavirus) emerged in 2019 in Wuhan, China, which led to an outbreak of COVID-19 disease globally. The emergence of highly pathogenic human coronaviruses, severe acute respiratory syndrome coronavirus (SARS-CoV), Middle East respiratory syndrome coronavirus (MERS-CoV), and SARS-CoV-2 has significantly impacted global health. These viruses share genomic similarities but exhibit unique epidemiological, clinical, and immunopathological characteristics. SARS-CoV emerged in 2002, MERS-CoV in 2012, and SARS-CoV-2 in 2019, leading to outbreaks of varying scales, with COVID-19 being a pandemic of unprecedented magnitude. The zoonotic origins of these viruses highlight the intricate relationship between humans, wildlife, and environmental changes (Kumari et al., 2022; Zhu et al., 2020; Wiersinga et al., 2020; Ghareeb et al., 2021). While the morbidity and mortality rates vary across SARS, MERS, and COVID-19, all three illnesses present severe respiratory and systemic complications in critical cases. The pathogenesis is largely driven by the interaction between the viral spike protein and host cell receptors, leading to complex immune responses that contribute to disease severity. Understanding these mechanisms has informed therapeutic strategies aimed at mitigating disease progression and mortality (Kumari et al., 2022; Zhu et al., 2020; Wiersinga et al., 2020; Ghareeb et al., 2021; Goswami et al., 2022). Pharmacological interventions for SARS-CoV, MERS-CoV, and SARS-CoV-2 have evolved from repurposing existing drugs to developing novel therapeutics, including antiviral agents, immunomodulators, and neutralizing antibodies. Drug-repurposing strategies identified candidates such as remdesivir and favipiravir, while targeted therapies against the viral spike protein have paved the way for monoclonal antibody treatments. For COVID-19, immunotherapies like tocilizumab and baricitinib have addressed hyperinflammatory states, while vaccines have played a critical role in pandemic control (Kumari et al., 2022; Zhu et al., 2020; Wiersinga et al., 2020; Ghareeb et al., 2021; Goswami et al., 2022). This chapter explores the pharmacological strategies, investigational drugs, and immunotherapies developed for SARS, MERS, and COVID-19. It examines the progression from initial outbreak responses to the development of evidence-based treatment protocols and novel therapeutics. By analyzing the lessons learned from these viral outbreaks, this chapter aims to provide a comprehensive understanding of the current state of coronavirus management and future therapeutic directions.

DOI: 10.1201/9781032721811-9

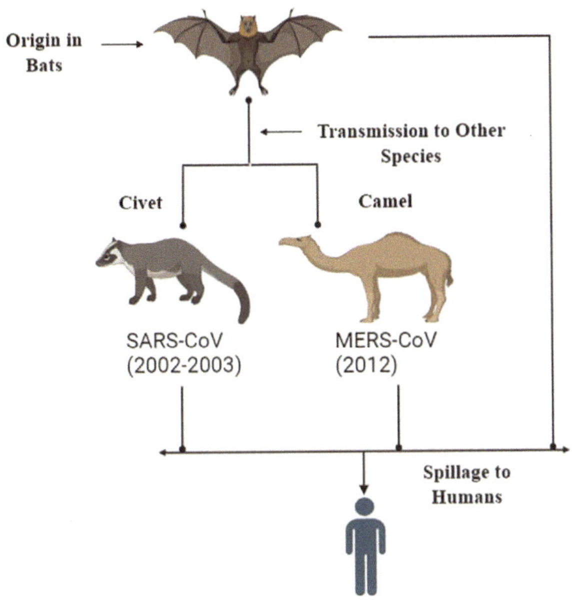

FIGURE 9.1 Flowchart representing zoonotic transmission of coronaviruses. Source: biorender.com.

9.2 SARS-CoV, MERS-CoV, AND SARS-CoV-2: THE FUNDAMENTALS

9.2.1 Severe Acute Respiratory Syndrome Coronavirus (SARS-CoV)

SARS-CoV, the causative agent of severe acute respiratory syndrome (SARS), is an enveloped, positive-sense single-stranded RNA virus from the Coronaviridae family. It belongs to the genus *Betacoronavirus*. The virus' genome encodes structural proteins such as the spike (S), membrane (M), envelope (E), and nucleocapsid (N) proteins, which play vital roles in its infectivity and immune response elicitation. The spike protein facilitates viral entry into host cells by binding to the angiotensin-converting enzyme 2 (ACE2) receptor, a process that involves significant structural changes and priming by host proteases like TMPRSS2. SARS-CoV primarily targets epithelial cells in the respiratory tract, leading to clinical manifestations ranging from mild respiratory symptoms to severe pneumonia and acute respiratory distress syndrome (ARDS). Its zoonotic origins are linked to bats as the natural reservoir, with intermediate hosts such as civets implicated in the spillover to humans. Molecular studies suggest that recombination events among bat coronaviruses likely contributed to the virus' emergence, enhancing its ability to infect human cells. The viral main protease (Mpro), essential for processing polyproteins into functional units, is a major target for antiviral drug development. Structural studies of Mpro have provided insights into its potential as a therapeutic target, aiding in the design of inhibitors aimed at halting viral replication. Understanding the biology of SARS-CoV has significantly informed research on related coronaviruses, including SARS-CoV-2, due to their genetic and structural similarities (Gao et al., 2023; Chapter 1: Understanding the Pathogen, 2023; Goswami et al., 2022).

The pathophysiology of SARS-CoV, the virus responsible for the severe acute respiratory syndrome (SARS) outbreak in 2002–2003, revolves around its interaction with the angiotensin-converting enzyme 2 (ACE2) receptor, which is abundant in the epithelial cells of the respiratory tract, heart, kidneys, and gastrointestinal tract. SARS-CoV primarily infects alveolar epithelial cells in the lungs, leading to diffuse alveolar damage, extensive inflammation, and impaired gas exchange. This viral replication triggers the release of pro-inflammatory cytokines, including IL-6, IL-8, and TNF-α, resulting in a cytokine storm, a hallmark of severe cases. This hyperinflammatory response can cause acute respiratory distress syndrome (ARDS), characterized by severe hypoxia and lung tissue destruction. The virus also exploits its structural proteins, particularly the spike (S) protein, to efficiently bind and enter host cells via endocytosis. Viral replication within host cells and the immune response contribute to endothelial damage and vascular leakage. This process not only exacerbates lung injury but may also lead to multi-organ dysfunction in severe cases. Infected individuals often display lymphopenia due to T-cell apoptosis, weakening the immune system's ability to control the infection. The systemic effects of this dysregulated immune response can include myocarditis, acute kidney injury, and hepatic dysfunction. Secondary bacterial infections further complicate the clinical progression of SARS-CoV, especially in individuals with preexisting health conditions (Zhu et al., 2023; Goswami et al., 2022).

9.2.2 Middle East Respiratory Syndrome Coronavirus (MERS-CoV)

The Middle East Respiratory Syndrome Coronavirus (MERS-CoV) is a zoonotic pathogen belonging to the genus *Betacoronavirus*. It was first identified in 2012 in Saudi Arabia. The virus has a single-stranded, positive-sense RNA genome of approximately 30.1 kb, the largest among its related coronaviruses, such as SARS-CoV and SARS-CoV-2. Its genome structure includes a 5′-end encoding non-structural proteins necessary for replication and a 3′-end encoding structural proteins like the spike (S), envelope (E), membrane (M), and nucleocapsid (N) proteins, essential for virus assembly and host interaction. MERS-CoV primarily infects humans through zoonotic transmission, with dromedary camels being the primary reservoir. The virus uses the dipeptidyl peptidase 4 (DPP4) receptor to enter host cells, a process mediated by its spike protein. The spike protein's receptor binding domain (RBD) is a critical determinant of host tropism and viral infectivity. Once inside the host, the virus replicates and assembles within the host cell's cytoplasm, leading to cell lysis and subsequent inflammation. Phylogenetic studies suggest that MERS-CoV originated from bats before adapting to camels as an intermediate host. It has a significant zoonotic potential, with periodic spillovers into human populations, particularly in regions where camel farming and human interactions are common (Figure 9.1). Despite its lower transmissibility compared to SARS-CoV-2, MERS-CoV poses a high fatality risk, with case fatality rates exceeding 35% in confirmed cases. Efforts to control MERS-CoV have included developing vaccines targeting the spike protein and enhanced surveillance in high-risk regions (Goswami et al., 2022; World Health Organization: WHO, 2019; Rabaan et al., 2021).

The pathophysiology of MERS-CoV (Middle East respiratory syndrome coronavirus) centers on its ability to infect human cells by binding to the dipeptidyl peptidase-4 (DPP4) receptor, primarily found on respiratory epithelial cells and certain immune cells. This interaction facilitates the viral entry and subsequent replication, leading to cellular damage and an inflammatory response. The infection predominantly affects the lower respiratory tract, often resulting in severe pneumonia and acute respiratory distress syndrome (ARDS). The disease progression involves a combination of direct viral cytopathic effects and an exaggerated immune response, including elevated levels of pro-inflammatory cytokines like IL-6 and TNF-α. This "cytokine storm" contributes to widespread tissue damage and multi-organ failure, particularly in patients with comorbid conditions such as diabetes, chronic kidney disease, or immunosuppression. Histopathological findings in MERS patients have revealed extensive lung injury, including alveolar damage and infiltration of mononuclear cells. In some cases, the infection extends beyond the respiratory system, affecting the kidneys, gastrointestinal tract, and cardiovascular system, potentially leading to complications such as acute kidney injury and septic shock. Mortality rates are influenced by patient age, comorbidities, and the timeliness of medical intervention, with a case fatality rate of approximately 35% (Zhang et al., 2021; World Health Organization: WHO, 2019; Rabaan et al., 2021).

9.2.3 Severe Acute Respiratory Syndrome Coronavirus 2 (SARS-CoV-2)

The basic biology of SARS-CoV-2, the virus responsible for COVID-19, highlights its complex structure, genome, and mechanisms of host interaction. SARS-CoV-2 is a single-stranded, positive-sense RNA virus belonging to the genus *Betacoronavirus* in the family Coronaviridae. Its genome, approximately 29.9 kb in length, encodes structural proteins, including the spike (S), envelope (E), membrane (M), and nucleocapsid (N) proteins, as well as non-structural and accessory proteins critical for replication and host interaction. The spike protein plays a pivotal role in the virus' pathogenicity by facilitating entry into host cells (Goswami et al., 2022; Le et al., 2023). It binds to the angiotensin-converting enzyme 2 (ACE2) receptor on host cells, primarily in the respiratory epithelium, with high affinity. Following receptor engagement, the spike protein undergoes proteolytic cleavage at the S1/S2 site, a process mediated by host proteases such as furin, enabling viral entry via membrane fusion or endocytosis. The presence of a polybasic cleavage site in SARS-CoV-2 distinguishes it from SARS-CoV and enhances its transmissibility and host range. The genome of SARS-CoV-2 is encapsulated by the N protein and surrounded by a lipid bilayer derived from the host cell membrane. The high A+U content of its genome and the stability conferred by structural elements such as the 5'-cap and 3'-poly-A tail facilitate efficient replication and translation (Goswami et al., 2022; Hardenbrook, 2022; Le et al., 2023). These adaptations, combined with mechanisms to evade host immune responses, such as inhibition of interferon signaling, contribute to its virulence and ability to cause widespread infection. SARS-CoV-2's genetic similarity to bat coronaviruses, particularly RaTG13, and its zoonotic origin underscore its evolutionary adaptability, which has resulted in various mutations and variants that affect transmissibility, immune evasion, and disease severity (Goswami et al., 2022; Le et al., 2023).

The pathophysiology of SARS-CoV-2, the causative agent of COVID-19, involves complex interactions between the virus, host immune responses, and various tissues. Following inhalation, SARS-CoV-2 primarily infects the respiratory epithelium via the ACE2 receptor, which is abundantly expressed in the lungs, heart, kidneys, and gastrointestinal tract. Viral replication within alveolar epithelial cells leads to direct cytopathic effects, resulting in tissue damage and triggering the host's innate immune response. Early in the infection, interferons and cytokines, including IL-6 and TNF-α, are released to combat the virus. However, an unregulated or excessive inflammatory response can result in severe pathology. The disease progression in severe cases is characterized by a "cytokine storm," where an overwhelming release of pro-inflammatory cytokines leads to widespread tissue damage and immune dysregulation. This state promotes endothelial dysfunction, vascular leakage, and the formation of microthrombi, which can culminate in acute respiratory distress syndrome (ARDS) and multi-organ failure. Neutrophil extracellular traps (NETs) have also been implicated in exacerbating tissue injury and thromboinflammation in severe COVID-19 cases. Additionally, SARS-CoV-2 downregulates ACE2 expression, resulting in increased angiotensin II levels, which exacerbates inflammation, vascular permeability, and fibrosis. These effects contribute to impaired oxygen exchange in the lungs and systemic complications such as cardiovascular stress and renal injury. Dysregulated adaptive immunity, including T-cell exhaustion and decreased regulatory T-cell populations, further complicates disease progression, impairing viral clearance and perpetuating inflammation. The systemic nature of SARS-CoV-2 infection explains its wide-ranging effects, including neurological symptoms, gastrointestinal disturbances, and long-term sequelae. This multifaceted pathophysiology underpins the heterogeneity of clinical manifestations and outcomes seen in COVID-19 patients (Goswami et al., 2022; Zhu et al., 2023; Mohandas et al., 2023).

9.3 CONVENTIONAL DRUG DESIGNING STRATEGIES FOR SARS, MERS, AND COVID-19

9.3.1 Designing Interventions Based on Receptor Binding Process

The conventional computer-aided drug designing against SARS/MERS viruses focuses on interfering with the S protein whose RBD (receptor binding domain) binds with the human ACE2 to bring about the viral infection. One of such approaches is the *in silico* pharmacophore modeling which involves construction of a 3D arrangement of a functional molecule to study its interactions just as they occur in an *in vivo* system (Shehroz et al., 2020). Another one is the molecular docking technique of drug design (Wu et al., 2020). Another approach could be repurposing existing drugs such that they become equipped to block the RBD-ACE2 interactions. Other ways to inhibit the mentioned binding is the inhibition of host proteases, by certain targeted drugs, which are responsible for cleavage and thereby activation of the S protein and also drugs which can promote/initiate S protein proteolysis by the host machinery (Goswami et al., 2022).

9.3.2 Designing Interventions Based on the Uncoating, Replication, and Translation of Viral Genetic Material

Another target for the drugs against these viruses are the components which are a part of the replication and translation machinery of the virus. One of the approaches targeting the viral replication is the 3CLpro as the target which is responsible for generation of 13 mature non-structural proteins playing an important role in viral replication (Zhang et al., 2022). Another target for such drugs could be PLpro which causes release of NSP1–3 post proteolytic cleavage, thereby inhibiting replication. The RNA-dependent RNA polymerase which plays a crucial part in both the viral replication and translation can also be targeted. Last but not least, effective approach would be to target some of the host proteins which aide the virus in its life cycle post–host hijacking (Goswami et al., 2022).

9.3.3 Designing Interventions Based on the Synthesis of Structural Proteins and Assembly of Viral Particles

One of the main structural proteins in SARS/MERS viruses is the spike protein and therefore could be an excellent target for the drugs. Some of the compounds which do have anti-S protein activity are some antihypertensive drugs, antifungal drugs, antibacterial drugs, and anticoagulant drugs. In addition to these, some natural flavonoids can also inhibit S protein. E protein/N proteins are other structural proteins which can also be a potent target. Many antibacterial, antitumor, antiasthmatic, and anti-inflammatory drugs have been found to show inhibition activity against these. Repurposing of drugs has proven to be a very effective strategy against SARS family. An example of this is use of emetine, an anti-amoebiasis drug which is capable of inhibiting viral protein synthesis by inhibiting activity of 40S ribosomal protein. including the synthesis of structural proteins (Shyr et al., 2020; Goswami et al., 2022).

9.4 STRATEGIES FOR INHIBITING THE RECEPTOR BINDING PROCESS

9.4.1 Strategies for Inhibiting the Receptor Binding Process in MERS-CoV

The receptor binding process in MERS-CoV is primarily mediated through the spike (S) protein, which interacts with the dipeptidyl peptidase-4 (DPP4) receptor on host cells. Drugs targeting this interaction aim to block the virus' entry into cells, preventing infection. Several strategies and agents have been developed to inhibit this critical step:

- *Monoclonal antibodies:* Neutralizing antibodies that specifically bind to the receptor binding domain (RBD) of the MERS-CoV S protein have shown promise. These antibodies can prevent the virus from attaching to the DPP4 receptor. For instance, monoclonal antibodies such as REGN3051 and REGN3048 have demonstrated effective neutralization by targeting unique epitopes on the RBD of the S protein, inhibiting viral entry into host cells (Rabaan et al., 2021; Ma et al., 2024; Labiod et al., 2023).
- *Nanobodies:* Nanobodies derived from camelids, like alpacas, offer an innovative approach. For example, nanobodies Nb9 and Nb14 have been shown to bind strongly to the

MERS-CoV RBD, blocking its interaction with the DPP4 receptor. These nanobodies have distinct binding affinities, and their combination can lead to enhanced neutralizing effects (Rabaan et al., 2021; Ma et al., 2024; Labiod et al., 2023).

- *Receptor decoys:* Decoy molecules mimicking the DPP4 receptor can competitively inhibit the binding of MERS-CoV to the actual cellular receptor. Recombinant forms of DPP4 or DPP4-immunoglobulin fusion proteins have been tested as potential therapeutics by acting as bait to trap the virus (Rabaan et al., 2021; Ma et al., 2024; Labiod et al., 2023).
- *Polysaccharide-based inhibitors:* Mannan, a polysaccharide derived from yeast, can inhibit C-type lectin receptors like DC-SIGN, which MERS-CoV utilizes for attachment and entry into certain cell types. Mannan competes with the virus for binding sites, acting as a decoy (Rabaan et al., 2021; Ma et al., 2024; Labiod et al., 2023).
- *Small-molecule inhibitors:* Small molecules that interfere with the spike protein-DPP4 interaction are under exploration. These compounds aim to block the conformational changes in the S protein necessary for receptor binding and subsequent membrane fusion (Rabaan et al., 2021; Ma et al., 2024; Labiod et al., 2023).

These receptor binding inhibitors represent a critical component of the therapeutic arsenal against MERS-CoV, providing a targeted approach to mitigate viral entry and limit disease progression (Rabaan et al., 2021; Ma et al., 2024; Labiod et al., 2023).

9.4.2 Strategies for Inhibiting the Receptor Binding Process in SARS-CoV

In SARS-CoV, the receptor binding process involves the spike (S) protein binding to the angiotensin-converting enzyme 2 (ACE2) receptor on host cells. Targeting this interaction is a critical strategy to prevent viral entry. Several drug classes and approaches have shown promise:

- *Monoclonal antibodies:* Antibodies targeting the receptor binding domain (RBD) of the SARS-CoV S protein effectively block ACE2 interaction. These antibodies are specifically engineered to neutralize viral entry into cells (Mohsen et al., 2023; Yin et al., 2021).
- *Small-molecule inhibitors:* Compounds designed to bind to the RBD or ACE2 can disrupt this interaction. Recent advances in *de novo* drug design have identified candidates that demonstrate high binding affinity and stability *in silico* and *in vitro* assays (Lopez et al., 2023; Mohsen et al., 2023; Yin et al., 2021).
- *Peptidomimetics:* These small peptides mimic the structure of ACE2, competitively inhibiting the S protein binding. They are highly specific and less likely to interfere with other cellular processes (Lopez et al., 2023; Mohsen et al., 2023; Yin et al., 2021).
- *Receptor decoys:* Soluble ACE2 proteins act as decoys, sequestering the virus and preventing its interaction with cellular ACE2 (Lopez et al., 2023; Mohsen et al., 2023; Yin et al., 2021).
- *Fusion inhibitors:* Drugs that prevent the conformational changes required for membrane fusion, such as heptad repeat peptides, are being explored for their efficacy in halting SARS-CoV entry (Yazdani et al., 2023; Yin et al., 2021).

These strategies, combined with advances in computational modeling and high-throughput screening, aim to provide a robust antiviral defense against SARS and related coronaviruses.

9.4.3 Drugs Inhibiting the Receptor Binding Process in SARS-CoV-2

The inhibition of receptor binding in COVID-19, particularly targeting the interaction between SARS-CoV-2 and its host receptor ACE2, has been a major focus in therapeutic drug development.

Drugs and molecules designed to block this interaction aim to prevent viral entry into host cells, thereby halting the replication cycle. Some of the important categories are as follows:

- *Monoclonal antibodies (mAbs):* Monoclonal antibodies such as bamlanivimab, etesevimab, and sotrovimab target the spike protein of SARS-CoV-2, specifically preventing its receptor binding domain (RBD) from engaging ACE2. These have shown efficacy in reducing hospitalization and severity when administered early in the disease course. However, the emergence of new variants with spike protein mutations has challenged their efficacy (Looi, 2023; Goswami et al., 2022; Backman, 2023).
- *ACE2 decoys and soluble ACE2:* Therapeutics like soluble recombinant ACE2 act as decoys to bind SARS-CoV-2, effectively sequestering the virus and reducing its ability to attach to actual cellular ACE2 receptors. These approaches have demonstrated promise in preclinical studies (Looi, 2023; Goswami et al., 2022; Backman, 2023).
- *Small molecules and peptides:* Small molecules and peptides are being explored to block the spike-ACE2 interaction. These compounds often work by binding to the ACE2 receptor or the spike protein, preventing their interaction. Examples include specific peptides derived from ACE2 that mimic its binding site, thereby competitively inhibiting viral entry (Looi, 2023; Goswami et al., 2022; Backman, 2023).
- *Broad-spectrum antivirals:* Drugs like camostat mesylate and nafamostat, originally developed as protease inhibitors, target the TMPRSS2 protease, which is essential for viral entry post-ACE2 binding. By inhibiting TMPRSS2, these drugs indirectly disrupt the receptor binding and subsequent viral entry processes (Looi, 2023; Goswami et al., 2022; Backman, 2023).
- *Innovative approaches:* Emerging therapies, including gene-editing tools and RNA-based inhibitors, aim to modify the expression of ACE2 or its affinity for SARS-CoV-2. These next-generation approaches hold potential for addressing the challenges posed by new variants (Looi, 2023; Goswami et al., 2022; Backman, 2023).

The ongoing refinement of these therapies emphasizes adaptability to overcome the viral evolution that challenges receptor binding inhibitors' efficacy. Such innovations aim to enhance accessibility and affordability for global use, particularly in areas where traditional treatments may face logistical or cost barriers.

9.5 STRATEGIES FOR INHIBITING THE REPLICATION OF VIRAL GENOME

9.5.1 Strategies for Inhibiting the Replication of Viral Genome in MERS-CoV

- *Remdesivir*, which happens to be a broad-spectrum antiviral drug and a nucleoside analog, has shown efficacy against the replication in MERS virus both in *in vitro* and *in vivo* studies. This drug is capable of both prophylactic and therapeutic activities (Goswami et al., 2022).
- *Ribavirin* is another nucleoside (guanosine) analog which inhibits the replication by interfering with the RNA polymerases. It can also inhibit the production of inosine monophosphate dehydrogenase, responsible for production of guanine required for replication.

9.5.2 Strategies for Inhibiting the Replication of Viral Genome in SARS-CoV

- *Berzosertib*, although not exclusively, is capable of inhibiting viral replication in SARS-CoV-1 (Garcia et al., 2021). This drug happens to be an ATR kinase inhibitor, an enzyme which plays an important role in sensing stressed DNA replication forks and destressing them.

- *Remdesivir* happens to be a broad-spectrum antiviral drug and a nucleoside analog has also shown efficacy against the replication in SARS-CoV in addition to the MERS virus of the Coronaviridae family.

9.5.3 Strategies for Inhibiting the Replication of Viral Genome in SARS-CoV-2

- *Favipiravir* is a prodrug and a broad-spectrum antiviral agent approved for use against influenzae virus. However, in its active form, favipiravir ribofuranosyl-triphosphate has been shown as an inhibitor of the RNA-dependent RNA polymerases in the COVID-19 virus.
- *Molnupiravir*, an orally available antiviral drug, is known to increase frequency of mutations in viral RNA and impair the replication in the process. This results when the viral RNA polymerase utilizes the active form of this drug-β-D-N4-hydroxycytidine (NHC) triphosphate as a substrate in place of cytidine triphosphate or uridine triphosphate. This ultimately leads to production of a mutated RNA (Kabinger et al., 2021).

9.6 STRATEGIES FOR INHIBITING THE VIRAL PROTEIN SYNTHESIS

9.6.1 Strategies for Inhibiting the Viral Protein Synthesis in MERS-CoV

- *Disulfiram* is an FDA-approved drug for alcohol aversion therapy. However, it has shown allosteric inhibition effect against papain-like proteases (PL^{pro}s) which are responsible for cleavage and therefore production of non-structural proteins in MERS virus (Lin et al., 2018).
- *Silvestrol*, which has shown broad antiviral activity, is an eIF4A inhibitor and inhibits expression of viral structural and non-structural proteins involved in infection (Liang et al., 2018).

9.6.2 Strategies for Inhibiting the Viral Protein Synthesis in SARS-CoV

- *Disulfiram* is an FDA-approved drug for alcohol aversion therapy. However, it has shown competitive inhibition effect against papain-like proteases (PL^{pro}s) in SARS-CoV (Lin et al., 2018) by the virtue of covalent inactivation of these proteases which are responsible for production of non-structural proteins.

9.6.3 Strategies for Inhibiting the Viral Protein Synthesis in SARS-CoV-2

- *Camostat mesylate*, a serine protease inhibitor, has been in use for treatment of pancreatitis and reflux disease in Japan. It can interfere with the TMPRSS2 gene activity resulting in inhibition of S protein priming and subsequent entry into the host cell (Su et al., 2021).
- *Plitidepsin*, which has been approved in Australia as a treatment against refractory multiple myeloma, can bind to the eEF1A and thereby brings about the inhibiting of viral protein translational elongation (Zhang et al., 2022).

9.7 IMMUNOMODULATORY AND ANTI-INFLAMMATORY AGENTS FOR MERS-CoV, SARS-CoV, AND SARS-CoV-2

- *IFNβ1b* is one of the more potent inhibitors of MERS-CoV replication as compared to other interferons. It does this by upregulation of the CD 73 in pulmonary endothelial cells resulting in secretion of anti-inflammatory adenosine and reduction in vascular leakage.

- *Systemic corticosteroids* (CS) can reduce the inflammation resulting from viral infection by inhibiting arachidonic acid, IL1, TNFα, and NF-κB. However, early studies showed that this drug also has adverse side effects on the body. Examples of such drugs are *thalidomide* and *methyl prednisolone*, which in combination with other antiviral drugs could treat completely.
- *Cyclosporine* is known to have interaction affinity to the N protein of SARS-CoV. It inhibits early phase of T-cell interaction and IL2–IL4, IL15, and INFγ production by inhibiting calcineurin.
- *Anakinra* happens to be an anti-inflammatory drug and primarily used as a therapeutic for rheumatoid arthritis (RA). Just like RA, overactivation of the immune system and the release of cytokines worsen the patient's situation in COVID-19 as well. IL-1, which is one such immune agent, can be regulated by this drug that functions by inhibiting the binding of IL-1 to its natural receptor.
- *Baricitinib* is a Janus kinase (JAK) 1/2 inhibitor drug also used for treatment of rheumatoid arthritis. By inhibiting this pathway, it brings about the reduction in anti-inflammatory response against the COVID-19 virus.

9.8 NON-ANTIVIRAL/ADJUNCTIVE AGENTS FOR MERS-COV, SARS-COV, SARS-COV-2

- Alternative therapeutics for MERS virus include cathepsin inhibitors, TMPRSS2 inhibitors, Furin inhibitors, kinase inhibitors and IFITM proteins (Liang et al., 2018). Other options include synthetic peptides such as the RVFV-6 peptide originally used against the Rift Valley fever virus and Kn2–7 which is the derivative from Scorpion venom and showed anti-HIV potential.
- Curcumin has shown broad-spectrum antiviral activity against enveloped viruses like COVID-19. It inhibits the infection by directly modifying the spike protein or the ACE2 receptor. Additionally, curcumin exerts immunomodulatory activity by blocking NF-κB, inflammasome, high-mobility group box 1 (HMGB1), and IL-6-driven inflammatory responses.
- Glycyrrhizic acid (GA), also known as glycyrrhizin, is a triterpene glycoside isolated from plants of *Glycyrrhiza* species (licorice). GA interacts with the spike protein and prevents its binding to the host receptor

9.9 CONCLUSION

Although a plethora of drugs, peptides, and compounds of other biochemical nature have been mentioned earlier, one cannot ignore the possibility that most of them may have adverse side effects which could be related to age or other factors. Every pathogen has been observed to exhibit a slightly different pathology in different individuals (infection may not be of the same degree of severity in each patient) and therefore there is no way to find out how an experimental drug will behave in a live model. One of the biggest challenges regarding drug design against this particular class of viruses is their rapidly changing structure. The number of variants and the speed at which COVID-19 generated them in such a short period of time poses a great threat to the existence of humanity. As virus like MERS do not have a very potent drug against them, it will continue to infect and cause loss of life unless throughput strategies are designed to block the virus in not only one aspect of its life cycle but multiple. All that being said, we must not forget that repurposing of the existing drugs can also be employed for the same and had shown promising results even in the case of highly contagious COVID-19.

REFERENCES

Backman, I. (2023, March 8). *New Class of Drugs Could Prevent Resistant COVID-19 Variants*. Yale School of Medicine. https://medicine.yale.edu/news-article/new-class-of-drugs-could-prevent-resistant-covid-19-variants/.

Chapter 1: Understanding the Pathogen. (2023, January 10). *GOV.UK*. https://www.gov.uk/government/publications/technical-report-on-the-covid-19-pandemic-in-the-uk/chapter-1-understanding-the-pathogen.

Gao, Y., et al. (2023). Structural biology of SARS-CoV-1/SARS-CoV-2 main protease. *Crystallography Reviews*, 29(2), 76–101. https://doi.org/10.1080/0889311X.2023.2222275.

Ghareeb, D. A., et al. (2021). Potential therapeutic and pharmacological strategies for SARS-CoV2. *Journal of Pharmaceutical Investigation*, 51, 281–296. https://doi.org/10.1007/s40005-021-00520-4.

Kabinger, F., Stiller, C., Schmitzová, J., Dienemann, C., Kokic, G., Hillen, H. S., Höbartner, C., & Cramer, P. (2021). Mechanism of molnupiravir induced SARS-CoV-2 mutagenesis. *Nature Structural & Molecular Biology*, 28, 740–746.

Kumari, M., et al. (2022). A critical overview of current progress for COVID-19: Development of vaccines, antiviral drugs, and therapeutic antibodies. *Journal of Biomedical Science*, 29(1), 68. https://doi.org/10.1186/s12929-022-00852-9.

Labiod, N., Luczkowiak, J., Tapia, M. M., Lasala, F., & Delgado, R. (2023). The role of DC-SIGN as a trans-receptor in infection by MERS-CoV. *Frontiers in Cellular and Infection Microbiology*, 13. https://doi.org/10.3389/fcimb.2023.1177270.

Le, K., Kannappan, S., Kim, T., Lee, J. H., Lee, H., & Kim, K. K. (2023). Structural understanding of SARS-CoV-2 virus entry to host cells. *Frontiers in Molecular Biosciences*, 10. https://doi.org/10.3389/fmolb.2023.1288686.

Liang, R., et al. (2018). Development of small-molecule MERS-CoV inhibitors. *Viruses*, 10(12), 721.

Lin, M. H., Moses, D. C., Hsieh, C. H., Cheng, S. C., Chen, Y. H., Sun, C. Y., & Chou, C. Y. (2018). Disulfiram can inhibit MERS and SARS coronavirus papain-like proteases via different modes. *Antiviral Research*, 150, 155–163.

Looi, M. (2023). What are the latest covid drugs and treatments? *BMJ*, 872. https://doi.org/10.1136/bmj.p872.

Ma, S., Zhang, D., Wang, Q., Zhu, L., Wu, X., Ye, S., & Wang, Y. (2024). Structure defining of ultrapotent neutralizing nanobodies against MERS-CoV with novel epitopes on receptor binding domain. *PLoS Pathogens*, 20(8), e1012438. https://doi.org/10.1371/journal.ppat.1012438.

Mohandas, S., et al. (2023). Immune mechanisms underlying COVID-19 pathology and post-acute sequelae of SARS-CoV-2 infection (PASC). *Elife*, 12, e86014. https://doi.org/10.7554/eLife.86014.

Rabaan, A. A., et al. (2021). MERS-CoV: Epidemiology, molecular dynamics, therapeutics, and future challenges. *Annals of Clinical Microbiology and* Antimicrobials, 20, 1–14.

Shehroz, M., Zaheer, T., & Hussain, T. (2020). Computer-aided drug design against spike glycoprotein of SARS-CoV-2 to aid COVID-19 treatment. *Heliyon*, 6(10).

Shyr, Z. A., Gorshkov, K., Chen, C. Z., & Zheng, W. (2020). Drug discovery strategies for SARS-CoV-2. *Journal of Pharmacology and Experimental Therapeutics*, 375(1), 127–138.

Su, H., Xu, Y., & Jiang, H. (2021). Drug discovery and development targeting the life cycle of SARS-CoV-2. *Fundamental Research*, 1(2), 151–165.

Wiersinga, W. J., Rhodes, A., Cheng, A. C., Peacock, S. J., & Prescott, H. C. (2020). Pathophysiology, transmission, diagnosis, and treatment of coronavirus disease 2019 (COVID-19): A review. *JAMA*, 324(8), 782–793. https://doi.org/10.1001/jama.2020.12839.

World Health Organization: WHO. (2019, November 8). *Middle East Respiratory Syndrome Coronavirus (MERS-CoV)*.https://www.who.int/health-topics/middle-east-respiratory-syndrome-coronavirus-mers#tab=tab_1.

Wu, D., Wu, T., Liu, Q., & Yang, Z. (2020). The SARS-CoV-2 outbreak: What we know. *International Journal of Infectious Diseases*, 94, 44–48. https://doi.org/10.1016/J.IJID.2020.03.004.

Yazdani, M., Jafari, A., Mahdian, S., Namazi, M., & Gharaghani, S. (2023). Rational approaches to discover SARS-CoV-2/ACE2 interaction inhibitors: Pharmacophore-based virtual screening, molecular docking, molecular dynamics and binding free energy studies. *Journal of Molecular Liquids*, 375, 121345. ISSN 0167-7322; https://doi.org/10.1016/j.molliq.2023.121345.

Yin, W., et al. (2021). Structural basis for inhibition of the SARS-CoV-2 RNA polymerase by suramin. *Nature Structural & Molecular Biology*, 28(3), 319–325. https://doi.org/10.1038/s41594-021-00570-0.

Zhang, A. R., et al. (2021). Epidemiology and evolution of Middle East respiratory syndrome coronavirus, 2012–2020. *Infectious Diseases of Poverty*, 10, 66.

Zhang, D., Zhu, L., Wang, Y., Li, P., & Gao, Y. (2022). Translational control of COVID-19 and its therapeutic implication. *Frontiers in Immunology*, 13, 857490.

Zhu, Y., Sharma, L., & Chang, D. (2023). Pathophysiology and clinical management of coronavirus disease (COVID-19): A mini-review. *Frontiers in Immunology*, 14, 1116131. https://doi.org/10.3389/fimmu.2023.1116131.

Zhu, Z., Lian, X., Su, X., Wu, W., Marraro, G. A., & Zeng, Y. (2020). From SARS and MERS to COVID-19: A brief summary and comparison of severe acute respiratory infections caused by three highly pathogenic human coronaviruses. *Respiratory Research*, 21, 1–14.

10 Vaccine-Based Approaches for Emerging and Re-Emerging Viral Diseases

Srijan Goswami, Rajesh Das, Palas Samanta, and Manab Deb Adhikari

10.1 INTRODUCTION

Emerging and re-emerging viral diseases pose significant threats to global public health due to their unpredictable nature and potential for rapid transmission. Emerging viral diseases are those that have newly appeared in a population or are rapidly increasing in incidence or geographic range, such as COVID-19 caused by SARS-CoV-2. Re-emerging viral diseases involve previously controlled infections, like measles, that resurge due to factors such as declining vaccination rates or ecological changes (Hakim et al., 2024; Nabity et al., 2023). Key factors driving the emergence of viral diseases include zoonotic spillover, environmental changes, globalization, and population growth. For example, pathogens like the Nipah virus and Marburg virus have emerged due to increased human interaction with wildlife, often driven by deforestation and agricultural expansion. Re-emergence often correlates with waning immunity in populations or breakdowns in public health infrastructure, as seen with measles and dengue fever outbreaks (Hakim et al., 2024; Harrison et al., 2023; Nabity et al., 2023). Emerging viruses, such as Ebola and Zika, have highlighted the vulnerabilities of healthcare systems. For instance, the 2014–2016 Ebola epidemic in West Africa revealed critical gaps in surveillance and response mechanisms. Meanwhile, re-emerging diseases like monkeypox have shown resilience, surviving on environmental surfaces, underscoring the need for improved sanitation and vaccination efforts (Hakim et al., 2024; Harrison et al., 2023; Nabity et al., 2023). To combat these threats, ongoing genomic surveillance, enhanced public health coordination, and robust vaccination campaigns are essential. Novel diagnostic tools and next-generation sequencing techniques have emerged as critical components in identifying and managing new viral strains (Hakim et al., 2024; Harrison et al., 2023; Nabity et al., 2023). Additionally, a One Health approach, integrating human, animal, and environmental health, is pivotal for preventing zoonotic spillovers.

Vaccines and immunomodulation are pivotal in combating viral diseases by directly preventing infections and regulating immune responses. Vaccines, such as those developed for COVID-19, have proven highly effective in reducing infection rates and severe disease outcomes, demonstrating efficacy levels of up to 91% after two doses against certain virus variants. These advances have been achieved through innovative vaccine platforms, including mRNA, viral vector, and protein subunit technologies. For example, mRNA vaccines, like Pfizer-BioNTech and Moderna, enable rapid adaptation to emerging variants, revolutionizing pandemic responses (Soheili et al., 2023). Vaccines train the immune system to recognize and neutralize pathogens by mimicking infection, typically without causing illness. Immunomodulators complement vaccines by either enhancing or suppressing immune responses, tailoring them to specific clinical contexts. For instance, immunostimulatory agents like interferons boost antiviral defenses, while immunosuppressive therapies mitigate hyperinflammatory states, such as cytokine storms in severe COVID-19 cases (Strzelec et al., 2023). The dual approach of vaccination and immunomodulation addresses diverse viral threats. In diseases like influenza, routine vaccination curtails annual outbreaks, whereas immunomodulators,

DOI: 10.1201/9781032721811-10

such as monoclonal antibodies, offer targeted therapies for high-risk patients. Similarly, emerging tools like anti-TNF agents and checkpoint inhibitors are expanding options for managing chronic viral infections and autoimmunity associated with viral triggers (Khatun et al., 2012). Despite successes, challenges remain, including vaccine hesitancy, equitable distribution, and the need for vaccines against difficult targets like HIV and hepatitis C. Advances in personalized medicine, such as tailoring immunotherapies based on genetic markers, promise improved outcomes. Additionally, integrating immunomodulatory strategies into pandemic preparedness frameworks can enhance resilience against future outbreaks (Haelle, 2024; Pardi et al., 2018; Fan et al., 2023).

10.2 UNDERSTANDING EMERGING AND RE-EMERGING VIRAL DISEASES

The terms "emerging" and "re-emerging" viral diseases are critical in the study of infectious diseases, particularly in public health and epidemiology. These concepts help identify patterns in disease occurrence and guide responses to outbreaks and pandemics. Emerging viral diseases are defined as infections caused by new or previously unrecognized viruses that appear in human populations or significantly increase in incidence. These pathogens often arise due to complex interactions between ecological, environmental, and human factors. A prominent example is SARS-CoV-2, the virus behind COVID-19, which emerged from zoonotic spillover, reflecting the close linkage between human activity and viral emergence (Hakim et al., 2024). Other examples include the Zika virus, Nipah virus, and H5N1 avian influenza, many of which result from environmental disruptions, urbanization, and global travel. Re-emerging viral diseases involve pathogens that were previously controlled or reduced to negligible levels but have resurfaced, often due to changes in host immunity, vector dynamics, or environmental factors. Examples include dengue fever, yellow fever, and measles. Dengue, now the most widespread arboviral infection, re-emerged largely due to vector adaptation and global warming (Buliva et al., 2017). The distinction between emerging and re-emerging diseases can sometimes blur. For instance, the re-emergence of monkeypox in non-endemic regions underscores the interplay between environmental shifts and viral adaptation, classifying it as both a re-emerging and emerging challenge in different contexts (Hakim et al., 2024). Emerging and re-emerging viral diseases share common etiological drivers:

1. *Zoonotic spillover:* Over 75% of these diseases originate from animals, emphasizing the need for robust surveillance of wildlife and livestock (Buliva et al., 2017).
2. *Climate change:* Altered temperature and precipitation patterns expand the habitats of vectors like mosquitoes and ticks, introducing pathogens such as dengue and Lyme disease to new regions (Buliva et al., 2017).
3. *Globalization and urbanization:* Increased global travel and urban density facilitate rapid disease spread (Buliva et al., 2017).
4. *Human activities:* Deforestation, habitat encroachment, and agricultural practices heighten human–animal interactions, increasing spillover risks (Buliva et al., 2017).

Understanding these definitions aids in anticipating disease trends and formulating prevention strategies. A multidisciplinary approach involving virologists, epidemiologists, ecologists, and public health professionals is essential to mitigate the risks posed by these diseases. Emerging and re-emerging viral diseases are driven by a combination of ecological, societal, and genetic factors that allow pathogens to adapt to new hosts or environments. The following are some of the most significant viral pathogens of concern in this context:

- *Coronaviruses:* Coronaviruses, particularly SARS-CoV-2 (the causative agent of COVID-19), exemplify how zoonotic spillovers can result in global pandemics. SARS-CoV and MERS-CoV have previously caused outbreaks with high case fatality rates. These viruses often emerge due to close contact between humans and animal reservoirs such as bats and camels, underscoring the role of zoonotic pathways in viral evolution (Rabaan et al., 2020; Viral Emerging and Re-emerging Infectious Diseases, 2023).

- *Arboviruses:* Arboviruses, transmitted by vectors like *Aedes* mosquitoes, are responsible for diseases with significant global morbidity. Dengue has seen dramatic increases in incidence due to urbanization and climate change, while the Zika virus outbreak in 2015 highlighted the risks of viral mutation and global dissemination (Hakim et al., 2024).
- *Influenza viruses:* Seasonal influenza remains a constant threat, but novel subtypes such as H5N1 (avian influenza) and H1N1 (swine flu) have emerged in the last two decades. Their ability to recombine and adapt in animal hosts makes them perennial concerns for pandemics (Influenza (FLU), 2024).
- *Filoviruses:* Ebola and Marburg viruses cause hemorrhagic fevers with high fatality rates and are often linked to outbreaks in Central and West Africa. These viruses are typically transmitted from wildlife (e.g., bats, non-human primates) to humans, with societal practices like bushmeat consumption exacerbating their spread (Trovato et al., 2020).
- *Henipaviruses:* Nipah virus is an emerging zoonotic pathogen with a history of outbreaks in South and Southeast Asia. It is transmitted through fruit bat reservoirs and poses risks of severe respiratory and neurological disease (Hakim et al., 2024).
- *Orthopoxviruses:* Monkeypox, a re-emerging orthopoxvirus, has garnered attention for outbreaks outside its traditional endemic regions in Africa. Changes in vaccination practices, such as the cessation of smallpox vaccination, have created immunological vulnerabilities (Hakim et al., 2024).
- *Retroviruses:* HIV, while long classified as a pandemic, continues to evolve and challenge public health systems, especially in low- and middle-income countries. New strains and drug-resistant forms pose persistent threats (Hakim et al., 2024).

Emergence and re-emergence are often linked to ecological changes such as deforestation, urbanization, and climate change. The globalization of trade and travel further facilitates the rapid spread of these viruses across continents. Moreover, socioeconomic inequalities and inadequate healthcare infrastructure in affected regions exacerbate the impact of these diseases (Arnold et al., 2003; McConnon et al., 2003; Brower et al., 2003; Jones et al., 2008). Emerging and re-emerging viral diseases have shown complex epidemiological patterns influenced by socioeconomic, ecological, and biological factors. These trends are pivotal in shaping global health responses and understanding their broader societal impacts (Arnold et al., 2003; McConnon et al., 2003; Brower et al., 2003; Jones et al., 2008). Emerging infectious diseases are often zoonotic in origin, with over 70% arising from animal reservoirs. Examples include the SARS-CoV-2 pandemic, which demonstrated rapid global spread facilitated by international travel and urbanization. Re-emerging diseases, such as measles and dengue, are linked to vaccine hesitancy, vector proliferation, and climate change (Arnold et al., 2003; McConnon et al., 2003; Brower et al., 2003; Jones et al., 2008). Vector-borne diseases like Zika, chikungunya, and dengue have expanded geographically due to rising global temperatures and changing precipitation patterns, which influence vector habitats. Meanwhile, diseases like monkeypox and avian influenza underline the influence of wildlife trade and agricultural practices in their resurgence (Farahat et al., 2022). Emerging viral diseases significantly strain healthcare systems, particularly in low- and middle-income countries, where resource limitations exacerbate morbidity and mortality. For instance, the COVID-19 pandemic revealed disparities in vaccine distribution and access to healthcare. Economic repercussions include disruptions in trade, tourism, and labor markets, with long-term impacts on global economic stability (Morens & Fauci, 2020; Arnold et al., 2003; McConnon et al., 2003; Brower et al., 2003; Jones et al., 2008). Key challenges include underreporting in surveillance systems, which hampers accurate disease burden estimation. Additionally, the emergence of drug-resistant pathogens complicates treatment strategies. Social factors such as vaccine misinformation and urbanization further exacerbate the spread and severity of these diseases (Emerging Infectious Diseases—CDC, n.d.; Arnold et al., 2003;

McConnon et al., 2003; Brower et al., 2003; Jones et al., 2008). Efforts to mitigate these impacts include the implementation of One Health approaches that integrate human, animal, and environmental health strategies. Strengthened international collaborations, as seen during the COVID-19 response, aim to improve early detection, preparedness, and equitable resource distribution (Emerging Infectious Diseases—CDC, n.d.; Arnold et al., 2003; McConnon et al., 2003; Brower et al., 2003; Jones et al., 2008)

10.3 IMMUNOLOGICAL BASIS OF VIRAL INFECTIONS

10.3.1 Host Immune Response to Viral Pathogens

The human immune response to viral infections involves a dynamic interplay between innate and adaptive immunity. This interaction plays a crucial role in limiting viral replication, controlling infection, and establishing immunological memory to prevent reinfection.

- *Innate immune response:* The innate immune system is the body's first line of defense. Pattern recognition receptors (PRRs), such as toll-like receptors (TLRs) and RIG-I-like receptors (RLRs), detect viral components like RNA or DNA. This detection triggers signaling cascades, activating transcription factors like NF-κB and interferon regulatory factors (IRFs). These factors induce the production of type I interferons and pro-inflammatory cytokines, which directly inhibit viral replication and recruit immune cells to the infection site (Sanyal, 2023). However, many viruses, including SARS-CoV-2, employ mechanisms to evade these innate responses, such as suppressing interferon signaling or modulating cytokine release (Emerging Viruses: Host Immunity and Novel Therapeutic Interventions, n.d).
- *Adaptive immune response:* The adaptive immune response involves antigen-specific reactions mediated by T cells and B cells. CD8+ cytotoxic T cells recognize and kill virus-infected cells, while CD4+ helper T cells assist in orchestrating broader immune responses. B cells produce neutralizing antibodies, essential for blocking viral entry into host cells. These adaptive responses not only control ongoing infections but also establish memory for faster response to future exposures (Kachuri et al., 2020).
- *Immune evasion mechanisms:* Viruses have evolved sophisticated strategies to evade host immunity. For instance, the influenza virus alters its surface proteins through antigenic drift and shift, allowing it to escape antibody-mediated neutralization. Similarly, HIV suppresses immune responses by targeting CD4+ T cells, undermining immune coordination (Emerging Viruses: Host Immunity and Novel Therapeutic Interventions, n.d.).

Emerging viruses like Zika, Ebola, and MERS-CoV are often zoonotic and capable of rapid mutation. This makes their detection and control more challenging. Furthermore, these viruses frequently exploit novel mechanisms to evade host defenses, underscoring the need for research into virus–host interactions and immune responses. Understanding these immune processes and their manipulation by pathogens is vital for developing effective vaccines and immunotherapeutics, especially in the face of evolving viral threats (Emerging Viruses: Host Immunity and Novel Therapeutic Interventions, n.d; Kachuri et al., 2020).

10.3.2 Mechanisms of Immune Evasion by Viruses

Viruses have developed sophisticated mechanisms to evade host immune defenses, ensuring their survival, replication, and persistence. Understanding these mechanisms is crucial for managing emerging and re-emerging viral diseases effectively.

- *Subversion of innate immunity:* The innate immune system is the first line of defense against viral infections, relying on pattern recognition receptors (PRRs) like Toll-like receptors (TLRs) and RIG-I-like receptors (RLRs) to detect viral components. Many viruses bypass this detection by modifying their RNA or DNA, suppressing the activation of key signaling pathways like the interferon (IFN) response. For instance, SARS-CoV-2 disrupts interferon signaling and employs double-membrane vesicles to shield its RNA from detection (Tay et al., 2022; Lu et al., 2024; Subbian, 2023).
- *Inhibition of antiviral pathways:* Viruses often encode proteins that directly inhibit antiviral pathways. For example, herpesviruses target the cGAS-STING pathway, a critical component of DNA virus detection, to modulate immune responses. Influenza viruses inhibit RIG-I and MAVS signaling, thereby attenuating IFN production (Tay et al., 2022; Lu et al., 2024; Subbian, 2023).
- *Decoy molecules:* Some viruses produce decoy molecules that mimic host components, interfering with immune recognition. Cytomegaloviruses (CMVs), for instance, encode MHC homologs to evade T-cell-mediated immunity. Similarly, hepatitis C virus (HCV) releases non-infectious particles to distract neutralizing antibodies (Tay et al., 2022; Lu et al., 2024; Subbian, 2023).
- *Evading adaptive immunity:* To avoid adaptive immunity, viruses rapidly mutate their surface proteins, a phenomenon known as antigenic variation. Influenza and HIV are notable examples, where mutations in hemagglutinin or envelope proteins help escape neutralizing antibodies. Viruses also downregulate MHC molecules, impairing antigen presentation to T cells (Tay et al., 2022; Lu et al., 2024; Subbian, 2023).
- *Immune exhaustion:* Chronic infections, such as those caused by HIV or hepatitis viruses, induce immune exhaustion by persistently activating T cells, leading to reduced functionality. This is often mediated by the upregulation of inhibitory receptors like PD-1 and CTLA-4 on immune cells (Tay et al., 2022; Lu et al., 2024; Subbian, 2023;).
- *Host factor exploitation:* Viruses manipulate host cellular machinery to enhance their replication while avoiding immune detection. For example, RNA viruses such as flaviviruses and coronaviruses remodel cellular organelles to create replication compartments, shielding their genetic material from immune sensors (Tay et al., 2022; Lu et al., 2024; Subbian, 2023).
- *Implications for therapeutics:* Targeting these immune evasion mechanisms offers a promising approach for antiviral therapies. Enhancing innate immune sensing, developing broad-spectrum antivirals, and designing vaccines that elicit robust and durable responses can mitigate the impact of these strategies (Tay et al., 2022; Lu et al., 2024; Subbian, 2023).

10.3.3 Implications for Vaccine and Immunotherapy Design

The rapid evolution and emergence of viral pathogens highlight critical challenges in vaccine and immunotherapy design, particularly for diseases with pandemic potential such as COVID-19, Nipah virus, or emerging strains of influenza. Designing effective interventions requires addressing viral diversity, immune evasion mechanisms, and rapid scalability in response to outbreaks (Hakim et al., 2024; Zhang et al., 2019; Seib et al., 2009).

- *mRNA and next-generation vaccines:* mRNA vaccines, such as those developed for SARS-CoV-2, have demonstrated remarkable adaptability, safety, and scalability. These vaccines leverage *in silico* methods like reverse vaccinology to identify antigenic targets within viral genomes. Techniques like self-amplifying mRNA and lipid nanoparticle delivery systems enhance immunogenicity and minimize cytotoxicity, offering platforms for rapid development against newly emerging pathogens (Pardi et al., 2018). Such advances are particularly relevant for RNA viruses like Zika and influenza, which require swift responses to mutations (Hakim et al., 2024; Zhang et al., 2019; Seib et al., 2009).

- *Genomic and reverse vaccinology approaches:* Reverse vaccinology and metagenomic analyses are transforming the discovery of vaccine targets by analyzing complete viral genomes. This approach has been instrumental in developing meningococcal vaccines, revealing conserved antigens capable of eliciting strong immune responses across diverse viral strains. For example, the factor H-binding protein (fHBP) was identified as a vaccine target for *Neisseria meningitidis*, offering insights for future vaccines aimed at highly variable pathogens (Hakim et al., 2024; Zhang et al., 2019; Seib et al., 2009).
- *Immunotherapy innovations:* Beyond vaccines, immunotherapies, including monoclonal antibodies and immune checkpoint inhibitors, are gaining traction. Strategies targeting viral entry points or immune checkpoints help neutralize the virus while modulating the immune response to prevent excessive inflammation, as observed in cytokine storms during severe infections. For diseases like Ebola, monoclonal antibody cocktails have shown efficacy, illustrating the potential for tailored immunotherapies in viral disease management (Hakim et al., 2024; Zhang et al., 2019; Seib et al., 2009).

Despite advancements, several challenges persist. Variants arising from viral mutations can evade immunity provided by existing vaccines. Strategies like multivalent vaccine formulations or universal vaccine approaches targeting conserved viral epitopes aim to address these hurdles. Additionally, designing interventions for re-emerging pathogens requires continuous genomic surveillance and adaptable platforms to manage outbreaks efficiently (Hakim et al., 2024; Zhang et al., 2019; Seib et al., 2009). The integration of genomics, mRNA technology, and immunotherapeutic innovations provides a robust framework for addressing the dynamic challenges posed by emerging and re-emerging viral diseases. These efforts are pivotal for global health security and pandemic preparedness.

10.4 ADVANCEMENTS IN VACCINE DEVELOPMENT

Vaccine development has made significant strides in addressing the threats posed by emerging and re-emerging viral diseases. These advancements have been crucial in combating pathogens like SARS-CoV-2, Ebola, and Zika, where rapid responses were essential to control outbreaks.

- *mRNA vaccine technology:* The COVID-19 pandemic highlighted the transformative potential of mRNA vaccine technology, enabling rapid development and deployment. mRNA vaccines work by using messenger RNA to instruct cells to produce viral antigens, which then stimulate an immune response without the risk of infection. This technology has shown promise for other diseases like Zika and influenza, where the fast mutation rates make traditional vaccine development slower and more challenging (Jackson et al., 2020).
- *Viral vector vaccines:* Viral vector vaccines, such as the Oxford-AstraZeneca COVID-19 vaccine, use a harmless virus to deliver genetic material from the target virus into cells. This approach has demonstrated efficacy in providing immunity against emerging viruses, including Ebola and MERS-CoV. The adaptability of viral vector platforms has spurred interest in applying them to other viral threats (Deng et al., 2022; Al-Halifa et al., 2019).
- *Nanoparticle and protein subunit vaccines:* Nanoparticle vaccines, such as those developed for respiratory syncytial virus (RSV), have shown that nanoparticle formulations can elicit a potent immune response while maintaining safety and efficacy. Additionally, protein subunit vaccines that use viral proteins to trigger immunity are being explored for pathogens like HIV and hepatitis C (Corti et al., 2021).
- *Universal vaccine approaches:* To combat viral mutations and the challenge of strain diversity, researchers are focusing on developing universal vaccines. For example, efforts

are underway to create a universal flu vaccine that targets conserved regions of the virus, ensuring effectiveness across different strains (Pica & Palese, 2013; Al-Halifa et al., 2019). These strategies aim to provide long-term protection against a broad range of viral variants.
- *Adjuvant development:* Adjuvants are substances that enhance the body's immune response to a vaccine. Recent advancements have improved adjuvant formulations, increasing vaccine efficacy and broadening their application, particularly in resource-limited settings where vaccine uptake is critical (Moni et al., 2023). Adjuvants also help induce stronger and longer-lasting immunity, which is essential for combatting emerging pathogens.

Despite these innovations, challenges remain in vaccine development. Rapid mutation rates, such as those seen in influenza or HIV, require continuous surveillance and adaptation of vaccine formulations. The emergence of new variants also underscores the need for global collaboration, early detection systems, and more flexible vaccine platforms. Moreover, global health disparities highlight the need for scalable and cost-effective vaccine solutions (Weinreich et al., 2021).

10.5 VACCINES FOR SARS, MERS, AND COVID-19

The development of vaccines for SARS-CoV, MERS-CoV, and SARS-CoV-2 has seen significant advancements, driven by both the urgency of global health crises and innovations in vaccine technology. The rapid pace of vaccine development for COVID-19 has been particularly notable, thanks to prior research on SARS and MERS, which provided valuable insights into the virus' spike protein as a key target for immunity (Jackson et al., 2020; Goswami et al., 2022; Chakraborty et al., 2023; Tai et al., 2022; Jiang et al., 2005; Li et al., 2020). SARS-CoV (2002) and MERS-CoV (2012) provided the groundwork for COVID-19 vaccine development. For example, early vaccine candidates for MERS-CoV, which targeted the spike protein and viral entry mechanisms, offered key insights that informed the development of mRNA and viral vector vaccines for COVID-19. Although no vaccine was licensed for MERS, research on these coronaviruses highlighted the importance of rapid vaccine development platforms (Jackson et al., 2020; Goswami et al., 2022; Chakraborty et al., 2023; Tai et al., 2022; Jiang et al., 2005; Li et al., 2020). The development of mRNA vaccines, such as the Pfizer-BioNTech and Moderna vaccines, revolutionized the fight against SARS-CoV-2. mRNA technology allows for rapid design and production of vaccines by encoding the virus' spike protein. The success of these vaccines, which demonstrated high efficacy in preventing severe COVID-19 cases and hospitalizations, has reshaped the landscape for vaccine development against future pandemics. This technology is also being explored for other diseases like MERS and Zika (Jackson et al., 2020; Goswami et al., 2022; Chakraborty et al., 2023; Tai et al., 2022; Jiang et al., 2005; Li et al., 2020).

The Oxford-AstraZeneca vaccine used a viral vector platform, where an adenovirus is modified to deliver the genetic material for the spike protein of SARS-CoV-2. This platform, developed with insights from MERS research, proved to be effective against COVID-19 and is also being adapted for use in ongoing vaccine trials against MERS and other emerging viruses (Jackson et al., 2020; Goswami et al., 2022; Chakraborty et al., 2023; Tai et al., 2022; Jiang et al., 2005; Li et al., 2020). Protein subunit vaccines, such as Novavax's NVX-CoV2373, use purified viral proteins to stimulate an immune response. This type of vaccine has been studied for both SARS-CoV and MERS-CoV, with the advantage of using a well-established technology platform that does not require the live virus, making it safer for production (Jackson et al., 2020; Goswami et al., 2022; Chakraborty et al., 2023; Tai et al., 2022; Jiang et al., 2005; Li et al., 2020). While vaccines for COVID-19 have been remarkably effective, the emergence of new variants of concern, such as Omicron, underscores the ongoing challenge of viral mutations. Future vaccine development must include strategies for broader protection against diverse SARS-like coronaviruses, potentially through universal vaccine approaches targeting conserved regions of the virus (Burki, 2021b). Additionally, improving vaccine access in low-resource settings and ensuring long-term immunity remain critical goals

(Jackson et al., 2020; Goswami et al., 2022; Chakraborty et al., 2023; Tai et al., 2022; Jiang et al., 2005; Li et al., 2020). The advancements in vaccine development driven by SARS, MERS, and COVID-19 research have demonstrated the power of modern vaccine technologies, particularly mRNA and viral vectors. These platforms are essential not only for future coronavirus variants but also for responding to other emerging infectious diseases (Jackson et al., 2020; Goswami et al., 2022; Chakraborty et al., 2023; Tai et al., 2022; Jiang et al., 2005; Li et al., 2020).

10.6 VACCINES FOR FILOVIRIDAE

Advancements in vaccine development for Ebola and Marburg viruses have been a focal point in the global fight against these highly lethal hemorrhagic fevers. Recent progress has emphasized the creation of vaccines that target the surface glycoproteins of these viruses, which are key to their ability to infect host cells. The Ebola virus (EBOV) has seen substantial success in vaccine development, notably with the rVSV-ZEBOV vaccine, which demonstrated high levels of protection in both clinical trials and during outbreak responses. Approved by the FDA in 2019, this vaccine has proven effective in preventing EBOV, which was particularly important during the 2013–2016 West Africa outbreak (Flaxman et al., 2024; Factsheet for Health Professionals about Marburg Virus Disease, 2024; Kallay et al., 2024; Ebola Virus Disease: Vaccines, n.d.; Ebola Vaccine Product Information, 2024; Everything You Need to Know About Candidate Vaccines and Treatments for Marburg Disease, 2024). Marburg virus (MARV), although rarer, also poses significant risks, with outbreaks occurring in regions overlapping with EBOV outbreaks. A promising approach for both EBOV and MARV has been the development of adenovirus-vectored vaccines. These vaccines, including the ChAd3-Marburg, have shown encouraging results in preclinical studies and some human trials. Multivalent vaccine strategies, combining protection against multiple pathogens, such as Ebola, Marburg, Sudan, and Lassa viruses, are currently in advanced development, highlighting the potential to provide broader protection in high-risk areas (Flaxman et al., 2024; Factsheet for Health Professionals about Marburg Virus Disease, 2024; Kallay et al., 2024; Ebola Virus Disease: Vaccines, n.d.; Ebola Vaccine Product Information, 2024; Everything You Need to Know About Candidate Vaccines and Treatments for Marburg Disease, 2024). These developments represent significant steps toward improving public health security, especially in endemic regions where these viral diseases continue to cause outbreaks with high mortality rates.

10.7 VACCINES FOR MOSQUITO-BORNE DISEASES

Advancements in vaccine development for mosquito-borne viral diseases such as chikungunya, dengue, and Zika have seen significant progress in recent years, particularly in response to the global burden these diseases place on public health. For chikungunya, despite the absence of specific antiviral treatments, there has been major progress in vaccine development. The first chikungunya vaccine, *Ixchiq* (VLA1553), was approved by the US FDA in November 2023, followed by EMA approval in 2024. This live-attenuated vaccine has shown high seroconversion rates and provides sustained immunogenicity after a single dose (Maure et al., 2024). This approval marks a crucial milestone in addressing the public health threat posed by chikungunya, with future goals aimed at broader implementation and surveillance to gauge long-term effectiveness (Maure et al., 2024). For dengue, a major focus has been on the issue of antibody-dependent enhancement (ADE), which complicates vaccine development. However, the recent introduction of vaccines such as *Dengvaxia* (Sanofi Pasteur) and ongoing research into alternative candidates aim to improve efficacy across all serotypes, which is a significant hurdle in global control efforts (Shukla et al., 2020). The development of these vaccines has been paralleled by efforts to refine strategies that reduce ADE risks, which remains a critical barrier in vaccine deployment (Screaton et al., 2015). For Zika virus, the rapid spread of the virus in recent years has catalyzed vaccine research, although there is still no fully licensed vaccine. Several candidates are in various stages of clinical trials, with mRNA vaccines being a promising area of investigation. These vaccines aim to offer protection without the risk of ADE, a challenge that has historically slowed progress in similar vector-borne viral diseases (Musso et al., 2015). While significant strides have been made, continued research is necessary to overcome challenges related to safety, efficacy, and delivery systems. The approval of vaccines like *Ixchiq* for chikungunya signals a new era in the fight against these diseases, but comprehensive strategies, including vector control and ongoing surveillance, remain key components of global health efforts (Maure et al., 2024; Musso et al., 2015).

10.8 VACCINES FOR LASSA FEVER AND NIPAH FEVER VIRUS

The development of vaccines for Lassa fever and Nipah virus (NiV) has made significant strides, especially as these viruses pose growing threats globally. Both are classified as emerging viral pathogens with high epidemic potential.

- *Lassa fever vaccine development:* Lassa fever, a hemorrhagic fever caused by the Lassa virus, is a major concern in West Africa, with up to 300,000 annual cases. In response, vaccine development has seen exciting progress, especially with the rVSVΔG-LASV-GPC vaccine, which is based on a recombinant vesicular stomatitis virus (rVSV) vector. This vaccine has entered advanced stages of clinical trials, with trials currently being conducted in Liberia and the United States (International AIDS Vaccine Initiative, 2024a). This rVSV vaccine expresses a key Lassa virus protein, essential for the virus' ability to infect cells. Early trials show promise in terms of safety and immune response, with ongoing evaluations set to further confirm its efficacy and potential for global distribution (International AIDS Vaccine Initiative, 2024a). The World Health Organization has recognized Lassa fever as a high-priority emerging disease, reinforcing the urgency for an effective vaccine (International AIDS Vaccine Initiative, 2024a).
- *Nipah virus vaccine development:* Similarly, Nipah virus, transmitted from bats and causing severe disease in humans, has prompted efforts toward vaccine development. In preclinical studies, several vaccine candidates have been tested, with promising results, including DNA vaccines and viral vectors. These have shown efficacy in animal models, with further trials being planned. As with Lassa fever, the emergence of global outbreaks, particularly in Asia, has highlighted the need for a rapid-response vaccine platform (Guillaume et al., 2004; International AIDS Vaccine Initiative, 2024b).

The advancements in vaccines for both Lassa and Nipah viruses demonstrate the importance of global collaborative efforts to address infectious diseases with high public health risks. These vaccines hold the potential to control and prevent future outbreaks, offering hope for regions most affected by these diseases.

10.9 VACCINES FOR HEMORRHAGIC FEVERS

Advancements in vaccine development for Crimean-Congo hemorrhagic fever virus (CCHFV) and Rift Valley fever virus (RVFV) have made significant developments in recent years, driven by the need for effective countermeasures against these potentially deadly zoonotic diseases. For CCHFV, a virus transmitted by ticks, the development of a virus-like replicon particle (VRP) vaccine has shown promise. This vaccine platform mimics the authentic viral infection process, inducing robust immune responses, including both humoral and cell-mediated immunity. In preclinical studies, the VRP vaccine provided complete protection against both homologous and heterologous CCHFV challenges in mice after a single dose. Notably, this vaccine demonstrated Fc-mediated antibody effector functions, such as complement activation and antibody-mediated phagocytosis, which may play a crucial role in protection. Despite these promising results, correlates of protection remain to be fully understood, and further studies are required to optimize this approach for human use (Scholte et al., 2023; Ozdarendeli et al., 2023). For RVFV, significant progress has been made with vaccines targeting the virus' glycoproteins. The virus, which primarily infects livestock but can also cause disease in humans, has led to the development of several vaccine candidates. These vaccines aim to induce strong neutralizing antibody responses and cell-mediated immunity, providing hope for preventing outbreaks in endemic regions. In addition to traditional inactivated vaccines, there is ongoing research into the use of mRNA and viral vector-based vaccines, reflecting the broader trend of leveraging new technologies to combat emerging viral threats (Kortekaas et al., 2011; Zhang et al., 2019). Both CCHFV and RVFV highlight the urgent need for vaccines that can be rapidly deployed in the face of outbreaks. Continued advancements in vaccine development, particularly in the areas of mRNA technology and viral vector platforms, hold promise for providing more effective and accessible vaccines to combat these deadly diseases in the future (Burki, 2021a).

10.10 CHALLENGES IN VACCINE DEVELOPMENT FOR EMERGING VIRUSES

The development of vaccines for emerging viruses presents significant challenges that complicate timely vaccine design and deployment. These viruses, often responsible for severe diseases and global health crises, require novel approaches due to the complexities of viral behavior, host interaction, and logistical constraints. Emerging viruses often exhibit high genetic variability, which poses a major challenge in vaccine development. For example, the rapid mutation rate of influenza viruses and SARS-CoV-2 complicates the creation of long-lasting vaccines. In addition, some viruses like HIV and the Zika virus are known to exhibit significant strain diversity, requiring vaccines to provide broad protection across various viral subtypes (Graham et al., 2018). This mutability can lead to a phenomenon called "antigenic drift" or "shift," where the virus changes its surface proteins, potentially rendering previously developed vaccines ineffective (Boni et al., 2008). A deep understanding of the host immune response to emerging viruses is often lacking at the outset. For example, with viruses such as Ebola and Zika, much of the initial research was focused on virus biology, but immune responses were not fully understood, complicating vaccine design. Some viruses, such as the Crimean-Congo hemorrhagic fever virus (CCHFV), have evolved sophisticated mechanisms to evade host immune responses, making it difficult to develop vaccines that stimulate protective immunity (Scholte et al., 2023). For others like dengue and chikungunya, antibody-dependent enhancement (ADE) presents a further complication, as immune responses may worsen disease symptoms instead of providing protection (Shukla et al., 2020). Another significant challenge is the absence of reliable animal models for many emerging viruses. While animal models are essential for testing vaccine candidates and understanding viral pathogenesis, for some emerging viruses, suitable models either do not exist or do not accurately replicate human infection. For instance, the lack of a perfect animal model for Zika virus or Ebola virus made it harder to test vaccine efficacy in preclinical stages (Wang et al., 2022). This not only delays vaccine development but also raises concerns about the safety and effectiveness of the vaccine candidates. Ensuring vaccine safety and efficacy is a critical barrier. The recent rollout of the COVID-19 vaccines, which utilized mRNA technology, highlighted the importance of rapid vaccine development but also underscored the potential for safety concerns, such as adverse reactions in certain populations (Beladiya et al., 2024). With emerging viruses, the risk of unforeseen adverse effects is particularly concerning due to the limited time for testing and the potential lack of long-term data on vaccine performance. Even once a vaccine is developed, producing it at scale presents substantial challenges. Vaccines for emerging viruses often need to be deployed in resource-limited settings with inadequate infrastructure for storage and distribution, especially in the case of vaccines requiring ultralow temperature storage, such as those for Ebola and COVID-19 (Wouters et al., 2021). Moreover, the rapid emergence of new variants may necessitate continual updates to vaccine formulations, further complicating manufacturing logistics. Finally, there are significant global health and regulatory barriers. Vaccine approval processes can be delayed by bureaucratic hurdles, and political and logistical factors can impede vaccine distribution. For instance, during the Ebola outbreak, the urgent need for a vaccine was met with challenges in securing funding and political will for large-scale trials (Malik et al., 2023). These barriers are particularly problematic in low-income countries, where emerging viral diseases tend to have the greatest impact. Thus, the development of vaccines for emerging viruses faces multiple hurdles, including viral diversity, understanding immune responses, lack of suitable animal models, safety concerns, and challenges related to manufacturing and distribution. While there have been significant advancements in vaccine technology and the use of platforms like mRNA and viral vectors, these challenges highlight the importance of global collaboration, investment in research, and the need for robust healthcare infrastructure to ensure that vaccines reach those most in need.

10.11 IMMUNE BOOSTING AGENTS FOR MANAGING EID-REID OF VIRAL ORIGIN

The rise of emerging and re-emerging infectious diseases (EID-REID) of viral origin has necessitated innovative approaches to enhance host immunity. Immunomodulatory strategies focusing on boosting the immune system with agents such as cytokines, adjuvants, and natural products aim to improve the host's resilience against viral infections.

- *Cytokines and interferons:* Cytokines, such as interferons (IFNs), are key regulators of the immune response to viral infections. Interferon-alpha (IFN-α) has demonstrated efficacy against several viruses, including hepatitis B virus (HBV) and hepatitis C virus (HCV). It activates innate immune pathways, particularly enhancing the activity of natural killer (NK) cells and macrophages (Ivashkiv & Donlin, 2014). Interferon-based therapies have been investigated for diseases like SARS-CoV-2 and Ebola, demonstrating the potential to modulate immune responses when administered early in the infection cycle (Decker et al., 2021).
- *Adjuvants:* Adjuvants are critical in vaccine formulations for amplifying immune responses. For example, AS03, an adjuvant used in influenza vaccines, enhances antigen-specific immune responses by stimulating Toll-like receptors (TLRs). Similarly, MF59 and CpG-1018 have been explored to boost immunity against viral infections like SARS-CoV-2 and dengue (Garçon & Van Mechelen, 2011). Adjuvants are particularly useful in vaccines designed for high-risk populations, such as the elderly, who often exhibit weaker immune responses.
- *Immunostimulatory compounds:* Immunostimulatory agents, including small molecules and nucleic acid–based agents, target immune pathways to combat viral infections. Imiquimod, a TLR-7 agonist, has shown efficacy in inducing antiviral responses in localized infections such as human papillomavirus (HPV) (Banchereau & Pascual, 2006). Similarly, polyinosinic-polycytidylic acid [poly(I:C)], a synthetic analog of viral double-stranded RNA, acts as an immune booster by mimicking viral infection and activating TLR-3 (Marrack et al., 2009).
- *Nutraceuticals and natural compounds:* Natural products like curcumin, resveratrol, and quercetin exhibit immunomodulatory properties. Curcumin, derived from *Curcuma longa*, has shown antiviral effects through the inhibition of viral entry and replication and modulation of inflammatory cytokine production (Ang et al., 2020). Resveratrol, a polyphenol found in grapes, modulates pro-inflammatory cytokine levels and enhances antiviral defense by upregulating interferon-stimulated genes (Abba et al., 2015).
- *Microbiome modulators:* Probiotics and prebiotics play a vital role in maintaining immune homeostasis, particularly in mucosal immunity. Certain strains of *Lactobacillus* and *Bifidobacterium* have shown efficacy in enhancing antiviral immunity by modulating gut-associated lymphoid tissue (GALT). These effects are particularly significant for respiratory viruses, as a healthy gut microbiota supports systemic immune responses (Belkaid & Hand, 2014).

- *Antioxidants:* Antioxidants, such as vitamins C and E, play a pivotal role in mitigating oxidative stress associated with viral infections. High doses of vitamin C have been studied for their potential to reduce the severity of respiratory viral infections, including influenza and COVID-19, by modulating inflammation and supporting immune cell function (Carr & Maggini, 2017).

Immunomodulatory approaches, particularly immune-boosting agents, provide a complementary strategy to vaccines and antiviral drugs in managing EID-REID. While these agents enhance innate and adaptive immunity, their application must be tailored to specific viruses, disease stages, and patient demographics to optimize efficacy and safety. Ongoing research into the mechanisms of action and the development of combination therapies holds promise for improving outcomes in viral epidemics.

10.12 CHALLENGES WITH VACCINE HESITANCY AND DISTRIBUTION

The effective management of emerging and re-emerging infectious diseases relies heavily on robust vaccination campaigns. However, vaccine hesitancy and issues with equitable distribution remain significant barriers to achieving global immunization goals. These challenges are influenced by a complex interplay of sociocultural, logistical, and political factors. Vaccine hesitancy refers to the reluctance or refusal to vaccinate despite the availability of vaccines, as defined by the World Health Organization (Trimble, 2019). This phenomenon is driven by several factors:

- *Spread of misinformation*: Social media platforms have amplified the dissemination of anti-vaccine rhetoric, undermining public confidence in vaccine safety and efficacy (Lazarus et al., 2021).
- *Historical distrust*: Marginalized communities often harbor mistrust due to historical medical injustices, as seen in hesitancy during the COVID-19 pandemic (Savoia et al., 2021).
- *Religious and cultural beliefs:* Religious objections and cultural norms can conflict with vaccination campaigns, particularly in regions with strong traditional practices (Kandiah et al., 2022).
- *Risk-benefit comparison:* Individuals may underestimate the severity of diseases or overestimate vaccine side effects, leading to reduced vaccine uptake (Dubé et al., 2013).

Vaccine distribution disparities exacerbate the global burden of infectious diseases. Key issues include the following:

- *Cold chain requirements*: Vaccines like the mRNA-based COVID-19 vaccines necessitate ultra-cold storage, posing challenges in low-resource settings (Ekwebelem et al., 2021).
- *Supply chain disruptions*: Global crises, including pandemics and geopolitical conflicts, disrupt vaccine production and distribution networks (COVAX Explained, 2022).
- *High costs*: Developing countries struggle with the financial burden of procuring vaccines, leading to inequitable distribution (So & Woo, 2020).
- *Dependency on donations*: Reliance on initiatives like COVAX can delay access for low-income nations compared to high-income countries with direct purchasing power (COVID-19 Vaccines: Everything You Need to Know, 2024).
- *Political conflicts and lack of international cooperation:* These can hinder vaccine distribution in conflict zones or politically unstable regions (Burki, 2021a).
- *Debates over prioritization:* Whether vaccines should go to high-risk populations first or be distributed equitably can delay implementation (Emanuel et al., 2020).

Addressing vaccine hesitancy and distribution inequities requires coordinated efforts like education campaigns, strengthening infrastructure, global cooperation, and so on. Tackling vaccine hesitancy and improving distribution systems is essential for the success of global immunization efforts. Addressing these barriers requires collaboration among governments, healthcare organizations, and communities, with a focus on education, infrastructure, and equitable practices.

10.13 INTEGRATION OF AI AND BIOINFORMATICS IN VACCINE DESIGN

The integration of artificial intelligence (AI) and bioinformatics has revolutionized vaccine design, offering innovative solutions to tackle emerging and re-emerging infectious diseases. By leveraging computational tools, machine learning algorithms, and vast biological datasets, these fields address critical challenges in vaccine development, including antigen identification, immune response prediction, and optimization of vaccine efficacy.

- *Antigen discovery and epitope prediction:* AI-powered tools accelerate the identification of potential vaccine targets by analyzing genomic and proteomic data of pathogens. Algorithms predict epitopes-specific regions of antigens recognized by the immune system ensuring precise vaccine formulation (Ong et al., 2020). For instance, AI platforms like DeepVacPred and Vaxign leverage deep learning to predict B-cell and T-cell epitopes with high accuracy (Rawat et al., 2023).
- *Accelerated vaccine candidate screening:* Machine learning models expedite the evaluation of vaccine candidates by simulating host–pathogen interactions and immune responses, reducing the time required for preclinical testing. AI systems like AlphaFold assist in protein structure prediction, crucial for developing vaccines targeting complex pathogens (Jumper et al., 2021).
- *Reverse vaccinology:* Bioinformatics tools enable reverse vaccinology, a method where vaccine candidates are identified by analyzing the pathogen's genome. This approach was instrumental in the development of the meningococcal B vaccine and has been widely adopted for viral pathogens, including SARS-CoV-2 (Rappuoli et al., 2016).
- *Immune response simulation:* Bioinformatics platforms, such as Immunoinformatics tools, model immune responses to predict vaccine efficacy and minimize adverse effects. These tools integrate omics data to design vaccines that elicit robust and targeted immune responses (Figure 10.1) (Tomar & De, 2014).

The convergence of AI and bioinformatics enhances the precision and efficiency of vaccine design:

- *Data integration*: AI algorithms analyze and integrate large-scale omics data, including genomics, transcriptomics, and proteomics, to identify novel vaccine candidates (Nandy & Basak, 2019).
- *Personalized vaccines*: By incorporating patient-specific genetic and immunological profiles, AI-driven bioinformatics tools pave the way for personalized vaccine design, improving effectiveness in diverse populations (Admin & Alkhaldi, 2024).

While AI and bioinformatics hold immense promise, challenges such as data standardization, computational cost, and algorithm transparency must be addressed. Collaborative efforts between computational scientists, immunologists, and public health experts are essential to fully harness these technologies' potential in global vaccine development (Rappuoli et al., 2016). Epitope prediction is a critical component of vaccine design, focusing on identifying regions of pathogens that can trigger immune responses. AI tools have revolutionized this process by leveraging machine learning

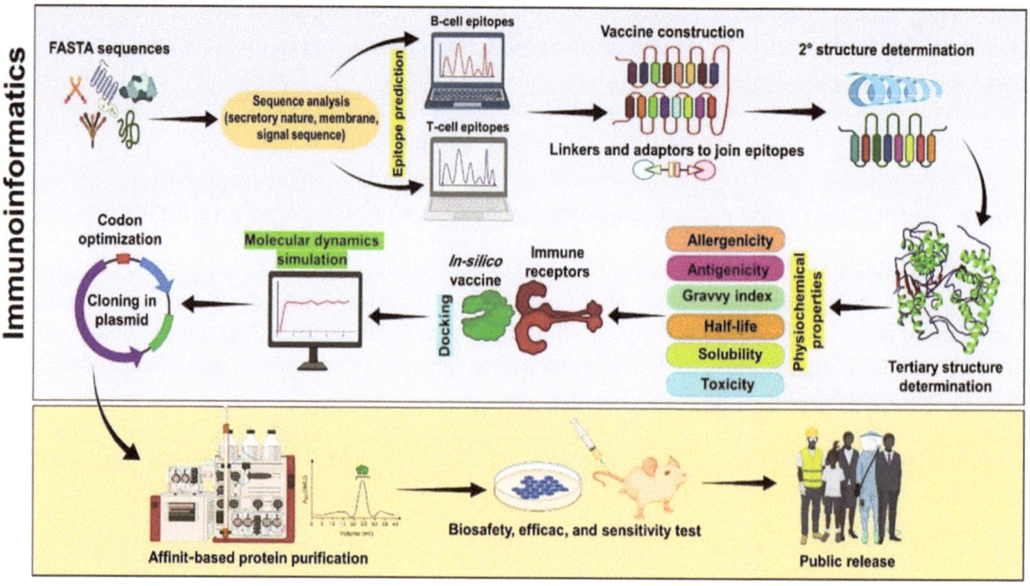

FIGURE 10.1 A detailed stepwise pipeline commonly used to construct a multi-epitope peptide vaccine through the Immunoinformatics approach. This figure provides a general idea of role of bioinformatics in vaccine designing.

Source: **Rawat et al., 2023. Available under Creative Commons Attribution (CC BY) license.**

TABLE 10.1
Key Tools for Epitope Prediction

VaxiJen	VaxiJen is a machine learning–based tool that predicts antigenicity without relying on alignment. It is particularly effective in distinguishing between potential vaccine candidates and non-candidates (Doytchinova & Flower, 2007)
DeepVacPred	DeepVacPred utilizes deep learning to predict both B-cell and T-cell epitopes, focusing on multi-epitope vaccine design for pathogens like SARS-CoV-2. Its high precision makes it a valuable resource in pandemic scenarios (Ong et al., 2020)
NetMHCpan	NetMHCpan is a widely used tool for predicting peptide-MHC binding across multiple HLA alleles. Its neural network-based approach has significantly improved accuracy in epitope mapping (Jurtz et al., 2017)

algorithms, deep learning, and large-scale biological datasets, enabling rapid and accurate epitope mapping. These tools streamline the identification of B-cell and T-cell epitopes, facilitating the design of targeted and effective vaccines. AI-based epitope prediction tools analyze vast genomic and proteomic data to identify immunogenic regions of pathogens. This approach enhances accuracy and efficiency compared to traditional experimental methods. The key tools for epitope prediction are presented in Table 10.1.

- *B-cell epitope prediction:* B-cell epitopes are regions on antigens that are directly recognized by antibodies. AI tools such as ABCpred and BepiPred use machine learning algorithms to predict linear and conformational B-cell epitopes based on amino acid sequences and structural features (Doytchinova et al., 2006).

- *T-cell epitope prediction:* T-cell epitopes are peptides presented by MHC molecules to T cells. AI platforms like NetMHC and DeepEpitope predict binding affinities between peptides and MHC molecules, improving the identification of T-cell epitopes for vaccine design (Jurtz et al., 2017).

AI-driven epitope prediction tools were pivotal in identifying immunogenic regions of the viral structural proteins, accelerating the development of mRNA vaccines. For example, SARS-CoV-2 vaccine from Pfizer and Moderna (Jumper et al., 2021). AI tools facilitate the design of multi-epitope vaccines by integrating predictions for both B-cell and T-cell epitopes. This approach enhances the immune response by targeting multiple aspects of the pathogen. AI tools are integral to reverse vaccinology, where potential antigens are identified computationally before experimental validation. This strategy has been employed in developing vaccines for pathogens such as *Neisseria meningitidis* and *Mycobacterium tuberculosis* (Rappuoli et al., 2016). Despite their success, AI tools face challenges such as limited availability of high-quality training data, algorithm transparency, and integration with experimental workflows. Future advancements in AI, coupled with improved datasets, will enhance the reliability and application of epitope prediction tools in vaccine design (Nandy & Basak, 2019).

Big data analytics has emerged as a transformative tool in immunology, providing comprehensive insights into immune system function, disease mechanisms, and therapeutic interventions. The integration of large-scale datasets with advanced computational tools enables researchers to uncover patterns, predict immune responses, and accelerate the development of vaccines and immunotherapies. Big data in immunology encompasses diverse datasets, including genomics, transcriptomics, proteomics, and clinical information. The ability to analyze these datasets allows for a deeper understanding of immune processes. Big data technologies facilitate detailed profiling of immune cell types, functions, and interactions. Techniques such as single-cell RNA sequencing generate high-resolution maps of immune cell heterogeneity and dynamics in health and disease (Luecken & Theis, 2019). By integrating omics data, big data analytics identifies immune-related pathways and biomarkers associated with diseases like autoimmune disorders, allergies, and cancer. This approach aids in understanding disease etiology and progression (Zhang et al., 2014). Machine learning algorithms analyze complex datasets to identify patterns and make predictions. For example, convolutional neural networks are employed to interpret imaging data of immune responses, while natural language processing tools analyze immunological literature for trend detection (Miotto et al., 2018). Platforms like Cytobank and ImmPort provide access to large immunological datasets and tools for visualization, analysis, and sharing of immune-related data (Bhattacharya et al., 2014). Cloud-based solutions offer scalable storage and computational power for analyzing massive immunological datasets. Google Cloud, Amazon Web Services (AWS), and Microsoft Azure have been instrumental in enabling collaborative immunological research (Hashem et al., 2015). Big data analytics identifies vaccine candidates by analyzing pathogen genomes, predicting immune responses, and modeling host–pathogen interactions. This approach was crucial in the rapid development of COVID-19 vaccines (Ong et al., 2022). Data-driven approaches guide the development of personalized cancer immunotherapies. Analysis of tumor microenvironment and immune signatures helps predict patient responses to immune checkpoint inhibitors (Fuchs et al., 2024). Integration of big data in autoimmune research aids in identifying genetic predispositions, environmental triggers, and potential therapeutic targets for conditions such as rheumatoid arthritis and lupus (Zhang et al., 2014). Advances in artificial intelligence, computational power, and data-sharing frameworks are expected to overcome current challenges. Integrating big data analytics with experimental immunology will enable the development of more effective and personalized therapies.

10.13.1 Precision Medicine in Vaccinology

Precision medicine in vaccinology refers to the development and application of vaccines tailored to individual genetic, immunological, and environmental factors. This approach aims to maximize vaccine efficacy, minimize adverse effects, and address variability in immune responses among different population groups. Advances in genomics, bioinformatics, and immunology have significantly propelled precision vaccinology, making it an essential frontier in combating infectious diseases. Precision vaccinology moves away from the "one-size-fits-all" model by considering the heterogeneity of immune responses influenced by genetic and non-genetic factors. It uses data-driven methodologies, including omics technologies (genomics, proteomics, and transcriptomics), to predict vaccine response at an individual or a population level (Pulendran & Ahmed, 2011). Genetic polymorphisms in immune system-related genes, such as those encoding HLA molecules, cytokines, and Toll-like receptors, significantly impact vaccine-induced immunity. For example, variations in the HLA-DRB1 gene influence immune responses to the hepatitis B vaccine (Linnik et al., 2022). Age-related changes such as immunosenescence in older adults or immature immune systems in neonates necessitate tailored vaccines. Additionally, sex-based differences in immune responses, often mediated by hormonal variations, can affect vaccine efficacy and adverse effects (Klein & Flanagan, 2016). Systems biology integrates multi-omics data to model and predict immune responses. Tools like immune signatures and transcriptional profiling are used to identify biomarkers that correlate with vaccine efficacy or adverse events (Nakaya et al., 2011). AI-driven models analyze large datasets to identify patterns and predict individual responses to vaccines. These tools are instrumental in designing vaccines for emerging infectious diseases, such as SARS-CoV-2 (Bai et al., 2021). Advances in CRISPR and sequencing technologies enable identification of genetic and epigenetic factors that influence vaccine responses. These insights are critical for developing vaccines targeting specific population subgroups. The rapid development of mRNA vaccines like Pfizer-BioNTech and Moderna for COVID-19 showcases the role of precision vaccinology. These vaccines were designed based on the genetic sequence of SARS-CoV-2 and tailored to elicit strong immune responses across diverse populations (Corbett et al., 2020). Precision approaches have been applied to develop universal influenza vaccines by targeting conserved regions of the virus, addressing antigenic drift and variability in immune responses (Sautto et al., 2018). Personalized cancer vaccines, based on tumor-specific neoantigens identified through genomic sequencing, are a growing area within precision vaccinology. These vaccines aim to enhance immune responses against specific cancer types. Integrating multi-omics data with clinical information poses significant computational and analytical challenges. The use of genetic and personal health data raises ethical issues related to privacy, consent, and equitable access to precision vaccines (Jalilian et al., 2023). Developing tailored vaccines requires significant investment, which may limit accessibility in low-resource settings. The integration of precision medicine in vaccinology is expected to grow with advancements in bioinformatics, AI, and omics technologies. Collaborative efforts between academia, industry, and policymakers will be critical in translating precision approaches into widespread vaccine deployment.

10.14 ETHICAL AND SOCIETAL CONSIDERATIONS

10.14.1 Vaccine Equity and Accessibility

Vaccine equity and accessibility are critical challenges in global public health, highlighting disparities in vaccine distribution, affordability, and availability across different regions and populations. The COVID-19 pandemic exposed stark inequities, with low-income countries receiving significantly fewer vaccines compared to high-income nations. Initiatives such as COVAX were launched to address these gaps, yet logistical challenges, geopolitical barriers, and limited funding hindered their effectiveness (COVAX Explained, 2022). Factors influencing vaccine inequity include inadequate healthcare infrastructure, vaccine nationalism, and patent restrictions that limit the production of affordable generics (Burki, 2021b). Additionally, sociocultural barriers such as misinformation and vaccine hesitancy exacerbate disparities in vaccine uptake, particularly in marginalized communities (Figueroa et al., 2021). Addressing these challenges requires a multifaceted approach, including strengthening local manufacturing capabilities, improving global collaboration, and implementing policies that prioritize equitable vaccine distribution. For instance, the World Health Organization's roadmap for equitable vaccine access emphasizes the need for robust international agreements and funding mechanisms to ensure vulnerable populations are not left behind (Global Vaccine Action Plan, 2020). Equity-focused strategies are essential not only to control pandemics but also to achieve long-term public health goals, such as the eradication of vaccine-preventable diseases.

10.14.2 Addressing Misinformation and Public Perception

Misinformation and negative public perception are significant barriers to vaccine acceptance and uptake, particularly in the context of emerging and re-emerging infectious diseases. The rapid spread of false or misleading information, especially via social media platforms, has fueled vaccine hesitancy and delayed vaccination efforts (Zimmerman et al., 2022). Common misconceptions include fears about vaccine safety, efficacy, and potential long-term side effects, which often outweigh factual information provided by health authorities (Larson et al., 2016). Furthermore, certain public figures and interest groups have contributed to vaccine skepticism by endorsing unproven claims, thus amplifying confusion and mistrust among the general public (Dror et al., 2020). Addressing vaccine misinformation requires a strategic, multifaceted approach that includes effective communication strategies, transparency, and engagement with communities. Health authorities must tailor messages to address specific concerns and cultural contexts, using trusted messengers to counteract misinformation and increase public confidence (Strengthen Vaccination Communications, 2024). Moreover, social media platforms and search engines have a critical role to play in regulating misleading content and promoting evidence-based vaccine information (Clark et al., 2022). Public health campaigns, public–private partnerships, and educational initiatives are also vital in building long-term trust and reducing the impact of misinformation.

10.14.3 Regulatory Frameworks and Safety Monitoring

Regulatory frameworks and safety monitoring are critical components in ensuring the efficacy and safety of vaccines. In most countries, vaccine development and approval are regulated by national health authorities such as the US Food and Drug Administration (FDA) and the European Medicines Agency (EMA), which follow rigorous scientific and ethical guidelines to evaluate the preclinical and clinical stages of vaccine trials. The regulatory process involves a series of phased trials (phases I, II, and III) to assess safety, immunogenicity, and efficacy before granting approval (Feikin et al., 2022; Mohammadi et al., 2024). These agencies rely on transparent data submission from pharmaceutical companies and independent reviews to ensure that vaccines meet established safety standards before they are introduced to the public. Post-marketing safety monitoring is another crucial aspect of vaccine regulation. Once vaccines are approved and distributed, they are continuously monitored through surveillance systems such as the Vaccine Adverse Event Reporting System (VAERS) in the United States and the European Medicines Agency's EudraVigilance database. These systems track adverse events following vaccination and allow for timely detection of potential safety concerns (Liu et al., 2021). In addition, regulatory bodies periodically review real-world data and adjust guidelines as needed to ensure continued safety, as seen with COVID-19 vaccines during the pandemic (Koritala et al., 2021). Such comprehensive regulatory frameworks and safety monitoring mechanisms are essential for maintaining public trust in vaccination programs and mitigating potential risks associated with new vaccines (Alcami & Koszinowski, 2000).

10.15 CONCLUSION

Advancements in vaccine delivery systems have significantly enhanced the efficiency, accessibility, and safety of immunization programs. Traditional vaccine delivery methods, typically via intramuscular or subcutaneous injection, have limitations, including the need for trained healthcare personnel and potential discomfort for recipients. New innovations, such as needle-free devices, microneedles, and oral vaccines, are designed to address these challenges and improve vaccine distribution globally. Needle-free systems, including jet injectors, use a high-pressure stream of liquid to deliver vaccines through the skin without a needle, reducing pain and increasing acceptance (Giudice & Campbell, 2006). Microneedle patches, which contain arrays of tiny needles that dissolve in the skin, offer the potential for self-administration and more effective delivery, especially in resource-poor settings (Jose et al., 2022). These technologies are being explored for their ability to improve vaccine stability, reduce storage requirements, and enhance immune responses by targeting the skin's immune cells (Shin et al., 2017). Additionally, oral vaccines have been a focus of research to overcome the logistical and hygienic barriers associated with injectable vaccines, particularly in rural and underserved regions (Kwong et al., 2023). The development of vaccines that can be delivered orally or through other non-injection routes could revolutionize immunization campaigns, making vaccines more accessible and cost-effective. Moreover, these innovations are vital in facilitating rapid responses to emerging infectious diseases, ensuring that vaccines can be distributed more efficiently during pandemics (Brisse et al., 2020). The continued research and development of innovative vaccine delivery systems are essential for increasing global vaccine coverage and ensuring that vaccines reach even the most remote populations.

Personalized medicine and genomics play an increasingly crucial role in the development and effectiveness of vaccines, offering the potential to tailor immunization strategies to individual genetic profiles and immune system variations. Genomic research enables the identification of genetic factors that influence vaccine response, helping to create more effective vaccines for diverse populations (Dugger et al., 2017). For example, variations in the major histocompatibility complex (MHC) can impact how individuals present antigens to their immune system, affecting the effectiveness of vaccines (Kimman et al., 2007). By utilizing genomic data, vaccine formulations can be optimized to target specific immune responses, ensuring that vaccines are more efficient for individuals with different genetic backgrounds. Moreover, personalized vaccine strategies may help address challenges such as vaccine efficacy in older adults or immunocompromised individuals. By analyzing the genetic determinants of immune aging, personalized vaccine approaches can be developed to enhance immunity in these populations (Chen et al., 2022). The application of genomics extends beyond vaccine development to monitoring vaccine responses, where genetic markers can identify individuals at higher risk of adverse reactions, allowing for more informed and targeted vaccination strategies (Rinaudo et al., 2009). Personalized medicine also aids in the creation of vaccines for rare diseases, where the genomic profiles of specific populations can inform the development of vaccines tailored to those groups' unique needs. Thus, the integration of genomics into vaccine development promises to revolutionize immunization practices, ensuring better outcomes and more precise targeting of vaccines (Corti et al., 2021; Morens & Fauci, 2013; Lu et al., 2020).

REFERENCES

Abba, Y., Hassim, H., Hamzah, H., & Noordin, M. M. (2015). Antiviral activity of resveratrol against human and animal viruses. *Advances in Virology*, 2015(1), 184241. https://doi.org/10.1155/2015/184241. Epub 2015 Nov 29. PMID: 26693226; PMCID: PMC4676993.

Admin, & Alkhaldi, N. (2024, March 20). *AI and Machine Learning in Bioinformatics—ITRex*. ITRex. https://itrexgroup.com/blog/ai-and-machine-learning-in-bioinformatics/#:~:text=Introduction%20to%20machine%20learning%20for,not%20explicitly%20programmed%20to%20handle.

Alcami, A., & Koszinowski, U. H. (2000). Viral mechanisms of immune evasion. *Immunology Today*, 21(9), 447–455. https://doi.org/10.1016/s0966-842x(00)01830-8. PMID: 10989308; PMCID: PMC7141568.

Al-Halifa, S., Gauthier, L., Arpin, D., Bourgault, S., & Archambault, D. (2019). Nanoparticle-based vaccines against respiratory viruses. *Frontiers in Immunology*, 10. https://doi.org/10.3389/fimmu.2019.00022.

Ang, L., Song, E., Lee, H. W., & Lee, M. S. (2020). Herbal medicine for the treatment of coronavirus disease 2019 (COVID-19): A systematic review and meta-analysis of randomized controlled trials. *Journal of Clinical Medicine*, 9(5), 1583. https://doi.org/10.3390/jcm9051583.

Arnold, J. C., Singh, K. K., Spector, S. A., & Sawyer, M. H. (2003) Human bocavirus: Prevalence and clinical spectrum at a children's hospital. *Clinical Infectious Diseases*, 43(3), 283–288. https://doi.org/10.1086/505399.

Bai, L., Zhao, Y., Dong, J., Liang, S., Guo, M., Liu, X., Wang, X., Huang, Z., Sun, X., Zhang, Z., Dong, L., Liu, Q., Zheng, Y., Niu, D., Xiang, M., Song, K., Ye, J., Zheng, W., Tang, Z., Tang, M., Zhou, Y., Shen, C., Dai, M., Zhou, L., Chen, Y., Yan, H., Lan, K., & Xu K. (2021). Coinfection with influenza A virus enhances SARS-CoV-2 infectivity. *Cell Res*, 31(4), 395-403. https://doi.org/10.1038/s41422-021-00473-1. Epub 2021 Feb 18. PMID: 33603116; PMCID: PMC7890106.

Banchereau, J., & Pascual, V. (2006). Type I interferon in systemic lupus erythematosus and other autoimmune diseases. *Immunity*, 25(3), 383–392. https://doi.org/10.1016/j.immuni.2006.08.010.

Beladiya, J., Kumar, A., Vasava, Y., Parmar, K., Patel, D., Patel, S., Dholakia, S., Sheth, D., Boddu, S. H. S., & Patel, C. (2024). Safety and efficacy of COVID-19 vaccines: A systematic review and meta-analysis of controlled and randomized clinical trials. *Reviews in Medical Virology*, 34(1), e2507. https://doi.org/10.1002/rmv.2507.

Belkaid, Y., & Hand, T. W. (2014). Role of the microbiota in immunity and inflammation. *Cell*, 157(1), 121–141. https://doi.org/10.1016/j.cell.2014.03.011.

Bhattacharya, S., et al. (2014). ImmPort: Disseminating data to the public for the future of immunology. *Immunologic Research*, 58, 234–239. https://doi.org/10.1007/s12026-014-8516-1.

Boni, M. F. (2008). Vaccination and antigenic drift in influenza. *Vaccine*, 26, C8–C14. https://doi.org/10.1016/j.vaccine.2008.04.011. PMID: 18773534; PMCID: PMC2603026.

Brisse, M., Vrba, S. M., Kirk, N., Liang, Y., & Ly, H. (2020, September 30). Emerging concepts and technologies in vaccine development. *Frontiers in Immunology*, 11, 583077. https://doi.org/10.3389/fimmu.2020.583077. PMID: 33101309; PMCID: PMC7554600.

Brower, J., & Chalk, P. (2003). *The Global Threat of New and Reemerging Infectious Diseases: Reconciling U.S. National Security and Public Health Policy*. Santa Monica, CA: RAND Corporation. https://www.rand.org/pubs/monograph_reports/MR1602.html.

Buliva, E., Elhakim, M., Minh, N. N. T., Elkholy, A., Mala, P., Abubakar, A., & Malik, S. M. M. R. (2017). Emerging and reemerging diseases in the World Health Organization (WHO) eastern mediterranean region—progress, challenges, and WHO initiatives. *Frontiers in Public Health*, 5. https://doi.org/10.3389/fpubh.2017.00276.

Burki, T. (2021a). Equitable distribution of COVID-19 vaccines. *The Lancet Infectious Diseases*, 21(1), 33–34. https://www.thelancet.com/journals/laninf/article/PIIS1473–3099(20)30949-X/fulltext.

Burki, T. (2021b). Global COVID-19 vaccine inequity. *The Lancet Infectious Diseases*, 21(7), 922–923. https://doi.org/10.1016/S1473–3099(21)00344–3.

Carr, A. C., & Maggini, S. (2017). Vitamin C and immune function. *Nutrients*, 9(11), 1211. https://doi.org/10.3390/nu9111211.

Chakraborty, C., Bhattacharya, M., & Dhama, K. (2023, March 17). SARS-CoV-2 vaccines, vaccine development technologies, and significant efforts in vaccine development during the pandemic: The lessons learned might help to fight against the next pandemic. *Vaccines (Basel)*, 11(3), 682. https://doi.org/10.3390/vaccines11030682. PMID: 36992266; PMCID: PMC10054865.

Chen, J., Deng, J. C., & Goldstein, D. R. (2022, December). How aging impacts vaccine efficacy: Known molecular and cellular mechanisms and future directions. *Trends in Molecular Medicine*, 28(12), 1100–1111. https://doi.org/10.1016/j.molmed.2022.09.008. Epub 2022 Oct 7. PMID: 36216643; PMCID: PMC9691569.

Clark, S. E., Bledsoe, M. C., & Harrison, C. J. (2022). The role of social media in promoting vaccine hesitancy. *Current Opinion in Pediatrics*, 34(2), 156–162. https://doi.org/10.1097/MOP.0000000000001111.

Corbett, K. S., et al. (2020). SARS-CoV-2 mRNA vaccine design enabled by prototype pathogen preparedness. *Nature*, 586(7830), 567–571. https://doi.org/10.1038/s41586-020-2622-0.

Corti, D., Purcell, L. A., Snell, G., & Veesler, D. (2021). Tackling COVID-19 with neutralizing monoclonal antibodies. *Cell*, 184(12), 3086–3108. https://doi.org/10.1016/j.cell.2021.05.005.

COVAX Explained. (2022, November 8). https://www.gavi.org/vaccineswork/covax-explained.

COVID-19 Vaccines: Everything You Need to Know. (2024, August 13). https://www.gavi.org/covid19-vaccines.

Decker, T. (2021, October 4). The early interferon catches the SARS-CoV-2. *Journal of Experimental Medicine*, 218(10), e20211667. https://doi.org/10.1084/jem.20211667. Epub 2021 Aug 23. PMID: 34424267; PMCID: PMC8404473.

Doytchinova, I. A., & Flower, D. R. (2007). VaxiJen: A server for prediction of protective antigens, tumour antigens and subunit vaccines. *BMC Bioinformatics*, 8, 4. https://doi.org/10.1186/1471-2105-8-4.

Doytchinova, I. A., Guan, P. & Flower, D. R. (2006). EpiJen: A server for multistep T cell epitope prediction. *BMC Bioinformatics* **7**, 131. https://doi.org/10.1186/1471-2105-7-131.

Dror, A. A., Eisenbach, N., Taiber, S., Morozov, N. G., Mizrachi, M., Zigron, A., Srouji, S., & Sela, E. (2020). Vaccine hesitancy: The next challenge in the fight against COVID-19. *European Journal of Epidemiology*, 35(8), 775–779. https://doi.org/10.1007/s10654-020-00671-y.

Dubé, E., Laberge, C., Guay, M., Bramadat, P., Roy, R., & Bettinger, J. (2013). Vaccine hesitancy: An overview. *Human Vaccines & Immunotherapeutics*, 9(8), 1763–1773. https://doi.org/10.4161/hv.24657.

Dugger, S. A., Platt, A., & Goldstein, D. B. (2018, March). Drug development in the era of precision medicine. *Nature Reviews Drug Discovery*, 17(3), 183–196. https://doi.org/10.1038/nrd.2017.226. Epub 2017 Dec 8. PMID: 29217837; PMCID: PMC6287751.

Ebola Vaccine Product Information. (2024, May 1). *Ebola*. https://www.cdc.gov/ebola/hcp/vaccines/index.html.

Ebola Virus Disease: Vaccines. (n.d.). https://www.who.int/news-room/questions-and-answers/item/ebola-vaccines.

Ekwebelem, O. C., Yunusa, I., Onyeaka, H., Ekwebelem, N. C., & Nnorom-Dike, O. (2021). COVID-19 vaccine rollout: Will it affect the rates of vaccine hesitancy in Africa? *Public Health*, 197, e18–e19. https://doi.org/10.1016/j.puhe.2021.01.010.

Emanuel, E. J., Persad, G., Kern, A., Buchanan, A., Fabre, C., Halliday, D., Heath, J., Herzog, L., Leland, R. J., Lemango, E. T., Luna, F., McCoy, M. S., Norheim, O. F., Ottersen, T., Schaefer, G. O., Tan, K. C., Wellman, C. H., Wolff, J., & Richardson, H. S. (2020). An ethical framework for global vaccine allocation. *Science (New York, N.Y.)*, 369(6509), 1309–1312. https://doi.org/10.1126/science.abe2803.

Emerging Infectious Diseases—CDC. (n.d.). *Emerging Infectious Diseases Journal*. https://wwwnc.cdc.gov/eid/.

Emerging Viruses: Host Immunity and Novel Therapeutic Interventions. (n.d.). *Frontiers*. https://www.frontiersin.org/research-topics/6725/emerging-viruses-host-immunity-and-novel-therapeutic-interventions/articles.

Everything You Need to Know about Candidate Vaccines and Treatments for Marburg Disease. (2024, October 8). https://www.gavi.org/vaccineswork/everything-you-need-know-about-candidate-vaccines-and-treatments-marburg-disease.

Factsheet for Health Professionals about Marburg Virus Disease. (2024, October 10). *European Centre for Disease Prevention and Control*. https://www.ecdc.europa.eu/en/infectious-disease-topics/marburg-virus-disease/factsheet-health-professionals-about-marburg-virus.

Fan, T., et al. (2023). Therapeutic cancer vaccines: Advancements, challenges and prospects. *Signal Transduction and Targeted Therapy*, 8, 450. https://doi.org/10.1038/s41392-023-01674-3.

Farahat, R. A., Abdelaal, A., Shah, J., Ghozy, S., Sah, R., Bonilla-Aldana, D. K., Rodriguez-Morales, A. J., McHugh, T. D., & Leblebicioglu H. (2022). Monkeypox outbreaks during COVID-19 pandemic: Are we looking at an independent phenomenon or an overlapping pandemic? *Ann Clin Microbiol Antimicrob*, 21(1), 26. https://doi.org/10.1186/s12941-022-00518-2. PMID: 35706004; PMCID: PMC9199472.

Feikin, D. R., Higdon, M. M., Abu-Raddad, L. J., Andrews, N., Araos, R., Goldberg, Y., Groome, M. J., Huppert, A., O'Brien, K. L., Smith, P. G., Wilder-Smith, A., Zeger, S., Deloria Knoll, M., & Patel, M. K. (2022). Duration of effectiveness of vaccines against SARS-CoV-2 infection and COVID-19 disease: Results of a systematic review and meta-regression. *Lancet (London, England)*, 399(10328), 924–944. https://doi.org/10.1016/S0140–6736(22)00152-0.

Figueroa, J. P., et al. (2021, September 13). Achieving global equity for COVID-19 vaccines: Stronger international partnerships and greater advocacy and solidarity are needed. *PLoS Medicine*, 18(9), e1003772. https://doi.org/10.1371/journal.pmed.1003772. PMID: 34516558; PMCID: PMC8475996.

Flaxman, A., Sebastian, S., Appelberg, S., Cha, K. M., Ulaszewska, M., Purushotham, J., Gilbride, C., Sharpe, H., Spencer, A. J., Bibi, S., Wright, D., Schmidt, I., Dowall, S., Easterbrook, L., Findlay-Wilson, S., Gilbert, S., Mirazimi, A., & Lambe, T. (2024). Potent immunogenicity and protective efficacy of a multi-pathogen vaccination targeting Ebola, Sudan, Marburg and Lassa viruses. *PLoS Pathogens*, 20(6), e1012262. https://doi.org/10.1371/journal.ppat.1012262.

Fuchs, T., Appelt, D., & Ellemunter, H. (2024). Long-term assessment of antibody response to COVID-19 vaccination in people with cystic fibrosis and solid organ transplantation. *Vaccines*, 12(1), 98. https://doi.org/10.3390/vaccines12010098.

Garçon, N., & Van Mechelen, M. (2011). Recent clinical experience with vaccines using MPL- and QS-21-containing adjuvant systems. *Expert Review of Vaccines*, 10(4), 471–486. https://doi.org/10.1586/erv.11.29.

Giudice, E. L., & Campbell, J. D. (2006). Needle-free vaccine delivery. *Advanced Drug Delivery Reviews*, 58(1), 68–89. https://doi.org/10.1016/j.addr.2005.12.003.

Global Vaccine Action Plan. (2020, August 31). https://www.who.int/teams/immunization-vaccines-and-biologicals/strategies/global-vaccine-action-plan.

Goswami, S., & Dey, C. (Eds.). (2022). *COVID-19 and SARS-CoV-2: The Science and Clinical Application of Conventional and Complementary Treatments (1st ed.). CRC Press.* https://doi.org/10.1201/9781003178514

Graham, B. S., Mascola, J. R., & Fauci AS. (2018). Novel vaccine technologies: Essential components of an adequate response to emerging viral diseases. *JAMA*, 319(14), 1431–1432. https://doi.org/10.1001/jama.2018.0345.

Guillaume, V., Contamin, H., Loth, P., Georges-Courbot, M. C., Lefeuvre, A., Marianneau, P., Chua, K. B., Lam, S. K., Buckland, R., Deubel, V., & Wild, T. F. (2004, January). Nipah virus: Vaccination and passive protection studies in a hamster model. *Journal of Virology*, 78(2), 834–840. https://doi.org/10.1128/jvi.78.2.834-840.2004. PMID: 14694115; PMCID: PMC368848.

Haelle, T. (2024). The staggering success of vaccines. *Nature*, 634(8035), S34–S39. https://doi.org/10.1038/d41586-024-03412-3.

Hakim, M. S., Wang, W., Goeijenbier, M., & Saxena, S. K. (2024). Editorial: Viral emerging and re-emerging diseases: Basic understanding and future intervention strategies. *Frontiers in Microbiology*, 15. https://doi.org/10.3389/fmicb.2024.1395742.

Harrison, R. E., Hamada, A., Haswell, N., Groves, A., Vihta, K. D., Cella, K., Garner, S., Walker, A. S., & Seale, A. C. (2023, October). Cycle threshold values as indication of increasing SARS-CoV-2 new variants, England, 2020–2022. *Emerging Infectious Diseases,* 29(10), 2024–2031. https://doi.org/10.3201/eid2910.230030. Epub 2023 Sep 7. PMID: 37678158; PMCID: PMC10521603.

Hashem, I. A. T., Yaqoob, I., Anuar, N. B., Mokhtar, S., Gani, A., & Khan, S. U. (2015). The rise of "big data" on cloud computing: Review and open research issues. *Information Systems*, 47, 98–115. ISSN 0306-4379; https://doi.org/10.1016/j.is.2014.07.006.

Huffman A, Ong E, Hur J, D'Mello A, Tettelin H, He Y. (2022 Jul 18). COVID-19 vaccine design using reverse and structural vaccinology, ontology-based literature mining and machine learning. *Brief Bioinform*, 23(4):bbac190. doi: 10.1093/bib/bbac190. PMID: 35649389; PMCID: PMC9294427.

Influenza (FLU). (2024, August 1). https://www.cdc.gov/flu/index.html.

International AIDS Vaccine Initiative. (2024a, January 10). *EDCTP and CEPI Funding Moves IAVI's Lassa Fever Vaccine Candidate into Advanced Clinical Development—IAVI. IAVI.* https://www.iavi.org/press-release/edctp-and-cepi-funding-moves-iavis-lassa-fever-vaccine-candidate-into-advanced-clinical-development/.

International AIDS Vaccine Initiative. (2024b, January 29). *IAVI Announces First Vaccinations at Liberia Site in Phase 1 Clinical Trial of Lassa Fever Vaccine Candidate—IAVI. IAVI.* https://www.iavi.org/press-release/iavi-announces-first-vaccinations-at-liberia-site-in-phase-i-clinical-trial-of-lassa-fever-vaccine-candidate/.

Ivashkiv, L. B., & Donlin, L. T. (2014). Regulation of type I interferon responses. *Nature Reviews in Immunology*, 14(1), 36–49. https://doi.org/10.1038/nri3581.

Jackson, L. A., et al. (2020). An mRNA vaccine against SARS-CoV-2—preliminary report. *The New England Journal of Medicine*, 383(20), 1920–1931. https://doi.org/10.1056/NEJMoa2022483.

Jalilian, H., Amraei, M., Javanshir, E., Jamebozorgi, K., & Faraji-Khiavi, F. (2023, March 14). Ethical considerations of the vaccine development process and vaccination: A scoping review. *BMC Health Services Research*, 23(1), 255. https://doi.org/10.1186/s12913-023-09237-6. PMID: 36918888; PMCID: PMC10013982.

Jiang, S., He, Y., & Liu, S. (2005, July). SARS vaccine development. *Emerging Infectious Disease*, 11(7), 1016–1020. https://doi.org/10.3201/1107.050219. PMID: 16022774; PMCID: PMC3371787.

Jones, K. E., Patel, N. G., Levy, M. A., Storeygard, A., Balk, D., Gittleman, J. L., & Daszak, P. (2008, February 21). Global trends in emerging infectious diseases. *Nature*, 451(7181), 990–993. https://doi.org/10.1038/nature06536. PMID: 18288193; PMCID: PMC5960580.

Jose, J., Khot, K. B., Shastry, P., Gopan, G., Bandiwadekar, A., Thomas, S. P., Muhammad, S. T., Ugare, S. R., Chopra, H., Priyanka, & Choudhary, O. P. (2022). Recent advancements in microneedle-based vaccine delivery. *International Journal of Surgery (London, England)*, 107, 106973. https://doi.org/10.1016/j.ijsu.2022.106973.

Jumper, J., et al. (2021). Highly accurate protein structure prediction with AlphaFold. *Nature*, 596(7873), 583–589. https://doi.org/10.1038/s41586-021-03819-2.

Jurtz, V., Paul, S., Andreatta, M., Marcatili, P., Peters, B., & Nielsen, M. (2017). NetMHCpan–4.0: Improved peptide-MHC class I interaction predictions integrating eluted ligand and peptide binding affinity data. *Journal of Immunology (Baltimore, MD: 1950)*, 199(9), 3360–3368. https://doi.org/10.4049/jimmunol.1700893.

Kachuri, L., Francis, S. S., Morrison, M. L., Wendt, G. A., Bossé, Y., Cavazos, T. B., Rashkin, S. R., Ziv, E., & Witte, J. S. (2020). The landscape of host genetic factors involved in immune response to common viral infections. *Genome Medicine*, 12(1), 93. https://doi.org/10.1186/s13073-020-00790-x.

Kallay, R., Doshi, R. H., Muhoza, P., Choi, M. J., Legand, A., Aberle-Grasse, E., Bagayoko, A., Hyde, T. B., Formenty, P., & Costa, A. (2024). Use of Ebola vaccines—worldwide, 2021–2023. *MMWR Morbidity and Mortality Weekly Report*, 73(16), 360–364. https://doi.org/10.15585/mmwr.mm7316a1.

Kandiah, S., Iheaku, O., Farrque, M., Hanna, J., Johnson, K. B., Wiley, Z., Franks, N. M., Carroll, K., Shin, S. R., Sims, K. M., & Kulshreshtha, A. (2022, May 3). CROSS collaborative. COVID-19 vaccine hesitancy among health care workers in four health care systems in Atlanta. *Open Forum Infectious Diseases*, 9(8), ofac224. https://doi.org/10.1093/ofid/ofac224. PMID: 36000002; PMCID: PMC9129176.

Khatun, S., Md Akram Hussain, S., Chowdhury, S., Ferdous, J., Hossain, F., Razia Begum, S., Jahan, M., Tabassum, S., Khatun, S., & Fazlul Karim, A. B. M. (2012). Safety and immunogenicity profile of human papillomavirus-16/18 AS04 adjuvant cervical cancer vaccine: A randomized controlled trial in healthy adolescent girls of Bangladesh. *Japanese Journal of Clinical Oncology*, 42(1), 36–41, https://doi.org/10.1093/jjco/hyr173.

Kimman, T. G., Vandebriel, R. J., & Hoebee, B. (2007). Genetic variation in the response to vaccination. *Community Genetics*, 10(4), 201–217. https://doi.org/10.1159/000106559.

Klein, S. L., & Flanagan, K. L. (2016). Sex differences in immune responses. *Nature Reviews in Immunology*, 16(10), 626–638. https://doi.org/10.1038/nri.2016.90.

Koritala, T., Hussain, A., Pleshkova, Y., Dondapati, L., Tirupathi, R., Rabaan, A. A., Al Mutair, A., Alhumaid, S., Al-Tawfiq, J. A., & Kashyap, R. (2021, September 10). A narrative review of emergency use authorization versus full FDA approval and its effect on COVID-19 vaccination hesitancy. *Le Infezioni in Medicina*, 29(3), 339. https://doi.org/10.53854/liim-2903-4. PMID: 35146338; PMCID: PMC8805497.

Kortekaas, J., Zingeser, J., de Leeuw, P., de La Rocque, S., Unger, H., & Moormann, R. J. (2011, September). Rift Valley fever vaccine development, progress and constraints. *Emerging Infectious Diseases*, 17(9), e1. https://doi.org/10.3201/eid1709.110506. PMID: 21888781; PMCID: PMC3322093.

Kwong, et al. (2023, July 12). Oral vaccines: A better future of immunization. *Vaccines*, 11(7), 1232. https://doi.org/10.3390/vaccines11071232. PMID: 37515047; PMCID: PMC10383709.

Larson, H. J., de Figueiredo, A., Xiahong, Z., Schulz, W. S., Verger, P., Johnston, I. G., Cook, A. R., & Jones, N. S. (2016). The state of vaccine confidence 2016: Global insights through a 67-country survey. *EBioMedicine*, 12, 295–301. https://doi.org/10.1016/j.ebiom.2016.08.042.

Lazarus, J. V., Ratzan, S. C., Palayew, A., Gostin, L. O., Larson, H. J., Rabin, K., Kimball, S., & El-Mohandes, A. (2021). A global survey of potential acceptance of a COVID-19 vaccine. *Nature Medicine*, 27(2), 225–228. https://doi.org/10.1038/s41591-020-1124-9.

Li, Y. D., et al. (2020). Coronavirus vaccine development: From SARS and MERS to COVID–19. *Journal of Biomedical Science*, 27, 104. https://doi.org/10.1186/s12929-020-00695-2.

Linnik, J., Syedbasha, M., Kaltenbach, H.-M., Vogt, D., Hollenstein, Y., Kaufmann, L., Cantoni, N., Ruosch-Girsberger, S., Müller, A. M. S., Schanz, U., Pabst, T., Stüssi, G., Weisser, M., Halter, J., Stelling, J., & Egli, A. (2022). Association of host factors with antibody response to seasonal influenza vaccination in allogeneic hematopoietic stem cell transplant patients. *The Journal of Infectious Diseases*, 225(8), 1482–1493, https://doi.org/10.1093/infdis/jiab391.

Liu, Y., Han, J., Li, X., Chen, D., Zhao, X., Qiu, Y., Zhang, L., Xiao, J., Li, B., & Zhao, H. (2021). COVID-19 Vaccination in People Living with HIV (PLWH) in China: A Cross Sectional Study of Vaccine Hesitancy, Safety, and Immunogenicity. *Vaccines*, 9(12), 1458. https://doi.org/10.3390/vaccines9121458

Lu, R. M., Hwang, Y. C., Liu, I. J., Lee, C. C., Tsai, H. Z., Li, H. J., & Wu, H. C. (2020). Development of therapeutic antibodies for the treatment of diseases. *Journal of Biomedical Science*, 27, 1–30. https://doi.org/10.1186/s12929-019-0592-z.

Lu, W., Wang, L., & Xing, J. (2024). Editorial: Antiviral innate immune sensing, regulation, and viral immune evasion. *Frontiers in Immunology*, 14. https://doi.org/10.3389/fimmu.2023.1358542.

Luecken, M. D., & Theis, F. J. (2019). Current best practices in single-cell RNA-seq analysis: A tutorial. *Molecular Systems Biology*, 15(6), e8746. https://doi.org/10.15252/msb.20188746.

Malik, S., Kishore, S., Nag, S., Dhasmana, A., Preetam, S., Mitra, O., León-Figueroa, D. A., Mohanty, A., Chattu, V. K., Assefi, M., Padhi, B. K., & Sah, R. (2023, January 26). Ebola virus disease vaccines: Development, current perspectives & challenges. *Vaccines (Basel)*, 11(2), 268. https://doi.org/10.3390/vaccines11020268. PMID: 36851146; PMCID: PMC9963029.

Marrack, P., McKee, A. S., & Munks, M. W. (2009). Towards an understanding of the adjuvant action of aluminium. *Nature Reviews in Immunology*, 9(4), 287–293. https://doi.org/10.1038/nri2510.

Maure, C., Khazhidinov, K., Kang, H., Auzenbergs, M., Moyersoen, P., Abbas, K., Santos, G. M. L., Medina, L. M. H., Wartel, T. A., Kim, J. H., Clemens, J., & Sahastrabuddhe, S. (2024). Chikungunya vaccine development, challenges, and pathway toward public health impact. *Vaccine*, 42(26), 126483. https://doi.org/10.1016/j.vaccine.2024.126483.

McConnon, P. J. (2003, September). The global threat of new and reemerging infectious diseases: Reconciling U.S. national security and public health policy. *Emerging Infectious Disease*, 9(9), 1189–1190. https://doi.org/10.3201/eid0909.030442. PMCID: PMC3016794.

Miotto, R., Wang, F., Wang, S., Jiang, X., & Dudley, J. T. (2018). Deep learning for healthcare: Review, opportunities and challenges. *Briefings in Bioinformatics*, 19(6), 1236–1246. https://doi.org/10.1093/bib/bbx044.

Mohammadi, D., Ghasemi, M., Manouchehrian, N., Zafarmand, M., Akbari, M., & Boroumand, A. B. (2024, November 6). COVID-19 vaccines: Current and future challenges. *Frontiers in Pharmacology*, 15, 1434181. https://doi.org/10.3389/fphar.2024.1434181. PMID: 39568586; PMCID: PMC11576167.

Moni, S. S., Abdelwahab, S. I., Jabeen, A., Elmobark, M. E., Aqaili, D., Ghoal, G., Oraibi, B., Farasani, A. M., Jerah, A. A., Alnajai, M. M. A., & Mohammad Alowayni, A. M. H. (2023, November 9). Advancements in vaccine adjuvants: The journey from alum to nano formulations. *Vaccines (Basel)*, 11(11), 1704. https://doi.org/10.3390/vaccines11111704. PMID: 38006036; PMCID: PMC10674458.

Monto, A. S. (2003, September 15). The global threat of new and reemerging infectious diseases: Reconciling U.S. national security and public health policy. *American Journal of Epidemiology*, 158(6), 606. https://doi.org/10.1093/aje/kwg196.

Morens, D. M., & Fauci, A. S. (2013). Emerging infectious diseases: Threats to human health and global stability. *PLOS Pathogens*, 9(7), e1003467. https://doi.org/10.1371/journal.ppat.1003467.

Morens, D. M., & Fauci, A. S. (2020). Emerging pandemic diseases: How we got to COVID-19. *Cell*, 182(5), 1077–1092. https://doi.org/10.1016/j.cell.2020.08.021.

Musso, D., Cao-Lormeau, V. M., & Gubler, D. J. (2015). Zika virus: Following the path of dengue and chikungunya? *Lancet (London, England)*, 386(9990), 243–244. https://doi.org/10.1016/S0140–6736(15)61273–9.

Nabity, S. A., et al. (2023). Characteristics of and deaths among 333 persons with tuberculosis and COVID-19 in cross-sectional sample from 25 jurisdictions, United States. *Emerging Infectious Diseases*, 29(10), 2016–2023. https://doi.org/10.3201/eid2910.230286.

Nakaya, H. I., Wrammert, J., Lee, E. K., Racioppi, L., Marie-Kunze, S., Haining, W. N., Means, A. R., Kasturi, S. P., Khan, N., Li, G. M., McCausland, M., Kanchan, V., Kokko, K. E., Li, S., Elbein, R., Mehta, A. K., Aderem, A., Subbarao, K., Ahmed, R., & Pulendran, B. (2011). Systems biology of vaccination for seasonal influenza in humans. *Nature Immunology*, 12(8), 786–795. https://doi.org/10.1038/ni.2067.

Nandy, A., & Basak, S. C. (2019). Bioinformatics in design of antiviral vaccines. *Encyclopedia of Biomedical Engineering*, 280–290. https://doi.org/10.1016/B978-0-12-801238-3.10878-5. Epub 2018 Sep 13; PMCID: PMC7149942.

Ong, E., Wang, H., Wong, M. U., Seetharaman, M., Valdez, N., & He, Y. (2020). Vaxign-ML: Supervised machine learning reverse vaccinology model for improved prediction of bacterial protective antigens. *Bioinformatics (Oxford, England)*, 36(10), 3185–3191. https://doi.org/10.1093/bioinformatics/btaa119.

Ozdarendeli, A. (2023, August 19). Crimean-Congo hemorrhagic fever virus: Progress in vaccine development. *Diagnostics (Basel)*, 13(16), 2708. https://doi.org/10.3390/diagnostics13162708. PMID: 37627967; PMCID: PMC10453274.

Pardi, N., et al. (2018). mRNA vaccines—a new era in vaccinology. *Nature Reviews Drug Discovery*, 17, 261–279. https://doi.org/10.1038/nrd.2017.243.

Pica, N., & Palese, P. (2013). Toward a universal influenza virus vaccine: Prospects and challenges. *Annual Review of Medicine*, 64, 189–202. https://doi.org/10.1146/annurev-med-120611-145115.

Pulendran, B., & Ahmed, R. (2011, June). Immunological mechanisms of vaccination. *Nature Immunology*, 12(6), 509–517. https://doi.org/10.1038/ni.2039. PMID: 21739679; PMCID: PMC3253344.

Rabaan, A. A., Al-Ahmed, S. H., Haque, S., Sah, R., Tiwari, R., Malik, Y. S., Dhama, K., Yatoo, M. I., Bonilla-Aldana, D. K., & Rodriguez-Morales, A. J. (2020). SARS-CoV-2, SARS-CoV, and MERS-COV: A comparative overview. *Le infezioni in medicina*, 28(2), 174–184.

Rappuoli, R., Bottomley, M. J., D'Oro, U., Finco, O., & De Gregorio, E. (2016, April 4). Reverse vaccinology 2.0: Human immunology instructs vaccine antigen design. *Journal of Experimental Medicine*, 213(4), 469–481. https://doi.org/10.1084/jem.20151960. Epub 2016 Mar 28. PMID: 27022144; PMCID: PMC4821650.

Rawat, S. S., Keshri, A. K., Kaur, R., & Prasad, A. (2023, January 19). Immunoinformatics approaches for vaccine design: A fast and secure strategy for successful vaccine development. *Vaccines (Basel)*, 11(2), 221. https://doi.org/10.3390/vaccines11020221. PMID: 36851099; PMCID: PMC9959071.

Rinaudo, C. D., Telford, J. L., Rappuoli, R., Seib, K. L. (2009, September). Vaccinology in the genome era. *Journal of Clinical Investigation*, 119(9), 2515–2525. https://doi.org/10.1172/JCI38330. PMID: 19729849; PMCID: PMC2735939.

Sanyal, S. (2023, October 1). Crossroads in virology: Current challenges and future perspectives in the age of emerging viruses. *Disease Models & Mechanisms*, 16(10), dmm050476. https://doi.org/10.1242/dmm.050476. Epub 2023 Sep 20. PMID: 37728628; PMCID: PMC10538290.

Sautto, G. A., Kirchenbaum, G. A., & Ross, T. M. (2018). Towards a universal influenza vaccine: Different approaches for one goal. *Virology Journal*, 15(1), 17. https://doi.org/10.1186/s12985-017-0918-y.

Savoia, E., Piltch-Loeb, R., Goldberg, B., Miller-Idriss, C., Hughes, B., Montrond, A., Kayyem, J., & Testa, M. A. (2021, July 9). Predictors of COVID-19 vaccine hesitancy: Socio-demographics, co-morbidity, and past experience of racial discrimination. *Vaccines (Basel)*, 9(7), 767. https://doi.org/10.3390/vaccines9070767. PMID: 34358184; PMCID: PMC8310049.

Scholte, F. E. M., Karaaslan, E., O'Neal, T. J., Sorvillo, T. E., Genzer, S. C., Welch, S. R., Coleman-McCray, J. D., Spengler, J. R., Kainulainen, M. H., Montgomery, J. M., Pegan, S. D., Bergeron, E., & Spiropoulou, C. F. (2023). Vaccination with the Crimean-Congo hemorrhagic fever virus viral replicon vaccine induces NP-based T-cell activation and antibodies possessing Fc-mediated effector functions. *Frontiers in Cellular and Infection Microbiology*, 13, 1233148. https://doi.org/10.3389/fcimb.2023.1233148.

Screaton, G., Mongkolsapaya, J., Yacoub, S., & Roberts, C. (2015). New insights into the immunopathology and control of dengue virus infection. *Nature Reviews in Immunology*, 15(12), 745–759. https://doi.org/10.1038/nri3916.

Seib, K. L., Dougan, G., & Rappuoli, R. (2009, October). The key role of genomics in modern vaccine and drug design for emerging infectious diseases. *PLoS Genet*, 5(10), e1000612. https://doi.org/10.1371/journal.pgen.1000612. Epub 2009 Oct 26. PMID: 19855822; PMCID: PMC2752168.

Shin, C. I., Jeong, S. D., Rejinold, N. S., & Kim, Y. C. (2017). Microneedles for vaccine delivery: Challenges and future perspectives. *Therapeutic Delivery*, 8(6), 447–460. https://doi.org/10.4155/tde-2017–0032.

Shukla, R., Ramasamy, V., Shanmugam, R. K., Ahuja, R., & Khanna, N. (2020). Antibody-dependent enhancement: A challenge for developing a safe dengue vaccine. *Frontiers in Cellular and Infection Microbiology*, 10, 572681. https://doi.org/10.3389/fcimb.2020.572681.

So, A. D., & Woo, J. (2020). Reserving coronavirus disease 2019 vaccines for global access: Cross sectional analysis. *BMJ*, 371, m4750. https://doi.org/10.1136/bmj.m4750.

Soheili, M., et al. (2023). The efficacy and effectiveness of COVID-19 vaccines around the world: A mini-review and meta-analysis. *Ann Clin Microbiol Antimicrob*, 22, 42. https://doi.org/10.1186/s12941-023-00594-y.

Strengthen Vaccination Communications. (2024, June 18). *IQIP*. https://www.cdc.gov/iqip/hcp/strategies/strengthen-vaccination-communications.html.

Strzelec, M., Detka, J., Mieszczak, P., Sobocińska, M. K., & Majka, M. (2023). Immunomodulation-a general review of the current state-of-the-art and new therapeutic strategies for targeting the immune system. *Frontiers in Immunology*, 14, 1127704. https://doi.org/10.3389/fimmu.2023.1127704.

Subbian, S. (2023). Editorial: Innate immune evasion strategies during microbial infection. *Frontiers in Cellular and Infection Microbiology*, 13. https://doi.org/10.3389/fcimb.2023.1332253.

Tai, W., Zhang, X., Yang, Y., Zhu, J., Du, L. (2022, April). Advances in mRNA and other vaccines against MERS-CoV. *Transl Res*, 242, 20–37. https://doi.org/10.1016/j.trsl.2021.11.007. Epub 2021 Nov 19. PMID: 34801748; PMCID: PMC8603276.

Tay, D. J. W., Lew, Z. Z. R., Chu, J. J. H., & Tan, K. S. (2022). Uncovering novel viral innate immune evasion strategies: What has SARS-COV-2 taught us? *Frontiers in Microbiology*, 13. https://doi.org/10.3389/fmicb.2022.844447.

Tomar, N., & De, R. K. (2014). Immunoinformatics: A brief review. *Methods in Molecular Biology (Clifton, N.J.)*, 1184, 23–55. https://doi.org/10.1007/978-1-4939-1115-8_3.

Trimble, M. (2019, January 16). WHO: Anti-Vaccine movement a top threat in 2019. *US News & World Report*. https://www.usnews.com/news/national-news/articles/2019-01-16/who-names-vaccine-hesitancy-as-top-world-threat-in-2019.

Trovato, M., Sartorius, R., D'Apice, L., Manco, R., & De Berardinis, P. (2020). Viral emerging diseases: Challenges in developing vaccination strategies. *Frontiers in Immunology*, 11. https://doi.org/10.3389/fimmu.2020.02130.

Wang, Y., Ling, L., Zhang, Z., & Marin-Lopez, A. (2022, October 28). Current advances in Zika vaccine development. *Vaccines (Basel)*, 10(11), 1816. https://doi.org/10.3390/vaccines10111816. PMID: 36366325; PMCID: PMC9694033.

Weinreich, D. M., et al. (2021). REGN-COV2, a neutralizing antibody cocktail, in outpatients with Covid–19. *The New England Journal of Medicine*, 384(3), 238–251. https://doi.org/10.1056/NEJMoa2035002.

Wouters, O. J., Shadlen, K. C., Salcher-Konrad, M., Pollard, A. J., Larson, H. J., Teerawattananon, Y., & Jit, M. (2021, March 13). Challenges in ensuring global access to COVID-19 vaccines: Production, affordability, allocation, and deployment. *Lancet*, 397(10278), 1023–1034. https://doi.org/10.1016/S0140-6736(21)00306-8. Epub 2021 Feb 12. PMID: 33587887; PMCID: PMC7906643.

Zhang, C., Maruggi, G., Shan, H., & Li, J. (2019). Advances in mRNA vaccines for infectious diseases. *Frontiers in Immunology*, 10. https://doi.org/10.3389/fimmu.2019.00594.

Zhang, G. L., Sun, J., Chitkushev, L., & Brusic, V. (2014). Big data analytics in immunology: A knowledge-based approach. *BioMed Research International*, 2014, 437987. https://doi.org/10.1155/2014/437987.

Zimmerman, T., Shiroma, K., Fleischmann, K. R., Xie, B., Jia, C., Verma, N., & Lee, M. K. (2023, January 4). Misinformation and COVID-19 vaccine hesitancy. *Vaccine*, 41(1), 136–144. https://doi.org/10.1016/j.vaccine.2022.11.014. Epub 2022 Nov 14. PMID: 36411132; PMCID: PMC9659512.

11 Clinical Examination, Immunoassays, and Molecular Diagnostic Techniques for Emerging and Re-Emerging Diseases of Viral Origin

Shayeri Das, Srijan Goswami, Sangita Agarwal, and Dibya Das

11.1 INTRODUCTION

Clinical diagnosis is a process through which healthcare practitioners identify a disease or disorder based on a patient's symptoms, medical history, and physical examination findings. This approach relies on clinical reasoning, where the practitioner gathers and interprets signs observed during the physical examination and symptoms reported by the patient to deduce the most probable cause of illness. Unlike laboratory-based diagnostics, clinical diagnosis is often preliminary and may guide further confirmatory tests if necessary. This method is foundational in healthcare as it allows timely decision-making and treatment planning, especially when rapid interventions are required. Clinical diagnosis is distinct but can be complemented by imaging and laboratory tests, especially in cases where symptom presentation alone cannot conclusively establish the diagnosis. It is a critical skill in medical practice, as accurate and efficient clinical diagnosis can significantly improve patient outcomes by enabling prompt and appropriate treatment. Clinical diagnosis is essential for pathogen identification as it allows healthcare professionals to recognize potential infections, guiding timely treatment quickly. Clinical assessments, which include evaluating symptoms and patient history, enable doctors to prioritize further laboratory and molecular tests specific to suspected pathogens. This is particularly important as early identification can lead to interventions that may limit the spread of contagious agents, an especially critical factor with high-risk, drug-resistant pathogens and emerging diseases. Clinical diagnosis also bridges the gap when advanced molecular or serological tests are not immediately available, offering an initial direction for disease management. Clinicians can accurately confirm pathogens' presence by combining clinical insights with laboratory techniques. Techniques like culture methods, PCR, and MALDI-TOF mass spectrometry have enhanced diagnostic precision by identifying pathogens based on unique proteins or genetic markers. These tools, integrated with clinical diagnosis, improve the speed and specificity of identifying disease-causing organisms, ultimately enhancing patient outcomes. Accurate diagnosis is not just critical; it is a compassionate act. An outbreak can be prevented if a disease or infection is diagnosed early. A correct diagnosis can significantly improve the effectiveness of treatments, thereby avoiding long-term complications and ensuring the patient's well-being. This emphasis on patient well-being is a testament to healthcare professionals' empathy and care, making them feel valued and appreciated for their work. If patients remain undiagnosed, they can unknowingly spread the infection. Properly explaining the symptoms to the patients relieves their mental stress and helps them decide on the right treatment path, showing them that their well-being is our utmost concern.

DOI: 10.1201/9781032721811-11

Proper diagnosis of the disease-causing pathogen is imperative since it leads to prevention strategies and the development of new treatment methods. Clinical decision-making comprises diagnosis, assessment of severity, and management (Berman, 2011). The main pillars of clinical decision-making are proper physical examination and patient history evaluation, which ensure precise diagnosis (Berman, 2011). In most situations, various diagnostic tests are performed. The correct diagnosis helps in understanding the severity of the disease or illness. The severity of magnitude is directly related to the treatment and management of the disease. It helps in deciding what treatment will be provided, the duration of the treatment, and whether the patient requires immediate treatment in a hospital or can be treated at home, followed by further patient monitoring. Appropriate clinical decision-making considers the need to make a precise diagnosis and the costs associated with inappropriate or indiscriminate use of diagnostics tests. The number of outbreaks of various viral infections has increased over the years. New diseases like COVID-19 have emerged, whereas preexisting diseases like malaria have re-emerged. These viral diseases pose a constant health threat to the global public, leading to massive outbreaks and population deaths. The correct diagnosis is crucial for outbreak prevention, disease control, and treatment of infected patients. The proper diagnosis can also explain the origin of the viral diseases.

11.2 CLINICAL EXAMINATION AND TRIAGE FOR EMERGING AND RE-EMERGING DISEASES OF VIRAL ORIGIN

Emerging and re-emerging diseases pose a severe risk to mankind and are one of the top reasons for human mortality and morbidity (Patel et al., 2021). Emerging infections are those that have never been recognized previously. Re-emerging infections are those that have existed previously and have come back again in a different form after their previous decline (Patel et al., 2021). These viral infections are deemed a public threat. The clinical features of patients infected with emerging or re-emerging viruses range from asymptomatic infection to severe symptoms and even death. The host immune response against the virus should be appropriate to control the disease without causing self-injury (Gogoi et al., 2024). Diagnosis of these infections is very crucial. The origin of the infections should also be determined (Mourya et al., 2019). Various kinds of viral diagnostic assays detect viral proteins in infected cells. There are different kinds of microscopic techniques. Genome quantification by PCR and serological tests is another infection identification technique. Typical targets for diagnostic tests are viral proteins (antigens), viral genomes, and antiviral antibodies. Some are designed to detect viral proteins, enzymes, or genomes directly from a patient sample (blood, throat swab), and these are useful in a clinical setting (Payne, 2017). Clinical examination and triage are foundational to managing emerging and re-emerging viral diseases. These processes are essential for early detection, risk assessment, and patient prioritization, which are critical for effective outbreak control and resource allocation in healthcare settings. Clinical examination involves systematically evaluating a patient's symptoms and history, which provides clues for identifying viral pathogens based on their epidemiological patterns, joint presentations, and transmission characteristics (Morens & Fauci, 2020). For instance, in cases of COVID-19, Ebola, or Zika, clinical indicators such as fever, respiratory symptoms, rashes, and bleeding are often primary red flags, prompting further laboratory or imaging tests. Triage, the process of prioritizing patients based on the severity of their condition, becomes particularly crucial during viral outbreaks due to the high number of cases that can overwhelm healthcare resources. Triage helps determine which patients require immediate medical attention, which can be managed with supportive care, and which need isolation to prevent transmission. This process is informed by clinical examination findings and risk factors, such as immunocompromised status or comorbidities, which may predispose patients to severe outcomes. Structured triage protocols, like the START (simple triage and rapid treatment) method, allow clinicians to quickly assess and categorize patients based on urgency, enhancing response efficiency and minimizing mortality in outbreak scenarios. Recent advancements in digital health and AI-driven tools have augmented the traditional clinical examination and triage process by facilitating the rapid processing of patient data. For example, wearable devices and telemedicine platforms enable real-time monitoring of vital signs and symptoms in suspected cases, potentially identifying critical cases sooner and reducing the risk of viral spread in clinical settings. However, challenges remain, such as the need for trained personnel and infrastructure, especially in low-resource settings where emerging and re-emerging viral diseases often occur. Clinical examination and triage are vital in managing viral diseases, particularly in rapidly evolving outbreaks. Adequate clinical examination and structured triage systems ensure timely patient care, optimize resource use and play a crucial role in controlling the spread of infections.

11.3 DIAGNOSTIC TESTS FOR EMERGING AND RE-EMERGING DISEASES OF VIRAL ORIGIN

Diagnostic testing for emerging and re-emerging viral diseases has become crucial for timely outbreak response and patient management. Rapid and accurate diagnostic tests help identify infections and guide intervention strategies, thus preventing disease spread. Molecular assays, such as reverse transcription-polymerase chain reaction (RT-PCR), are among the primary diagnostic tools for detecting viral RNA in diseases like COVID-19 and Ebola due to their high sensitivity and specificity (World Health Organization, 2022). Immunoassays, including enzyme-linked immunosorbent assays (ELISA) and lateral flow tests, are also widely used for detecting viral antigens or antibodies, providing faster results and ease of use in point-of-care settings. Advanced technologies like next-generation sequencing (NGS) and CRISPR-based diagnostics have enhanced pathogen detection, enabling the identification of novel viral strains and mutations, which is essential in monitoring viral evolution and guiding vaccine development. Despite these advancements, challenges remain, such as ensuring accessibility in low-resource settings and maintaining diagnostic accuracy amid rapidly mutating viruses.

Polymerase chain reaction (PCR) is a molecular technique that amplifies specific DNA sequences, enabling their detection and analysis. The PCR process relies on a cyclic series of temperature changes to produce millions of copies of a target DNA segment, making it detectable even in small quantities. The process begins with denaturation, where the double-stranded DNA is heated to around 94–98°C to separate it into two single strands. Next is the annealing phase, where the temperature is lowered to around 50–65°C, allowing short DNA primers to bind to their complementary sequences on the target DNA strands. This step is critical for synthesizing the DNA segment of interest. In the final extension phase, the reaction temperature is raised to about 72°C, optimal for the DNA polymerase enzyme, which synthesizes a new DNA strand by adding nucleotides to the primer-bound sequences. Repeating these steps for 20–40 cycles exponentially amplifies the target DNA, allowing it to be analyzed for various applications, including pathogen detection, genetic research, and forensic analysis (World Health Organization, 2022; Tayyeb & Basit, 2023). Polymerase chain reaction (PCR) is a powerful tool for diagnosing viral diseases and detecting viral genetic material with high sensitivity and specificity. PCR enables the identification of viruses by amplifying small amounts of viral DNA or RNA to detectable levels, a crucial step in diagnosing infections early, even when viral load is low. For RNA viruses like SARS-CoV-2, reverse transcription PCR (RT-PCR) is particularly valuable as it converts RNA to DNA before amplification, providing a reliable diagnostic method used worldwide during the COVID-19 pandemic (World Health Organization, 2022). PCR's applications extend to detecting viral pathogens, including HIV, influenza, and hepatitis viruses, allowing timely and precise diagnosis and guiding effective treatment and infection control measures. Beyond diagnostics, PCR supports viral load monitoring, resistance testing, and epidemiological studies, making it indispensable in public health surveillance and managing outbreaks.

11.3.1 Serological Tests: Working Principle, Purpose, and Application

Serological tests are critical for diagnosing viral diseases, as they detect specific antibodies or antigens associated with viral infections. These tests assess the immune response rather than the virus, providing insight into past and recent infections. They play a precious role in identifying viruses that may not be detectable by molecular methods due to low viral loads, such as in the later stages of infection (World Health Organization, 2022). Common types of serological tests include enzyme-linked immunosorbent assay (ELISA) and lateral flow immunoassays, which detect antibodies like IgM and IgG that develop in response to viral infections. For instance, serology has been crucial in tracking and diagnosing infections with HIV, hepatitis B, and, more recently, SARS-CoV-2, helping to determine previous exposure and potential immunity. Serological testing has extensive applications in epidemiology, enabling public health officials to monitor the spread of infection, assess population immunity, and evaluate vaccine efficacy. While these tests are valuable for detecting immune responses, they do not provide immediate infection status, as antibodies can take days to weeks to develop. As a result, they are often used alongside PCR tests for comprehensive diagnosis and monitoring of viral disease spread (Vainionpää et al., 2015). Serological tests for viral disease diagnosis include several key types, each with distinct applications based on the target analyte—antibodies or antigens—and the test format. The enzyme-linked immunosorbent assay (ELISA) is one of the most commonly used types, ideal for detecting specific antibodies (such as IgM and IgG) and quantifying them in a patient's serum. This test is widely used in diagnosing viral infections like HIV, hepatitis B, and, recently, COVID-19, as it provides high sensitivity and specificity. Another prevalent type is the lateral flow immunoassay (LFA), a rapid, point-of-care test often used in settings where quick results are essential. LFAs are widely applied in field settings, as they require minimal equipment and provide results within minutes, making them valuable for tracking outbreaks (World Health Organization, 2022). Additionally, chemiluminescent immunoassays (CLIA) have gained popularity for their enhanced sensitivity and automation capabilities, particularly in laboratory settings where high-throughput testing is necessary. Neutralization assays, which measure the ability of antibodies to prevent viral infection in cell cultures, are essential for assessing immunity, vaccine efficacy, and the detection of specific neutralizing antibodies in research and clinical contexts. These diverse serological testing methods allow for comprehensive surveillance and diagnosis, supporting various stages of disease monitoring and management.

11.3.2 Virus Isolation and Culture: Working Principle, Purpose, and Application

Virus isolation and culture involve growing viruses in controlled environments, such as cell cultures, to study their biological properties and support diagnosis. Virus isolation begins by inoculating a sample suspected of containing the virus into a suitable host cell culture, which provides the virus with the necessary nutrients and environmental conditions to replicate. This approach is commonly conducted in biosafety laboratories to prevent contamination and ensure safe handling. The cells are observed for cytopathic effects (CPE)—visible changes in cell morphology or cell death that indicate viral infection. If these effects are observed, they confirm the presence of an active virus, allowing researchers to proceed with further characterization, including genetic sequencing and drug susceptibility testing. Culturing viruses requires specific host cells that align with the virus' natural tropism; for example, respiratory viruses like influenza are often grown in respiratory tract cells, while enteric viruses are cultured in gastrointestinal cells. Although virus culture remains a gold standard for identifying novel pathogens, it is time-consuming and labor-intensive. However, it remains essential for applications such as vaccine development, antiviral testing, and studying viral mutations, making it a vital tool in virology research and clinical diagnostics (WHO, 2022; Corona et al., 2018). Virus isolation and culture aim to obtain live virus samples for further study, diagnostics, and development of antiviral therapies and vaccines. Virus isolation is crucial for identifying pathogens, particularly newly emerging or re-emerging, as it enables the cultivation of viruses in cell cultures or animal models. By growing viruses in controlled conditions, researchers can observe viral replication and morphological changes in infected cells, essential for understanding viral pathogenesis and transmission. One of the primary applications of virus isolation is in diagnostic virology, where cultured viruses help confirm infections, especially when molecular tests like PCR are unavailable or inconclusive. It also aids in identifying viral strains, determining their infectivity, and assessing their susceptibility to drugs and vaccines. In addition to diagnostic applications, virus isolation, and culture are fundamental in vaccine development. Isolated viruses can produce viral antigens or attenuated strains for vaccine production (World Health Organization, 2022). Furthermore, virus culture plays a role in studying viral genetics and mutation patterns, which is essential for tracking viral evolution and monitoring resistance to antiviral treatments. While labor-intensive, virus isolation and culture remain indispensable tools in virology.

11.3.3 Next-Generation Sequencing: Working Principle, Purpose, and Application

Next-generation sequencing (NGS) and CRISPR-based techniques have revolutionized viral diagnostics by offering high-throughput, precise, and versatile methods for identifying viral genomes. NGS works by sequencing DNA or RNA in parallel, allowing for the rapid detection and characterization of viral genomes from various pathogens, even when the virus is novel or highly mutagenic. The process involves fragmenting viral genetic material sequenced by advanced sequencing platforms, producing vast amounts of data that can be analyzed to identify viral sequences, mutations, and genetic variations. NGS has been pivotal in diagnosing infections like COVID-19 and Zika by facilitating the identification of new viral strains and their mutations, which is essential for epidemiological tracking and the development of vaccines. On the other hand, CRISPR-based diagnostics utilize the genome-editing capabilities of CRISPR-Cas systems to identify specific viral sequences with high sensitivity. The CRISPR-Cas9 system targets and cleaves viral DNA or RNA sequences, and this activity can be detected through various methods, such as fluorescence or color change, making it a promising tool for rapid, point-of-care viral diagnostics. The CRISPR-based diagnostic platform, known as SHERLOCK (specific high-sensitivity enzymatic reporter UnLOCKing), has successfully detected low levels of viral nucleic acids, including in samples from patients with asymptomatic infections. NGS and CRISPR technologies represent a significant leap forward in viral diagnostics, enabling faster, more accurate detection of a wide range of viral pathogens (Cheng et al., 2023).

Next-generation sequencing (NGS) and CRISPR-based techniques have transformed the landscape of viral diagnostics, offering advanced capabilities for detecting and analyzing viral pathogens. NGS enables the sequencing of entire viral genomes in a high-throughput manner, allowing for the identification of novel and evolving viral strains. This technology has been instrumental in tracking viral mutations, monitoring epidemics, and identifying emerging pathogens, such as SARS-CoV-2, by providing comprehensive genomic data that can inform vaccine development and therapeutic strategies. NGS is beneficial for detecting viruses that cannot be cultured or are challenging to detect using traditional methods, including rare or genetically diverse viruses. CRISPR-based diagnostics, particularly CRISPR-Cas systems, offer a compassionate and rapid alternative for detecting viral nucleic acids. The CRISPR-Cas9 system can be engineered to recognize specific viral sequences, and its action can be measured by colorimetric or fluorescent readouts, allowing for quick detection of infections even in low-resource settings. These systems, such as SHERLOCK, have demonstrated outstanding potential for point-of-care diagnostics, enabling the detection of viruses like SARS-CoV-2, Zika, and HIV with minimal equipment. Applying NGS and CRISPR technologies in viral diagnostics provides a powerful toolkit for real-time surveillance, accurate diagnosis, and timely outbreak response.

11.3.4 Point of Care Testing (POCT): Working Principle, Purpose, and Application

Point-of-care testing (POCT) for virus diagnosis operates on the principle of rapid detection of viral antigens, nucleic acids, or antibodies at or near the site of patient care, bypassing the need for centralized laboratory infrastructure. POCT devices typically utilize immunoassays, lateral flow assays, or nucleic acid amplification technologies, such as isothermal amplification, to detect specific viral markers quickly and accurately. For example, lateral flow assays detect viral antigens by binding antibodies to target proteins, generating a visible signal in minutes. Similarly, nucleic acid-based POCT systems amplify viral RNA or DNA using loop-mediated isothermal amplification (LAMP), providing results in under an hour. The critical components of POCT include user-friendly interfaces, minimal sample preparation, and portability, making these systems particularly useful in resource-limited settings, outbreak areas, or emergencies (World Health Organization, 2022). During the COVID-19 pandemic, POCT devices such as antigen-based rapid diagnostic tests and molecular platforms significantly enhanced testing capacity and accessibility. Their ability to deliver timely and actionable results plays a crucial role in early disease detection, treatment initiation, and containment of viral spread (Khan et al., 2023).

Point-of-care testing (POCT) enables rapid and accessible diagnosis of viral infections directly at or near the site of patient care without requiring advanced laboratory infrastructure. Its primary goal is to provide timely diagnostic results to facilitate early detection, prompt treatment, and effective containment of viral diseases, particularly in resource-limited or emergency settings. POCT is widely applied in detecting viral infections such as HIV, influenza, and COVID-19. During the COVID-19 pandemic, antigen-based rapid diagnostic tests and molecular POCT platforms proved invaluable in expanding testing capacities and alleviating the burden on centralized laboratories. Applications of POCT extend to epidemiological surveillance, outbreak management, and monitoring disease progression. In low-resource areas, POCT plays a pivotal role in diagnosing diseases like dengue and hepatitis, where timely results are essential for reducing morbidity and mortality. These systems are also increasingly integrated into public health strategies for controlling emerging and re-emerging diseases by enabling decentralized and mass testing efforts (World Health Organization, 2022). Overall, POCT enhances diagnostic accessibility, improves patient outcomes, and contributes significantly to global health initiatives (Bhattacharya et al., 2020).

11.3.5 Imaging Techniques: Working Principle, Purpose, and Application

Imaging techniques play a crucial role in diagnosing viral infections by enabling visualization of viral particles, their structures, and interactions with host cells. Electron microscopy (EM), one of the most commonly used imaging methods, operates on the principle of electron beam transmission or scanning to generate high-resolution images of viral morphology and ultrastructure. Transmission electron microscopy (TEM) can directly visualize viral particles in clinical specimens, facilitating the identification of viruses based on size, shape, and structural features. On the other hand, scanning electron microscopy (SEM) provides detailed surface topography of infected cells or viral particles. Advanced imaging techniques, such as cryo-electron microscopy (cryo-EM), further enhance diagnostic capabilities by allowing near-atomic resolution imaging of viral capsids and spike proteins without requiring staining or fixation. Additionally, fluorescence and confocal microscopy are applied to detect labeled viral antigens or nucleic acids within cells, often using specific dyes or tagged antibodies for localization studies. These imaging modalities are particularly valuable in studying viral pathogenesis, monitoring disease progression, and evaluating the efficacy of antiviral agents (van Oostveen & de Lange, 2021). Imaging techniques in viral diagnosis serve the dual purpose of identifying viruses and understanding their structural and functional properties, which are essential for accurate diagnosis and research. These techniques, such as electron microscopy (EM), enable direct visualization of viruses, providing insights into their morphology, size, and surface features, which are critical for distinguishing viral species. Cryo-electron microscopy (cryo-EM) extends these capabilities by capturing high-resolution, near-native images of viral structures, including envelope proteins and capsid architecture, aiding vaccine development and antiviral drug design. Fluorescence microscopy and confocal imaging, on the other hand, are widely used for detecting viral antigens or nucleic acids in infected tissues or cells, often utilizing labeled antibodies or probes to localize the virus within host systems. Applications of imaging techniques range from routine diagnostics, such as identifying viral particles in clinical specimens, to advanced research exploring viral–host interactions and pathogenesis. For instance, during outbreaks of emerging viruses like Ebola or SARS-CoV-2, EM has been pivotal in confirming the presence of the virus in patient samples. Imaging also supports therapeutic research by visualizing the effects of antiviral agents on viral replication and morphology (Su et al., 2023). These tools remain indispensable in virology for both clinical and investigative purposes (Bunaciu et al., 2015; Pontone et al., 2021; Chien et al., 2020).

11.3.6 Mass Spectrometry (MS): Working Principle, Purpose, and Application

Mass spectrometry (MS) measures ionized molecules' mass-to-charge ratio (*m*/*z*) to identify and characterize viral components such as proteins, lipids, and nucleic acids. In virus diagnosis, MS involves ionizing viral particles or their biomarkers, such as capsid proteins, using methods like electrospray ionization (ESI) or matrix-assisted laser desorption/ionization (MALDI). The generated ions are then separated based on their *m*/*z* in a mass analyzer, such as time-of-flight (TOF) or quadrupole, and detected to produce a mass spectrum unique to the virus. This technique allows for identifying viral species and strains by analyzing their proteomic or lipidomic signatures, offering high sensitivity and specificity. For instance, MALDI-TOF MS has been used to identify influenza viruses and SARS-CoV-2 by detecting specific protein profiles directly from clinical samples. MS is also applied in viral metagenomics, where it helps profile host–pathogen interactions and discover novel viruses. Its rapid and accurate detection capabilities make MS valuable in routine diagnostics, outbreak investigation, and therapeutic monitoring (Shuken, 2023). Mass spectrometry (MS) is a versatile tool in virus diagnosis, serving the dual purpose of identifying viral pathogens and analyzing their structural and functional components. Its application lies in its ability to detect viral proteins, lipids, and nucleic acids with high specificity and sensitivity, enabling the rapid identification of known and novel viruses. For instance, MALDI-TOF MS has been widely adopted to identify viruses like influenza, dengue, and SARS-CoV-2 from clinical samples, offering a time-efficient alternative to traditional diagnostic methods. Moreover, MS is instrumental in studying viral proteomes, discovering biomarkers for early disease detection, and monitoring treatment efficacy. MS can quickly profile viral particles in an outbreak, facilitating real-time epidemiological surveillance and containment efforts. Beyond diagnosis, it is used in vaccine development and antiviral research by elucidating virus–host interactions and the structural dynamics of viral components. Advanced MS technologies, such as tandem MS (MS/MS), further enhance its applications by enabling detailed characterization of viral genomes and posttranslational modifications, making it indispensable in modern virology and public health strategies (Sorelle et al., 2020; Su et al., 2023).

11.3.7 Loop-Mediated Isothermal Amplification: Working Principle, Purpose, and Application

Loop-mediated isothermal amplification (LAMP) is a nucleic acid amplification technique designed to amplify viral DNA or RNA rapidly and efficiently under isothermal conditions. The method relies on four to six primers recognizing six to eight regions of the target sequence and a strand-displacing DNA polymerase. Unlike polymerase chain reaction (PCR), LAMP operates at a constant temperature (usually 60–65°C), eliminating the need for thermal cycling. The reaction proceeds through initial DNA strand displacement followed by self-primed DNA synthesis, resulting in large quantities of amplified DNA with a characteristic looped structure. The amplification can be detected in real-time using fluorescence dyes, turbidity, or colorimetric indicators, enabling rapid visualization of results without advanced equipment. LAMP's robustness against inhibitors and minimal equipment requirements makes it particularly suitable for virus diagnostics in resource-limited settings. It has been widely applied for detecting viruses like SARS-CoV-2, dengue, Zika, and hepatitis viruses due to its speed, specificity, and sensitivity. These features position LAMP as a critical tool for rapid and point-of-care viral diagnostics, especially during outbreaks (Shirshikov & Bespyatykh, 2022).

Loop-mediated isothermal amplification (LAMP) provides a rapid, cost-effective, and accessible method for detecting viral pathogens. It is precious for point-of-care testing and resource-limited settings due to its simplicity and efficiency. LAMP's high sensitivity and specificity make it a robust tool for diagnosing viral infections, including SARS-CoV-2, Zika, dengue, and hepatitis viruses. Unlike traditional polymerase chain reaction (PCR), LAMP operates under isothermal conditions, allowing the reaction to proceed without complex thermal cycling, thus significantly reducing the need for sophisticated laboratory equipment. The application of LAMP extends to early disease detection, surveillance during outbreaks, and monitoring of treatment outcomes. Its rapid turnaround time, often within 30 minutes, makes it ideal for emergency diagnostics during pandemics. In addition, LAMP can be integrated with portable detection devices, making it feasible for use in remote areas. Colorimetric and fluorescence-based detection further enhance its applicability by enabling quick and visual interpretation of results, thereby aiding healthcare providers in making timely and informed decisions.

11.4 DIAGNOSIS OF CCHFV

Crimean-Congo hemorrhagic fever (CCHF) is a rare, severe viral disease spread to humans by tick bites or infected animal blood (Frank et al., 2024). Symptoms usually start 1–14 days after being exposed to the virus. RT-PCR detects antibodies in the blood. Laboratory tests used to diagnose CCHF include reverse transcriptase (RT)-PCR, immunofluorescence assay (IFA), antibody (IgG, IgM) and antigen-capture ELISA, and virus isolation. RT-PCR primarily diagnoses patients suspected of CCHF as these assays provide the highest detection sensitivity to active infection at the earliest time (Mazzola & Kelly-Cirino, 2019a; Raabe, 2020):

- *Molecular diagnostics:* RT-PCR-based techniques typically target the nucleoprotein gene region in the S segment, which is the more conserved region of the CCHFV genome across geographical isolates. Quantitative real-time RT-PCR (qRT-PCR) has better performance over conventional RT-PCR or nested RT-PCR, with a lower contamination rate, higher sensitivity and specificity, and better time-effectiveness (Mazzola & Kelly-Cirino, 2019a; Raabe, 2020).
- *Serological assays:* Serological assays are sensitive to antigenic variation but less impacted by genetic variation. Most assays target the CCHFV N protein, which induces an early, strong, and long-lasting immune response in humans. Active CCHFV infection can be detected by IgM or a significant increase in IgG titer following the acute phase of infection four to nine days after symptom onset; however, severe and fatal cases often do not mount a detectable antibody response. For CCHFV, capture ELISA is more sensitive than IFA or neutralization assay. Virus neutralization assays are less functional for diagnosis since CCHFV elicits relatively low levels of neutralizing antibodies, but they can be helpful for epidemiology and vaccine research. CCHFV neutralization is generally performed using plaque reduction neutralization, with results lasting five to seven days (Mazzola & Kelly-Cirino, 2019a; Raabe, 2020). Several commercial ELISA (IgG, IgM) and IFA kits are available, although primarily marketed as RUO.
- *Rapid diagnostic tests:* RDTs can leverage the same antibody/antigen capture agents as an ELISA but in a lateral flow strip format, with minimal specimen processing (blood, plasma, and swabs). However, this results in a faster time with a lower detection sensitivity than ELISA, partly due to reduced sample volume. RDTs are ideal screening tests for field testing and low-infrastructure settings. However, follow-up confirmatory testing is often required. The primary challenge is detection sensitivity, as the IgG/IgM serological response is typically detectable only five days post-infection and often undetectable in severe and fatal infections (Mazzola & Kelly-Cirino, 2019a; Raabe, 2020).

11.5 DIAGNOSIS OF RIFT VALLEY FEVER VIRUS

Rift Valley fever (RVF) is a viral zoonosis that primarily affects animals but can infect humans. It is transmitted by mosquitoes and blood-feeding flies (WHO Newsroom Fact Sheet, 2018). The interval from infection to onset of symptoms for Rift Valley fever (RVF) varies from two to six days (WHO Newsroom Fact Sheet, 2018). Most human cases are relatively mild; however, a small percentage of patients develop a much more severe form of the disease. Clinical diagnosis is usually difficult in the early course of the disease since the symptoms are unspecific and varied. Suspected RVF cases have high containment levels. Limited capacities at the laboratories in the endemic regions have posed a significant hurdle in the timely diagnosis of RVF, leading to major outbreaks. Rift Valley fever virus infections can only be definitively diagnosed in the dedicated reference laboratory by trained biomedical staff using the following tests (Petrova et al., 2020; Cobo, 2012; Reta et al., 2020):

- Reverse transcriptase polymerase chain reaction (RT-PCR) assay
- IgG and IgM antibody enzyme-linked immunosorbent assay (ELISA)
- Virus isolation by cell culture

The selection of an optimal assay depends on the timing of sampling relative to disease progression and the ability to detect antigenic (isolated virus, viral RNA) or immunological markers (IgM and IgG). If the timing of infection is unknown, a combination of molecular and serological assays is usually needed to confirm RVFV cases (Petrova et al., 2020; Cobo, 2012; Reta et al., 2020).

- *Molecular tests:* It is most beneficial two to four days after infection and up to eight days after the onset of symptoms. Quantitative Polymerase Chain Reaction methods are preferred to ensure diagnosis, as the viral load is correlated with disease severity. Several commercially available molecular testing kits are based on RT-PCR and enable the detection of the RVF virus as a single test (Petrova et al., 2020; Cobo, 2012; Reta et al., 2020).
- *Serological tests:* These assays help assess the active infection in ongoing diseases or previous exposure to the virus. The presence of virus-specific IgM and IgG antibodies can detect the presence of active antigens or prior exposure to the virus. These tests are based on ELISA. ELISA tests are usually based on recombinant nucleocapsid protein (NP) and offer high specificity and simple sample processing. Diagnosis can also be made by immunofluorescent antibody (IFA) assay. Virus antigens are undetectable after four to five days of infection. A robust use of serological assays includes a combination of tests for detecting viral antigens and serum IgM. ELISA methods can detect responses to one antigen at a time, whereas RVF virus-specific antibodies can be detected using bead-based assays (Petrova et al., 2020; Cobo, 2012; Reta et al., 2020).
- *Rapid diagnostic tests:* These tests are critical in detecting potential RVF virus outbreaks early, especially in areas with remote laboratory access. If the rapid diagnostic tests (RDTs) results are accurate and their sensitivity and specificity are comparable to RT-PCR, this will pave the pathway for potential vaccine trials. Loop-mediated isothermal amplification and RPA-PCR form the base for several methods for rapid molecular testing (Petrova et al., 2020; Cobo, 2012; Reta et al., 2020).
- *Sequencing analysis:* RVF virus genome sequencing has been used to monitor the epidemiological spread of the virus and the geographic distribution of different strains. Whole genome sequencing of clinical specimens helps identify new strains and map virus evolution and transmission. Fifteen RVF virus lineages have been identified, and differences in virulence and pathogenicity in animal models have been observed. Genetic diversity should be considered when designing molecular diagnostic assays to ensure optimal assay sensitivity across all virus lineages (Petrova et al., 2020; Cobo, 2012; Reta et al., 2020).

11.6 DIAGNOSIS OF LASSA VIRUSES

Lassa fever is an infectious disease caused by the Lassa virus (LASV), part of the Arenaviridae family, known for causing hemorrhagic fevers. Symptoms can range from mild, such as fever, headache, sore throat, and gastrointestinal issues, to severe, including low blood pressure, swelling, and respiratory distress. They can potentially progress to a hemorrhagic phase with multi-organ failure and high mortality (Raabe & Koehler, 2017). Lassa fever symptoms resemble most other joint diseases, such as malaria and typhoid fever, which are challenging clinical diagnoses. Testing options are also limited. Patients with a high viral load followed by unexplained disease of any organ system are suspected of having Lass fever. Liver tests, urinalysis, serologic tests, and a complete blood count should then be done. Proteinuria is common and may be massive (Mazzola & Kelly-Cirino, 2019b). The gold standard for diagnosing LASV is polymerase chain reaction (PCR). PCR is highly accurate and requires advanced laboratory facilities and skilled personnel. This limits its accessibility in remote areas where Lassa fever often strikes massive (Mazzola & Kelly-Cirino, 2019b). Samples like blood, throat swabs, urine, and spinal taps are used for preliminary diagnosis of the disease. The viral lineage is detected through PCR. The LASSA virus' lineage diversity should be considered when choosing the PCR primers and serology antigens. These tests must be revalidated as more mutations occur, leading to more lineages. LASSA virus can be found in many body fluid compartments besides blood, including urine, semen, and cerebrospinal fluid. The virus persists in the central nervous system, urine, and semen long after viral clearance in the blood. The disease outcome is predicted based on the titer value of the virus. Fatality has been linked to suppressed or diminished IgM response throughout infection. This supports the hypothesis that early detection is critical to surviving infection.

- *Molecular diagnostics:* Polymerase chain reaction, loop-mediated isothermal amplification, and similar assays are molecular diagnostics tests designed to detect highly conserved regions of the pathogen genome. These are the most commonly used sensitive methods to detect active infection. These tests target the S genome segment encoding the GPC or the NP regions. Due to the virus' high mutation rate and genetic diversity, molecular diagnostic tests must encompass the sequence diversity. If not, this will risk false negatives in mutated or uncovered strains (Mazzola & Kelly-Cirino, 2019b; Cobo, 2012; Reta et al., 2020).
- *Serological and antigen detection assays:* Serological tests can detect IgM and IgG antibodies raised against LASV antigens and directly capture and detect LASV antigens. IgM and antigen tests are used to detect infection. However, not all patients have detectable IgM at the acute stage. In several cases, both IgM and IgG antibodies can be immunosuppressed. IgG is usually used only for surveillance in endemic settings. LASV antigen detection has been demonstrated to be a robust method for detecting active infection. While genetic sequence diversity and minor genetic mutations typically impact protein sequence less, early antigen tests have shown some variation in LASV lineage sensitivity (Mazzola & Kelly-Cirino, 2019b; Cobo, 2012; Reta et al., 2020).
- *Rapid diagnostic tests:* LASV RDTs leverage the same antibody/antigen capture agents as an ELISA but are packaged in a stripped-down lateral flow format. This is essential for patient care and outbreak response in outlying laboratories and clinics. The monoclonal capture agents developed for this LASV RDT showed reduced sensitivity to LASV lineages II and III. The improvements to the assay using a polyclonal approach suggest increased pan-lineage antigen sensitivity (Mazzola & Kelly-Cirino, 2019b; Cobo, 2012; Reta et al., 2020).

11.7 DIAGNOSIS OF NIPAH VIRUSES

Nipah virus is the causative agent of Nipah virus infection. It is a type of RNA virus in the genus *Henipavirus*. It naturally circulates among fruit bats of the genus *Pteropus*. The symptoms presented are fever, cough, headache, and shortness of breath. In worse situations, it can lead to coma, resulting in mortality (Mazzola & Kelly-Cirino, 2019c; Cobo, 2012; Reta et al., 2020). The symptoms of Nipah virus infection are similar to other febrile diseases. Early diagnosis is crucial to contain the disease and facilitate proper patient care (Mazzola & Kelly-Cirino, 2019c; Cobo, 2012; Reta et al., 2020). The laboratory tests include nucleic acid simplification testing, such as polymerase chain reaction, IgG/IgM/antigen ELISA, immunofluorescence assay, histopathology, virus isolation, and neutralization (Mazzola & Kelly-Cirino, 2019c; Cobo, 2012; Reta et al., 2020). PCR is considered the most effective method for diagnosing active infection. However, Nipah virus–specific IgM ELISA is regarded as an alternative serological approach (Banerjee et al., 2019). Histopathology (immunochemistry) confirms Nipah virus infection in fatal cases. Virus isolation and neutralization techniques are usually used for confirmation, but these are mostly restricted to Biosafety Level 4 laboratories (Mazzola & Kelly-Cirino, 2019c; Cobo, 2012; Reta et al., 2020). In the early stages of infection, specimens are collected from the throat, nasal swabs, cerebrospinal fluid, urine, and blood.

PCR is the most preferred method for detecting active viral infection since the infection is susceptible. This is available in high-infrastructure hospitals and unsuitable in endemic settings. These tests lose susceptibility if the viral genome undergoes significant genetic mutation or new strains arise. Reverse-transcriptase PCR (RT-PCR) tests for the Nipah virus have targeted the conserved N, M, or P genome. Several types of PCR tests have been developed, such as RT-PCR, nested RT-PCR, and real-time RT-PCR. Real-time RT-PCR is also known as quantitative PCR or qPCR. Real-time PCR has proven to be 1,000 times more sensitive than conventional PCR. Additionally, next-generation sequencing and deep sequencing enable a direct read of the viral genome, allowing virus and clade identification without prior knowledge of the composition. This approach is highly complex, expensive, and impractical for screening more samples.

Serological assays can directly detect the Nipah virus antigens and IgM and IgG antibodies raised against these antigens. These tests are generally less sensitive to minor genetic variation. IgG tests are primarily used for epidemiological studies and surveillance. IgM ELISA is the first-line serological diagnostic test for the Nipah virus, followed by serum neutralization or PCR as a confirmatory test. Recombinant proteins and monoclonal or polyclonal antibodies are capture agents for increased specificity. ELISA combinations of polyclonal antibodies can enable pan-henipavirus (NiV and HeV) detection and differentiation between NiV and HeV. However, comparing assay sensitivity is challenging due to differences in sample preparation and quantitation units. ELISA test reagents can be translated into a lateral flow format to create a rapid diagnostic test (RDT). This can be implemented in a non-laboratory environment, typically with minimal specimen processing (blood, plasma, swabs). The lateral flow format enables a faster time to result (10–30 minutes), although it has a lower detection sensitivity than its ELISA counterparts.

11.8 DIAGNOSIS OF EBOLA AND MARBURG VIRUSES

Ebola is infectious, and the causative agent is a group of viruses known as Orthoebolaviruses. Orthoebolavirus was formerly known as the Ebola virus. The origin of Orthoebolaviruses is primarily in sub-Saharan Africa. Orthoebolaviruses can cause severe disease, and if untreated, it can be fatal. There is a high mortality rate of 80–90%. Four orthoebolaviruses cause illness in people:

- Ebola virus (species *Orthoebolavirus zairense*) causes Ebola virus disease.
- Sudan virus (species *Orthoebolavirus sudanense*) causes Sudan virus disease.
- Taï Forest virus (species *Orthoebolavirus taiense*) causes Taï Forest virus disease.
- Bundibugyo virus (species *Orthoebolavirus bundibugyoense*) causes Bundibugyo virus disease.

It may take 2–21 days for the symptoms of Ebola disease to appear after someone is infected with the virus. On average, eight to ten days are required for the symptoms to show after exposure.

At the initial stage, symptoms are generic and are known as "dry" symptoms. Some dry symptoms are fever, muscle and joint pain, severe headache, weakness, fatigue, and sore throat. Since the symptoms at the initial stage are generic and commonly observed for other diseases, it often becomes difficult for healthcare professionals to identify the Ebola disease. The symptoms at a later stage of disease progression at four to five days are called "wet" symptoms. Some wet symptoms are loss of appetite, unexplained breathing, nausea, abdominal pain, diarrhea, and vomiting. Severe symptoms may include chest pain, shortness of breath, confusion, red eyes, skin rash, hiccups, and seizures. People who have severe symptoms at the early onset of the disease and left untreated often expire. In contrast, those treated at the early onset may begin improving about six days after symptoms start. Two vaccines, Inmazeb and Ebanga, are FDA-approved treatments currently available to treat Ebola disease caused by *Orthoebolavirus zairense* (US Center for Disease Control and Prevention, 2024a).

Polymerase chain reaction (PCR) is one of the most commonly used diagnostic methods because it can detect low virus levels. Four species of *Ebolavirus* are known to cause disease in humans. The *Zaire ebolavirus* has been the leading cause of infections, so laboratory diagnostics have focused on identifying that species (NETEC, 2022). Tests to confirm suspected Ebola cases are often performed in reference laboratories. This is because there are few laboratories in territories where Ebola viruses are endemic and a limited number of surveillance programs. This measure provides a guarantee regarding the accuracy of results. However, shipping samples to a centralized facility where the diagnostic test is performed takes significant time, leading to increased costs and delays in the test (Bettini et al., 2023). In recent outbreaks, molecular tests have been the most frequently used tool in Ebola diagnosis. The Ebola virus should be managed in Biosafety Level 4 facilities as it is categorized as a high-hazard pathogen. These laboratories have high-efficiency particulate air (HEPA) filtered air-handling systems with negative pressure, and all waste material is inactivated. The staff working at these laboratories are highly skilled. Mobile diagnostic laboratories provide an option to carry out rapid diagnosis. In a mobile laboratory, there are three separate areas: places for inactivation and extraction, reagent preparation, and quantitative reverse-transcription polymerase chain reactions (RT-PCR). In the first area (High-Risk area), the samples are inactivated, another area is present for reagent preparation, and RT-PCR execution, which is considered a Low-Risk area, and a third area is used as office space, with locker rooms and a shower room (Bettini et al., 2023).

- *Reverse transcription–polymerase chain reaction (RT-PCR):* RT-PCR is considered the gold standard in Ebola diagnosis due to its specificity and sensitivity. RT-PCR can be of two types: qualitative (RT-PCR) or quantitative RT-PCR. The viral RNA should be purified by extraction steps in both cases. Most RNA extraction methods are based on chaotropic methods, which use, prevalently, guanidium-isothiocyanate and a column with glass fiber filters or magnetic beads for RNA isolation. The reverse transcriptase step can be carried out separately from PCR or in a "one-step method," where all the reagents necessary for reverse transcription and amplification of cDNA are located in the same tube. The primers used in PCR must target the conserved regions of a virus. PCR products are checked in an ethidium bromide gel by electrophoresis. Even though this method is inexpensive, it is rarely used because it is time-consuming. A second PCR round (nested PCR) is often required to improve the sensitivity and specificity. Moreover, separate rooms are required for RNA extraction, PCR reagent preparation, and gel electrophoresis (Bettini et al., 2023).
- *Rapid diagnostic test (RDT):* Commercially available RDT include ReEBOV, SD Zaire Ag, OraQuick, Ebola Antigen System, and QuickNavi-Ebola. The ReEBOV test received emergency use authorization from the FDA and the WHO during the 2013–2016 Ebola outbreak. However, in 2018, after Zalgen Laboratories had acquired the company and failed to reproduce the claimed test accuracy of ReEBOV, the FDA revoked this authorization. These RDTs can be performed on capillary blood, whole blood with EDTA, plasma, serum, or saliva. The use of RDTs has a specificity and sensitivity of 99%. The WHO's

ASSURED criteria for point-of-care devices in November 2014 are affordable, selective, user-friendly, rapid, equipment-free, and deliverable to users. Point-of-care rapid diagnostic tests significantly reduce the requirement of trained personnel, laboratory infrastructures that hinder diagnostics in an outbreak setting, and the requirement of frequently sending samples to a reference center to be analyzed with a delay in the availability of results (Bettini et al., 2023).

- *Differential diagnosis*: Ebola virus disease has non-specific symptoms, and they pose a significant problem in isolation efforts at Ebola Treatment Centers. Viral hemorrhagic fevers (VHF) and malaria have overlapping clinical characteristics, making differential diagnosis a challenge. A new immunoassay called surface-enhanced Raman scattering (SERS) has been developed, which detects antigens from Ebola, Lassa, and malaria within a single blood sample. Results are provided within 30 minutes for each sample. Results from a trial setting indicate the potential of the SERS technology as an essential tool for outbreak management in low-resource settings. However, in real-time PCR, a few probe sets can be included, which is a limitation in differential diagnosis.
- *Enzyme-linked immunosorbent assays (ELISA):* Antigen ELISA assay is a traditional serological diagnostic tool for Ebola virus-infected patients. Though the body produces an IgM response in the early stages of infection, the anti-Ebola IgM titer decreases one month after the onset of infection, eventually becoming undetectable. The Ebola virus has a high mortality rate, and there is no proper humoral immune response. This is suggestive that an antibody-based detection system for Ebola may not be helpful in Ebola diagnosis. A new ELISA test was developed to detect filovirus species-specific antibodies. However, not much data is available to assess the sensitivity and specificity of the ELISA test (Bettini et al., 2023).
- *Immunofluorescent assay*: Biosafety Laboratories Level 4 has an indirect immunofluorescent assay (IFA), which uses virus-infected Vero cells to detect Ebola virus antibodies. Sample testing is done at a fixed dilution of 1:20 for IgM and IgG detection. The positive samples can also be tested at limiting dilution. Commercial Immunofluorescent assay IgG and IgM detection tests are unavailable (Bettini et al., 2023).
- *Viral isolation:* In the past, cultures of live agents were considered the gold standard for viral detection, but currently, PCR and next-generation sequencing are becoming increasingly relevant in diagnostics. Ebola virus isolation can be performed from the patient's serum, but this assay is not used for primary diagnosis since it is less sensitive than PCR and requires a BSL4 facility. However, viral isolation is helpful for virulence or pathogenesis studies (Bettini et al., 2023).
- *Electron microscopy:* Electron microscopy (EM) can visualize virus particles in specimens directly from clinical materials or viral culture and was a commonly used technique to diagnose filovirus technique in past decades. The immuno-EM method is a feasible method to perform for the detection of a filovirus that is serologically related to EBOV. An indirect immuno-electron microscopy method has been successfully applied to identify Ebola Reston (EBOV-R) particles in serum and tissue culture fluid specimens. The immuno-EM method also differentiated antigenically different filoviruses (Bettini et al., 2023).

11.9 DIAGNOSIS OF CORONAVIRUS DISEASES

Coronavirus struck the world at the end of 2019 and soon spread across the globe. The disease outbreak became a global health concern and posed a severe threat. Coronaviruses are related to RNA viruses that cause respiratory tract infections in mammals and humans. The causative agent of COVID-19 was severe acute respiratory syndrome coronavirus 2 (SARS-CoV-2), a positive-sense single-stranded RNA virus. SARS-CoV-2 is a strain of the species *Betacoronavirus pandemicum*. SARS-CoV-2 is of zoonotic origin, and its close genetic similarity to bat coronaviruses suggests it emerged from such a bat-borne virus. The disease was transmitted primarily via respiratory droplets from coughs and sneezes within about 1.8 meters (6 feet). COVID-19 affects people in different ways. The onset of symptoms usually takes five to six days up to 14 days from the day of infection. Most infected people develop mild to moderate illness and recover without hospitalization. The most common symptoms are fever, cough, tiredness, and loss of taste or smell. Less common symptoms are sore throat, headache, aches and pain, diarrhea, red or irritated eyes, and a rash on the skin or discoloration of fingers or toes. The most severe symptoms are chest pain, loss of speech or mobility, and difficulty breathing or shortness of breath. The symptoms of COVID-19 are similar to those of other diseases, such as making diagnostic tests essential for identifying people with COVID-19. The test to determine the illness should be taken immediately if the person has COVID-19 symptoms. A person should wait for at least five days if the person has been exposed to COVID-19 but is asymptomatic. COVID-19 tests use samples from the nose, throat, or saliva.

- *Laboratory tests:* Laboratory tests explained the potential effect of COVID-19 on lymphocytes, especially T cells. Routine blood tests, C-reactive protein, coagulation function, arterial blood gas, and tissue function tests played a significant role in understanding the impact of COVID-19. It was observed that nearly half of the patients infected with COVID-19 had reduced WBC count and reduced lymphocytes (lymphopenia) or low blood platelet count (thrombocytopenia) with increased activated thromboplastin time. C-reactive protein levels were increased in most patients, but procalcitonin concentrations were average. High levels of serum ferritin were reported in some patients (Abbasi-Oshaghi et al., 2020).
- *Imaging:* Radiological evaluations have been critical in diagnosing, managing, and following COVID-19 infections. The most commonly used radiological test is a thin-slice chest computed tomography (CT) scan. If an imaging abnormality is diagnosed early, it can be suspected of pneumonia. Lung imaging often shows the severity of the infection and possible lung abnormalities. If COVID-19 pneumonia is recognized early, it provides quick management and follow-ups. A high-resolution computed tomography (HRCT) scan can detect ground glass capacity (GGO) in the lungs more quickly in the early stages. The usual imaging features of COVID-19 pneumonia include interstitial inflammation, extensive consolidation with multifocal bilateral GGOs, bilateral involvement, noticeable peripheral or subpleural distribution, posterior or lower lobe predilection, and multiple lesions (Abbasi-Oshaghi et al., 2020). It is worth noting that the CT scan results of COVID-19 overlap with other viral pneumonia.
- *RT-PCR test:* The RT-PCR assay is highly recommended for rapid detection and treatment. This assay uses RNA isolated from respiratory specimens such as sputum, swabs from the nose or saliva, and deep tracheal aspirate. Presently, it is the most commonly used method to confirm the infection. However, this assay had high false-negative RT-PCR results and was inapplicable in the initial phase of the disease, which limited the rapid diagnosis of infected subjects. The low effectiveness is owed to low viral load, inappropriate samples, variation in the diagnosis rate among different kits, and undeveloped technology for nucleic acid detection. However, the positive rate of RT-PCR for throat swab specimens was about 60% with correct sample collection transportation and kit quality (Abbasi-Oshaghi et al., 2020).
- *Antigen tests:* Rapid tests such as antigen or at-home tests produce quicker results. However, antigen tests are less likely to detect the virus if no symptoms are present than nucleic acid amplification tests. This test looks for current infection and provides results within a few minutes. It is highly recommended that individuals with symptoms repeat the test twice, and those with no symptoms repeat it thrice at 48-hour intervals to confirm that the test is negative (American Lung Association, 2024).
- *Antibody tests:* Antibody tests detect past infections. In an antibody test, blood is drawn, and a laboratory test is performed to check for antibodies and see if there was a past

infection of SARS-CoV-2. An antibody test must not be used to determine the current infection, as the body can take sufficient time to make antibodies. This test is not recommended for immunity determination or vaccination requirements (American Lung Association, 2024).

11.10 DIAGNOSIS OF MOSQUITO-BORNE VIRAL INFECTIONS

Mosquito-borne diseases are transmitted to people through the bite of an infected mosquito. Parasites or viruses can cause these diseases. Vector-borne diseases account for more than 17% of all infectious diseases, causing more than 700,000 deaths annually (WHO Fact Sheet, 2024). Vectors are living organisms transmitting infectious pathogens between humans or from animals to humans. These diseases are highest in tropical and subtropical areas. Malaria, dengue, West Nile virus, chikungunya, yellow fever, and Zika are common mosquito-borne diseases.

- *Zika virus disease:* Zika fever is a mosquito-borne disease caused by the Zika virus. The causative agent is a virus transmitted primarily by *Aedes aegypti* mosquitoes. These mosquitoes bite during the day, usually just after sunrise and around sunset. The common symptoms are fever, rash, conjunctivitis, headache, muscle and joint aches. Most people infected with this virus are asymptomatic; however, if symptoms are shown, they last for two to seven days. Women infected by Zika virus during pregnancy may bear children with severe health conditions, including microcephaly, which can cause life-long disability. Laboratory tests and molecular tests can diagnose the disease. Laboratory tests of blood samples or other body fluids can confirm the infection. These tests are usually referred to as antibody tests. They check the blood sample for antibodies such as IgG or IgM that fight against the virus. The test should be done within a week after symptoms appear for accurate results. They should be differentiated from cross-reactive related flaviviruses such as dengue, as the patient could have been exposed to or previously vaccinated against them (WHO, 2022). Molecular tests such as the nucleic acid amplification test (NAAT) look for the genetic material from the Zika virus. PCR is a widespread molecular test that checks for the virus in a blood or urine sample. This is highly recommended for pregnant women. However, the period of detection of the RNA following infection is limited.
- *Yellow fever:* Yellow fever is a viral disease transmitted to humans by the bites of infected mosquitoes. It is prone to epidemics and can be prevented with a vaccine. The causative agent is a virus transmitted primarily by *Aedes* or *Haemagogus* species mosquitoes. Sylvatic or jungle yellow fever, intermediate yellow fever, and urban yellow fever are three types of transmission cycles (WHO Fact Sheet, 2024). Urban yellow fever may cause large epidemics when the virus spreads in heavily populated areas with high mosquito density (WHO Fact Sheet, 2024). Yellow fever has a wide range of symptoms and severity. In most cases, people are asymptomatic, but once symptoms appear, the most common are fever, muscle pain with backache, headache, loss of appetite, and nausea or vomiting. The body's incubation period is three to six days. Symptoms disappear after three to four days. There is a small percentage of patients who enter a second, more toxic phase within 24 hours of recovering from initial symptoms. In such cases, body systems like the liver and kidneys are affected. Patients tend to develop jaundice, which leads to yellowing of the skin and eyes, thus causing the name yellow fever. These patients have a high mortality rate. There are rapid detection methods, which include the detection of yellow fever antigen using monoclonal enzyme immunoassay in serum specimens and the detection of viral genome sequences using polymerase chain reaction (PCR) assay (Simon et al., 2023). ELISA and serology titers of antibodies are also some of the diagnostic tests. Other investigations depend on the affected organ. Laboratory tests with blood samples may reveal leukopenia with elevated transaminase levels. The coagulation profile will be abnormal if the liver

is impacted (Simon et al., 2023). The differential diagnosis profile of yellow fever has a broad spectrum, and careful travel history is crucial. Some differential diagnoses are viral hemorrhagic fevers, viral hepatitis, malaria, Lassa fever, the Ebola virus, and dengue fever (Simon et al., 2023).

- *Chikungunya:* Chikungunya is a viral disease transmitted to humans primarily via infected *Aedes aegypti* mosquitoes. These mosquitoes usually bite during the day, just after sunrise and around sunset. The typical symptoms are fever, joint pain and swelling, headache, muscle pain, and rash. Some clinical signs among chikungunya, Zika, and dengue lead to potential misdiagnosis. Usually, the patient's condition improves within a week. However, the joint pain may occasionally last for months or even years. The laboratory test is the primary diagnostic method for detecting chikungunya virus or viral RNA. The various types of laboratory diagnosis are serologic methods, virus isolation, and viral RNA detection by RT-PCR. The serum for this laboratory test is collected within six days after the onset of the illness. Two commercially available real-time reverse transcription polymerase chain reaction (RT-PCR) kits are available, but the validity of the data is limited (Johnson et al., 2016).
- *West Nile virus:* West Nile virus (WNV) is a member of the *Flavivirus* genus and belongs to the Japanese encephalitis antigenic complex of the family Flaviviridae. It is mainly transmitted to people through the bites of infected mosquitoes. The principal vectors of the virus are mosquitoes belonging to the genus *Culex*, particularly *Culex pipiens*. This virus causes fatal neurological disease in humans. The majority of people infected with West Nile virus are asymptomatic. A significantly smaller percentage develop symptoms such as fever, headache, tiredness, body aches, nausea, vomiting, occasional skin rash, and swollen lymph glands. This population who exhibits symptoms develops West Nile fever. The incubation period is usually 3–14 days. Very rarely, infected persons develop symptoms of severe neuroinvasive disease such as West Nile encephalitis, meningitis, or West Nile poliomyelitis (WHO Newsroom Fact Sheet, 2017). The symptoms are headache, high fever, neck stiffness, stupor, disorientation, coma, tremors, convulsions, and paralysis. Several different tests can diagnose West Nile virus (WHO Newsroom Fact Sheet, 2017).
    - IgG antibody sero-conversion (or significant increase in antibody titers) in two serial specimens collected at a one-week interval by enzyme-linked immunosorbent assay (ELISA).
    - IgM antibody capture enzyme-linked immunosorbent assay (ELISA).
    - Neutralization assays.
    - Viral detection by reverse transcription polymerase chain reaction (RT-PCR) assay. Virus isolation by cell culture.

  IgM can be detected in nearly all cerebrospinal fluid (CSF) and serum specimens from WNV-infected patients. Serum IgM antibodies may persist for over a year (WHO Newsroom Fact Sheet, 2017).
- *Dengue:* Dengue fever is a mosquito-borne disease caused by the dengue virus. The virus is transmitted primarily by *Aedes aegypti* mosquitoes, which bite during the daytime, usually just after sunrise and around sunset. The global incidence of dengue has significantly increased recently, with more than half of the world's population at risk, and has a high mortality rate. Dengue symptoms usually appear three to 14 days after infection and last two to seven days. Common symptoms are high fever, headache, vomiting, rash, and muscle and joint pain. Some people with dengue develop complications that can result in internal bleeding, shock, and even death. Severe dengue requires careful monitoring in the hospital. Molecular tests are the best for symptomatic patients. Dengue RNA virus can usually be detected by molecular tests for the first zero to seven days. These tests lose their sensitivity after seven days. A combination of NAAT, such as RT-PCR and an IgM antibody test, is highly recommended. NS1 can detect dengue virus infection in the

first zero to seven days. NS1 ELISA tests can be as sensitive as molecular tests during symptoms' first zero to seven days. However, it loses sensitivity after seven days. Serum is the preferred sample for dengue testing. A positive NAAT test does not require further confirmatory testing (US Center for Disease Control and Prevention, 2024a). Serologic testing by IgM antibody testing can identify the most recent dengue infection after three days of illness. These tests must be performed on samples with negative NS1 ELISA and PCR results, especially after three days of illness. Serologic testing by IgG in a single specimen sample is not preferred for diagnosing acute dengue in patients. Interpreting positive results from these tests is complicated as they may detect antibodies from dengue infections or other flavivirus infections that have occurred in the past (US Center for Disease Control and Prevention, 2024a). Plaque reduction neutralization tests (PRNT) can resolve false-positive IgM antibody results caused by non-specific reactivity and, in some cases, help identify the infecting virus. However, in areas with a high prevalence of dengue and Zika virus-neutralizing antibodies, PRNT may not confirm a significant proportion of IgM-positive results (US Center for Disease Control and Prevention, 2024b).

- *Malaria:* Malaria is a life-threatening disease primarily prevalent in tropical countries. Unlike others, it is not contagious; the disease is transmitted through the bites of female *Anopheles* mosquitoes. Five parasites can cause malaria in humans, and two of these species—*Plasmodium falciparum* and *Plasmodium vivax*—pose the greatest threat. Common symptoms are fever, headache, and chills. The first symptoms usually begin 10–15 days after an infected mosquito bite. People who have developed partial immunity in malaria-endemic areas may have asymptomatic infections. Severe malaria may cause multi-organ failure in adults, while children suffer from anemia and respiratory distress. The two main diagnostic tests are a microscopic examination of blood smears and rapid diagnostic tests. Light microscopy is the diagnostic standard against which other diagnostic methods have traditionally been compared. It allows the detection of different malaria-causing parasites (*Plasmodium falciparum*, *Plasmodium vivax*, *Plasmodium malariae*, *Plasmodium ovale*, and *Plasmodium knowlesi*), the various parasite stages, including gametocytes, and the quantification of parasite density to monitor response to treatment. Giemsa is the recommended stain for malaria microscopy, and diagnosis requires the examination of both thin and thick films from the same patient (WHO Fact Sheet, 2024). Malaria rapid diagnostic tests (RDTs) could potentially improve the quality of malaria infection management. They are simple to perform and interpret, provide results rapidly, and require limited training, allowing diagnosis at the community level. Malaria RDTs detect specific antigens (proteins) malaria parasites produce in infected individuals' blood. Some RDTs detect a single species (*Plasmodium falciparum* or *Plasmodium vivax*). Some detect multiple species (*Plasmodium falciparum*, *Plasmodium vivax*, *Plasmodium malariae*, and *Plasmodium ovale*), and some further distinguish between *Plasmodium falciparum* and non-*Plasmodium falciparum* infection, or between specific species (WHO Fact Sheet, 2024). Nucleic acid amplification–based diagnostics can sensitively detect low-density malaria infections, which microscopy and RTDs cannot. The use of NAATs can be broadly divided into four roles (WHO Fact Sheet, 2024):
    - Qualitative or quantitative parasite detection, determination of the multiplicity of infections
    - Genotyping to distinguish recrudescence from reinfection
    - Detection of drug resistance mutations

  PCR, including nested (n), quantitative (q) or RT-PCR, loop-mediated isothermal amplification, and quantitative nucleic acid sequence–based amplification (QT-NASBA), are among the critical NAATs developed to detect malaria. NAATs can offer quantitative as well as qualitative output, and many tests have also been developed to enable specific detection of both sexual and asexual stage parasites. The 18S ribosomal RNA gene has unique sequences that enable the identification of all five malaria species infecting humans, namely, *Plasmodium falciparum*, *Plasmodium vivax*, *Plasmodium malariae*, *Plasmodium ovale*, and *Plasmodium knowlesi* and is, therefore, commonly targeted for amplification (WHO Fact Sheet, 2024). NAATs may differ according to required equipment and infrastructure, pre-amplification sample processing, reaction temperature, time to result, ease of use, limit of detection, or cost. Most require a nucleic acid extraction step followed by amplification, which may occur at a single temperature or by cycling through different temperatures (WHO Fact Sheet, 2024).

REFERENCES

Abbasi-Oshaghi, E., Mirzaei, F., Farahani, F., Khodadadi, I., & Tayebinia, H. (2020). Diagnosis and treatment of coronavirus disease 2019 (COVID-19): Laboratory, PCR, and chest CT imaging findings. *Int J Surg*, 79, 143–153. https://doi.org/10.1016/j.ijsu.2020.05.018.

American Lung Association. (2024, October 30). *COVID-19 Testing and Diagnosis*. https://www.lung.org/lung-health-diseases/lung-disease-lookup/covid-19/covid-19-testing.

Banerjee, S., Gupta, N., Kodan, P., Mittal, A., Ray, Y., Nischal, N., Soneja, M., Biswas, A., & Wig, N. (2019). Nipah virus disease: A rare and intractable disease. *Intractable Rare Dis Res*, 8(1), 1–8. https://doi.org/10.5582/irdr.2018.01130. PMID: 30881850; PMCID: PMC6409114.

Berman, S. (2011). *Berman's Pediatric Decision Making* (5th ed.). Mosby. https://doi.org/10.1016/B978-0-323-05405-8.00010-3.

Bettini, A., Lapa, D., & Garbuglia, A. R. (2023). Diagnostics of ebola virus. *Frontiers in Public Health*, 11, 1123024. https://doi.org/10.3389/fpubh.2023.1123024.

Bhattacharya, M., Bose, P., Jana, A., & Chandra, S. (2020). Rapid and cost-effective COVID-19 diagnostic test kit. *International Journal of Research in Pharmaceutical Sciences*, 11(Suppl. 1). https://doi.org/10.26452/ijrps.v11ispl1.4198.

Bunaciu, A. A., Udriştioiu, E. G., & Aboul-Enein, H. Y. (2015). X-ray diffraction: Instrumentation and applications. *Critical Reviews in Analytical Chemistry*, 45(4). https://doi.org/10.1080/10408347.2014.949616.

Cheng, C., Fei, Z., & Xiao, P. (2023). Methods to improve the accuracy of next-generation sequencing. *Frontiers in Bioengineering and Biotechnology*, 11. https://doi.org/10.3389/fbioe.2023.982111.

Chien, A., Weaver, J. S., Kinne, E., & Omar, I. (2020). Magnetic resonance imaging of the knee. *Polish Journal of Radiology*, 85(1). https://doi.org/10.5114/pjr.2020.99415.

Cobo, F. (2012). Application of molecular diagnostic techniques for viral testing. *Open Virol J*, 6, 104–114. https://doi.org/10.2174/1874357901206010104. Epub 2012 Nov 30. PMID: 23248732; PMCID: PMC3522074.

Corona, T. F., Böger, B., da Rocha, T. C., Svoboda, W. K., & Gomes, E. C. (2018). Comparative analysis of mouse inoculation test and virus isolation in cell culture for rabies diagnosis in animals of Parana, Brazil. *Revista Da Sociedade Brasileira de Medicina Tropical*, 51(1). https://doi.org/10.1590/0037-8682-0303-2017.

Frank, M. G., Weaver, G., & Raabe, V. (2024). Crimean Congo hemorrhagic fever virus for clinicians—virology, pathogenesis, and pathology. *Emerging Infectious Diseases*, 30(5), 847–853. https://doi.org/10.3201/eid3005.231646.

Gogoi, D., Baruah, P. J., & Narain, K. (2024). Immunopathology of emerging and re-emerging viral infections: An updated overview. *Acta Virologica*, 68. https://doi.org/10.3389/av.2024.12108.

Johnson, B. W., Russell, B. J., & Goodman, C. H. (2016). Laboratory diagnosis of Chikungunya virus infections and commercial sources for diagnostic assays. *J Infect Dis*, 214(Suppl. 5), S471–S474. https://doi.org/10.1093/infdis/jiw274.

Khan, A. I., Pratumvinit, B., Jacobs, E., Kost, G. J., Kary, H., Balla, J., Shaw, J., Milevoj Kopcinovic, L., Vaubourdolle, M., Oliver, P., Jarvis, P. R. E., Pamidi, P., Erasmus, R. T., O'Kelly, R., Musaad, S., & Sandberg, S. (2023). Point-of-care testing performed by healthcare professionals outside the hospital setting: Consensus-based recommendations from the IFCC committee on point-of-care testing (IFCC C-POCT). *Clinical Chemistry and Laboratory Medicine*. https://doi.org/10.1515/cclm-2023–0502.

Mazzola, L. T., & Kelly-Cirino, C. (2019a). Diagnostic tests for Crimean-Congo haemorrhagic fever: A widespread tickborne disease. *BMJ Glob Health*, 4, e001114. https://doi.org/10.1136/bmjgh-2018-001114.

Mazzola, L. T., & Kelly-Cirino, C. (2019b). Diagnostics for Lassa fever virus: A genetically diverse pathogen found in low-resource settings. *BMJ Glob Health*, 4, e001116. https://doi.org/10.1136/bmjgh-2018-001116.

Mazzola, L. T., & Kelly-Cirino, C. (2019c). Diagnostics for Nipah virus: A zoonotic pathogen endemic to Southeast Asia. *BMJ Global Health*, 4, e001118. https://doi.org/10.1136/bmjgh-2018-001118.

Morens, D. M., & Fauci, A. S. (2020). Emerging pandemic diseases: How we got to COVID-19. *Cell*, 182(5), 1077–1092. https://doi.org/10.1016/j.cell.2020.08.021.

Mourya, D. T., Yadav, P. D., Ullas, P. T., Bhardwaj, S. D., Sahay, R. R., Chadha, M. S., Shete, A. M., Jadhav, S., Gupta, N., Gangakhedkar, R. R., Khasnobis, P., & Singh, S. K. (2019). Emerging/re-emerging viral diseases and new viruses on the Indian horizon. *Indian J Med Res*, 149(4), 447–467. https://doi.org/10.4103/ijmr.IJMR_1239_18.

National Emerging Special Pathogens Training and Education Center. (2022, October 11). *Laboratory Testing for Ebola*. https://netec.org/2022/10/11/laboratory-testing-for-ebola/.

Patel, M., Goel, A. D., Bhardwaj, P., Joshi, N., Kumar, N., Gupta, M. K., Jain, V., Saurabh, S., & Patel, K. (2021). Emerging and re-emerging viral infections in India. *J Prev Med Hyg*, 62(3), E628–E634. https://doi.org/10.15167/2421-4248/jpmh2021.62.3.1899. PMID: 34909490; PMCID: PMC8639133.

Payne, S. (2017). Methods to study viruses. *PubMed Central*, 37–52. https://doi.org/10.1016/B978-0-12-803109-4.00004-0. PMCID: PMC7149989.

Petrova, V., Kristiansen, P., Norheim, G., & Yimer, S. A. (2020). Rift valley fever: Diagnostic challenges and investment needs for vaccine development. *BMJ Global Health*, 5(8), e002694. https://doi.org/10.1136/bmjgh-2020–002694. PMID: 32816810; PMCID: PMC7437696.

Pontone, G., Scafuri, S., Mancini, M. E., Agalbato, C., Guglielmo, M., Baggiano, A., Muscogiuri, G., Fusini, L., Andreini, D., Mushtaq, S., Conte, E., Annoni, A., Formenti, A., Gennari, A. G., Guaricci, A. I., Rabbat, M. R., Pompilio, G., Pepi, M., & Rossi, A. (2021). Role of computed tomography in COVID–19. *Journal of Cardiovascular Computed Tomography*, 15(1). https://doi.org/10.1016/j.jcct.2020.08.013.

Raabe, V. N. (2020). Diagnostic testing for Crimean-Congo hemorrhagic fever. *Journal of Clinical Microbiology*, 58. https://doi.org/10.1128/jcm.01580–19.

Raabe, V. N., & Koehler, J. (2017). Laboratory diagnosis of Lassa fever. *Journal of Clinical Microbiology*, 55. https://doi.org/10.1128/jcm.00170–17.

Reta, D. H., Tessema, T. S., Ashenef, A. S., Desta, A. F., Labisso, W. L., Gizaw, S. T., Abay, S. M., Melka, D. S., & Reta, F. A. (2020). Molecular and immunological diagnostic techniques of medical viruses. *International Journal of Microbiology*, 2020, 19. https://doi.org/10.1155/2020/8832728.

Shirshikov, F. V., & Bespyatykh, J. A. (2022). Loop-mediated isothermal amplification: From theory to practice. *Russian Journal of Bioorganic Chemistry*, 48(6). https://doi.org/10.1134/S106816202206022X.

Shuken, S. R. (2023). An introduction to mass spectrometry-based proteomics. *Journal of Proteome Research*, 22(7). https://doi.org/10.1021/acs.jproteome.2c00838.

Simon, L. V., Hashmi, M. F., & Torp, K. D. (2023). Yellow fever. In *StatPearls [Internet]*. Treasure Island, FL: StatPearls Publishing. https://www.ncbi.nlm.nih.gov/books/NBK470425/.

Sorelle, J. A., Patel, K., Filkins, L., & Park, J. Y. (2020). Mass spectrometry for COVID–19. *Clinical Chemistry*, 66(11). https://doi.org/10.1093/clinchem/hvaa222.

Su, R., Yang, T., Zhang, X., Li, N., Zhai, X., & Chen, H. (2023). Mass spectrometry for breath analysis. *TrAC—Trends in Analytical Chemistry*, 158. https://doi.org/10.1016/j.trac.2022.116823.

Tayyeb, A., & Basit, Z. (2023). Polymerase chain reaction. In *Genetic Engineering: Volume 1: Principles Mechanism, and Expression*. https://doi.org/10.5005/jp/books/12238_22.

US Center for Disease Control and Prevention. (2024a, August 26). *Clinical Testing Guidance for Dengue*. https://www.cdc.gov/dengue/hcp/diagnosis-testing/.

US Center for Disease Control and Prevention. (2024b, April 23). *Ebola Disease Basics*. https://www.cdc.gov/ebola.

Vainionpää, R., Waris, M., & Leinikki, P. (2015). Diagnostic techniques: Serological and molecular approaches☆. In *Reference Module in Biomedical Sciences*. Elsevier. https://doi.org/10.1016/B978-0-12-801238-3.02558-7.

Van Oostveen, W. M., & de Lange, E. C. M. (2021). Imaging techniques in Alzheimer's disease: A review of applications in early diagnosis and longitudinal monitoring. *International Journal of Molecular Sciences*, 22(4). https://doi.org/10.3390/ijms22042110.

World Health Organization (WHO). (2022, July 14). *Laboratory Testing for Zika Virus and Dengue Virus Infections: Interim Guidance*. https://www.who.int/publications/i/item/WHO-ZIKV_DENV-LAB–2022.1.

World Health Organization (WHO) Fact Sheet. (2024, January 12). *Global Malaria Programme*. https://www.who.int/teams/global-malaria-programme/case-management/diagnosis/microscopy.

World Health Organization (WHO) Newsroom Fact Sheet. (2017, October 3). *West Nile Virus*. https://www.who.int/news-room/fact-sheets/detail/west-nile-virus.

World Health Organization (WHO) Newsroom Fact Sheet. (2018, February 19). *Rift Valley Fever*. https://www.who.int/news-room/fact-sheets/detail/rift-valley-fever.

World Health Organization (WHO) Newsroom Fact Sheet. (2024, September 26). *Vector-Borne Diseases*. https://www.who.int/news-room/fact-sheets/detail/vector-borne-diseases.

12 Preventive and Social Medicine in Managing Emerging and Re-Emerging Diseases of Viral Origin

Sonali Sain

12.1 INTRODUCTION

Preventive medicine is a branch of medicine which deals with prevention of disease and promotion of health. The main achievement of preventive medicine during COVID-19 pandemic is its role in control by vaccination and quarantine. Early success has been achieved in this field by invention of bacterial vaccines and antisera at the turn of the century which led to the wide spectrum conquest of specific diseases. Later development of antiviral vaccines include smallpox vaccine (led to eradication in 1977) and polio vaccines (1955, 1960). Discovery of better and newer vaccines (e.g., malaria, leprosy, syphilis and other parasitic diseases, and even cancer) continues in recent times. Successful decline has been observed in morbidity and mortality from diphtheria, tetanus, typhoid fever, and others. Discoveries in nutrition enable to combat specific deficiency disorders (Park's Textbook, 28th Edition 2024). Preventive medicine deals with all aspects of life which helps in prevention and control of disease, both communicable and non-communicable. Maintenance of hygiene, discovery of vaccines, healthy housing, control of disease transmission by various modes, etc. proved helpful for control of communicable diseases. Control of emergence of risk factors like lifestyle modification, prevention of addiction, healthy diet, control of pollution, etc. can halt the emergence and progression of non-communicable diseases. Consideration of nutrition, physical activity, occupation, control addiction, genetics, screening, etc. all contribute to healthy living. Multidisciplinary approach like maternal health, child health, adolescent health, reproductive health, and geriatric health constitute holistic approach to health. This chapter deals with the various aspects of emerging and re-emerging diseases of viral origin, so that the information may provide insight into dealing with control of these diseases in case of any future outbreak. Emerging infections are the conditions which has increased or has potential to increase in recent past or in near future. These are newly appearing infection or infections with newer presentation in new geographical areas or known persistent diseases that are difficult to control. Commonly found emerging diseases are coronavirus infections, such as COVID-19, SARS (severe acute respiratory syndrome), and MERS (Middle East respiratory syndrome), HIV infections, Lyme disease, *Escherichia coli* (*E. coli*) O157:H7, Hantavirus, West Nile virus, Zika virus etc. Re-emerging diseases are those infections that have come back after major decline. In this era of antimicrobial resistance, there are resurgence of new strains or mutation of existing pathogens. These may be due to change in people's behavior, decline in host immunity, decline in immunization status, and irrational use of antibiotics. Commonly found re-emerging diseases are malaria, tuberculosis, cholera, pertussis, influenza, pneumococcal disease, gonorrhea, etc. (*Emerging Infectious Diseases*, 2024). Important factors responsible for emergence and re-emergence of infectious diseases are unplanned urbanization with poor sanitation and inadequate public health infrastructure, overcrowding and rapid

DOI: 10.1201/9781032721811-12

population growth, increased vector population and reservoirs of infection, microbial genetic mutation and antibiotic resistance, and international travel.

12.2 SURVEILLANCE AND EARLY DETECTION OF EMERGING AND RE-EMERGING VIRAL DISEASES

Surveillance is the systematic collection, analysis, and interpretation of outcome-specific data for planning, implementing, and evaluating public health policies and practices. It is of three types: active surveillance, passive surveillance, and sentinel surveillance. It has two functions: early warning of potential threats to public health and program-monitoring functions. It is essential for country's health security. The International Health Regulations (IHR) 2005 emphasized global security. It requested all Member States to establish and implement effective surveillance systems to detect and contain any public health threats of national and international importance. Primary goal of surveillance system is eradication or elimination of diseases like measles and surveillance of eradicated diseases like polio by searching for acute flaccid paralysis. Surveillance systems also monitor trends of endemic diseases and progress toward disease control objectives, as well as provide information regarding impact of disease control programs (Figure 12.1).

A national surveillance plan should consider the diversity in the uses of data and approaches (Davis JR, Lederberg J, 2000). Integrated Disease Surveillance Project (IDSP) was launched by the Department of Health & Family Welfare in November 2004. Now it is known as Integrated Health Information Platform (IHIP). A Central Surveillance Unit (CSU) was established in Delhi, State Surveillance Units (SSU) in all State/UT headquarters, and District Surveillance Units (DSU) in all Districts. The objectives of IDSP are to strengthen/maintain decentralized laboratory-based IT-enabled disease surveillance system for epidemic-prone diseases, to monitor disease trends, and to detect and respond to outbreaks in early rising phase through trained Rapid Response Teams (RRTs). Program component includes integration and decentralization of surveillance activities through establishment of surveillance units at center, state, and district levels, human resource development, and training of state surveillance officers, district surveillance officers, rapid response team, and other medical and paramedical staff on principles of disease surveillance.

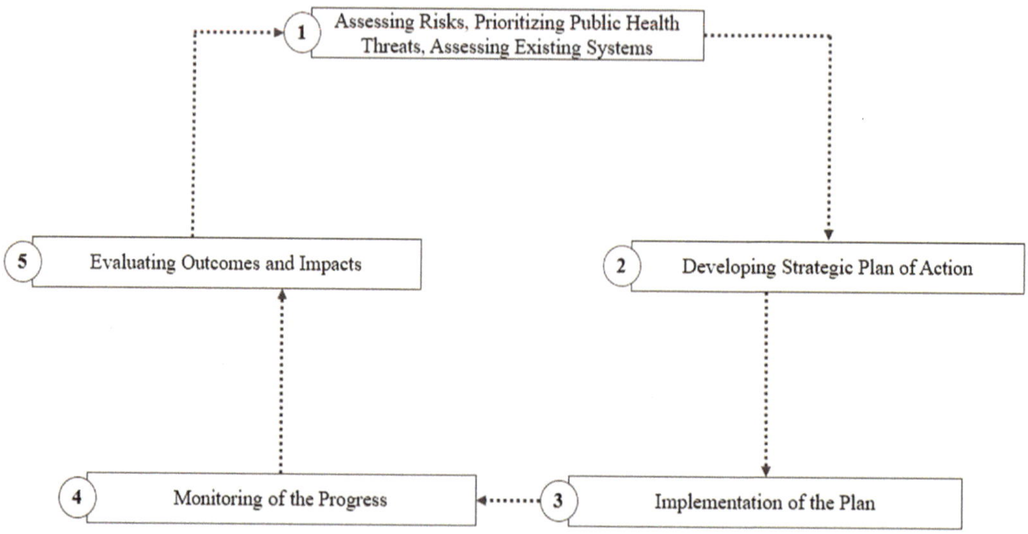

FIGURE 12.1 Cycle illustrating surveillance systems strengthening activities (WHO_CDS_EPR_LYO_2006_2_eng).

Use of information communication technology for collection, collation, compilation, analysis, and dissemination of data along with strengthening of public health laboratories and intersectoral coordination for zoonotic diseases are also important components. Data is collected on epidemic-prone diseases on weekly basis (Monday–Sunday) under IDSP. The information is collected on three specified reporting formats, namely, "S" (suspected cases), "P" (presumptive cases), and "L" (laboratory-confirmed cases) filled by health workers, clinicians, and laboratory staff, respectively, and provides information of disease trends and seasonality of diseases. Rapid response teams (RRT) investigate a rising trend of illnesses in any area to diagnose and control the outbreak. Data analysis and following actions are undertaken by respective State/District Surveillance Units (Integrated Disease Surveillance Programme—IDSP). Sentinel surveillance is a type of passive surveillance done for inflow of a large number of cases at selected health facility. It is a prearranged reporting sources to report all cases of defined conditions, which might indicate trends in the entire target population. When properly implemented, these systems offer an effective method of using limited resources and enable prompt monitoring and investigation of suspected public health problems. Examples of sentinel surveillance are networks of private practitioners reporting cases of influenza or a laboratory-based sentinel system reporting cases of certain bacterial infections among children. Sentinel surveillance is excellent for detecting large public health problems like HIV and AIDS. It may be insensitive to rare events and early emergence of a new disease, because these infections may emerge anywhere in the population (Nsubuga P et al., 2006).

Sentinel surveillance provides primary source of essential information regarding trends and dynamics of epidemic among different at-risk groups of country. For example, sentinel surveillance of HIV infection is conducted biennially in India which is calibrated according to the latest WHO guidelines. The Indian HIV Sentinel Surveillance (HSS) is one of the largest in the world and conducted among eight population groups like pregnant women, single male migrants, long distance truckers, prisoners, female sex workers, men who have sex with men, hijra/transgender people, and injecting drug users. The National AIDS Control Organization (NACO) implements HIV Sentinel Surveillance (HSS). A network of testing and reference laboratory provides laboratory support. There is a network of service delivery points at national, regional, state, and district levels (HIV Sentinel Surveillance 2016–17). National notification and surveillance efforts led to the rapid identification of the foci of Ebola infection and to effective containment of Ebola outbreak in Uganda in 2000.

12.3 INVESTIGATION OF EMERGING AND RE-EMERGING DISEASE OUTBREAK

Epidemiology is defined as the study of the occurrence and distribution of health-related events, states, and processes in specified populations, including the study of the determinants influencing such processes, and the application of this knowledge to control relevant health problems (IEA, 2014).

Epidemiological research is conducted to explore about disease. It investigates etiology, mode of transmission, spectrum of clinical presentation, and prevention and effective management. There are different types of epidemiological studies: descriptive, analytical, and experimental. Descriptive study is for description of the occurrence of disease or health events in a population. An analytical study analyzes relationship between health status and other variables. Experimental or intervention study involves an active attempt to change a disease determinant or the progress of disease (Park's Textbook, 27th Edition 2023). Natural history of disease is investigated by a type of analytical study, known as cohort study. In this study, a group of people are followed from exposure of risk factors, early onset of symptoms, till the end point. Thus, it helps to explore the source, risk factors, mode of transmission, and progress of disease. Doll and Hill study was conducted among British medical doctors in 1951. Association of smoking and lung cancer was established by this study (Di Cicco et al., 2016). The Framingham heart study was organized by the United States Public Health Service in 1948. It was conducted to establish relationship between risk factors like weight, smoking, blood pressure, serum cholesterol, etc. and the development of cardiovascular diseases over 20 years (Mahmood et al., 2014). Royal College of General Practitioners conducted a cohort study of oral contraceptives and health in England in 1968. It was conducted among 23,000 pill users aged 15–49 years together with a similar number of controls using other methods or no method of contraception to assess the risks and benefits of oral contraceptive use. The study observed an increased cardiovascular-related mortality among the pill users (Beral et al., 1999). Rapid response team (RRT) is for early and rapid intervention in case of medical emergency. It identifies and prevents significant medical deterioration leasing to cardiopulmonary arrest. It was originally conceptualized for the management strategies of severe trauma. Key elements are early detection of condition and rapid response (Al-Qahtani et al., 2010). Rapid response team (RRT) is a multidisciplinary team which is to be activated in response to disease outbreak, natural or manmade disasters, and other urgent health threats. RRT includes clinicians, epidemiologists, surveillance officers, laboratory technicians, environmental health officers, and social officers. They are trained in wide range of knowledge and skill required to deal with medical emergency and adapting solutions to specific scenario. They help to investigate wide range of health emergencies and collect field data for public health action. They also mobilize and engage community for public health response. RRT played an important role in testing, case investigation, contact tracing, risk assessment, infection prevention and control, and community mitigation along with surveillance and epidemic investigation during COVID-19 pandemic throughout the world.

12.4 COMMUNITY EDUCATION AND AWARENESS

Health education is a continuous, dynamic, complex, and planned teaching–learning process throughout the lifespan and in different settings that is implemented through an equitable and negotiated client and health professional "partnership" to facilitate and empower the person to promote/initiate lifestyle-related behavioral changes that promote positive health status outcomes. Health education considers individuals' or groups' internal and external factors that improve knowledge, skills, attitudes, and beliefs regarding their health-related needs and behavior, within a positive health paradigm (Figure 12.2) (Alissa et al., 2024; Hossain & Luies, 2017).

Social media can be used in health education because of its ability to remove physical barriers which impede access to healthcare service traditionally. Thus, health educators should be competent

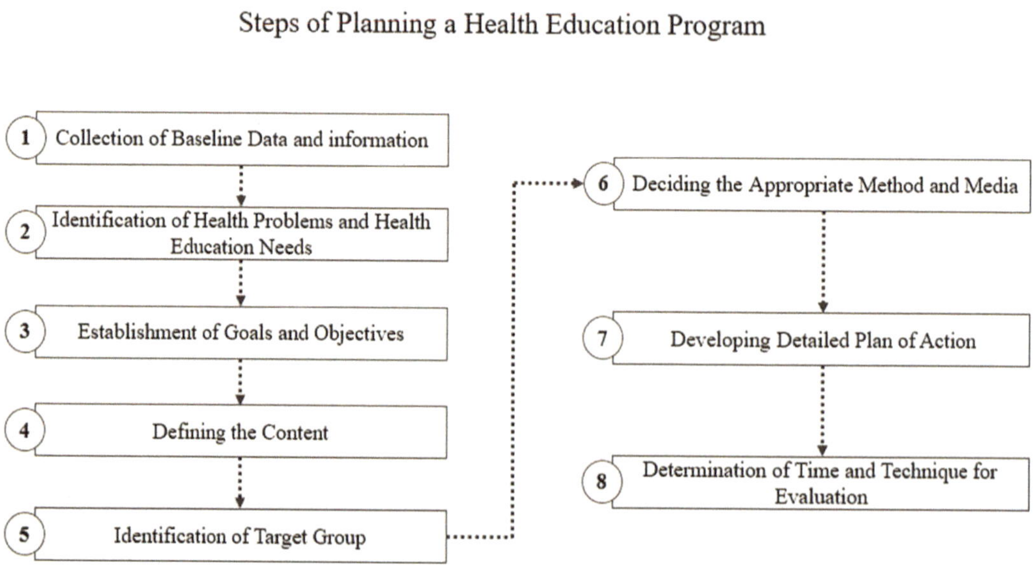

FIGURE 12.2 Steps of planning a health education program (Hossain & Luies, 2017).

in computer-mediated contexts for both online and offline beneficiary health experiences (Stellefson et al., 2020). It is essential to consider limitation of technical information during health education, because recipient might find it difficult to internalize, motivate, and change attitude (Knittel et al., 1971). Involvement of community in public health intervention is a complex process. It is to be substantiated by well-built community processes, where range of social and cultural factors would be taken into account. Involving community in health education program has pragmatic impact on community health. Short-term health outcome, that is, changes in community member's health status, are estimated as effectiveness of health education program. Social effects in community following organizational intervention are regarded as intermediate social effects. It involves changes in knowledge, attitudes, and behaviors of community, thus leading to community development and sociocultural and environmental improvements. Long-term effects are described as larger sociopolitical influences and stakeholder perception. Community empowerment is an important outcome of public health intervention (Haldane et al., 2019).

12.5 VACCINATION PROGRAMS FOR EMERGING AND RE-EMERGING DISEASES OF VIRAL ORIGIN

Vaccination is a basic human right and best investment of money in health. It is one of the important components of primary prevention which can prevent 3.5–5 million deaths every year from diseases like diphtheria, tetanus, pertussis, influenza, and measles. COVID-19 pandemic can be successfully controlled by different health measures, including vaccination. Our world can be measles-free only by vaccination (Vaccines and Immunization—WHO, n.d.). Vaccines help the body to develop defense mechanism by imitating the infection, without full-blown infection. It provides protection for few months, years, or even for lifetime. Some viral infections can have minimal or no symptoms but can have long-term consequences. For example, human papillomavirus never show any significant symptoms of infection but may result in aggressive, life-threatening cervical cancer. Children below 5 years of age do not die due to measles infection, but due to its complications (*Explaining How Vaccines Work*, 2024). Appearance of new infectious diseases has been apprehended even before the discovery of infectious agents. During the era of jet age, intercontinental travel and global interdependence put in complexity to the outbreak containment in spite of advances in medical science. Emerging infectious diseases (EIDs) and re-emerging infectious diseases (REIDs) are new threats to global health and mankind. Coronavirus epidemic, like SARS-COV-2 pandemic, is one such example. Population vulnerability because of growth and complexities are amenable to devastating outbreak and pandemics due to emerging and re-emerging infectious diseases (Excler et al., 2021). India's Routine Immunization program is the largest public health programs in the world. It is the most cost-effective public health intervention for reduction of vaccine-preventable diseases in children below 5 years of age and pregnant mothers. Vaccination program planning requires leadership, clinical positions, clinic location, and layout and planning with all stakeholders (*Vaccine Administration Protocols | CDC, n.d.*). CoWIN portal was developed for planning, implementation, monitoring, and evaluation during COVID-19 vaccination drive during COVID-19 pandemic in India. The experience of CoWIN encourages to develop a similar platform for routine immunization in India. This will help to register and track the beneficiaries for vaccination. This real-time activity will ensure compliance of children toward vaccination. This will help to tackle the challenges of low immunization coverage in migratory and urban population. The real-time monitoring of vaccine coverage data is derived by Health Management Information System (HMIS). The targeted vaccine campaigns are conducted to strengthen and re-energize the program so that full immunization coverage for all children and pregnant mothers can be achieved rapidly (*Universal Immunization Programme (UIP) | Ministry of Health and Family Welfare | GOI*, n.d.). Vaccine hesitancy is defined as delay in acceptance or refusal of vaccines despite availability of vaccination services. Vaccine hesitancy is complex and context-specific varying across time, place, and vaccines. It includes complacency, convenience, and confidence as modifying factors. Despite being identified as most successful public health measures, vaccination is considered as unsafe and unnecessary by many. Lack of confidence in vaccines is a threat to the successful vaccination programs. Vaccine hesitancy is believed to be responsible for poor vaccine coverage and heightened risk of vaccine-preventable disease outbreaks and epidemics (Sain et al., 2023). Hesitancy for vaccination was significantly higher among lesser educated person. Challenges perceived were lesser among geriatric participants with comorbidities (Karmakar et al., 2023). Communication is a powerful tool for influencing public opinion and attitudes toward various health-related issues. It should be well planned and properly conducted. Advocacy for vaccine hesitancy can alter beneficiary behavior and can improve vaccine acceptance. Social marketing techniques can be effective for improving vaccine uptake. This was evident in South Korea which had achieved high COVID-19 vaccine coverage during pandemic. However, social marketing strategies may not be helpful in all circumstances. The crucial factors for advocacy of immunization program are high degree of proactiveness, credibility, fighting misinformation, emphasizing social norms, prosocial behavior, and coherence (Hong, 2023).

12.6 CONTROLLING THE VECTOR ASSOCIATED WITH VIRAL DISEASES

Integrated vector management (IVM) is the strategic vector control approach advocated by WHO. It is defined as "a rational decision-making process for the optimal use of resources for vector control." Prime elements of IVM are as follows:

- *Advocacy social mobilization and legislation:* Development of policies of all organizations and civil society; the establishment or strengthening of regulatory and legislative controls for public health; and the empowerment of communities (Control of Neglected Tropical Diseases (NTD), 2009)
- *Collaboration within the health sector and with other sectors:* Consideration for collaboration within and between public and private sectors; planning and decision-making delegated to all administrative levels; and strengthening communication among policymakers, managers of programs, and other key partners (Control of Neglected Tropical Diseases (NTD), 2009)
- *Integrated Approach to Disease Control:* Ensuring the rational use of available resources through the application of a multi-disease control approach; integration of non-chemical and chemical vector control methods; and integration with other disease control measures (Control of Neglected Tropical Diseases (NTD), 2009)
- *Evidence-Based Decision-Making:* Adaptation of strategies and interventions to local vector ecology, epidemiology, and resources, guided by operational research and routine monitoring and evaluation (Control of Neglected Tropical Diseases (NTD), 2009)
- *Capacity-Building:* The development of essential infrastructure, financial resources, and adequate human resources at national and local levels to manage IVM programs, based on a situation analysis (Control of Neglected Tropical Diseases (NTD), 2009)

Environmental management involves processes that minimize humanity's impact on its surrounding environment. The main objective is to create and maintain sustainable environment. It has become a mandatory practice for several governments and organizations to ensure its successful implementation since the 1970s. Environmental management helps to identify degradation factors and implement strategies to mitigate, to predict future impacts of environmental degradation, and to initiate processes to minimize the effects (Agyemang-Badu et al., 2023). The World Health Organization also plays important roles in environmental management. Different principles of environmental management are as follows:

(a) *Prevention principle:* Preventive measures to anticipate and avoid environmental damage.
(b) *Precautionary principle:* Adequate risk management system to minimize environmental harm.

(c) *Integration principle:* Environmental management policy for ensuring sustainable development across all areas.
(d) *Polluter pays:* Party that causes pollution should be accountable for the harm and bear the cost of remedy.

However, environmental management system can be employed as a supplementary method to integrate vector control methods. Effective collaboration among key stakeholders are necessary for successful implementation (Agyemang-Badu et al., 2023). Vector surveillance is the analysis and interpretation of systematically collected entomological data for facilitating appropriate vector control (User, n.d.). The purpose of vector surveillance is to reduce the risk of vector-borne disease transmission. This helps to maintain vector-free status and reduce transmission of pathogens and invasion and dispersal of vectors to and from local environments. National or regional intersectoral cooperation is required for successful implementation of integrated vector management (IHR POE, n.d.).

12.7 QUARANTINE AND ISOLATION MEASURES

Different public health measures are adopted for prevention of disease transmission to others. Isolation is the separation and restriction of movement of sick persons suffering from contagious disease. It is done in a hospital setting, at home, or in a special facility. Quarantine refers to the restriction of movement or separation of apparently healthy persons who have been exposed to a contagious disease, before appearance of symptoms. It may be applied at the individual level or to a group or community of exposed persons. Quarantine is usually organized in the home. It is done in conjunction with disease surveillance and monitoring for symptoms in persons quarantined, rapid diagnosis and timely referral of those who developed symptoms, and provision of preventive interventions, including vaccination or prophylactic antibiotics—for example, containment strategies employed during the recent COVID-19 pandemic. Contact surveillance is the process of monitoring of signs and symptoms in persons who have been exposed to a contagious disease. Surveillance can be active or passive. The principal strategies of containment during pandemic were education of high-risk populations, early detection of suspected and probable cases, and rapid implementation of isolation and other infection control tools (Cetron et al., 2004; *Emergency Preparedness and Response*, 2023). Following Holocaust of six million Jews, United Nations Declaration on Human Rights in 1948 and World Medical Association's Helsinki Declaration were announced. Any public health activity must follow these ethical issues. It might be mandatory immunization of children or may be isolation or quarantine during pandemic (Tulchinsky et al., 2018). During COVID-19 pandemic, governments around the world imposed lockdown in the community. Individual freedom was compromised to control disease transmission. It was necessary for control of disease transmission during pandemic at the cost of people's freedom susceptible to human rights violations, especially in *marginalized populations.* Person with underlying medical conditions may experience difficult access to essential healthcare. It can result in delayed diagnosis, treatment, and prevention of diseases. Other human rights compromise impacted by public health interventions were the right to protection of individual privacy and the right to peaceful assembly, which sometimes resulted in resistance and protest globally (Zweig et al., 2021). Thus, a balance should be maintained between human rights, ethical issues, and public health intervention.

12.8 INTERNATIONAL COLLABORATION

World Health Organization launched a global network for control of infectious disease threat by providing a platform with the help of pathogen genomics. The International Pathogen Surveillance Network connects different countries throughout the world, helps in collection and genomic analysis of sample for genetic code of viruses, bacteria, and other disease-causing organisms, utilization of data in public health decision-making, and sharing of information among themselves before the outbreak turns into epidemic and pandemic. This broader disease surveillance method can play a vital role in development of vaccine and treatment guidelines during disease threat. It helps to optimize routine disease surveillance program by accessing to pathogen genomic sequencing and analytics. COVID-19 pandemic highlighted the critical role of rapid pathogen genomics sequencing of the SARS-COV-2 genome. Thus, quick development of vaccine and early identification of variants of the transmissible virus were possible. Genomics is the core of effective epidemic and pandemic preparedness and response. Thus, international collaboration in this regard has been effective for early intervention. Similar effort has been initiated in foodborne diseases, influenza, tuberculosis, and HIV AIDS and monitoring of drug resistance (World Health Organization: WHO, 2023b). Leadership and governance were two most important factors for international collaboration among countries. Knowledge- and information-sharing could prevent experience of repetitive negative situations. COVID-19 pandemic was controlled globally because of sharing of resources and appropriate communication between countries, and thus changed the trajectory of a pandemic. Facilitating international cooperation was especially helpful among low- and middle-income countries with limited resources and expertise. International collaboration is guided by the following categories—political, structure, infrastructure, leadership and governance, knowledge- and information-sharing, community engagement, and process/action (Hameed et al., 2022; Fanning et al., 2021). Pandemics can increase morbidity and mortality over a wide geographic area leading to significant economic, social, and political disruption. Severe acute respiratory syndrome (SARS) pandemic in 2003 and threat of avian influenza led to devise pandemic plans in several countries. International Health Regulations (IHR) issued directives to meet specific standards for detecting, reporting on, and responding to outbreaks. But significant gaps and challenges exist in global pandemic preparedness, especially in low- and middle-income countries. A country's expected ability to contain pandemic can be expressed using a preparedness index developed by Oppenheim and others. This index illustrates global variation in preparedness to detect and respond to a large-scale outbreak of infectious disease and draws on core capacity metrics and other publicly accessible several indicators. It was found that many countries have been unable to meet basic requirements for pandemic preparedness (Madhav et al., 2017). Emerging infectious diseases (EIDs) are driven by socioeconomic, environmental, and ecological factors and can cause significant burden on global economy and public health. There was emergence of 335 infectious diseases between 1940 and 2004 throughout the world, maximum in the 1980s, due to the emergence of HIV/AIDS pandemic. EID emergence has highlighted major threat of viral pathogens (especially RNA viruses). The majority (60.3%) of EID events were caused by zoonotic pathogens. Climate change may lead to the emergence of vector-borne diseases sensitive to changes in environmental conditions such as rainfall, temperature, and severe weather events. Thus, climate anomalies during the 1990s led to significant rise in EID events. The number of EID events caused by pathogens originating in wildlife has increased significantly with time, due to the increasing contact between wildlife and humans. A latitudinal spatial gradient in human pathogen species concentration increased toward the equator. The highest concentration of EID events was found between 30° and 60° North and between 30° and 40° South latitudes. Socioeconomic factors like human population density, antibiotic drug use, and agricultural practices are major determinants of the spatial distribution along with ecological or environmental factors. EID events caused by drug-resistant microbes (20.9%) are related to a corresponding rise in antimicrobial drug use, particularly in high-latitude developed countries (Jones et al., 2008). Well-prepared countries have ability to detect and manage disease outbreaks, including surveillance, mass vaccination, and risk communications, due to strong economies and adequate investment in the public health. Poorly prepared countries may suffer from political instability, weak public administration, inadequate public health resources, and gaps in outbreak detection and response systems. Thus, emphasis of pandemic preparedness should be given to those countries with risk of EID events and poorly prepared countries.

12.9 HEALTHCARE INFRASTRUCTURE AND PREPAREDNESS

Public health surveillance provides scientific information for informed decision-making and appropriate public health action. The public health surveillance system makes successful interventions and determines the design and implementation. This surveillance system provides rapid early warning information on pathogens from clinics and laboratories, for example, Integrated Disease Surveillance Programme in India. In chronic diseases, health-related behaviors can affect the disease outcome. For example, tobacco-smoking is considered as high-risk behavior for respiratory morbidity. A surveillance system also monitors the effects of infection control procedure in susceptible population (Nsubuga et al., 2006). Infection control refers to the policy and procedures required to control healthcare-associated infections (HAIs) and minimize the dissemination of infections in all healthcare delivery sites. Infection control activity is implemented to control infection and to improve outcomes (Habboush et al., 2024). The Preparedness and Resilience for Emerging Threats Initiative (PRET) is a new initiative by WHO during the COVID-19 pandemic and other recent public health emergencies. It incorporates the latest tools and approaches for shared learning and collective action. PRET addresses technical guidance and support for promoting and strengthening integrated preparedness and response, as outlined in World Health Assembly resolutions. WHO will use a mode of transmission approach in pandemic planning of the member countries (World Health Organization: WHO, 2023a). It is important to integrate pandemic preparedness into national emergency preparedness frameworks. The goal of situation analysis and assessment is to collect, interpret, and disseminate information about the risk of an outbreak before it occurs. It is necessary to monitor pandemic progression pattern and characteristics once in the midst of outbreak (Pandemic Influenza Preparedness and Response, 2009).

12.10 RESEARCH AND INNOVATION

Pandemic preparedness requires investment in research in myriad disciplines, including basic biomedical research, clinical research, and anthropological research. These help to identify the natural history of disease and social and cultural factors related to disease as well as develop new vaccines and drugs. Readiness for outbreak containment can be achieved by appropriate PPE, point-of-care diagnostics, novel therapeutic agents, assessment of antimicrobial resistance (AMR), and vaccine constructs. Research on population, policy, and implementation is required to assess the population factors, policies, and delivery systems for appropriate biomedical interventions. Social research is required to be conducted at vulnerable hotspots to anticipate potential outbreaks and generate vital information about the causes of infection. A comprehensive approach of design and implementation with common goals will succeed with the contributions of various stakeholders in future emergencies (National Academy of Medicine, Secretariat, 2016). Genomic surveillance is the process of constantly monitoring pathogens and analyzing their genetic similarities and differences. However, genomics surveillance has been applied to monitoring infectious diseases widely by sequencing the genomes of bacteria, parasites, and viruses, which needs global collaboration. Thus, WHO has proposed the development of ten-year (2022—2032) global genomic surveillance strategy for pathogens with pandemic and epidemic potential. The vision is to unite and strengthen genomic surveillance activities globally. Research and Diagnosis unit was comprised of academia, government, industry, and civil society. It acts as cohesive entity to determine the necessary biomedical interventions during global health emergency (Morton et al., 2023). Recent advancements in pathogen detection and typing were made by genomic sequencing technology and data analytics. Genomic analyses of infectious diseases are rapidly recognized as the standard for pathogen analysis and control. Genome sequencing managed a pivotal role in hospital outbreaks investigation. Genomic-based AMR data can provide rapid, reliable, and highly accurate predictions of resistance phenotypes of specific strains. It played a valuable role in surveillance of emerging infectious diseases (Bianconi et al., 2023). A range of digital data sources were being used during COVID-19 pandemic to interpret key epidemiological data like infection transmission with respect to time, place, and person, and identifying risk factors obtained by public health authorities. Data dashboards were used extensively for collating real-time public health data, including confirmed cases, deaths, and testing figures, to redefine interventions during COVID-19 pandemic (Budd et al., 2020).

12.11 CONCLUSION

Successful regulation of morbidity and mortality from infections of public health importance was achieved by invention of vaccines and antisera, along with early detection and effective management protocol of disease. Outbreak investigation helps to identify the origin, mode of transmission, and the factors responsible for spread of disease. These help the epidemiologists to contain the epidemic and will be helpful even in future. Future epidemic of various emerging and re-emerging diseases would be successfully managed by epidemiological investigation and community awareness generation. Recent COVID-19 pandemic provides practical experience of epidemic investigation, contact tracing, early diagnosis and management, complication minimization, and adoption of different preventive measures, including vaccine invention. These experience can help to combat future pandemics of various emerging and re-emerging diseases.

REFERENCES

Agyemang-Badu, S. Y., Awuah, E., Oduro-Kwarteng, S., Dzamesi, J. Y. W., Dom, N. C., Kanno, G. G. (2023, January 3). Environmental management and sanitation as a malaria vector control strategy: A qualitative cross-sectional study among stakeholders, Sunyani municipality, Ghana. *Environ Health Insights*, 17. https://doi.org/10.1177/11786302221146890. PMID: 36620305; PMCID: PMC9817013.

Alissa, N. A., & Alwargash, M. (2024, January–December). Health promotion and education. Awareness and attitudes toward health education and promotion among physicians and nurses: Implications for primary health care. *Inquiry*, 61. https://doi.org/10.1177/00469580241248127. PMID: 38654504; PMCID: PMC11041534.

Al-Qahtani, S., & Al-Dorzi, H. M. (2010, January). Rapid response systems in acute hospital care. *Ann Thorac Med*, 5(1), 1–4. https://doi.org/10.4103/1817–1737.58952. PMID: 20351953; PMCID: PMC2841802.

Beral, V., Hermon, C., Kay, C., Hannaford, P., Darby, S., & Reeves, G. (1999, January 9). Mortality associated with oral contraceptive use: 25 year follow up of cohort of 46 000 women from Royal College of general practitioners' oral contraception study. *BMJ*, 318(7176), 96–100. https://doi.org/10.1136/bmj.318.7176.96. PMID: 9880284; PMCID: PMC27684.

Cetron, M., et al. (2004). Isolation and quarantine: Containment strategies for SARS 2003. In S. Knobler et al. (Eds.). *Institute of Medicine (US) Forum on Microbial Threats: Learning from SARS: Preparing for the Next Disease Outbreak: Workshop Summary*. Washington, DC: National Academies Press. https://www.ncbi.nlm.nih.gov/books/NBK92450/.

Control of Neglected Tropical Diseases (NTD). (2009, April 21). *Dengue Guidelines, for Diagnosis, Treatment, Prevention and Control*. https://www.who.int/publications/i/item/9789241547871.

Dengue.health.gov-vector-surveillance-and-integrated-vector-management; https://www.dengue.health.gov.lk/web/index.php/en/activities-and-services/vector-surveillance-and-integrated-vector-management#.

Di Cicco, M. E., Ragazzo, V., & Jacinto, T. (2016, September). Mortality in relation to smoking: The British doctors study. *Breathe (Sheff)*, 12(3), 275–276. https://doi.org/10.1183/20734735.013416. PMID: 28210302; PMCID: PMC5298160.

Emergency Preparedness and Response. (2023, November 15). *Emergency Preparedness and Response*. https://www.cdc.gov/emergency/index.html.

Emergency, Quarantine and Isolation Strategies CDC. https://emergency.cdc.gov/newsletters/epic/012522.htm.

Emerging Infectious Diseases. (2024, June 20). *Johns Hopkins Medicine*. https://www.hopkinsmedicine.org/health/conditions-and-diseases/emerging-infectious-diseases.

Excler, J. L., et al. (2021). Vaccine development for emerging infectious diseases. *Nat Med*, 27, 591–600. https://doi.org/10.1038/s41591-021-01301-0.

Explaining How Vaccines Work. https://www.cdc.gov/vaccines/hcp/conversations/understanding-vacc-work.html.

Explaining How Vaccines Work. (2024, August 10). *Vaccines & Immunizations*. https://www.cdc.gov/vaccines/basics/explaining-how-vaccines-work.html.

Haldane, V., Chuah, F. L. H., Srivastava, A., Singh, S. R., Koh, G. C. H., Seng, C. K., & Legido-Quigley, H. (2019, May 10). Community participation in health services development, implementation, and evaluation: A systematic review of empowerment, health, community, and process outcomes. *PLoS One*, 14(5), e0216112. https://doi.org/10.1371/journal.pone.0216112. PMID: 31075120; PMCID: PMC6510456.

HIV Sentinel Surveillance. (2016–17). *Technical Brief*. NACO. https://naco.gov.in › sites › default › files › HIV Sentinel Surveillance.

Hong, S. A. (2023). COVID-19 vaccine communication and advocacy strategy: A social marketing campaign for increasing COVID-19 vaccine uptake in South Korea. *Humanit Soc Sci Commun*, 10(1), 109. https://doi.org/10.1057/s41599-023-01593-2. Epub 2023 Mar 16. PMID: 36942012; PMCID: PMC10018596.

Hossain, M. T., & Luies, S. K. (2017). Designing a school based health education program to prevent nutritional anaemia among the adolescent girls in a rural school in Bangladesh. *Journal of Health Education Research & Development*, 5, 229. https://doi.org/10.4172/2380–5439.1000229.

IEA. (2014). *A Dictionary of Epidemiology* (6th ed., M. Porta, Ed.). Oxford University Press.

IHR POE. (n.d.). https://ihpoe.mohfw.gov.in/vector.php.

Institute of Medicine (US) Forum on Emerging Infections, Davis, J. R., & Lederberg, J. (Eds.). (2000). *Public Health Systems and Emerging Infections: Assessing the Capabilities of the Public and Private Sectors: Workshop Summary* (Vol. 3). Washington, DC: National Academies Press. https://www.ncbi.nlm.nih.gov/books/NBK100249/.

IntegratedDisease Surveillance Programme (IDSP). https://idsp.mohfw.gov.in › Old Site › HWM.

Integrated-Vector-Management. https://ihpoe.mohfw.gov.in/vector.php.

Karmakar, T., Sain, S., Dey, S. P., Sarkar, A. P., & Biswas, A. (2023, November). Perception of beneficiaries on motivating factors and challenges of COVID-19 vaccination: An institute-based study. *J Family Med Prim Care*, 12(11), 2916–2921. https://doi.org/10.4103/jfmpc.jfmpc_687_23. Epub 2023 Nov 21. PMID: 38186813; PMCID: PMC10771184.

Knittel, R. E., Child, R. C., & Hobgood, J. (1971, August). Role and training of health education aides. *Am J Public Health*, 61(8), 1571–80. https://doi.org/10.2105/ajph.61.8.1571. PMID: 5563275; PMCID: PMC1530204.

Mahmood, S. S., Levy, D., Vasan, R. S., & Wang, T. J. (2014, March 15). The Framingham Heart Study and the epidemiology of cardiovascular disease: A historical perspective. *Lancet*, 383(9921), 999–1008. https://doi.org/10.1016/S0140–6736(13)61752-3. Epub 2013 Sep 29. PMID: 24084292; PMCID: PMC4159698.

Morton, L., Creppage, K., Rahman, N., Early, J., Hartman, L., Hydrick, A., & Kasper, M. (2023). Challenges and opportunities in pathogen agnostic sequencing for public Health surveillance: Lessons learned from the global emerging infections surveillance program. *Health Security*, 22(1), 16–24. https://doi.org/10.1089/hs.2023.0068.

Nsubuga, P., et al. (2006). Chapter 53: Public health surveillance: A tool for targeting and monitoring interventions. In D. T. Jamison et al. (Eds.), *Disease Control Priorities in Developing Countries* (2nd ed.). Washington, DC: The International Bank for Reconstruction and Development/The World Bank; New York: Oxford University Press. https://www.ncbi.nlm.nih.gov/books/NBK11770/.

Park, K. (2023). *Park's Textbook of Preventive and Social Medicine* (27th ed.). B. Bhanot Hopkins Medicine. https://www.hopkinsmedicine.org/health/conditions-and-diseases/emerging-infectious-diseases.

Rapid Response Team Training. https://www.fieldepiinaction.com/our-programs-blog/rrt.

Sain, S., Chattopadhyay, A., & Haldar, D. (2023). Awareness, acceptance, and hesitancy about COVID-19 vaccine among residents of urban and rural health training center field practice area of a medical college at Kolkata: A cross-sectional survey. *Asian Journal of Medical Sciences*, 14, 3–8. https://doi.org/10.3126/ajms.v14i9.54891.

Stellefson, M., Paige, S. R., Chaney, B. H., & Chaney, J. D. (2020, February 12). Evolving role of social media in health promotion: Updated responsibilities for health education specialists. *International Journal of Environmental Research and Public Health,* 17(4), 1153. https://doi.org/10.3390/ijerph17041153. PMID: 32059561; PMCID: PMC7068576.

Tulchinsky, T. H. (2018). Ethical issues in public health. *Case Studies in Public Health*, 277–316. https://doi.org/10.1016/B978-0-12-804571-8.00027-5. Epub 2018 Mar 30. PMCID: PMC7149338.

Universal Immunization Program in India | Ministry of Health and Family Welfare. https://main.mohfw.gov.in › sites › default › files.

Universal Immunization Programme (UIP) | Ministry of Health and Family Welfare | GOI. (n.d.). https://www.mohfw.gov.in/?q=Major-Programmes/universal-immunization-programme-uip.

User, S. (n.d.). *NDCU—Vector Surveillance and Integrated Vector Management*. https://www.dengue.health.gov.lk/web/index.php/en/activities-and-services/vector-surveillance-and-integrated-vector-management.

Vaccination Clinic Planning Activities: Centers for Disease Control and Prevention. https://www.cdc.gov › vaccines › hcp › admin › planning.
Vaccine Administration Protocols | CDC. (n.d.). https://www.cdc.gov/vaccines/hcp/admin/admin-protocols.html.
Vaccines and Immunization—World Health Organization (WHO). https://www.who.int/health-topics/vaccines-and-immunization#tab=tab_1.
WHO_CDS_EPR_LYO_2006_2_eng—World Health Organization (WHO). https://iris.who.int › bitstream › handle › WHO.
World Health Organization: WHO. (2023a, April 26). WHO launches new initiative to improve pandemic preparedness. *World Health Organization—News Release*. https://www.who.int/news/item/26-04-2023-who-launches-new-initiative-to-improve-pandemic-preparedness
World Health Organization: WHO. (2023b, May 20). WHO launches global network to detect and prevent infectious disease threats. *World Health Organization—News Release*. https://www.who.int/news/item/20-05-2023-who-launches-global-network-to-detect-and-prevent-infectious-disease-threats.
World Health Organization (WHO) and the Special Programme for Research and Training in Tropical Diseases (TDR). (2009). *Dengue: Guidelines for Diagnosis, Treatment, Prevention and Control* (New ed., Vol. 3). Geneva: World Health Organization/Vector Management and Delivery of Vector Control Services. https://www.ncbi.nlm.nih.gov/books/NBK143163/.
Zweig, S. A., Zapf, A. J., Beyrer, C., Guha-Sapir, D., & Haar, R. J. (2021, December). Ensuring rights while protecting health: The importance of using a human rights approach in implementing public health responses to COVID-19. *Health and Human Rights*, 23(2), 173–186.

13 Emerging and Re-Emerging Viral Diseases

Perspectives from Ayurvedic System of Medicine

Banani Ray Chowdhury, Debopriyo Pal, Mihika Bhattacharjee, Shreya Sen, Pallabi Roy Chowdhury, Abichal Chattopadhyay, Tirtha Pratim Mandal, and Sayari Bhunia

13.1 INTRODUCTION

Both conventional wisdom and modern thought regard Ayurveda as a suitable cure for chronic illnesses. In the current environment, it is never considered or tested for any potential effects on severe, life-threatening, or urgent situations (Sabila Afzal et al., 2018). In recent times, numerous therapeutic approaches have been introduced to the world, such as the study of Ayurvedic medicinal techniques, with the aim of finding effective treatments for viral infections. This chapter delves into the fascinating area of Ayurvedic intervention for viral therapy, offering a complete approach that blends traditional wisdom with state-of-the-art science. An extensive history of herbal remedies, natural remedies, and holistic approaches forms the foundation of Ayurveda, the age-old Indian medical science (Pekosz & Glass, 2008).

13.2 AYURVEDIC DIAGNOSIS OF VIRAL DISEASES

The age-old comprehensive medical system known as Ayurveda is becoming more popular worldwide due to its emphasis on natural healing techniques and preventative healthcare. Ayurvedic home treatments for fever, cough, and other typical viral infection symptoms shed light on complementary medicine practices, which are the source of Ayurveda's healing properties. The foundation of Ayurveda is based on the three *doshas* theory. This theory forms the foundation of Ayurvedic physiology, pathology, diagnostics, medications, and treatments. It is believed that each *dosha* has unique innate qualities in a person's physiological, psychological, and physical attributes. The Ayurvedic books from the Middle Ages provide a detailed explanation of how to identify *dosha* qualities based on indications and symptoms that lead to *Prakriti* and the appearance of disease (Vrinda et al., 2013; Jaiswal & Williams, 2016).

Methods based on these threefold systems of diagnosis:

Pulse examination method: Pulse diagnosis (*nadi pariksha*) is a non-harmful ancient technique of ayurvedic disease diagnosis by measuring the pulse. It accurately identifies the physical, mental, and emotional imbalances of the patient. Pulse examination is performed by placing the index, middle, and ring fingers at the root of the thumb of the patients.

Tongue examination method: Physicians evaluate the level of coating on the tongue. There are three types of tongue coating: thin coating (*alpasama*), thick, sticky coating (*sama jivha*), and no coating (*niramjivha*).

DOI: 10.1201/9781032721811-13

Prakriti examination method: The *prakriti* has distinct physiological, psychological, and physical qualities based on *dosha* traits. To detect the patient's *prakriti*, Ayurvedic practitioner examine, question, and palpate these characteristics (Sharoni Narang et al., 2018)

13.3 STRATEGIES FOR AYURVEDIC INTERVENTIONS

When treating chronic disorders, where active care of a clinical state is not necessary, Ayurvedic therapies are frequently considered to be helpful. In no critical care condition requiring ongoing care have Ayurvedic therapies been examined. Scientists, researchers, and health officials realized during the pandemic that Ayurveda could be helpful in diagnosing illnesses, particularly those caused by viruses such as COVID-19, Ebola, influenza, adenoviruses, etc. The World Health Organization estimates that 70–80% of people worldwide use complementary and alternative therapies, primarily extracted from herbal sources. The public's interest in complementary and substitute medicine is primarily driven by factors such as rising synthetic drug side effects, inadequate therapy for multiple chronic conditions, prohibitively high costs of new pharmaceuticals, microbial resistance, developing diseases, etc. (Girija P. et al., 2022).

13.3.1 Treating COVID-19

Fever (*jwara*) is a condition covered in Ayurveda and is the subject of the first chapter of therapy (treatment) in two of the field's canonical books. It focuses on classification, management, medications, food, prognosis, pathophysiology (*samprapti*), diagnosis (*nidanam*), and treatment. Typically, the diagnosis of fever is made based on the symptoms that are present, such as a mostly *vata–kapha* condition that needs to be appropriately managed. The patient subsequently tested positive for COVID-19 is taken for examining the efficiency of Ayurveda. A thorough and precautionary Ayurvedic checkup is performed for present complaints. The following conditions are examined as *kapha–vata* defects: fatigue (*vata* domination), fever (*kapha–vata*-dominated *jwara*), discomfort in the throat (*kapha* vitiation), myalgia (*kapha* and *vata* vitiation), and hypertension in most cases (*vata pitta* vitiation). *Vata pitta prakriti* is the body composition; *agnimandya* is the mind *sattva* (psychic condition) (Gundeti et al., 2020).

13.3.2 Treating Influenza

One of the primary causes of morbidity and mortality in low- and middle-income countries (LMICs) in tropical regions like India is influenza, with young and older persons having the highest estimates of influenza-like illness (ILI) and influenza-associated ILI. According to the description in the texts of Ayurveda, the clinical manifestation of ILI is comparable to *vata–kapha jvara* (fever with high *vata* and *kapha dosha*). Through *langhana* (fasting/reduced calorie intake) and formulations (single herb polyherbas) made of ingredients with *tikta rasa*, *ushna virya*, *katu vipaka*, and *amapachaka* (anti-inflammatory) qualities, *jvara* is treated with the goal of breaking down *ama* (partially digested metabolic waste). In the CCRAS peripheral, AYUSH-64, a polyherbal formulation, has been used since the 1980s to treat a variety of ailments, including fevers, inflammation, and joint discomfort. By measuring the ILI symptoms—such as fever, pain in head, muscle pain, runny nose, nasal obstruction/congestion, cough, sore throat, weariness, chills/aversion to cold, and sweating—the capacity of the medication was evaluated. Using the VAS, the symptoms were assessed on a range of 0–100 cm, where 0 denotes no symptoms and 100 denotes the highest possible degree of symptoms. Exams were conducted on these symptoms on days 3, 7, 14, 21, and 28. The time interval between the initial dosage of the study drug and the point at which the patient's body temperature dropped to 37.4 °C or lower and remained there for at least 24 hours was used to quantify the improvement in fever (Gundeti et al., 2020).

13.3.3 Treating Ebola

Numerous outbreaks have already been caused by the lethal Ebola virus. Although there isn't a known cure for Ebola, Ayurveda has been researched for treating the disease's symptoms. Natural remedies are used in Ayurveda to treat a variety of ailments. It has been demonstrated that certain Ayurvedic remedies, such as *haritaki*, *guduchi*, *amalaki*, *haridra*, and *tulasi*, have antiviral properties and may be helpful in treating Ebola. Ayurveda can also be used to treat Ebola hemorrhagic fever (EHF), using the *Ayurvedokta* variant of *jwara—Raktasthivi Jwara Sannipaat* (Deshpande, 2018).

Even though Ayurveda appears to be a promising cure for fatal diseases like the Ebola virus sickness, further research is needed to ascertain the illness's effectiveness. It is critical that we remember the benefits of Ayurveda, as we attempt to comprehend how the system has intervened in the medical sciences. The following is a list of benefits:

- Ayurveda is acceptable for daily usage since it seamlessly integrates with alternative therapies and emphasizes lifestyle choices that, when followed, send a wave of general well-being into everyday life.
- Ayurveda promotes health through well-balanced dietary recommendations, healthy sleeping patterns, herbal medicines, daily and seasonal routines, yoga, and exercising sessions.
- The use of Ayurvedic medicines and supplements improves immunity, appetite, and digestion.
- The basal tenets of Ayurveda include leading a disciplined lifestyle, which includes eating a light dinner, not skipping breakfast, and avoiding too much sleeping, meaning stay awake and have a healthy day.

Thus, improving the Ayurvedic system for medicinal sciences holds vast opportunities for the future. Also, educating people about the benefits of Ayurveda will not only cure diseases but also produce a whole generation of healthy individuals (Anagha et al., 2014).

13.4 STRATEGIES FOR AYURVEDA-BASED IMMUNE SYSTEM ENHANCEMENT

Numerous published acts have demonstrated the immune-boosting effects of various Ayurvedic medications. This ancient practice that dates back to the 2nd century BC focuses on harmonizing the three humors—*tridoshas*, five elements—*pancha mahabhootas*, and seven tissues—*saptadhatus*—to promote overall wellness by prescribing a plethora of treatments for diverse ailments. Ayurvedic therapy continues to be studied for its benefits on both healthy and ill individuals. The majority of these remedies utilize only natural ingredients (Aggarwal et al., 2016). All reliable texts have sporadic descriptions of the clinical picture of AIDS as it is found in Ayurveda. *Ojus* has been used as an explanation for immunity. The works of Charaka, Sushruta, and Vagbhata offer intriguing insights into clinical attributes of *ojus*, including its genesis and characteristics of destabilization such as *visramsa*, *vyapat*, and *kshaya*. The stages of AIDS are similar to Ayurvedic treatments such as *udanavrata prana vata*, *ojanirodhaka jvara*, *shosha*, *madhumeha*, *asadhya pandu*, and *hatoujas*, a complex fever. Here, *ojo kshaya* and AIDS-related opportunistic illnesses have been linked. When the virus invades the body, it decreases both the quantity and quality of the *rasadi sapta dhatus*. In Ayurveda, *ojus* is considered the essential element of all *dhatus*, determining an individual's resistance to illness (known as *vyadhibala virodhitvam*) and their ability to prevent future risk factors (known as *vyadhi utpadaka pratibandhakatvam*) (Ahmad et al., 2021). As stated by the principles of Ayurveda, *rasayana* therapy is a unique and effective way to rejuvenate the body. The potent ingredients found in *rasayana* medications target the levels of *agni dhatu* and *srotas*. It is vital to maintain a balanced *dosha* and *dhatu* system with the aim to achieve overall well-being and the *prashasta dhatus* play a crucial role in this process trusted by renowned Ayurvedic experts like Sushruta and

Charaka. *Shilajatu* is highly regarded for its ability to address imbalances in physical strength *bala* and vital energy *ojus* in order to determine their impact on immunity. Esteemed *rasayana* herbs including *shilajatu mandukaparni guduchi* and *amalaki* were meticulously chosen for the study (Ajala et al., 2016).

13.5 EXPLORING THE AYURVEDIC CONCEPT OF *OJUS* IN RELATION TO HIV INFECTION AND AIDS

After carefully examining and obtaining consent, 20 individuals living with HIV who were accepting ART therapy were identified from the Outpatient and Inpatient departments at Kayachikitsa, S.S. Hospital, IMS, BHU. The inclusion and exclusion criteria are mentioned in Table 13.1. When diagnosing HIV, healthcare professionals look for certain markers such as a history of blood transfusions, a history of multiple sexual partners, and a particular occupation. Additionally, there are clinical manifestations that strongly indicate the presence of HIV. These include abnormal biochemical results like a low CD4 count and a positive ELISA test for HIV I/II. At the BHU's IMS Rasa Shastra Department, the experimental medication was meticulously crafted. The renowned ingredient, *shilajatu*, was purified seven times with *gomutra* before being finely ground into a powder. This powder was then carefully divided into 500 mg capsules and stored after undergoing seven rounds of trituration with the potent *swarasa* of *amalaki*, *guduchi*, and *mandukaparni* (Alajmi et al., 2018).

Over the month-long study, the initial two follow-ups were scheduled every 15 days. As the month progressed, an additional two follow-ups were also completed. The effectiveness of the put to test narcotics were sizing up through (I) assessing relief of symptoms on a point scale, and (II) monitoring any changes in biochemical markers. The results were categorized as follows:

- *Improved:* The patient's clinical signs and symptoms had improved, as had the biochemical findings (LFT, renal profile), and the value of CD4 had been gradually rising.
- *Unchanged:* The clinical characteristic and CD4 count remain unchanged.

A total of 20 individuals diagnosed with HIV were selected for the trial, presenting a diverse range of symptoms. At randomization, 12 participants received a combination of ART and *shilajatu rasayana* (known as the treated group), while the remaining eight received HAART therapy alone. Two individuals from each group were excluded from the study due to non-compliance with medication, despite the researcher's recommendations. ELISA testing was utilized to confirm the presence of both HIV I and HIV II. Further evaluation revealed that out of the 20 documented cases, 18 were males with *vata-pittaj prakriti*, between the ages of 31 and 40, belonging to the labor class (including coolies and drivers), all of whom were married and had a history of multiple sexual partners.

TABLE 13.1
Inclusion and Exclusion Criteria of the Study

Inclusion Criteria	Exclusion Criteria
The patient is not experiencing numerous problems or recurrent infections (age between 20 and 50 years)	Patients in the age category of under 20 and over 40
A CD4 count of at least 150/μL	Patients with problems such as diabetes, hypertension, asthma, and other opportunistic infections on several occasions
Informed consent is given by patients who are willing to take part	Ladies who are nursing or pregnant

All of the aforementioned medications function as immune-boosters and adaptogenics, acting on the body and the mind to increase a person's physical ability to fight off allergies and opportunistic infections as well as to address mood disorders associated with AIDS.

13.6 CURRENT SCENARIO OF AYURVEDA IN TREATING VIRAL DISEASES

Viruses and related diseases have long been seen as humanity's biggest problem and the center of the scientific, agricultural, and healthcare systems. These days, many viral illnesses are under control and extremely uncommon, such as measles, poliomyelitis, rabies, and yellow fever. This is a result of immunizations and other public health initiatives. On the other hand, a lot of new viruses appear on a regular basis and cause terrible harm to both individuals and society (Enquist, 2009). Coronaviruses, also known as CoVs, are a type of RNA virus that can cause disorder in mammals, including mankind. They are just one of many different types of viruses that exist.

Present medical history: On the initial day of feeling unwell, the patient promptly reached out to their Ayurvedic Vaidya in Chennai through a telephonic consultation. Following the Vaidya's advice, the patient began treatment with Ayurvedic remedies for their symptoms. As a precaution, the patient chose to self-isolate due to their New York residence and potential exposure to COVID-19. The presenting symptoms included fever, body aches, coughing, deprivation of vapor and scent, and midriff discomfort. The pyrexia persisted for a period of a week. With a proficient understanding of fever, particularly in relation to COVID-19, the Vaidya administered appropriate Ayurvedic management, in accordance with the principles of *jwara* (Table 13.2) (Gundeti et al., 2020).

A Scanning of the Sufferer's Condition Utilizing Ayurveda

- *Diagnosis:* Understanding the sufferer's indication and the flow of year, our initial diagnosis pointed toward a *nija jwara*, caused by an imbalance of the body *doshas*. This fever exhibited a predominance of *vata-kapha*, for which we prescribed appropriate medication, dietary changes, and a specific regimen. At the time, COVID-19 had not been diagnosed, prompting our treatment plan for the observed symptoms. The extensive body of publicly available material on COVID-19 and the sufferer's stated indications were used to arrive at the final detection. We are able to comprehend all forms of novel and emerging fevers by categorizing fevers based on the aggravating *doshas* (disease-causing components).

TABLE 13.2
Details of the Symptoms, Number of Days It Lasted, Tests Performed, and Medicines Recommended

Date/Day	Symptoms	Test/Result
29.03.2020/Day 1	Experiencing severe anatomical pains at an intensity of 8 out of 10, accompanied by mild to balanced midriff suffering at a 2 to 3 out of 10 level. Currently, running a temperature of 100°F and also to accord in the company of a deprivation of vapor and scent	NA
30.03.2020/Day 2	Once the Ayurvedic medicines were taken, the abdominal pain became easily manageable. However, the body ache still persisted and the temperature remained at 101°F. Additionally, there was a continued loss of taste and smell and some mild coughing present	NA
31.03.2020/Day 3	The individual experienced utmost anatomical suffering go along with a far-up degree of 103°F, as well as ongoing deprivation of vapor and scent. They also suffered from intense bouts of coughing, which were characterized as infrequent, ineffectual, and exacerbated in the evenings	NA

Source: Ayurvedic treatment of COVID-19/SARS-CoV-2: A case report

- *Pathophysiology:* Fever, respiratory distress, and cough, as recognized by *Ashtanga Hrdayam*, are the primary manifestations of COVID-19 and are closely linked to the *abhyantara* pathway of disease. Among the three *roga margas*, or routes of illness, described in this ancient medical text, *abhyantara* holds a significant place. This condition is known to cause *pranavaha sroto dushti*, which is characterized by severe respiratory distress in addition to other symptoms that can occasionally result in death. The *ura* (chest area) is the primary site of affliction for this condition. According to the information above, coronavirus illness is associated with *agantuja sannipataja jwara*, which is primarily *vata–kapha*. Because *bhoota abhishanga* aggravates all three *doshas*, this *jwara* might be categorized as *agantu* (external). *Sannipata*, known for its ability to aggravate all three *doshas*, is a prominent term in the world of Ayurveda. This ancient system of medicine also offers the *bhoota abhishanga* classification, which can provide valuable insight into the ways in which the virus spreads and impacts the body during *jwara*. *Agantu jwara* is to be treated as *nija jwara* due to an imbalance of *doshas*, according to Ayurveda.
- *Etiology:* As per the latest findings from MWM, the ongoing outbreak is caused by a novel CoV that has now been pointed out as SARS-CoV-2. In line with Ayurvedic principles, this virus exhibits all attributes of a *janapadodhwamsa vikara* and is classified as a *vata-kapha*-type fever. What makes this disease particularly concerning is its high level of contagiousness and the various routes through which it can spread from person to person. Current learning has been seen that T lymphocyte, particularly CD4+ and CD8+, participate in a pivotal role in fighting vigorous suppurations and are distinctly affected by SARS-CoV-2. However, there is a possibility that an excessive immunity could lead to inflammation or autoimmunity. While CD8+ T lymphocytes have the ability to demolish germ-contaminated lymphocytes, CD4+ T lymphocytes are vital in the production of viral-specific antibodies through T-cell-dependent B cells. Furthermore, in individuals infected with SARS-CoV, CD8+ T lymphocytes are responsible for both causing damage in the lung tissue and eliminating the virus (16). T helper (Th) lymphocytes participate in a pivotal role in defending against vigorous contaminations, for example, CoV-2. This defense is made possible through the activation of nuclear factor kappa-light-chain-enhancer of activated B cells (NF-kB) signaling, which leads to the production of pro-inflammatory cytokines. The significance of the T-cell response cannot be underestimated in safeguarding against these types of infections. During an infection, the body's immune system activates several cytokines, including tumor necrosis factor (TNF)-β, monocyte chemoattractant protein-1 (MCP-1), interleukin (IL)-17, and others, to recruit monocytes and neutrophils to the site of infection (Bunte & Beikler, 2019). However, prior research has shown that the presence of CD8+ and CD4+ T cells is crucial for combating SARS-CoV. Furthermore, CD4+ T helper cells play a vital role in producing SARS-CoV-specific neutralizing antibodies. However, in some cases, an excessive immune response may lead to severe inflammation and the destruction of host cells. This cascade of cytokine release then causes cytokine release syndrome (CRS), a potentially fatal illness. It is more severe in individuals who also have other concomitant conditions such as heart problems, diabetes, arterial hypertension, and kidney or liver problems. Investigating the precise strategy where a customized immune response is necessary becomes crucial, as does determining the precise method and cause of the cytokine storm-causing hyperbolic and aberrant immune response that can exist in some circumstances (Abruzzese et al., 2016; Gundeti et al., 2020). The ancient practice of traditional medicine takes a thorough proceed toward preventing disease by focusing on three key principles *nidanaparivarjana*, *apakarshana* and *prakritivighata*. These strategies involve not only avoiding the root causes of illness but also actively removing them from the body and creating an environment that is not conducive to the growth of harmful agents. Furthermore, traditional medicines strategy includes *rasayana* therapy, a combination of immune-boosting methods, nutritive control, and manner of living adjustments to

strengthen the body's defenses against a variety of illnesses. Current research has shown that herbs specifically targeted toward the respiratory tract have proven effective in treating viral respiratory infections such as dengue whether through eliminating disease-causing agents or bolstering overall immunity. These powerful herbs play a vital part in encouraging a holistic state of well-being (Nair & Deep, 2024).

13.7 SAFETY AND EFFICIENCY OF AYURVEDIC INTERVENTIONS

Carefully conducted clinical studies have started to assess medications derived from Ayurveda. The purpose of the experiments, which are frequently placebo-controlled, is typically to prove superiority.

- In order to estimate the favorable outcomes of traditional medicines aid and yoga in rehabilitating independently with lingering effects of COVID-19, the WHO's recovery Self-Management has been evaluated.
- In order to assess the protection of Ayur remedy to the persons who are in pain from long-term effects of CoV-2, a thorough evaluation is necessary (Murthy, 2001).

The intervention and comparator are both crucial aspects that are worth considering:

- *Intervention Arm (Group-I):* As per AYUSH'S bureau, a cutting-edge national clinical protocol rooted in the principles of Ayurveda and yoga has been developed to effectively address COVID-19. This protocol includes a variety of interventions such as dedicated yoga sessions twice a day for 45 minutes as well as the daily use of *agastya haritaki* and *ashwagandha* supplements in doses of 6 grams and 500 mg, respectively, with warm water after meals. This comprehensive plan should be consistently followed for a span of 90 days specifically designed to aid in post-COVID-19 recovery. In comparison, the second group in the study will receive standard care for 90 days.
- *Comparator Arm (Group II):* A 90-day rehabilitation self-management program is provided by the World Health Organization. All necessary study medications have been sourced from a reputable pharmaceutical company with GMP certification (Paul, 1993).

Principal Results

Main outcome: The San Diego breathless dyspneic panting short-winded, 6-minute walk test, and pulmonary function test will be utilized to assess any fluctuations in respiratory function.

Secondary Results

- Enhancement of the chest HRCT through modification.
- The fatigue score of the Modified Fatigue Impact Scale exhibits notable variation.
- The Hospital Anxiety and Depression Scale Anxiety Score underwent a revision.
- The Pittsburgh Sleep Quality Index proves to be a highly valuable resource for tracking changes in sleep patterns and overall quality of rest.
- The COV19-QoL scale's evaluation of caliber of existence has been modified.

In order to guarantee the potency and harmlessness of our mediations, we will manage a thorough evaluation using a variety of measures. This will involve analyzing data gathered from biochemical and hematological tests as well as comparing results before and succeeding the mediation period. Additionally, we will carefully examine any reported adverse events or drug reactions through diligent analysis of this information. We can confidently determine the overall safety of our interventions (Rothman & Warren, 1994).

TABLE 13.3
Four Combinations of Safety and Efficacy in the Trade-Off-Type Matrix

	Effective	Not Effective
Safe	Efficient and risk-free	Not only is it ineffective, but also provides minimal safety
Not safe	Even though it may not be completely safe, this method has proven to be effective	These methods are neither safe nor effective

Source: Sanjeev Sarmukaddam et al., 2023

Timelines for Evaluating Outcomes

Throughout the study period, which spans 90 days, subjective and clinical evaluations will be conducted on days 15, 30, 60, and 90. Additionally, laboratory tests such as CBC, LFT, KFT, HbA1c, Hs-CRP, and D-dimer, along with pulmonary function tests and HRCT chest scans, will be administered at both the beginning and conclusion of the study.

Randomization: To generate the random number sequences, the Statistical Package for Social Sciences (SPSS) version 15.0 is utilized. The process of randomization will follow a 1:1 ratio to assign volunteers to one of two study groups.

Masking or blinding: This study employs an open-label design, with the outcome assessor being kept unaware of the participants' research group assignment. Despite this, the hypothesis that "Ayurvedic interventions are as effective as conventional medicine but safer" could be used as a basis for designing trials.

Even though some studies seemed to show promising results individually, it was difficult to draw any conclusions about the safety or effectiveness of Ayurvedic treatments for sinusitis due to the extremely low certainty of total impact (downgraded twice for RoB and once for each of inconsistency, indirectness, and imprecision). Well-planned, well-executed clinical investigations with reported outcomes that are clinically meaningful are required (Table 13.3).

13.8 CASE STUDIES AND CLINICAL TRIALS

CASE STUDY

An extensive investigation was conducted by Gundeti et al. to look at the performance and security of Ayush 64, a multi-herbal ayurvedic medicine, on patients who had signs of influenza-like illness (ILI). The formulation comprises four potent ingredients like powder of *C. crista* L. or *kuberaksha* seed, *S. chirataPexbex. karst (kiratatikta)* whole-plant aqueous extract, *P. Kurroa Royle ex. Benth (katuki)* root aqueous extract, and *A. scholarisR. Br (saptaparna)* bark aqueous extract. Ayush 64 was given to each of the 38 individuals who fulfilled the eligibility requirements and received a clinical diagnosis of ILI. They were then closely observed for 21 days with the improvement of ILI symptoms: nasal discharge being one of the most notable outcomes, headaches, sore muscles, fatigue from fever, swollen nose, and hoarse voice. Cough sensitivity to cold and shivering was assessed at the start of the study and again on days 3, 7, 14, 21, and 28. The overall severity of the illness was also determined. The findings demonstrated a noteworthy enhancement in ILI symptoms starting from the third day. A remarkable 18 participants experienced complete relief from fever on the third day, while 10 participants achieved the same result on the seventh day of treatment. Furthermore,

TABLE 13.4
Grouping of the Patients, and Interventions (Rais et al., 2022)

Group	No. of Patients	Experimental Drug	Quantity	Time Period
A	40	Vyaghryadi Kashaya	50 mL two times daily before breakfast and in evening, with 250 mg of long pepper powder	Ten days
B	40	Samshamani vati	Two 500 mg tablets, two times daily	Ten days
		Shunthi in powdered form	After meals, 2 grams twice a day with heated water	
		Liquified Rasona and Kalka	1 gram with heated water per day.	
C	40	Vitamin C tablet	500 mg twice daily	Ten days
		Acetaminophen tablet	A 500 mg SOS dosage	

the data revealed a consistent decrease in the severity of illness, throughout the entire intervention. Safety was also thoroughly examined, with Ayush 64 proving to have no significant adverse effects on participants. Furthermore, during the course of the treatment, a number of measurements, including blood parameters, kidney and liver functioning tests, stayed within standard ranges. Crucially, the investigation verified that none of the subjects experienced any subsequent ILI-related issues. The reports' findings evaluated together suggest using Ayush 64 in addition to conventional medical treatments may substantially decrease ILI symptoms and promote a quicker recovery. It was also shown that the intervention was well-accepted and had no unfavorable side effects. It should be remembered that this was only the initial analysis with a small number of participants, and that in order to confirm the clinical value of Ayush 64 in treating ILI, larger, more rigorous trials would be preferable (Gundeti et al., 2022).

CLINICAL TRIALS

A study conducted in a hospital in Lucknow, India, examined the efficacy of Ayurvedic therapy for patients with moderate to no symptoms from coronavirus. A total of 120 patients were separated into groups: Group A, Group B, and Group C (Group C is the control group). Group A and Group B were given Vyaghryadi Kashaya and Shunthi Churna with Rasona Kalka, respectively, while Group C served as the control group, as illustrated in Table 13.4. Vital indicators such as oxygen level, body heat, heart rate, BP, and respiration rate were continuously checked on the patients. The nasal and saliva swabs were obtained for real-time PCR testing, which is thought to be the most accurate way for diagnosing COVID-19, on prearranged days (5th, 7th, and 10th). The preparation of Vyaghryadi Kashaya, an Ayurvedic treatment medication, involved equal parts boiling of *shunthi*, *guduchi*, and *kantakari* in drinkable water. The decoction was reduced and strained before being administered twice a day to patients in Group A in addition to powdered *pippali*. Shunthi Churna and Rasona Kalka were given to Group B, and vitamin C tablets were given to Group C, which was the control group. This all-encompassing strategy guaranteed careful patient monitoring and care.

On the fifth day, the majority of patients in Groups A and B reported clear for COVID-19. and On the seventh day of the intervention, the participants recovered faster in contrast to the control group. Moreover, patients in Group A, who were administered Vyaghryadi Kashaya, showed faster recovery rates than those in Group B. The Ayurvedic treatments were well-tolerated, with only a few minor side effects such as a subtle discomfort in the abdominal region that had been conveniently managed. Figure 13.1 shows the steps involved in conducting a research study: After evaluating 200

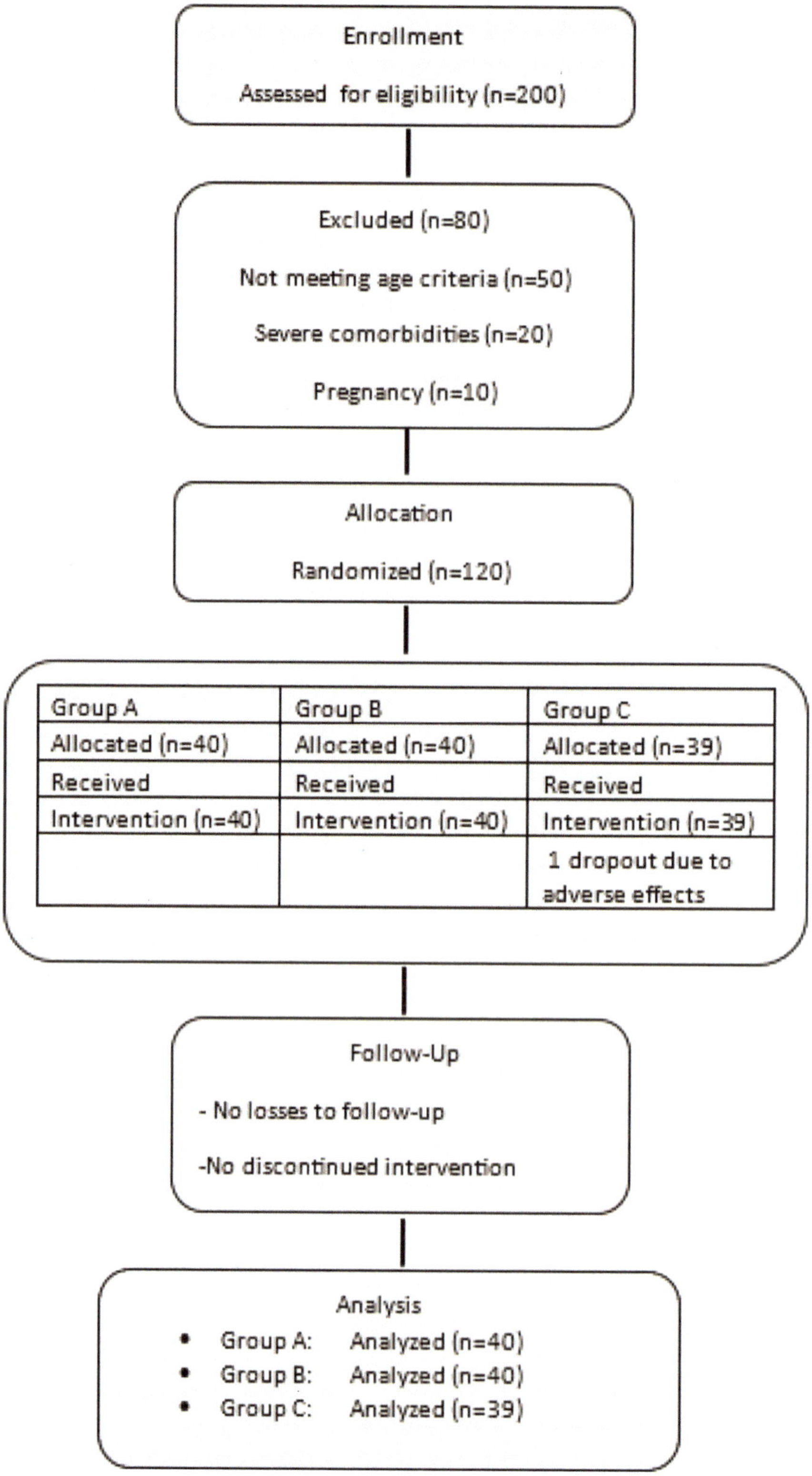

FIGURE 13.1 Graphical presentation of the trial.

participants for eligibility, 80 were disqualified. A total of 120 individuals were randomly assigned to one of three groups: Group A ($n = 40$), Group B ($n = 40$), and Group C ($n = 39$). Following up, none of the groups had any losses or stopped using interventions. According to the final study, there are 40 people in Group A, 40 in Group B, and 39 in Group C.

In addition to these findings, the study emphasized the importance of early detection and prompt management of COVID-19 cases to prevent its spread. The Ayurvedic interventions, by improving digestive health and strengthening the body's natural defense mechanisms, played a pivotal role in controlling disease progression and enhancing the overall prognosis. In this study, various statistical tests were utilized to examine the data. Particularly noteworthy, the Kruskal Wallis test unveiled a significant distinction in the length of time between two subsequent negative real-time PCR tests within the three groups. Subsequently, post hoc tests were conducted to compare the results between each group pairing. Moving on to the discussion, this study offers valuable insights into the Ayurvedic approach to disease management. Ayurveda views disease as a disparity of the biohumors (*doshas*) and a disruption of the digestive fire, also known as *mandagni*, which leads to compromised immunity. As per this approach, the treatment involves correcting the digestive fire and bolstering immunity. The medications used in the study were selected with the intent of managing these key aspects. The research brought attention to how crucial preliminary identification of COVID-19 patients is to stopping the diseases development. It revealed that early infection recognition and treatment might have a major effect on the healing process. It is noteworthy to observe that patients in the Kashaya category recovered faster than patients in the groups receiving control Shunthi Churna and Rasona Kalka treatments. Similarly, on day 7, majority of Group A patients viral elimination tests showed negative findings. *Inference:* this investigation provides strong proof that traditional Ayurvedic treatments work well for treating moderate and silent COVID-19 patients. The herbal remedies that were utilized in the research showed good efficacy in terms of symptom relief, improved digestion, and more rapid healing from the illness. Ayurveda can be applied in combination with current pharmaceutical treatments as an alternative way to address COVID-19. Because of its complete methodology, extensive studies and investigations are necessary to confirm the outcomes and explore Ayurveda's possibility in COVID-19 treatment (Wardani et al., 2023).

13.9 AYURVEDA VERSUS MODERN APPROACHES

Considering the current COVID-19 epidemic, the subject of discussion is fascinating and relevant viral illnesses which can affect many different body systems and cause symptoms like a high body temperature, throat irritation, coughing, and stomach problems are caused by a variety of viruses. Without appropriate medical attention, some viral illnesses can be severe and even fatal, causing skin rashes blisters and tiredness. Ayurveda, an ancient and holistic medical system, has been continuously practiced in India for thousands of years. Modern medicine, also known as allopathy, is a relatively new medical field. Both systems of medicine have different approaches to treating viral diseases. Modern medicine is a relatively recent and scientific system of medicine that is based on empirical evidence and experimentation.

Here are some of the main differences between Ayurveda and modern medicine in case of viral disease treatment:

- *Philosophy:* Our constitution is particular as per Ayurveda, and so our treatment plan should be adapted to our specific *prakriti* and *vikriti*. Ayurveda uses organic methods such as food consumption, changes in habits, herbal medicines, and therapeutic practices to balance the three *doshas* in both the physical and mental bodies. Furthermore, Ayurveda emphasizes sickness prevention by improving the body's immunity, a concept known as *vyadhikshamatva*. In contrast, contemporary medicine treats viral diseases in a more standardized and universal manner. It is largely concerned with identifying the precise type and strain of the virus as well as giving drugs or vaccines to battle both the virus and its symptoms in modern medicine. Advanced technologies are also used to identify and cure viral illnesses (Pandey et al., 2013).

- *Treatment:* Ayurveda offers several ways to manage viral infections:
  - *Herbal remedies:* Ayurveda capitalizes on the therapeutic potential of diverse herbs renowned for their robust anti-inflammatory, antioxidant, antiviral, and immunomodulatory qualities to counter viral infections. A selection of noteworthy examples encompasses oregano, olive leaf, echinacea, St. John's wort, licorice, cranberry, lemon balm, elderberry, turmeric, astragalus, ginger, garlic, honey, *tulsi*, lemon, *kudzu*, sage, and *ashwagandha*. These potent botanicals can be employed in various formulations, such as teas, decoctions, powders, capsules, syrups, or pastes, offering relief from symptoms and fortifying the immune system (Payyappallimana & Venkatasubramanian, 2016).
  - *Dietary guidelines:* To battle infections caused by viruses, Ayurveda suggests a wholesome food rich in fruits and veggies, as well as cereal grains, nuts, seeds, dairy, and antioxidant-rich herbs. Fruits with citrus, berries, spinach, yogurt, coconut milk, and warm beverages are touted as immune-boosting nutrients. For improved overall well-being and immunity against infectious threats, restrict junk food, sugary products, caffeine, alcoholic beverages, deep-fried dishes, and spicy meals (Pandey et al., 2013; Wallace et al., 2020).
  - *Lifestyle modifications:* Ayurveda advises adopting a healthy lifestyle that promotes physical and mental well-being and prevents diseases. Some lifestyle modifications that can help prevent and treat viral infections are regular exercise, yoga, meditation, *pranayama* (breathing exercises), adequate sleep, proper hygiene, stress management, avoiding smoking, and exposure to pollutants (Wallace, 2020; Razonable, 2011).

Modern medicine offers several ways to treat viral infections:

- *Drugs:* Various medications are used in modern times to either kill or inhibit virus replication or activity—antivirals (like acyclovir for herpes), antibiotics (like azithromycin for bacterial co-infections), antipyretics (like paracetamol for fever), analgesics (like ibuprofen for pain), antihistamines (like cetirizine for allergies), decongestants (like pseudoephedrine for nasal congestion), expectorants (like guaifenesin for cough), antiemetics (e.g., ondansetron for nausea), antidiarrheals (e.g., loperamide for diarrhea), and so on. Depending on the severity of the infection, these drugs can be administered through either oral ingestion or intravenous injection (Rahmatullah et al., 2022; Singh et al., 2021).
- *Vaccines:* In today's advanced medical practices, the use of vaccines has become a crucial tool in preventing or minimizing the chance of contracting various viral illnesses. Vaccines, which are made up of attenuated or annihilated variants of a pathogen or its elements, promote the body's defenses to develop antibodies and fight off possible infections. Common examples of viral vaccines include those for hepatitis, influenza, polio, rubella, chickenpox, mumps, measles, and HPV. Typically, these vaccines are administered through injections or oral drops (Singh et al., 2021).

- *Benefits and limitations:* Ayurvedic and modern approaches to treating viral diseases have their own benefits and limitations:
  - *Ayurvedic benefits:* Ayurvedic treatment for viral diseases has several benefits over conventional treatment methods:
    - *It is natural and safe:* Ayurvedic medicines are made with natural components that have little or no negative effects compared to synthetic drugs that may cause adverse reactions or complications.
    - *It is holistic and personalized:* Ayurvedic treatment considers the whole person and not just the disease. It takes into account the individual's constitution, imbalance, symptoms, preferences, environment, and other factors to provide a customized treatment plan that suits their needs (Kaur & Talwar, 2022).

        - *It is preventive and curative:* Ayurvedic treatment aims to not only cure the disease but also prevent its recurrence by enhancing the immunity and restoring the balance of the body and mind.
- *Ayurvedic limitations:* Ayurvedic treatment for viral diseases also has some limitations that need to be addressed:
    - *It is not standardized or regulated:* Ayurvedic remedies are not standardized or regulated by any authority, which may lead to variations in quality, potency, dosage, and safety. There may also be a lack of scientific evidence or clinical trials to support the efficacy and safety of some Ayurvedic remedies.
    - *It is not widely available or accessible:* Ayurvedic treatment may not be widely available or accessible to everyone, especially in remote or rural areas, where there may be a lack of qualified Ayurvedic practitioners, facilities, or resources. There may also be a lack of awareness or acceptance of Ayurveda among some people, who may prefer conventional treatment methods or have misconceptions. While Ayurvedic medicine can enhance health and quality of life, it is crucial to note that it is not a substitute for conventional medical care. This is especially critical in the event of serious or life-threatening illnesses such as HIV/AIDS, hepatitis, and herpes, where more intensive and prompt medical attention may be required. Nevertheless, when integrated as a supplementary or complementary approach, Ayurvedic medicine has the potential to augment overall outcomes and contribute to the overall well-being of individuals.
- *Modern benefits:* Modern treatment for viral diseases has several benefits over traditional treatment methods:
    - *It is fast and effective:* Modern drugs and vaccines can quickly and effectively target and eliminate the virus and its symptoms. They can also prevent or reduce the risk of complications or transmission of the disease.
    - *It is based on scientific evidence:* Modern treatment is based on empirical evidence and experimentation that can prove its efficacy and safety. It also follows strict standards and regulations that ensure quality and consistency (Kumar, 2021).
    - *It is widely available and accessible:* Modern treatment is widely available and accessible to everyone, especially in urban areas, where there are plenty of qualified doctors, hospitals, pharmacies, laboratories, etc. There is also a high level of awareness and acceptance of modern medicine among most people.
- *Limitations of modern system:* Modern treatment for viral diseases also has some limitations that need to be addressed. Some of these limitations are:
    - *It is synthetic and harmful:* Modern drugs and vaccines may contain synthetic chemicals or biological agents that may have harmful side effects or complications. They may also interact with other drugs or conditions that may worsen the situation. They may also cause resistance or mutation of the virus that may make it more challenging to treat.
        - *It is symptomatic and temporary:* Modern treatment focuses on relieving the symptoms of the disease rather than curing it. This approach fails to adequately target the main source of the illness or the fundamental disharmony within the body and psyche. Moreover, it may only yield short-term alleviation that can diminish over time.
        - *It is expensive and dependent:* Modern treatment can be very expensive and unaffordable for some people who may not have access to health insurance or subsidies. It may also create a dependency on drugs or vaccines that may reduce the natural immunity or self-healing ability of the body.

Therefore, it is not possible to say which approach is better or superior in treating viral diseases. Both Ayurveda and modern medicine have their strengths and weaknesses and can complement each other in providing holistic and effective care for patients. The most effective strategy can vary depending on a multitude of factors, including the nature and intensity of the viral illness, the patient's age, overall health, personal preferences and beliefs, the accessibility, cost and suitability of potential treatments, as well as the knowledge, skills, and track record of healthcare professionals (Upadhayay et al., 2022).

13.10 FUTURE PROSPECTS AND CHALLENGES OF AYURVEDA IN TREATING VIRAL DISEASES

The future prospect of Ayurveda in treating viral diseases is very promising and optimistic. Ayurveda has much potential to offer effective and safe solutions for viral diseases that can complement and enhance the modern medical system. The following are some of the reasons why Ayurveda has a bright future in treating viral diseases:

- *Integrative and complementary therapy:* Ayurveda can help cure viral disorders in conjunction with contemporary therapy. It can be used in addition to traditional antiviral drugs to boost immunity, lessen symptoms, and promote general health.
- *Increasing immunity:* Ayurveda places a high value on having a robust immune system. Herbal treatments, dietary adjustments, and lifestyle adjustments are all examples of Ayurvedic practices that can strengthen immunity and increase resistance to viral infections.
- *Preventive measures:* Ayurveda offers advice on methods to preserve good health and lessen the likelihood of contracting illnesses. This has the potential to lower the risk of viral infections.
- *Herbal formulations:* Ayurveda provides a variety of herbal remedies that may have antiviral effects. Investigations into these compositions' efficacy may result in the creation of Ayurvedic remedies designed especially to address certain viral infections.
- *Handling of post-viral complications:* Ayurvedic medicine offers techniques for handling weariness, weakness, and respiratory problems that arise from viral infections. Those recuperating from viral illnesses may find their quality of life enhanced by these methods.
- *Holistic approach:* Ayurvedic medicine treats the full person, taking into account mental and emotional wellness in addition to physical health. This strategy may help control the tension and worry brought on by viral infections.
- *Revival of traditional knowledge:* Research into and validation of the efficacy of traditional medical systems, such as Ayurveda, in treating viral infections may expand due to the rising interest in these systems.
- *Customized care:* Ayurveda bases its treatment plans on the unique constitution (Prakriti) of each patient. Individualized treatment may be more appealing to patients who are looking for comprehensive care.
- *Interest on a global scale:* Ayurveda's fame has spread beyond India. Ayurveda may gain traction as a more mainstream method of treating viral infections as interest in complementary and alternative medicine increases.
- *Regulatory framework:* In order to guarantee the efficacy, safety, and quality control of Ayurvedic therapies, certain nations are striving to create regulatory frameworks for traditional medicine, which includes Ayurveda.
- *Research and collaboration:* The creation of evidence-based protocols for Ayurvedic therapies in the management of viral diseases can result from collaborative efforts between contemporary medical researchers and Ayurvedic practitioners.
- *Public education and awareness:* Raising public knowledge of the possible advantages and restrictions of Ayurveda in the treatment of viral infections might enable people to make well-informed healthcare decisions (Singh et al., 2021; Talwar et al., 2020).

Ayurveda, an ancient system of traditional medicine, faces several challenges when it comes to treating viral diseases. These challenges are primarily due to the differences between Ayurvedic principles and the nature of viral infections. Here are some of the key challenges:

- *Absence of particular antiviral drugs:* Herbal medicines, dietary adjustments, and lifestyle alterations are the mainstays of Ayurvedic treatment for a variety of ailments. It does not, however, include any particular antiviral drugs that can target and eliminate viruses directly. This restricts its ability to quickly suppress viral infections.
- *Variation in diagnosis:* The doshas (*vata*, *pitta*, and *kapha*) form the foundation for Ayurvedic diagnosis. When using these ideas to identify viral illnesses, treatment approaches may differ and interpretations may become subjective. When dealing with viral diseases that are quickly developing, the lack of a uniform diagnosis might provide difficulties.
- *Opposition to conventional medicine:* Patients may occasionally put off or refuse to get conventional medical care in favor of Ayurvedic treatments. Negative effects may result from improper or delayed treatment for severe viral infections.
- *The challenge to conventional medicine:* Patients may occasionally put off or refuse to get conventional medical care in favor of Ayurvedic treatments. Serious viral infections may not respond well to early or improper treatment (Upadhayay et al., 2022).
- *Limited knowledge of viral biology:* In contrast to contemporary virology, Ayurveda has a very limited grasp of viral biology. The development of focused therapeutics for viral infections may be hampered by this ignorance.
- *Viral mutations:* Viruses frequently undergo fast mutations, which give rise to new strains with distinct traits. It is possible that Ayurvedic remedies can't be modified to successfully fight these novel virus strains.
- *Lack of standardization:* The efficacy and consistency of therapies may be impacted by variations in the manufacturing and quality control of Ayurvedic medications. The level of standardization and quality control procedures is not as high as it is in contemporary medicine.
- *Ethical and Regulatory Challenges:* The integration of Ayurvedic remedies with contemporary medicine and adherence to regulatory regulations might provide ethical and regulatory challenges. It takes work to ensure that Ayurvedic treatments are both safe and effective (Chopra et al. 2010).
- *Danger of overreliance:* Relying too much on Ayurveda to treat viral illnesses might result in missed opportunities for prompt and efficient conventional medical therapies, such antiviral medications and vaccinations.
- *Availability of vaccination:* Unlike contemporary medicine, Ayurveda does not recommend vaccination as a prophylactic strategy against viral illnesses. The spread of several viral diseases has been largely controlled by vaccination

Even though Ayurveda may be able to cure viral infections in the future, evidence-based research and integration with contemporary medical techniques are crucial. When traditional and contemporary medicine work together, they can offer more complete and efficient treatments for the treatment and prevention of viral infections (Goyal, 2019; Chauhan et al., 2015).

13.11 CONCLUSION

To battle with several infectious diseases which have the capability to re-emerge and infect people, Ayurvedic intervention can be included for treating these diseases. Ayurvedic remedies include pure natural herbs which are effective in preventing viral diseases. Moreover, the herbal medicines are used to relieve symptoms along with boosting immune system against viral attack. With awareness and ease of availability, people today have realized that prescribed chemical drugs have failed in offering mechanisms to build-up immunity and prevent chronic issues within the body. Thus, Ayurveda demand will surge in the coming years and individuals will understand that the only way to avoid long-term health issues is to work with 3R formula of healing (Remove, Renew, and Restore) and tools of diet, home remedies, detox, herbs used in Ayurveda which can combat with emerging and re-emerging viral diseases.

REFERENCES

Abruzzese, M. P., et al. (2016). Inhibition of bromodomain and extra-terminal (BET) proteins increases NKG2D ligand MICA expression and sensitivity to NK cell-mediated cytotoxicity in multiple myeloma cells: Role of cMYC-IRF4-miR-125b interplay. *Journal of Hematology & Oncology*, 9, 134. https://doi.org/10.1186/s13045-016-0362-2.

Aggarwal, L. M., Chacko, M. K., & Kuruvilla, T. B. (2016). Systematic and comprehensive investigation of the toxicity of curcuminoid-essential oil complex: A bioavailable turmeric formulation. *Molecular Medicine Reports*, 13, 592–604. https://doi.org/10.3892/mmr.2015.4579.

Ahmad, S., et al. (2021). Indian medicinal plants and formulations and their potential against COVID-19 – preclinical and clinical research. *Frontiers in Pharmacology*, 11, 2470. https://doi.org/10.3389/fphar.2020.578970.

Ajala, T. O., et al. (2016). The physicochemical, safety, and antimicrobial properties of Phyllanthus amarus herbal cream and ointment. *Journal of Pharmaceutical Investigation*, 46, 169–178. https://doi.org/10.1007/s40005-015-0226-8.

Alajmi, M. F., Mothana, R. A., Al-Rehaily, A. J., & Khaled, J. M. (2018). Antimycobacterial activity and safety profile assessment of Alpinia galanga and Tinospora cordifolia. *Evidence-Based Complementary and Alternative Medicine*, 2018, 2934583–2934512. https://doi.org/10.1155/2018/2934583.

Anagha, K., Manasi, D., Priya, L., & Meera, M. (2014). Scope of Glycyrrhiza glabra (Yashtimadhu) as an antiviral agent: A review. *Int J Curr Microbiol App Sci*, *3*(1), 657–65.

Bunte, K., & Beikler, T. (2019). Th17 cells and the IL-23/IL-17 axis in the pathogenesis of periodontitis and immune-mediated inflammatory diseases. *International Journal of Molecular Sciences*, 20(14), 3394.

Chauhan, A., Semwal, D. K., Mishra, S. P., & Semwal, R. B. (2015). Ayurvedic research and methodology: Present status and future strategies. *Ayu*, 36(4), 364–369. https://doi.org/10.4103/0974–8520.190699.

Deshpande, M. (2018). *Review on Ayurvedic Perspective of Ebola Hemorrhagic Fever (EHF)-(Raktashthiivi Sannipat)*. ResearchGate.

Enquist, L. W. (2009). Virology in the 21st century. *Journal of Virology*, 83(11), 5296–5308.

Girija, P., Sivan, N., Murugavel, Y. A., Naik, P., Mukundan, T., & Durai, K. M. (2022). Stand-alone Ayurvedic intervention with home quarantine in COVID-19 – outcomes of clinical practice. *Journal of Ayurveda and Integrative Medicine*, 13(2), 100445.

Goyal, M. (2019). Threats and challenges of emerging viral diseases and scope of Ayurveda in its prevention. *Ayu*, 40(2), 67–68. https://doi.org/10.4103/ayu.AYU_18_20.

Gundeti, M. S., Bhurke, L. W., Mundada, P. S., Murudkar, S., Surve, A., Sharma, R., . . . Dhiman, K. S. (2022). AYUSH 64, a polyherbal ayurvedic formulation in influenza-like illness – results of a pilot study. *Journal of Ayurveda and Integrative Medicine*, *13*(1), 100325.

Gundeti, M. S., et al. (2020). AYUSH 64, a polyherbal Ayurvedic formulation in Influenza-like illness—results of a pilot study. *Journal of Ayurveda and Integrative Medicine*, 13(1), 100325. https://doi.org/10.1016/j.jaim.2020.05.010.

Jaiswal, Y. S., & Williams, L. L. (2016). A glimpse of Ayurveda—The forgotten history and principles of Indian traditional medicine. *Journal of Traditional and Complementary Medicine*, 7(1), 50–53. https://doi.org/10.1016/j.jtcme.2016.02.002.

Kaur, N., & Talwar, K. K. (2022). Role of integrative wellness in the care of cardiometabolic disorders. *Journal of Cardio-diabetes and Metabolic Disorders*, 2(2), 47–52.

Kumar, D. S. (2021). Ayurveda vs. modern medicine: Superiority debate is counterproductive to people. *Health World from The Economic Times*.

Murthy, P. H. C. (2001). *Sarngadhara Samhita of Sarngadharacarya* (1st ed.). Chowkhamba Sanskrit Series Office.

Nair, P. G., & Deep, V. C. (2024). Unveiling the life and works of Manakoddam Keshavan Vaidyan: Expert in Panchakarma and Ayurvedic narrative medicine approach. *Journal of Research in Ayurvedic Sciences*, 8(Suppl 1), S64-S69.

Pandey, M. M., Rastogi, S., & Rawat, A. K. (2013). Indian traditional Ayurvedic system of medicine and nutritional supplementation. *Evidence-Based Complementary and Alternative Medicine*, 2013, 376327. https://doi.org/10.1155/2013/376327.

Paul, K. (1993). Trials and errors: Alternative thoughts on the methodology of clinical trials. *BMJ*, 306, 1706–1707.

Payyappallimana, U., & Venkatasubramanian, P. (2016). Exploring ayurvedic knowledge on food and health for providing innovative solutions to contemporary healthcare. *Frontiers in Public Health*, 4, 57.

Pekosz, A., & Glass, G. E. (2008). Emerging viral diseases. *Md Med*, 9(1), 11, 13–16. https://doi.org/10.1093/medmed.

Rahmatullah, M., Jahan, R., Nissapatorn, V., Pereira, M. D., & Wiart, C. (2022). Editorial: Emerging and old viral diseases: Antiviral drug discovery from medicinal plants. *Frontiers in Pharmacology*, 13, 976592.

Rais, A., et al. (2022). A randomized open-label parallel-group pilot study to evaluate the efficacy of Ayurveda interventions in the management of asymptomatic and mild COVID-19 patients. *Journal of Ayurveda and Integrative Medicine*, 13(2), 100393.

Razonable, R. R. (2011). Antiviral drugs for viruses other than human immunodeficiency virus. *Mayo Clinic Proceedings*, 86(10), 1009–1026.

Rothman, J. E., & Warren, G. (1994). Implications of the SNARE hypothesis for intracellular membrane topology and dynamics. *Current Biology*, *4*(3), 220–233.

Sabila Afzal, S., et al. (2018). Zoonotic diseases and Ayurvedic treatment: A review. *Global Scientific Journals*, 6(11). ISSN 2320-9186.

Sharoni Narang, S. P., Batwal, O., & Khandagale, M. (2018). Ayurveda based disease diagnosis using machine learning. *Inetnational Research Journal of Engineering and Technology*, 5(3), 3704–3707.

Singh, R., Goel, S., Bourgeade, P., Aleya, L., & Tewari, D. (2021). Ayurveda Rasayana as antivirals and immunomodulators: Potential applications in COVID–19. *Environmental Science and Pollution Research International*, 28(40), 55925–55951.

Talwar, S., Sood, S., Kumar, J., Chauhan, R., Sharma, M., & Tuli, H. S. (2020). Ayurveda and allopathic therapeutic strategies in coronavirus pandemic treatment. *Current Pharmacology Reports*, 6(6), 354–363. https://doi.org/10.1007/s40495-020-00245-2.

Upadhayay, A., Patel, G., Pal, D., & Kumar, A. (2022). Frequently used allopathic and traditional medicine for COVID-19 treatment and feasibility of their integration. *Chinese Journal of Integrative Medicine*, 28(11), 1040–1047.

Vrinda, K., Bilgrau, A. E., Waagepetersen, R., Toft, E., & Prasad, R. (2013). Interrater reliability of diagnostic methods in traditional Indian Ayurvedic medicine. *Evidence-Based Complementary and Alternative Medicine*, 2013, 658275.

Wallace, C. L., Wladkowski, S. P., Gibson, A., & White, P. (2020). Grief during the COVID-19 pandemic: Considerations for palliative care providers. *Journal of Pain and Symptom Management*, 60(1), e70–e76.

Wallace, R. K. (2020). The microbiome in health and disease from the perspective of modern medicine and Ayurveda. *Medicina*, 56(9), 462. https://doi.org/10.3390/medicina56090462.

Wardani, R. S., et al. (2023, February 12). Treatment of the common cold with herbs used in Ayurveda and Jamu: Monograph review and the science of ginger, liquorice, turmeric, and peppermint. *Drugs in Context*, 12.

14 Emerging and Re-Emerging Viral Diseases (Influenza-Like Illnesses)

Perspectives from Homoeopathic System of Medicine

Srijan Goswami, Ushmita Gupta Bakshi, and Chiranjeeb Dey

14.1 INTRODUCTION

Emerging infectious diseases (EIDs) and re-emerging infectious diseases (REIDs) represent a significant and growing threat to global public health. EIDs are infections that have recently appeared within a population or whose incidence or geographic range is rapidly increasing, while REIDs are infections that were once under control but have resurged due to changes in human behavior, microbial evolution, climate, or environmental disruptions (Morens et al., 2004). These diseases, such as severe acute respiratory syndrome (SARS), Middle East respiratory syndrome (MERS), Zika virus, Ebola, and the ongoing coronavirus disease (COVID-19), have demonstrated the catastrophic impact of viral infections on healthcare systems, economies, and societies worldwide. Despite remarkable advancements in modern medicine, the conventional approaches to managing EIDs and REIDs often face substantial challenges. These include the high cost and lengthy timelines of vaccine and antiviral drug development, limited availability of effective therapies for novel pathogens, and the growing issue of antimicrobial resistance (AMR) (Fauci & Morens, 2012). Moreover, the unpredictability of viral evolution, combined with gaps in healthcare accessibility in low-resource settings, underscores the need for complementary and integrative strategies. In light of these challenges, exploring alternative therapeutic approaches becomes imperative. The Homoeopathic System of Medicine, a well-established complementary healthcare system, offers a promising perspective in addressing the EID-REID crisis. Homoeopathy, based on the principle of "like cures like," emphasizes individualized treatment and the stimulation of the body's innate healing mechanisms (Bell et al., 2013). Its approach considers not only the physiological but also the psychological and environmental factors contributing to disease susceptibility, thus promoting holistic health. The inclusion of homoeopathy as a complementary tool in combating viral diseases has gained interest due to its potential in mitigating symptoms, enhancing immunity, and reducing the burden on conventional healthcare systems. Evidence from clinical and epidemiological studies has highlighted its efficacy in managing certain viral infections, such as influenza-like illnesses, dengue, and chikungunya (Mathie et al., 2013). Furthermore, its cost-effectiveness and safety profile make it particularly attractive in resource-limited settings. This chapter aims to examine the role of homoeopathy in addressing the global challenge of EIDs and REIDs. By providing insights into its therapeutic potential and exploring its integration with conventional medicine, the discussion seeks to present homoeopathy as a viable component of a comprehensive public health response to emerging viral threats.

DOI: 10.1201/9781032721811-14

14.2 HOMOEOPATHIC SYSTEM OF MEDICINE: THE FUNDAMENTALS

Homoeopathy is a system of medicine that was founded in the late 18th century by Samuel Hahnemann, a German physician, who developed the core principles and practices of this alternative healthcare system. The term "homoeopathy" is derived from two Greek words: *homoios* (meaning similar) and *pathos* (meaning suffering), reflecting its foundational principle of "like cures like." According to this principle, a substance that causes symptoms in a healthy person can, when administered in extremely diluted forms, treat similar symptoms in a sick person (Hahnemann, 2002). Homoeopathic treatments are prepared through a process called potentization, which involves serial dilution and succussion (vigorous shaking) of the substance. This process is believed to enhance the healing properties of the substance, despite the fact that the final preparation may not contain any measurable molecules of the original substance (Teixeira, 1992). The system treats the whole person, addressing physical, mental, and emotional symptoms, aiming to stimulate the body's innate healing mechanisms to restore balance and health. The scientific logic underlying homoeopathy is based on several interconnected principles, although it remains a topic of ongoing debate and research within the medical and scientific communities. Below are the core scientific concepts that support homoeopathic practice:

- *Principle of similars (like cures like):* The core idea of homoeopathy emphasizes that a substance capable of producing certain symptoms in a healthy person can be used to treat similar symptoms in a sick person. This principle is founded on observations of symptom patterns rather than specific molecular effects. While conventional pharmacology focuses on biochemical interactions and dose–response relationships, homoeopathy emphasizes the concept of "energetic healing," where the remedy is thought to stimulate the body's healing energy in response to the disease (Bell et al., 2013). Although the mechanisms are not fully understood, this concept resonates with the idea of "biological resonance" in certain alternative medicine theories, where an energetic imprint of a substance might influence physiological processes.
- *Dilution and potentization:* Homoeopathic remedies are typically prepared through serial dilution followed by succussion (vigorous shaking), which is believed to potentiate the therapeutic properties of the remedy. This process results in extremely diluted solutions, often beyond the point where any molecules of the original substance are likely to remain. The scientific challenge has been to explain how such highly diluted remedies could exert a therapeutic effect, given that traditional pharmacology relies on measurable amounts of active substances. Some researchers suggest that homoeopathic remedies may interact with water's molecular structure in ways not yet fully understood. The concept of "water memory," proposed by Benveniste (2004), speculates that water might retain an "imprint" of a substance, even after it is no longer present, which could explain the biological activity of highly diluted remedies (Thomas, 2007).
- *The vital force:* Hahnemann's concept of the "vital force" is central to homoeopathy. The vital force refers to a dynamic, immaterial energy that maintains the body's health and equilibrium. According to homoeopathy, illness occurs when the vital force is disturbed, and treatment aims to restore this balance. This concept is aligned with some holistic and energetic approaches in modern medicine, such as biofields and energy medicine (Moss & Green, 2009). While the vital force remains an abstract concept without a direct scientific equivalent, its application in homoeopathy encourages a whole-person approach to treatment, considering the interplay of mind, body, and environment.
- *Individualization of treatment:* Homoeopathy is characterized by its individualized approach to treatment. Unlike conventional medicine, which often targets a disease or symptom in a standardized way, homoeopathy takes into account the patient's unique

physical, emotional, and psychological profile. This holistic approach is based on the understanding that health is a dynamic balance and that disease results from the disruption of this balance. The homoeopathic practitioner aims to choose a remedy that matches the patient's total symptom picture, rather than treating just one specific symptom (Bell et al., 2013). This individualized treatment plan can be seen as an early form of personalized medicine, which is gaining prominence in modern healthcare.

Homoeopathy has been a subject of criticism, particularly due to the lack of consensus regarding its scientific plausibility. Critics argue that the highly diluted nature of homoeopathic remedies exceeds Avogadro's number, meaning no molecules of the original substance should remain (Ernst, 2002). Despite this, several clinical trials and systematic reviews suggest that homoeopathy may offer therapeutic benefits, especially for chronic conditions or as a complementary treatment in conjunction with conventional therapies. A systematic review by Shang et al. (2005) found that homoeopathic treatments appeared to have a certain effect beyond placebo, while a study by Mathie et al. (2013) reported positive outcomes for conditions such as influenza-like illnesses. Homoeopathy offers a distinctive approach to healing, rooted in the idea of stimulating the body's innate capacity to restore health. While its scientific plausibility remains debated, homoeopathy provides an individualized, holistic model that may complement conventional medicine, especially for managing chronic and complex conditions. The integration of homoeopathy into mainstream healthcare requires further investigation, particularly through rigorous clinical trials and exploration of its underlying biological mechanisms. Homoeopathy is based on several fundamental principles that distinguish it from conventional medicine. The core principle, *like cures like* (*similia similibus curentur*), suggests that a substance capable of producing symptoms in a healthy person can cure those same symptoms in a sick person when administered in highly diluted form (Hahnemann, 2002). Another key principle is the concept of *individualization*, where treatment is tailored to the patient's unique symptom profile, encompassing physical, emotional, and mental aspects. Homoeopathy also emphasizes the *vital force*, an innate energy that maintains health by regulating physiological functions. Illness occurs when this vital force is disturbed, and homoeopathic remedies are believed to restore balance by stimulating the body's healing energy (Bell et al., 2013). Additionally, homoeopathy relies on *potentization*, a process of serial dilution and succussion, to enhance the remedy's therapeutic properties, even at extreme dilutions where no molecules of the original substance are likely to remain (Teixeira, 1992). In contrast, conventional medicine primarily operates on a mechanistic understanding of disease, focusing on diagnosing specific conditions and treating them with drugs or surgical interventions aimed at altering the biochemical or physiological processes involved (Fauci & Morens, 2012). Conventional medicine tends to target symptoms or pathogens directly, often through standardized, one-size-fits-all treatments. Homoeopathy, by contrast, treats the person as a whole, emphasizing the body's ability to self-heal and the importance of individualized care. While both systems aim to restore health, homoeopathy's holistic, energy-based approach contrasts sharply with the reductionist, symptom-targeted nature of conventional medicine. Homoeopathy offers several advantages that make it a suitable complementary or alternative option to conventional medicine, particularly in cases where conventional treatments may fall short or involve adverse effects. One of the primary benefits of homoeopathy is its individualized approach, focusing on the person as a whole rather than just the disease. This personalized care considers the physical, emotional, and psychological aspects of the patient, which aligns with the growing emphasis on holistic and patient-centered care in modern healthcare (Bell & Koithan, 2012). Additionally, homoeopathic remedies are typically prepared through highly diluted solutions, which are generally considered safe with minimal side effects, making them especially suitable for patients seeking a gentler alternative to potent pharmaceutical drugs, particularly in cases of chronic conditions or for vulnerable populations such as children and the elderly (Teixeira, 1992). Furthermore, homoeopathy has a lower risk of drug interactions, which is particularly important for patients on polypharmacy or those with complex health conditions (Mathie et al., 2013). In contrast, conventional medicine often relies on drugs that target specific symptoms or pathogens, which may be accompanied by side effects or long-term complications. Homoeopathy's non-invasive, symptom-matching approach can provide a safer and more comprehensive treatment, especially when combined with conventional therapies for a more integrative model of care.

14.3 INFECTIOUS DISEASES AND HOMOEOPATHY: THE CONNECTION

Infectious diseases are illnesses caused by pathogenic microorganisms, such as bacteria, viruses, fungi, and parasites that can spread directly or indirectly from person to person, animal to person, or environment to person. These diseases remain a major cause of morbidity and mortality worldwide, particularly in regions with inadequate healthcare infrastructure (Fauci & Morens, 2012). The traditional conceptual framework for understanding infectious diseases is the *epidemiological triad*, which consists of three components: the host, the agent (microorganism), and the environment. This model assumes that disease results from the interaction of these three elements, with the agent being the central cause of infection. However, this model oversimplifies the complexity of disease processes and fails to fully account for the role of the host's immune system, genetics, and broader environmental factors, such as socioeconomic conditions and lifestyle, which influence disease outcomes (Moss & Green, 2009). The *germ theory* of disease, proposed by Louis Pasteur and Robert Koch in the 19th century, revolutionized the understanding of infectious diseases by emphasizing that microorganisms are the primary cause of infection. While the germ theory has been instrumental in advancing medical science and treatments, it has been criticized for focusing too narrowly on the pathogenic agents and neglecting the host's susceptibility, resilience, and the dynamic interactions between the host and the pathogen (Moss & Green, 2009). Conventional treatments typically target the elimination of the pathogen using antibiotics, antivirals, or vaccines. However, this approach can be limited by issues such as antimicrobial resistance, side effects, and a failure to address the root causes of host vulnerability, including stress, lifestyle, and environmental factors (Fauci & Morens, 2012). This conceptual mismatch between the germ theory, the epidemiological triad, and conventional treatments suggests that a more holistic approach, one that considers the host's entire context and potential interactions with alternative therapies, may be necessary to address the complexity of infectious diseases more effectively. The epidemiological triad, Rudolf Virchow's ideology, and the homoeopathic system of treatment share a common emphasis on the interconnectedness of factors influencing health and disease. The epidemiological triad highlights the interaction between the host, agent, and environment, recognizing that disease arises not solely from pathogens but also from the host's susceptibility and environmental conditions (Moss & Green, 2009). Similarly, Rudolf Virchow, regarded as the father of modern pathology, emphasized the role of social determinants, such as poverty, nutrition, and sanitation, in the development of disease. He famously asserted that "medicine is a social science," emphasizing the broader systemic factors that impact health (Virchow, 1848/2006). Homoeopathy aligns with these perspectives by considering the individual as a whole—encompassing physical, mental, and environmental factors—in the treatment process. It acknowledges that a person's susceptibility to disease is influenced by a combination of internal and external factors and seeks to restore balance by addressing these underlying causes (Bell & Koithan, 2012). Moreover, homoeopathy's principle of individualized care mirrors Virchow's understanding of how disease manifests differently depending on the unique circumstances of the individual. Both frameworks, along with the triad, move beyond the reductionist focus on the pathogen to address the broader context of health, offering an integrative and holistic approach to understanding and managing disease. This convergence highlights the need for healthcare systems to incorporate multidimensional perspectives that acknowledge biological, social, and environmental determinants of health.

14.4 EMERGING AND RE-EMERGING INFECTIOUS DISEASES: CAUSES FROM HOMOEOPATHIC PERSPECTIVE AND MIASM CONCEPT

In homoeopathy, the concepts of *fundamental cause*, *exciting cause*, and *maintaining cause* are integral to understanding disease susceptibility, particularly in the context of infectious diseases. The *fundamental cause* refers to the underlying predisposition or inherited tendency of an individual to develop certain diseases, which is often linked to genetic, constitutional, or miasmatic factors. This aligns with the notion of host susceptibility in infectious disease epidemiology, where some individuals are inherently more vulnerable to infection (Hahnemann, 2002). The *exciting cause* encompasses external triggers that initiate the disease process, such as exposure to a specific pathogen, aligning with the agent in the epidemiological triad. However, homoeopathy views this trigger as secondary, emphasizing that disease arises only when the host's internal susceptibility allows it (Bell & Koithan, 2012). Finally, the *maintaining cause* includes environmental, lifestyle, or psychological factors that perpetuate the disease state, such as poor hygiene, stress, or inadequate nutrition. These causes often correspond to the environmental factors in the triad that exacerbate infectious diseases (Moss & Green, 2009). By addressing all three causes, homoeopathy seeks to not only treat the symptoms of the disease but also reduce susceptibility and remove the conditions that sustain illness. This comprehensive approach contrasts with conventional medicine, which often focuses primarily on eradicating the pathogen without addressing the deeper, systemic vulnerabilities of the host. The miasm theory in homoeopathy, introduced by Samuel Hahnemann, provides a framework for understanding chronic disease predisposition and susceptibility to infectious diseases. According to this theory, miasms are deep-seated, inherited or acquired dyscrasias (constitutional weaknesses) that predispose individuals to specific patterns of illness, including susceptibility to infections. Hahnemann identified three primary miasms, namely, psora, syphilis, and sycosis, each corresponding to distinct constitutional tendencies and disease manifestations (Hahnemann, 2002). Psora is linked to conditions of deficiency and hypersensitivity, syphilis to destructive and degenerative processes, and sycosis to excessive growth or suppression. In the context of infectious diseases, the miasm theory suggests that an individual's susceptibility is determined not just by exposure to pathogens but also by the constitutional miasmatic imprint, which influences how the body responds to infections (Banerjee, 2010). This perspective complements modern understandings of host–pathogen interactions by emphasizing the role of inherited and chronic predispositions in disease vulnerability. For example, individuals with a predominant psoric miasm may be more prone to conditions characterized by hypersensitivity, such as allergic or inflammatory responses during infections. The miasm theory underscores the importance of addressing these deep-seated predispositions to enhance the body's resilience and reduce susceptibility, offering a unique perspective that goes beyond the pathogen-centric focus of conventional medicine.

14.5 EMERGING AND RE-EMERGING INFECTIOUS DISEASES: THE APPROACH OF HOMOEOPATHIC CASE TAKING

Case taking is a foundational process in homoeopathic practice, emphasizing a comprehensive understanding of the patient to identify an individualized remedy. Unlike conventional medical histories that focus predominantly on physical symptoms, homoeopathic case taking involves a detailed exploration of the patient's physical, mental, and emotional state, as well as lifestyle, environmental factors, and hereditary influences (Hahnemann, 2002). This holistic approach ensures that the selected remedy matches the patient's unique symptom profile, in accordance with the principle of *similia similibus curentur* (like cures like). The process begins with an open-ended and empathetic dialogue, allowing the patient to describe their condition in their own words. The practitioner actively listens to identify the main complaint and associated symptoms, including their onset, duration, intensity, and modalities (factors that aggravate or ameliorate symptoms). Special attention is paid to peculiar, individualizing symptoms that distinguish the patient's case, as these often guide the selection of the remedy (Teut et al., 2022). Beyond physical symptoms, homoeopathic case taking delves into the patient's mental and emotional state, capturing aspects such as fears, dreams, temperament, and coping mechanisms. For instance, anxiety, irritability, or sadness associated with the illness are documented, as these factors are integral to selecting the remedy that aligns with the patient's overall condition (Bell & Koithan, 2012). Environmental and lifestyle factors, such as diet, sleep patterns, work stress, and relationships, are also considered, as they may contribute to the underlying cause or perpetuation of the disease. Additionally, the practitioner explores the patient's medical history, family history, and any past treatments or medications to identify potential miasmatic influences and hereditary predispositions (Banerjee, 2010). After gathering comprehensive data, the practitioner analyzes the case to identify the most characteristic symptoms and their underlying pattern. These are matched to the homoeopathic *materia medica* or repertory to find the remedy that closely resembles the patient's symptom profile. This individualized approach ensures that the selected remedy not only addresses the presenting complaint but also aims to restore overall health by stimulating the body's self-healing capacity (Hahnemann, 2002). Case taking is thus a dynamic, patient-centered process that distinguishes homoeopathy from conventional medicine, focusing on the totality of symptoms rather than isolated disease manifestations.

Homoeopathic case taking is a structured yet flexible process aimed at gathering comprehensive information about the patient to facilitate individualized treatment. The steps involved include the following:

- *Initiation of rapport:* The practitioner begins by creating a welcoming and empathetic environment to encourage the patient to share their concerns openly. Establishing trust is crucial, as it ensures the patient feels comfortable discussing sensitive aspects of their physical, mental, and emotional health.
- *Elicitation of the chief complaint:* The primary focus is on understanding the patient's main reason for seeking treatment. Details about the chief complaint, including its onset, duration, intensity, progression, and associated symptoms, are explored. Questions often focus on what aggravates or alleviates the condition, known as *modalities*.
- *Detailed symptom analysis:* The practitioner delves deeper into the symptoms, seeking unique and peculiar aspects that differentiate the patient's case. This includes physical symptoms (e.g., pain characteristics, location, and triggers), mental symptoms (e.g., stress, fears, mood), and emotional responses (e.g., anxiety, grief).
- *Assessment of the patient's generalities:* Information about general tendencies, such as appetite, food preferences, sleep patterns, perspiration, and sensitivity to environmental factors like temperature or weather, is gathered. These generalities provide important clues for remedy selection.
- *Exploration of mental and emotional state:* The patient's psychological state is assessed, including their temperament, coping mechanisms, and personality traits. Insights into their fears, dreams, and emotional reactions are also noted, as these are vital in understanding the individual holistically.
- *Evaluation of lifestyle and environmental factors:* Lifestyle aspects such as daily routines, occupational stress, habits, and relationships are examined. The practitioner identifies potential maintaining causes, such as stress, poor nutrition, or environmental triggers that may perpetuate the illness.
- *Review of medical and family history:* A thorough history of past illnesses, treatments, and family medical history is documented. This helps identify hereditary patterns, miasmatic influences, and chronic predispositions that may impact susceptibility to disease.
- *Holistic synthesis of the case*: After collecting all relevant information, the practitioner synthesizes the data to identify the totality of symptoms. Key symptoms, including peculiar, prominent, and constitutional characteristics, are prioritized for remedy selection.

- *Selection of the remedy*: The practitioner matches the synthesized symptom profile to the homoeopathic *materia medica* or repertory to identify the most suitable remedy. The remedy is chosen based on its ability to closely mimic the patient's unique symptom pattern and stimulate the body's healing response.
- *Follow-up and monitoring*: Regular follow-up sessions are scheduled to assess the patient's response to the remedy and make any necessary adjustments. This iterative process ensures the treatment remains aligned with the patient's evolving condition.

This meticulous and individualized approach distinguishes homoeopathic case taking from conventional diagnostic methods, focusing on the person as a whole rather than just the disease.

14.6 EMERGING AND RE-EMERGING INFECTIOUS DISEASES: THE APPROACH OF SELECTING HOMOEOPATHIC REMEDY

Homoeopathic medicine selection for infectious diseases is an individualized process grounded in the principles of holistic care and the law of similars (*similia similibus curentur*). It involves matching the totality of the patient's symptoms with a remedy from the homoeopathic *materia medica* that most closely resembles their unique symptom pattern. This selection process follows the comprehensive case taking process and emphasizes both the acute symptoms of the infection and the patient's constitutional and miasmatic predispositions (Hahnemann, 2002).

- *Analysis of the case:* After completing case taking, the practitioner begins by analyzing the totality of symptoms. This includes identifying the primary, secondary, and auxiliary symptoms of the infectious disease. Special attention is given to *peculiar* and *individualizing* symptoms, which are crucial for remedy differentiation. For example, in a case of influenza, the specific nature of the fever, associated body aches, and unique mental or emotional states may guide remedy selection (Teut et al., 2022).
- *Classification of symptoms:* The collected symptoms are classified into three categories:
  - *Characteristic symptoms:* These are peculiar and uncommon symptoms that distinguish the case. For example, a patient's desire for cold drinks despite having a sore throat may indicate *phosphorus.*
  - *Common symptoms:* General symptoms shared by most patients with the disease, such as fever or cough, are considered less important unless they present in a peculiar manner.
  - *General symptoms:* These include broader aspects such as appetite, thirst, sleep, and sensitivity to temperature, providing an overall picture of the patient's condition (Banerjee, 2010; Bhattacharjee et al., 2021).
- *Repertorization:* The symptoms are then repertorized using a homoeopathic repertory, a systematic index of symptoms and corresponding remedies. This step involves cross-referencing the patient's symptom profile with the repertory to generate a list of potential remedies (Boericke, 2007). Modern software tools often assist practitioners in streamlining this process.
- *Selection of a similimum:* The practitioner selects the remedy that most closely matches the patient's total symptom profile, known as the *similimum.* For infectious diseases, the chosen remedy must address both acute pathological symptoms and the underlying susceptibility. For example, in a case of typhoid fever, the remedy *Baptisia tinctoria* might be indicated based on symptoms of stupor, high fever, and a sensation of body fragmentation (Hahnemann, 2002; Bhattacharjee et al., 2021).
- *Consideration of miasmatic influence:* Miasmatic tendencies like psora, sycosis, or syphilis are evaluated to ensure the remedy addresses chronic predispositions that may influence susceptibility to the infectious disease. For instance, patients with a psoric miasmatic

background may require remedies like *sulphur* or *Calcarea carbonica* to trigger their resistance to infection (Banerjee, 2010; Bhattacharjee et al., 2021).

- *Potency and dose selection:* The potency (strength) of the remedy is selected based on the patient's vitality, the intensity of the disease, and the presence of chronic predispositions. For acute infectious diseases, higher potencies (e.g., 200°C or 1 M) may be used in dynamic and robust patients, while lower potencies (e.g., 6°C or 30°C) may be suitable for more sensitive or weakened individuals (Teut et al., 2022; Bhattacharjee et al., 2021).
- *Administration and follow-up:* The selected remedy is administered, and the patient is monitored for responses such as improvement in symptoms, aggravation, or new symptom emergence. Follow-up sessions ensure that the remedy continues to match the evolving symptom profile. Adjustments in potency or remedy may be made as required, based on Hering's law of cure, which observes symptom resolution from within outward and from the most vital organs to the least (Hahnemann, 2002).
- *Integration with acute and prophylactic remedies:* In infectious disease outbreaks, homoeopaths may use "genus epidemicus" remedies those identified based on the symptom pattern common to most affected individuals during the epidemic. This approach provides both therapeutic and prophylactic benefits. For example, *Arsenicum album* was widely recommended during the COVID-19 pandemic as a prophylactic remedy (Chaudhary et al., 2022).

Homoeopathic medicine selection for infectious diseases is thus a dynamic process that integrates individualized care, an understanding of disease susceptibility, and the broader principles of holistic health. By addressing both the immediate symptoms and underlying predispositions, homoeopathy aims to enhance the patient's resistance and promote long-term wellness.

14.7 CONCLUSION

Homoeopathy offers a unique, patient-centered approach to the treatment and management of infectious diseases by focusing on individualized care, holistic understanding of symptoms, and addressing susceptibility through constitutional and miasmatic considerations. The process of case taking and remedy selection emphasizes the totality of the patient's symptoms, integrating physical, mental, and emotional aspects to identify the most appropriate similimum. Unlike the conventional approach, which often targets pathogens directly, homoeopathy seeks to restore the body's inherent healing capacity, addressing both acute symptoms and underlying predispositions. Through its systematic methodology, which includes detailed case analysis, repertorization, and potency selection, homoeopathy provides a dynamic framework for treating infectious diseases. Additionally, the integration of "genus epidemicus" remedies offers potential benefits during epidemics, supporting both therapeutic and prophylactic needs. By addressing the fundamental, exciting, and maintaining causes of illness, homoeopathy complements modern epidemiological principles while offering a more comprehensive understanding of host–pathogen dynamics. This holistic and individualized approach positions homoeopathy as a valuable alternative or adjunct in managing infectious diseases, particularly for those seeking treatments that go beyond symptom suppression to enhance overall health and resilience.

REFERENCES

Banerjee, S. (2010). *Chronic Diseases: Theory of Miasms*. B. Jain Publishers.

Bell, I. R., & Koithan, M. (2012). A model for homeopathic remedy effects: Low dose nanoparticles, allostatic cross-adaptation, and time-dependent sensitization in a complex adaptive system. *BMC Complement Altern Med*, 12, 191. https://doi.org/10.1186/1472-6882-12-191. PMID: 23088629; PMCID: PMC3570304.

Bell, I. R., Koithan, M., & Brooks, A. J. (2013). Testing the nanoparticle-allostatic cross-adaptation-sensitization model for homeopathic remedy effects. *Homeopathy: The Journal of the Faculty of Homeopathy*, 102(1), 66–81. https://doi.org/10.1016/j.homp.2012.10.005.

Benveniste, Watts G. J. (2004 Nov 27). *BMJ*, 329(7477), 1290. PMCID: PMC534457.

Bhattacharjee, B., Roja, V., & Khurana, A. (2021). Homoeopathic medicine arsenicum album, the prophylactic/immune booster in COVID-19 pandemic. *Indian J Res Homoeopathy*, 15(3). https://doi.org/10.53945/2320–7094.1028.

Boericke, W. (2007). *Pocket Manual of Homoeopathic Materia Medica & Repertory* (9th ed.). B. Jain Publishers.

Chaudhary, A., Nayak, D., Pandey, S. R. R., Khurana, A., & Study Group, C. (2022). Homoeopathic medicine arsenicum album 30 C for Covid-19 in 2020: A retrospective analysis from mass-level data. *Alternative Therapies in Health and Medicine*, 28(7), 72–79.

Ernst, E. (2002, December). A systematic review of systematic reviews of homeopathy. *Br J Clin Pharmacol*, 54(6), 577–582. https://doi.org/10.1046/j.1365–2125.2002.01699.x. PMID: 12492603; PMCID: PMC1874503.

Fauci, A. S., & Morens, D. M. (2012). The perpetual challenge of infectious diseases. *The New England Journal of Medicine*, 366(5), 454–461. https://doi.org/10.1056/NEJMra1108296.

Hahnemann, S. (2002). *Organon of Medicine* (6th ed.). B. Jain Publishers.

Mathie, R. T., Baitson, E. S., Frye, J., Nayak, C., Manchanda, R. K., & Fisher, P. (2013). Homeopathic treatment of patients with influenza-like illness during the 2009 A/H1N1 influenza pandemic in India. *Homeopathy: The Journal of the Faculty of Homeopathy*, 102(3), 187–192. https://doi.org/10.1016/j.homp.2013.04.001.

Morens, D. M., Folkers, G. K., & Fauci, A. S. (2004). The challenge of emerging and re-emerging infectious diseases. *Nature*, 430(6996), 242–249. https://doi.org/10.1038/nature02759.

Moss, D., & Green, B. (2009). *Energy Medicine: The Scientific Basis*. Churchill Livingstone.

Shang, A., Huwiler-Müntener, K., Nartey, L., Jüni, P., Dörig, S., Sterne, J. A., Pewsner, D., & Egger, M. (2005). Are the clinical effects of homoeopathy placebo effects? Comparative study of placebo-controlled trials of homoeopathy and allopathy. *Lancet (London, England)*, 366(9487), 726–732. https://doi.org/10.1016/S0140–6736(05)67177–2.

Teixeira, M. Z. (2024, May 3). "Homeopathy is not placebo effect": Proof of the scientific evidence for homeopathy. *Rev Assoc Med Bras*, 70(4), e20231438. https://doi.org/10.1590/1806–9282.20231438. PMID: 38716949; PMCID: PMC11068375.

Teut, M., van Haselen, R. A., Rutten, L., Lamba, C. D., Bleul, G., & Ulbrich-Zürni, S. (2022, February). Case reporting in homeopathy-an overview of guidelines and scientific tools. *Homeopathy*, 111(1), 2–9. https://doi.org/10.1055/s-0041–1731313. Epub 2021 Sep 14. PMID: 34521146; PMCID: PMC8803476.

Thomas, Y. (2007). The history of the memory of water. *Homeopathy: The Journal of the Faculty of Homeopathy*, 96(3), 151–157. https://doi.org/10.1016/j.homp.2007.03.006.

Virchow, R. C. (2006, December). Report on the typhus epidemic in Upper Silesia. *Am J Public Health*, 96(12), 2102–2105. https://doi.org/10.2105/ajph.96.12.2102. PMID: 17123938; PMCID: PMC1698167. (Original work published 1848)

15 Emerging and Re-Emerging Viral Diseases

Perspective from Holistic Medical Nutrition Therapy

Sahana Kamaraj, Mandke Bhakti Ashok, Ulas Acaroz, Mohamad Fawzi Mahomoodally, and Ayesha Noor

15.1 INTRODUCTION TO MEDICAL NUTRITION THERAPY

Food, in addition to being the source of energy and the building blocks of the body, has the potential to prevent or assist in the treatment of diseases. Since ancient times, food has been regarded as a therapeutic agent by both Eastern and Western medicine. Food, nutrition, and lifestyle have a significant influence on our health, standard of living and longevity. Medical nutrition therapy (MNT) is an evidence-based treatment that involves understanding the patient's nutritional status and providing an individualized nutrition plan to help control or treat a medical condition. MNT has been employed to treat various conditions such as malnutrition, diabetes, cardiovascular diseases, obesity, and gastrointestinal disorders. MNT is practiced by Registered Dietician Nutritionists at the suggestion of the patient's healthcare provider (Hicks-Roof et al., 2021). In addition to meeting the energy requirements, a balanced nutritious diet also helps in protecting the body from infectious diseases.

In the past few years, not only did new diseases begin to emerge but also existing infectious diseases began to resurface with full effects causing serious outbreaks. Emerging infectious diseases (EIDs) are those diseases that haven't affected humans previously; or that have only affected a small population of humans in remote places; or that have affected humans in the past but only now have been documented as distinct diseases caused due to an infectious agent. Re-emerging infectious diseases (REIDs) refer to those diseases that have once caused a major impact on the health of a particular geographical region or globally and then went dormant for a while and have now resurfaced causing serious impacts on human health (Spernovasilis et al., 2022).

15.2 FACTORS CONTRIBUTING TO THE EMERGENCE OF VIRAL DISEASES

There are three main factors responsible for the emergence of EIDS and REIDs, namely, viral factors, human factors, and ecological factors.

15.2.1 Viral Factor

EIDs are mostly caused due to RNA viruses that have high mutation rates. They evolve and adapt to the new environment and host very quickly. They adopt one of these mechanisms to mutate: point mutation, recombination, and reassortment (Ka-Wai Hui, 2006).

DOI: 10.1201/9781032721811-15

15.2.2 Human Factor

Human behavior such as unsafe sexual practices and drug abuse have contributed to the rise of EIDs and REIDs. Some of the other human factors include overpopulation, international travel, the introduction of exotic species that carry parasitic viruses, and healthcare-related transmission (Ka-Wai Hui, 2006).

15.2.3 Ecological Factor

If the changing climate due to global warming favors the environmental condition of the ecological niche of the vector, there will be a rise in the population of vector-borne pathogens (Ka-Wai Hui, 2006).

15.3 NUTRIENTS POSSESSING ANTIVIRAL PROPERTIES

Some of the nutrients and non-nutrients possessing antiviral activity are shown in Table 15.1 along with their sources, functions, and possible modes of action.

15.4 POSSIBLE MECHANISMS INVOLVED IN THE TREATMENT OF VIRAL INFECTIONS USING DIETARY NUTRIENTS

15.4.1 Role of Oxidative Stress in Viral Pathogenesis

Free radicals like O_2, OH, and NO are produced because of oxidative stress, which occurs due to imbalance between nitrogen species and reactive oxygen. Viral infections also causes oxidative stress, promoting viral replication and triggering macrophages and monocytes to release molecules like ROS, TNF, IFN, NF-κB, and IL-1, causing damage (Li et al., 2017). Minerals like Fe, Zn, Se, Cu, and Mg and vitamins E, D, C, A are essential for treating viral diseases due to their antioxidant properties. Antioxidants like catalase, superoxide dismutase, and glutathione peroxidase neutralize ROS, protect cells, while proteolytic enzymes like peptidases and proteinases synthesize new antioxidants and aid in tissue repair. Selenium, a mineral with strong antiviral properties, is a crucial trace element in mammalian redox biology, as it covalently attaches to organic molecules (Guillin et al., 2019). Hence nutritional supplements or maintaining a balanced diet can help combat infectious diseases.

15.4.2 Inflammation in Viral Infections

Pattern recognition receptors (PRR) bind to pathogen associated molecules and initiate innate immune responses, involving cytokines, chemokines, cell maturation, and adaptive responses, with anti-inflammatory cytokines playing a crucial role in viral processes (Chen & Ichinohe, 2015). Inflammasomes are crucial for immune response, but viruses can manipulate their activation and reproduction through changes in molecular expression, PRR detection, immune pathway activation, and effector molecule binding. Therefore, nutritional supplements can be considered as therapy to improve the host immune system and treat viral diseases. Micronutrients significantly enhance immune function by supporting innate immunity through physical barriers, antimicrobial activity, protein synthesis, cell proliferation, differentiation, phagocytosis, cell death, and inflammatory recovery. Vitamin A regulates cell-mediated and humoral immunity via surface IgA and cytokine synthesis, while vitamin D reduces inflammatory cytokine expression and increases antimicrobial peptide production. Inadequate protein consumption leads to malnutrition and lower immunity. Lactoferrin and active milk proteins exhibit antiviral and immune-regulating properties and stimulate the formation of anti- and proinflammatory cytokines like IL-6, IL-1, IL-18, IL-8, IL-ɣ, and TNF-cytokines (Singh et al., 2021).

TABLE 15.1
Antiviral Properties of Different Nutrients

Nutrient	Source	Function	Antiviral Activity: Nutrient Subtype (If Any)	Antiviral Activity: Virus (If Any in Particular)	Antiviral Activity: Possible Mode of Action	References
15.3.1 Polysaccharides						
Polysaccharides derived from plants and mushrooms	Plant cell walls, seaweed, mushrooms	Prevents hypoglycemia, immune system regulation, anti-tumor activity	Polysaccharides from mushroom	HIV	Inhibits reverse transcriptase and proteases	Claus-Desbonnet et al. (2022)
			Astragalus polysaccharides (*Astragalus membranaceus*)	NA	Strengthens Th1 and Th2 response	
Polysaccharides derived from animals	Organs and tissues of animals such as marine fish, shellfish, sponges, molluscs	Heparin anticoagulant and antithrombotic agent. Chitosan-biomaterials, pharmaceuticals	Heparin	SARS-CoV-2	Binds to the spike protein of SARS-CoV-2 and prevents the virus from attaching to the ACE-2	Claus-Desbonnet et al. (2022)
			Chitosan	NA	Stimulates immune cells, enhances NO, increases IgG and Ig A response, promotes migration of neutrophils	
			Chondroitin sulfate	HIV	Inhibits viral entry	
Sulfated algal-derived polysaccharides	Marine algae (Red, Brown, Green), some bacteria; Carrageenan (Red algae), Fucoidan (Brown algae)	Supports and protects the cell wall of algae	Iota carrageenan	HPV	Prevents virions from binding to the cell	Hans et al. (2021)
			λ-Carrageenan and hybrid μ/ι carrageenan	HSV	Inhibits the HSV glycoprotein from linking with the heparin sulfate on the cell	
			Fucoidan	HIV	Inhibits reverse transcriptase, prevents viral entry	
15.3.2 Proteins						
Whey proteins	Constitutes 20 % of the milk proteins; (includes α lactalbumin, β-lactoglobulin, lactoferrin, immunoglobulins)	Enhances muscle strength and bone growth, lowers cholesterol levels, immunomodulatory effect	Lactoferrin	H1N1(influenza), H3N2 (influenza)	Inhibits viral replication	Wakabayashi et al. (2014)
				PIV-2	Inhibits viral entry	
				HSV-1	Inhibits cytopathic effects	
Lectins	Raw legumes, whole grains	Antioxidant activity, prevents sugar spikes	Lectin	HIV	Prevents viral entry into cells by binding on to glycoproteins on the virus	Thirumdas et al. (2021)
				HCV	Prevents vial entry into cells by binding on to the virus' E1 and E2 glycosylation sites	

(*Continued*)

TABLE 15.1
Antiviral Properties of Different Nutrients

Nutrient	Source	Function	Antiviral Activity: Nutrient Subtype (If Any)	Antiviral Activity: Virus (If Any in Particular)	Antiviral Activity: Possible Mode of Action	References
15.3.3 Vitamins						
Vitamin A	Leafy vegetables, carrots, sweet potato, papaya, milk, oily fish, egg yolks, and cheese	Antioxidant activity, supports vision, cell differentiation, maintains the immune system	Vitamin A	NA	During acute inflammation, the active retinol derivative produces IFN (α and β)	Fath et al. (2022)
Vitamin D	Synthesized by the body on exposure to sunlight. Dietary source: oily fish, cod liver oil, fortified dietary products	Stimulates absorption of Ca from the gut, supports the functioning of the nervous system, helps in bone growth	Vitamin D	RSV	Production of β-defensin 2 and antimicrobial peptide LL 3	Siddiqui et al. (2020)
				Dengue virus	Prevents the virus from attaching to the host cell	
Vitamin E	Vegetable oils such as soybean, corn, sunflower, peanut	Inhibits oxidation of PUFA	Vitamin E	NA	Decreases NO production which leads to regulation of prostaglandin E2 and COX-2 (cyclooxygenase-2) inhibition	Lee & Han (2018)
Vitamin C	Citrus fruits, green vegetables, tomatoes, peppers	Antioxidant activity, plays a role in collagen formation and neurotransmitter synthesis	Vitamin C	NA	Antiviral mechanism may involve the formation of cytokines (IFN-α/β), production of free radicals or by directly binding to the virus	Bae and Kim (2020)
15.3.4 Minerals						
Zinc	Meat, poultry, seafood	Serves as a cofactor and structural element, influences the immune system	Zinc	RNA viruses	Inhibits RNA polymerase	Toledano et al. (2023)
Magnesium	Leafy vegetables, legumes, whole grains, seeds, nuts	Cofactor for enzymes, regulates nerve and muscle function	Magnesium	NA	Causes decrease in NF-κB activation, cytokine production, and systemic inflammation. Suppresses IL-6 and c-reactive protein levels	Wallace (2020)

Selenium	Seafood, muscle and organ meat, garlic, cabbage	Part of many selenoprotein/ selenoenzymes, anticancer activity	Selenium supplementation	Poliovirus (vaccination)	Accelerates the antiviral immune response by producing IFN-γ and IL-10	Steinbrenner et al. (2015)
15.3.5 Non-nutrients						
Probiotics	Fermented foods such as yogurt, cheese, kefir, and dietary supplements	Helps maintain balance of intestinal microbes, activates and regulates immune system	*Enterococcus*	H1N1, H3N2	Enterocin, an AMP produced by the bacteria restricts cytopathic effect of virus	Wang et al. (2022)
			Lactobacillus acidophilus	HIV	Production of H_2O_2, inhibits viral replication	
			Lactobacilli (vaginal mucosa)	HIV, HSV	Lactic acid produced	
			Bifidobacterium bifidum	H1N1	Production of antibodies, IL-4, IL-12, IFN-γ	
Polyphenols	Spices, fruits, vegetables, seeds, drinks and oils	Potent antioxidant, anti-inflammatory property, immunomodulators	Epigallocatechin-3-gallate	HBV	Inhibits viral entry into the cell	Chojnacka et al. (2021); Ninfali et al. (2020)
				HIV	Targets reverse transcriptase and inhibits viral replication	
			Quercetin	HCV	Inhibits viral replication and virion production	
			Vitexin	H1N1	Reduces inflammatory injury and increases IFN-β levels	
			Pinocembrin	Zika	Inhibits synthesis of viral envelope protein	

Note: HBV: hepatitis B virus; HCV: hepatitis C virus; HIV: human immunodeficiency virus; HPV: human papillomavirus; HSV: herpes simplex virus; NA: not available; PIV: parainfluenza virus; RSV: respiratory syncytial virus; SARS-CoV-2: severe acute respiratory syndrome coronavirus 2.

15.4.3 Improving the Gut Microbiome

Viral infections in the gastrointestinal system can be influenced by probiotic and commensal bacteria by modifying leukocyte activity, secreting antimicrobial peptides (AMPs), blocking viral attachment, and increasing mucosal barrier function (Harper et al., 2021). *Lactobacilli* and *Bifidobacterium*, a probiotic bacterium in fermented foods, have antiviral properties, particularly useful in respiratory tract infections. Probiotics suppress viral infectivity by boosting immune system and secreting macrophages and natural killer cells which initiate interferon gamma signaling and IL-12 cytokines.

15.5 TREATMENT OF VIRAL DISEASES USING NUTRITION THERAPY

Maintaining a healthy immune system and a balanced diet are crucial for preventing viral infections, and nutritional therapy may play a significant role in treating viral diseases by reducing ROS/RNS and inflammation in host immune system by regulating the levels of pro- and anti-inflammatory markers (Figure 15.1).

15.5.1 SARS-CoV-2: COVID-19

The SARS-CoV-2 is an enveloped positive-sense RNA virus that attacks the respiratory tract and is the cause of COVID-19. Aerosols or contact with afflicted individuals can spread the virus. By adhering to the angiotensin-converting enzyme 2 receptor on alveolar epithelial cells, it stimulates immune cells. Severe cases can cause acute respiratory syndrome, cardiac arrest, organ failures, and septic shock (Fernández-Quintela et al., 2020). Further, zinc is an anti-inflammatory trace element which increases cytotoxicity of NK cells, targeting plasma membrane proteins, which may destroy microbes. It also, promotes regulatory T-cell growth, reducing inflammatory Th17 and Th9 cell proliferation and boosting IgG antibody production (Cervantes-Pérez et al., 2020).

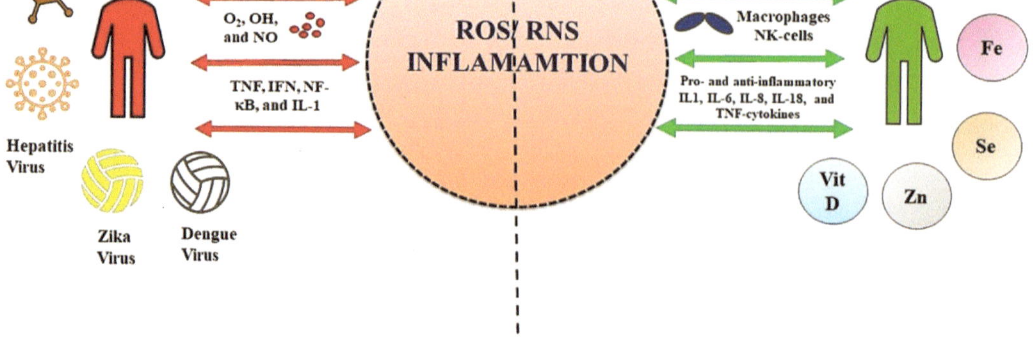

FIGURE 15.1 Dietary constituents in combating viral infections.

15.5.2 HIV Virus

HIV, a human immunodeficiency virus, attacks CD4+ T cells, causing immunological abnormalities and a decline in CD4+ T cells and it peaks around day 30 when the body develops HIV antibodies. Antiretroviral therapy (ART) has been used for over 20 years but often resurfaces. According to Coelho et al. (2015), the standardized vitamin D3 supplementation can increase 25(OH)D levels, positively correlated with efavirenz, and improves immunologic recovery in HIV-infected individuals.

15.5.3 Influenza Virus

Influenza A virus, a common flu infection, is an eight-segmented ssRNA negative-sense virus with various structural, non-structural, and functional proteins, consisting of PB1, PB2, PA, HA, NA, NP, M (1,2), and NS (1,2). Pathogenicity is determined by virus–host cell protein interaction, binding to epithelium cells. Vitamin C and red ginseng administration raised CD25 and CD69 expression in peripheral blood mononuclear cells and natural killer cells, decreased BCBL-1 lytic expression, and increased IFN- and NKp46 production in Gulo mice, according to Kim et al. (2016).

15.5.4 Zika Virus

The flavivirus Zika is a positive-sense single-stranded RNA virus that is spread by mosquitos. It encodes three structural proteins (precursor membrane protein, capsid protein, and NS1–NS5) as well as seven non-structural proteins (NS1–NS5). The virus is transmitted through female mosquito *Aedes aegypti* or *Aedes albopictus* which suck blood in humans through perinatal, sexual, and breast milk feeding routes. Nutraceutical substances like lemon balm, rosamarinic acid, kanzonol, isoquercitine, curcumin, cotton seeds' gossypol, harringtonine, palmatine, and indole have shown antiviral efficacy against the Zika virus, inhibiting virus entry, adhesion, and enhancing host cell health (Muthuraj et al., 2023).

15.5.5 Dengue Virus

Dengue fever is a potentially dangerous viral infection. It is a member of the Flaviviridae family and has four serotypes, from DV-1 to DV-4. The virus is mostly transmitted by the mosquitos *Aedes aegypti* and *Aedes albopictus*. High vitamin D3 supplementation may reduce proinflammatory cytokines IL-2 responses in dengue fever patients, and monocyte-derived dendritic cells showed increased IL-10 anti-inflammatory cytokines (Martínez-Moreno et al., 2020).

15.5.6 Hepatitis Virus

Hepatotropic viruses like hepatitis E, C, B, and A cause severe illness and death, targeting the liver, causing impaired metabolism, decreased food intake, nutritional deficiencies, and severe cases of anorexia and dysgeusia (Praharaj & Anand, 2023). Vitamin E protects cells from reactive oxygen species and inflammatory cytokines, aiding in hepatitis B virus treatment. Nutrients from milk, coffee, tea, spinach, and broccoli can create personalized management food strategies to cure the disease.

15.6 CONTROL OF EID AND REID USING LIFESTYLE CHANGES AND PUBLIC AWARENESS

The COVID-19 pandemic emphasizes the value of having a strong immune system, with probiotics, nutrition, trace elements, and herbal therapies boosting immunity against various kinds of viral infections. Further, raising public awareness about viral illnesses, promoting health practices like handwashing, exercise, and hygiene, and ensuring regular checkups are crucial for controlling virus spread and preventing disease prognosis.

15.7 CONCLUSION

A viral infection triggers a state of oxidative stress in the body and activates immune response. Supplementation of certain nutrients is found to help treat infectious diseases, decrease its severity, and have an overall protective effect. These nutrients can directly interfere with the virus by binding to it, inhibiting its entry, restricting its replication, and producing antimicrobial products such as AMPs, lactic acid, and H_2O_2. They can also act on the body and helps to regulate the immune system by stimulating certain signaling pathways and modulating immune cell proliferation, differentiation, and function. A proper diet rich in the right amount of macronutrients, micronutrients, and probiotics along with good hygiene practices and regular health checkups can help to prevent and control viral infections.

ACKNOWLEDGMENT

The authors thank the management of Vellore Institute of Technology (VIT) for providing the required facilities.

REFERENCES

Bae, M., & Kim, H. (2020). The role of vitamin C, vitamin D, and selenium in immune system against COVID–19. *Molecules*, 25(22). https://doi.org/10.3390/MOLECULES25225346.

Cervantes-Pérez, E., Cervantes-Guevara, G., Martínez-Soto Holguín, M. C., Cervantes-Pérez, L. A., Cervantes-Pérez, G., Cervantes-Cardona, G. A., González-Ojeda, A., Fuentes-Orozco, C., & Ramírez-Ochoa, S. (2020). Medical nutrition therapy in hospitalized patients with SARS-CoV-2 (COVID-19) infection in a non-critical care setting: Knowledge in progress. *Current Nutrition Reports*, 9(4), 309–315. https://doi.org/10.1007/S13668-020-00337-X.

Chen, I. Y., & Ichinohe, T. (2015). Response of host inflammasomes to viral infection. *Trends in Microbiology*, 23(1), 55–63. https://doi.org/10.1016/J.TIM.2014.09.007.

Chojnacka, K., Skrzypczak, D., Izydorczyk, G., Witek-Krowiak, A., Mikula, K., & Szopa, D. (2021). Antiviral properties of polyphenols from plants. *Foods (Basel, Switzerland)*, 10(10). https://doi.org/10.3390/FOODS10102277.

Claus-Desbonnet, H., Nikly, E., Nalbantova, V., Karcheva-Bahchevanska, D., Ivanova, S., Pierre, G., Benbassat, N., Katsarov, P., Michaud, P., Lukova, P., & Delattre, C. (2022). Polysaccharides and their derivatives as potential antiviral molecules. *Viruses*, 14(2). https://doi.org/10.3390/V14020426.

Coelho, L., Cardoso, S. W., Luz, P. M., Hoffman, R. M., Mendonça, L., Veloso, V. G., Currier, J. S., Grinsztejn, B., & Lake, J. E. (2015). Vitamin D3 supplementation in HIV infection: Effectiveness and associations with antiretroviral therapy. *Nutrition Journal*, 14(1). https://doi.org/10.1186/S12937-015-0072-6.

Fath, M. K., Naderi, M., Hamzavi, H., Ganji, M., Shabani, S., ghahroodi, F. N., Khalesi, B., Pourzardosht, N., Hashemi, Z. S., & Khalili, S. (2022). Molecular mechanisms and therapeutic effects of different vitamins and minerals in COVID-19 patients. *Journal of Trace Elements in Medicine and Biology: Organ of the Society for Minerals and Trace Elements (GMS)*, 73. https://doi.org/10.1016/J.JTEMB.2022.127044.

Fernández-Quintela, A., Milton-Laskibar, I., Trepiana, J., Gómez-Zorita, S., Kajarabille, N., Léniz, A., González, M., & Portillo, M. P. (2020). Key aspects in nutritional management of COVID-19 patients. *Journal of Clinical Medicine*, 9(8), 1–24. https://doi.org/10.3390/JCM9082589.

Guillin, O. M., Vindry, C., Ohlmann, T., & Chavatte, L. (2019). Selenium, selenoproteins and viral infection. *Nutrients*, 11(9). https://doi.org/10.3390/NU11092101.

Hans, N., Malik, A., & Naik, S. (2021). Antiviral activity of sulfated polysaccharides from marine algae and its application in combating COVID-19: Mini review. *Bioresource Technology Reports*, 13. https://doi.org/10.1016/J.BITEB.2020.100623.

Harper, A., Vijayakumar, V., Ouwehand, A. C., ter Haar, J., Obis, D., Espadaler, J., Binda, S., Desiraju, S., & Day, R. (2021). Viral infections, the microbiome, and probiotics. *Frontiers in Cellular and Infection Microbiology*, 10. https://doi.org/10.3389/FCIMB.2020.596166.

Hicks-Roof, K., Xu, J., Fults, A. K., & Latortue, K. Y. (2021). Beyond the clinical walls: Registered dietitian nutritionists providing medical nutrition therapy in the home setting. *Nutrition Research and Practice*, 15(6), 789. https://doi.org/10.4162/NRP.2021.15.6.789.

Ka-Wai Hui, E. (2006). Reasons for the increase in emerging and re-emerging viral infectious diseases. *Microbes and Infection*, 8(3), 905. https://doi.org/10.1016/J.MICINF.2005.06.032.

Kim, H., Jang, M., Kim, Y., Choi, J., Jeon, J., Kim, J., Hwang, Y. I., Kang, J. S., & Lee, W. J. (2016). Red ginseng and vitamin C increase immune cell activity and decrease lung inflammation induced by influenza A virus/H1N1 infection. *The Journal of Pharmacy and Pharmacology*, 68(3), 406–420. https://doi.org/10.1111/JPHP.12529.

Lee, G. Y., & Han, S. N. (2018). The role of vitamin E in immunity. *Nutrients*, 10(11). https://doi.org/10.3390/NU10111614.

Li, Z., Xu, X., Leng, X., He, M., Wang, J., Cheng, S., & Wu, H. (2017). Roles of reactive oxygen species in cell signaling pathways and immune responses to viral infections. *Archives of Virology*, 162(3), 603–610. https://doi.org/10.1007/S00705-016-3130-2.

Martínez-Moreno, J., Hernandez, J. C., & Urcuqui-Inchima, S. (2020). Effect of high doses of vitamin D supplementation on dengue virus replication, toll-like receptor expression, and cytokine profiles on dendritic cells. *Molecular and Cellular Biochemistry*, 464(1–2), 169–180. https://doi.org/10.1007/S11010-019-03658-W.

Muthuraj, P. G., Krishnamoorthy, C., Anderson-Berry, A., Hanson, C., & Natarajan, S. K. (2023). Novel therapeutic nutrients molecules that protect against Zika virus infection with a special note on palmitoleate. *Nutrients*, 15(1). https://doi.org/10.3390/NU15010124.

Ninfali, P., Antonelli, A., Magnani, M., & Scarpa, E. S. (2020). Antiviral properties of flavonoids and delivery strategies. *Nutrients*, 12(9), 1–19. https://doi.org/10.3390/NU12092534.

Praharaj, D. L., & Anand, A. C. (2023). Nutrition in viral hepatitis. *Current Hepatology Reports*, 22, 9–23. https://doi.org/10.1007/s11901-023-00596-6.

Siddiqui, M., Manansala, J. S., Abdulrahman, H. A., Nasrallah, G. K., Smatti, M. K., Younes, N., Althani, A. A., & Yassine, H. M. (2020). Immune modulatory effects of vitamin D on viral infections. *Nutrients*, 12(9), 1–16. https://doi.org/10.3390/NU12092879.

Singh, S., Kola, P., Kaur, D., Singla, G., Mishra, V., Panesar, P. S., Mallikarjunan, K., & Krishania, M. (2021). Therapeutic potential of nutraceuticals and dietary supplements in the prevention of viral diseases: A review. *Frontiers in Nutrition*, 8. https://doi.org/10.3389/FNUT.2021.679312.

Spernovasilis, N., Tsiodras, S., & Poulakou, G. (2022). Emerging and re-emerging infectious diseases: Humankind's companions and competitors. *Microorganisms*, 10(1). https://doi.org/10.3390/MICROORGANISMS10010098.

Steinbrenner, H., Al-Quraishy, S., Dkhil, M. A., Wunderlich, F., & Sies, H. (2015). Dietary selenium in adjuvant therapy of viral and bacterial infections. *Advances in Nutrition (Bethesda, Md.)*, 6(1), 73–82. https://doi.org/10.3945/AN.114.007575.

Thirumdas, R., Kothakota, A., Pandiselvam, R., Bahrami, A., & Barba, F. J. (2021). Role of food nutrients and supplementation in fighting against viral infections and boosting immunity: A review. *Trends in Food Science & Technology*, 110, 66–77. https://doi.org/10.1016/J.TIFS.2021.01.069.

Toledano, J. M., Puche-Juarez, M., Moreno-Fernandez, J., Ochoa, J. J., & Diaz-Castro, J. (2023). Antioxidant and immune-related implications of minerals in COVID-19: A possibility for disease prevention and management. *Antioxidants (Basel, Switzerland)*, 12(5). https://doi.org/10.3390/ANTIOX12051104.

Wakabayashi, H., Oda, H., Yamauchi, K., & Abe, F. (2014). Lactoferrin for prevention of common viral infections. *Journal of Infection and Chemotherapy*, 20(11), 666–671. https://doi.org/10.1016/J.JIAC.2014.08.003.

Wallace, T. C. (2020). Combating COVID-19 and building immune resilience: A potential role for magnesium nutrition? *Journal of the American College of Nutrition*, 39(8), 685–693. https://doi.org/10.1080/07315724.2020.1785971.

Wang, Y., Moon, A., Huang, J., Sun, Y., & Qiu, H. J. (2022). Antiviral effects and underlying mechanisms of probiotics as promising antivirals. *Frontiers in Cellular and Infection Microbiology*, 12. https://doi.org/10.3389/FCIMB.2022.928050.

16 Role of Systems Biology in Emerging and Re-Emerging Viral Diseases

Saurav Kumar Mishra, Nikita Krishna, Somenath Dutta, and John J. Georrge

16.1 INTRODUCTION

In the present situation, different emerging infectious diseases and re-emerging infectious diseases (EID and REID) are occurring daily; this generates a large amount of complex biological data. Understanding host–pathogen interaction with complex biological processes is also becoming difficult. With the advancement of computer-assisted study, system biology came as new hope with precise and accurate steps that can easily explore and uncover the key target and molecular interaction between the host–pathogen interaction (McArthur, 2019; Mishra et al., 2023; Tariq et al., 2021; Zhu et al., 2021). Systems biology is one of the emerging and successful approaches applied in various biological systems, such as identifying potential proteins in viral infection, responsible key molecules, and potential drug targets and how these are intercorrelated with the target pathways. Understanding that a biological system's distinctive features and behaviors are derived from the interactions among its components is one of the core concepts of systems biology (Ben-Othman et al., 2020; Tavassoly et al., 2018). The field of systems biology aims to shed light on the fundamental ideas driving these complicated behaviors, which can range from the coordinated reactions of an entire organism in response to external stimuli to the molecular interactions happening within a cell. System biology considered bioinformatics, statistical calculation, genomics, and transcriptomics concepts to reveal host–pathogen interactions based on protein–protein interaction. However, the selection and data representation are crucial for the system biology study (Sameni et al., 2023; Zou & Laubichler, 2018). The emergence of high-throughput methods, especially transcriptomics, proteomics, and genomes, has resulted in considerable data that constitutes the foundation of systems biology investigation. Although system biology has a long history, at the beginning of 2000, researchers took a keen interest in it and applied it in various ways to solve complex biological problems (Kitano, 2002; Tavassoly et al., 2018). System biology requires experts from different areas, such as chemists, researchers, biologists, and mathematicians, to simplify the biology of complicated biological structures by combining carefully constructed mathematical models with various rigorous molecular techniques for measurement and uncovering the molecular and key target interaction. In EID and REID, the involvement of system biology was more suitable based on the successful study. Computer-assisted research mainly focuses on a single or double target; however, system-biology research considers a huge amount of targets which have the accuracy to be more specific and emphasize the interpretation of their results in different forms, such as molecular and cellular networks (Abouelwafa et al., 2023; Chaudhary et al., 2023; Naimo et al., 2019; Patel et al., 2018). Analysis of biological systems during infection, intricate relationships between host–pathogen, prediction of system dynamic behavior and their evolution and mutational landscape for the revolving mechanism help to understand and bridge the gap between the infection and the clinical outcomes. As viruses and pathogens emerge daily, the advancement and the available information are essential to understand for further research (Malange et al., 2022; Tariq et al., 2021). So, this

DOI: 10.1201/9781032721811-16

study aims to summarize the available information in the context of system biology and its role in EID and REID, along with the list of available tools and databases crucial to performing the system biology-assisted study. This study also included the successful selective study conducted through the system biology approach that can help readers understand the involvement of system biology in basic and translational research.

16.2 ROLE OF SYSTEM BIOLOGY IN EID AND REID

EID and REID are challenging worldwide problems; for controlling and reducing these infections, drug and vaccine developments, which can combat the infection, are also challenging (Mishra et al., 2023; Vinjoda et al., 2024). However, as the host–pathogen interaction and their complete mechanism is a complex process to understand and uncover the most potential target for the drug design and candidate vaccine development, system biology is a new hope that can quickly uncover the complex biological and host–pathogen interaction in a network form, and also helps to get the more potential target based on their assisted step (Hu et al., 2022; Zhang et al., 2012). The target and genes identified through system biology are promising as they are identified and selected through the specific relation, that is, the target's potential role in infection and correlated with the specific possible pathways (Naimo et al., 2019; Vinjoda et al., 2024; Zhu et al., 2021). Moreover, system biology involves many steps to handle EID and REID. These aspects cover the system biology–assisted study in emerging and re-emerging infections (Figure 16.1).

Host–pathogen interaction: In the case of EID and REID, it is essential to understand how the infection interacts with the host cell. There have been outbreaks of some of the re-emerging infectious diseases such as COVID-19; to overcome this, system biology plays a significant role as it helps to clear the molecular interaction between the infection and host cell and also helps to find the target pathways that are involved in the host–pathogen interaction (Mishra et al., 2023; Tariq et al., 2021; Zhu et al., 2021).

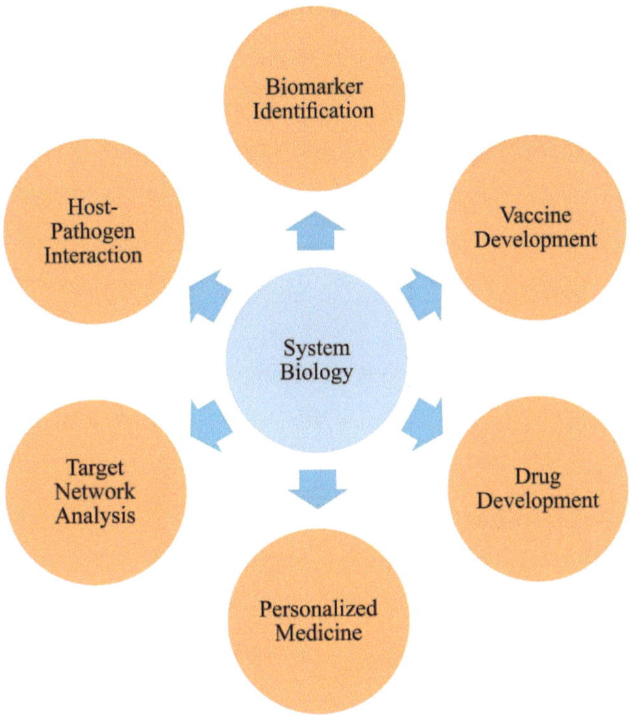

FIGURE 16.1 Representation of involvement of system biology in different areas.

Biomarker identification: In the case of emerging and re-emerging infectious diseases, it is essential to find the molecular indicator or marker that can help to get an idea of the severity of the progression. Applying system biology in this aspect can quickly identify the associated biomarker with disease susceptibility, progression, and severity. One of the significant roles of biomarkers is to detect the infection's early stage and monitor its severity (Peedicayil, 2019; Zhang et al., 2012).

Vaccine developments: In the case of vaccine development, the most antigenic protein target is essential to develop a vaccine with a major role in host–pathogen interaction. The optimization of vaccine development, especially identifying antigenic sites and understanding immune responses, has been made easier by systems biology. It also helps to understand the complex interaction between the pathogen and host receptor through network analysis of complex biological processes (Ben-Othman et al., 2020; Tariq et al., 2021).

Drug development: In the case of EID and REID, the identification of more potential inhibitors and drug development is essential. For this, through the system biology-assisted study, the researchers utilized and validated the drug repurposing approaches based on the crucial step as it can provide the potential for quick therapeutic development (Sameni et al., 2023; Vaghasia et al., 2024; Vakhariya Sakina et al., 2023).

Personalized medicine: With the help of system biology, researchers can easily identify the genetic variation, environmental factors, immune response, and many others which are the essential parts of personalized medicine, and based on the detailed information, personalized medicine is easy for the more specific treatments based on the obtained information of patient and pathogen characteristics (Naimo et al., 2019; Tavassoly et al., 2018).

Target network analysis: Selecting a potential target is essential to develop an effective drug and vaccine candidate. Identifying and selecting the most potential target, key node, and associated pathways that can lead to novel drug and vaccine developments through the system-biology-assisted biological network is easy (Farooq et al., 2020).

16.3 TOOLS AND DATABASES IN SYSTEMS BIOLOGY

As system biology was involved in novel research, finding the different tools and databases is one of the most important aspects of this approach in revolutionizing novel discovery. The involvement of these tools and databases helps to perform a more precise study with regard to precision and efficiency of time. A variety of platforms and tools have been established and are presently being employed by scientists across the globe to understand and uncover complex biological systems. This section will help to understand the available tools and their significant role, that is, how they are helpful. Some of the useful tools and databases are provided in Table 16.1.

TABLE 16.1
List of Useful Tools and Databases in System Biology

Sl. No.	Name	Urls	Reference
1.	STRING	https://string-db.org/	(Szklarczyk et al., 2023)
2.	STITCH	http://stitch.embl.de/	(Szklarczyk et al., 2016)
3.	GeneCards	https://www.genecards.org/	(Stelzer et al., 2016)
4.	Cytoscape	https://cytoscape.org/	(Otasek et al., 2019)
5.	BioGRID	https://thebiogrid.org/	(Stark et al., 2006)
6.	BioCarta	https://maayanlab.cloud/Harmonizome/dataset/Biocarta+Pathways	(Rouillard et al., 2016)
7.	DAVID	https://david.ncifcrf.gov/	(Sherman et al., 2022)
8.	DIP	https://dip.doe-mbi.ucla.edu/dip/Main.cgi	(Salwinski et al., 2004)
9.	KEGG	https://www.genome.jp/kegg/	(Kanehisa & Goto, 2000)
10.	Gene Ontology	https://geneontology.org/	(Ashburner et al., 2000)

In the system biology–assisted study, the network analysis and protein–protein interaction are significant aspects, and the STRING database is one of the helpful tools that can help generate network between proteins, which is useful to identify the most potential targets for the novel drug discovery and vaccine developments (Szklarczyk et al., 2023). STITCH is an open-access database that can help generate and predict the interaction between substance and targets (Szklarczyk et al., 2016). GeneCards is one of the most important tools in the system biology–assisted study as it can assist in obtaining comprehensive data on every labeled and anticipated human gene (Stelzer et al., 2016). Cytoscape is an open-access network analysis and visualizing tool that can help generate the network between the genes and their associated data, such as molecular information and biological pathways. An open-access database named the biological general repository for interaction datasets (BioGRID) provides information on interactions about genetic data and the proteins that go along with it (Stark et al., 2006). BioCarta is an online tool that can help with the knowledge of the graphical models of proteomics, genomics, their associated information, and the specific pathways (Rouillard et al., 2016). The omics-assisted open-access essential database, that is, database for annotation, visualization, and integrated discovery (DAVID), can help understand the functional interpretation of genes selected and predicted through proteomics, microarrays, and genomic studies (Sherman et al., 2022). Database of interacting proteins—DIP—is an open-access curated database that contains information on experimentally identified proteins and is useful for understanding protein–protein interactions (Salwinski et al., 2004). Kyoto encyclopedia of genes and genomes (KEGG) is an open access to pathways analysis tools that can help get information about the association between the target gene and respective pathways. Getting detailed ideas and information regarding systems, genomics, chemicals, and health is also helpful (Kanehisa & Goto, 2000). Gene ontology is one of the most important tools as it can help understand the function of genes at three specific levels (biological, cellular, and molecular) that help identify the more potential genes among the selected and predicted ones (Ashburner et al., 2000).

16.4 SYSTEMS BIOLOGY ASSISTED IN SUCCESSFUL STUDIES

In the current situation, system biology is emerging as a new hope to overcome the host–pathogen complex disease mechanism for better understanding. However, it is essential to understand how the worldwide research community is utilizing the system biology for novel discovery. This section will help readers learn the involvement of system biology–assisted applications and pipelines to overcome the EID and REID (Table 16.2).

The influenza A virus, or IAV, is a hazardous respiratory infection that poses a global hazard to public health, and the first case with a human was reported in Hong Kong in 1997. Humans are still susceptible to influenza multiple times in their lifetimes, even after acquiring vaccinations and becoming vaccinated through previous infections. This required more significant details to understand the potential target for the drug design. To overcome this, a study performed by Farooq et al. on the influenza A virus applied the system biology–assisted strategy to recognize the protein–protein connectivity in the human host and the virus. The authors used Cytoscape to construct the network between the infectious agent and the recipient, identifying 1,477 interacting proteins between the virus and host. To prioritize the essential ones, the identified proteins were analyzed using KEGG and GO enrichment analyses. Finally, the authors found that NS1 has a more significant contribution than others and potentially serves as a therapeutic candidate for therapeutic development (Farooq et al., 2020).

Another study, assisted by Ben-Othman et al., applied rigorous system biology approaches targeting the blood and tissue to understand the vaccine immune response in adults. Based on integrating wet lab experiments and system biology–assisted approaches, the authors suggested the feasibility of recruiting healthy adult samples (blood and tissue) for multiple vaccine administrations and complete retention. Also, they suggested that these approaches can apply to other designed vaccines for viral infection to understand the immune response better (Ben-Othman et al., 2020).

TABLE 16.2
List of Studies to Understand the Role of System Biology

Sl. No.	Target Virus	Remarks	Reference
1	Influenza A virus	Construction of host and influenza A virus protein interaction employing system biology	(Farooq et al., 2020)
2	Hepatitis B	To understand the immune response of vaccines on Hepatitis B employing system biology	(Ben-Othman et al., 2020)
3	COVID-19	To regulate the immune mechanism associated with COVID-19, targeting vitamin C, curcumin, and glycyrrhizic acid	(Chen et al., 2020)
4	COVID-19	Identification of pathophysiological links between COVID and neurological diseases to understand the severity	(Rahman et al., 2021)
5	ZIKA and DENV	Identification of host machinery proteasome for the ZIKA and DENV viruses	(Song et al., 2021)
6	Chikungunya (CHIKV)	Identification of molecular signature to combat infection	(Zhu et al., 2021)
7	COVID-19	To comprehend the molecular mechanisms that govern RA and COVID in order to discover possible treatments	(Hu et al., 2022)
8	SARS-CoV-2	To gain insight into the cellular processes underlying the infection of SARS-CoV-2 through the implementation of a system biology framework	(Sameni et al., 2023)
9	Influenza A (H5N1)	To comprehend the essential biological processes and genes that are linked to influenza A (H5N1)	(Chaudhary et al., 2023)
10	Monkeypox	Identification of potential hub genes along with identification of potential inhibitor targeting F13	(Singhvi et al., 2023)

Another study by Chen et al. employed system biology approaches to regulate the immune response associated with COVID-19, targeting the combination of different compounds. In this study, the authors utilized different compounds to understand the modulation within the path and site of the defensive and inflammatory reaction. The overall finding shows that these combinations can be helpful for the regulation of those associated with the site of the defensive and inflammatory reaction (Chen et al., 2020).

Another study performed by Rahman et al. focused on exploring the mechanism of COVID-19 and neurological disease interaction and their severity impacts. The authors employed bioinformatics, network-based approaches, and transcriptomic analysis for the hub gene identification in this study. Finally, the author found that there is a maximum correlation with Parkinson's disease and a minimum with multiple sclerosis, and there is a maximum score (Semantic) with Huntington's disease and a minimum with epilepsy disease (Rahman et al., 2021).

Further, another study led by Song et al. employed a system biology approach for identifying the ZIKA and DENV proteasomes, as this virus exhibits a similar pathway for the replication process. In this study, the authors employed various systems biology–assisted applications and considered protein–protein interaction and RNAi screening, and they performed high-throughput screening to identify bioactive compounds. The overall finding suggested that the multiple complexes were identified through the protein array analysis, and 10,415 responsible genes were identified. Through the high-throughput screening, 134 potential compounds along with 6 proteasome inhibitors were identified that can be further leads for the novel discovery (Song et al., 2021).

Further, a study by Zhu et al. targeted chikungunya as a target and employed the system biology–assisted rigorous approach for identifying responsible key genes. In this study, the authors utilized the microarray data of chikungunya along with three infections and one disorder, which was extracted from the GEO database. They performed various steps such as network analysis, pathway analysis, and key gene identification. The overall finding suggested that 12 hub genes are more associated with chikungunya infection (Zhu et al., 2021).

Similarly, another study was performed by Hu et al. to analyze the responsible molecule and identify the potential drug employing the bioinformatics and system biology approaches. In this study, the authors utilized the two differentially expressed genes (DEGs) (SE171110 and GSE1775544), which were investigated along with the gene ontology analysis and drug interaction analysis. Moreover, the overall study suggested that among the 103 DEG, the top 10 hub genes can be potential leads for drug development, and these identified genes are correlated with the mutational aspects of COVID-19 that can be helpful for the new drug target (Hu et al., 2022).

As the SARS-CoV-2 pandemic is not yet over and due to the mutation, their host–pathogen interaction is also getting complex, to overcome this, a study performed by Sameni et al. targeted the GRN motifs to understand the pathogenesis. In this study, the DEG (differentially expressed genes) data of the targets, including the SARS-CoV, were retrieved to identify motifs along with pathway analysis. Further, the drug–genes interaction was performed. The overall study found that some transcription factors are responsible for similar pathways, and six drugs were identified through the drug–gene network analysis (Sameni et al., 2023).

Similarly, another study by Chaudhary et al. employs system biology to predict hub gene and their restricted pathways. In this study, the authors retrieved genes that are involved in influenza A (H5N1) in humans from the NCBI gene database, employing advanced search; apart from that, they have also applied STRING 11.5 for protein–protein network, KEGG for pathways analysis, and Cytoscape for network construction. Based on the rigorous steps of system biology, six hub genes were identified as having significant contributions in inflammation and many more, and the authors suggest that these genes can be a potential lead for the therapeutics development (Chaudhary et al., 2023).

A recently re-emerged virus, that is, monkeypox virus, is increasing daily, and to understand the different aspects of this infection, a study was performed by Singhvi et al. They employed system biology approach along with the comparative genomics to understand the different molecular mechanisms (mutational hotspots, functional patterns) and potential drug target. In this study, ten hub genes were identified, and further F13 (having significant role in host–pathogen interaction) was considered a target; some promising potential inhibitors were identified, and the authors suggested that these ligands can be helpful in therapeutic developments (Singhvi et al., 2023).

16.5 CONCLUSION

Systems biology is one of the emerging technologies that have the capability to understand and interpret complex biological systems easily. Due to the frequent occurrence and the reappearance of viral illnesses, with the complex replication system, the application of the system can be helpful in understanding and determining the potential target that has a more significant contribution to the host–pathogen interaction. However, more studies are required to be performed through the system biology approach on other diseases as there are fewer studies available, and it can revolutionize drug and vaccine developments.

REFERENCES

Abouelwafa, M., Ibrahim, T. M., El-Hadidi, M. S., Mahnashi, M. H., Owaidah, A. Y., Saeedi, N. H., Attia, H. G., Georrge, J. J., & Mostafa, A. (2023). Using CADD tools to inhibit the overexpressed genes FAP, FN1, and MMP1 by repurposing ginsenoside C and Rg1 as a treatment for oral cancer. *Frontiers in Molecular Biosciences*, 10, 1248885. https://doi.org/10.3389/fmolb.2023.1248885.

Ashburner, M., Ball, C. A., Blake, J. A., Botstein, D., Butler, H., Cherry, J. M., Davis, A. P., Dolinski, K., Dwight, S. S., Eppig, J. T., Harris, M. A., Hill, D. P., Issel-Tarver, L., Kasarskis, A., Lewis, S., Matese, J. C., Richardson, J. E., Ringwald, M., Rubin, G. M., & Sherlock, G. (2000). Gene ontology: Tool for the unification of biology. The gene ontology consortium. *Nat Genet*, 25(1), 25–29. https://doi.org/10.1038/75556.

Ben-Othman, R., Cai, B., Liu, A. C., Varankovich, N., He, D., Blimkie, T. M., Lee, A. H., Gill, E. E., Novotny, M., Aevermann, B., Drissler, S., Shannon, C. P., McCann, S., Marty, K., Bjornson, G., Edgar, R. D., Lin, D. T. S., Gladish, N., Maclsaac, J., Amenyogbe, N., Chan, Q., Llibre, A., Collin, J., Landais, E., Le, K., Reiss, S. M., Koff, W. C., Havenar-Daughton, C., Heran, M., Sangha, B., Walt, D., Krajden, M., Crotty, S., Sok, D., Briney, B., Burton, D. R., Duffy, D., Foster, L. J., Mohn, W. W., Kobor, M. S., Tebbutt, S. J., Brinkman, R. R., Scheuermann, R. H., Hancock, R. E. W., Kollmann, T. R., & Sadarangani, M. (2020). Systems biology methods applied to blood and tissue for a comprehensive analysis of immune response to hepatitis B vaccine in adults. *Frontiers in Immunology*, 11, 580373. https://doi.org/10.3389/fimmu.2020.580373.

Chaudhary, R. K. L. A., Patil, P., Mateti, U. V., Sah, S., Mohanty, A., Rath, R. S., Padhi, B. K., Malik, S., Jassim, K. H., Al-Shammari, M. A., Waheed, Y., Satapathy, P., Barboza, J. J., Rodriguez-Morales, A. J., & Sah, R. (2023). System biology approach to identify the hub genes and pathways associated with human H5N1 infection. *Vaccines (Basel)*, 11(7). https://doi.org/10.3390/vaccines11071269.

Chen, L., Hu, C., Hood, M., Zhang, X., Zhang, L., Kan, J., & Du, J. (2020). A novel combination of vitamin C, curcumin and glycyrrhizic acid potentially regulates immune and inflammatory response associated with coronavirus infections: A perspective from system biology analysis. *Nutrients*, 12(4). https://doi.org/10.3390/nu12041193.

Farooq, Q. U. A., Shaukat, Z., Aiman, S., Zhou, T., & Li, C. (2020). A systems biology-driven approach to construct a comprehensive protein interaction network of influenza A virus with its host. *BMC Infect Dis*, 20(1), 480. https://doi.org/10.1186/s12879-020-05214-0.

Hu, H., Tang, N., Zhang, F., Li, L., & Li, L. (2022). Bioinformatics and system biology approach to identify the influences of COVID-19 on rheumatoid arthritis. *Frontiers in Immunology*, 13, 860676. https://doi.org/10.3389/fimmu.2022.860676.

Kanehisa, M., & Goto, S. (2000). KEGG: Kyoto encyclopedia of genes and genomes. *Nucleic Acids Res*, 28(1), 27–30. https://doi.org/10.1093/nar/28.1.27.

Kitano, H. (2002). Systems biology: A brief overview. *Science*, 295(5560), 1662–1664. https://doi.org/10.1126/science.1069492.

Malange, V. N. E., Hedermann, G., Lausten-Thomsen, U., Hoffmann, S., Voldstedlund, M., Aabakke, A. J. M., Eltvedt, A. K., Jensen, J. S., Breindahl, M., Krebs, L., Christiansen, M., & Hedley, P. L. (2022). The perinatal health challenges of emerging and re-emerging infectious diseases: A narrative review. *Frontiers in Public Health*, 10, 1039779. https://doi.org/10.3389/fpubh.2022.1039779.

McArthur, D. B. (2019). Emerging infectious diseases. *Nursing Clinics*, 54(2), 297–311.

Mishra, S. K., Priya, P., Rai, G. P., Haque, R., & Shanker, A. (2023). Coevolution based immunoinformatics approach considering variability of epitopes to combat different strains: A case study using spike protein of SARS-CoV–2. *Comput Biol Med*, 163, 107233. https://doi.org/10.1016/j.compbiomed.2023.107233.

Naimo, G. D., Guarnaccia, M., Sprovieri, T., Ungaro, C., Conforti, F. L., Ando, S., & Cavallaro, S. (2019). A systems biology approach for personalized medicine in refractory epilepsy. *Int J Mol Sci*, 20(15). https://doi.org/10.3390/ijms20153717.

Otasek, D., Morris, J. H., Boucas, J., Pico, A. R., & Demchak, B. (2019). Cytoscape automation: Empowering workflow-based network analysis. *Genome Biol*, 20(1), 185. https://doi.org/10.1186/s13059-019-1758-4.

Patel, C. N., Georrge, J. J., Modi, K. M., Narechania, M. B., Patel, D. P., Gonzalez, F. J., & Pandya, H. A. (2018). Pharmacophore-based virtual screening of catechol-o-methyltransferase (COMT) inhibitors to combat Alzheimer's disease. *J Biomol Struct Dyn*, 36(15), 3938–3957. https://doi.org/10.1080/07391102.2017.1404931.

Peedicayil, J. (2019). Identification of biomarkers in neuropsychiatric disorders based on systems biology and epigenetics. *Front Genet*, 10, 985. https://doi.org/10.3389/fgene.2019.00985.

Rahman, M. H., Rana, H. K., Peng, S., Kibria, M. G., Islam, M. Z., Mahmud, S. M. H., & Moni, M. A. (2021). Bioinformatics and system biology approaches to identify pathophysiological impact of COVID-19 to the progression and severity of neurological diseases. *Comput Biol Med*, 138, 104859. https://doi.org/10.1016/j.compbiomed.2021.104859.

Rouillard, A. D., Gundersen, G. W., Fernandez, N. F., Wang, Z., Monteiro, C. D., McDermott, M. G., & Ma'ayan, A. (2016). The harmonizome: A collection of processed datasets gathered to serve and mine knowledge about genes and proteins. *Database (Oxford)*, 2016. https://doi.org/10.1093/database/baw100.

Salwinski, L., Miller, C. S., Smith, A. J., Pettit, F. K., Bowie, J. U., & Eisenberg, D. (2004). The database of interacting proteins: 2004 update. *Nucleic Acids Res*, 32(Database issue), D449–451. https://doi.org/10.1093/nar/gkh086.

Sameni, M., Mirmotalebisohi, S. A., Dehghan, Z., Abooshahab, R., Khazaei-Poul, Y., Mozafar, M., & Zali, H. (2023). Deciphering molecular mechanisms of SARS-CoV-2 pathogenesis and drug repurposing through

GRN motifs: A comprehensive systems biology study. *3 Biotech*, 13(4), 117. https://doi.org/10.1007/s13205-023-03518-x.

Sherman, B. T., Hao, M., Qiu, J., Jiao, X., Baseler, M. W., Lane, H. C., Imamichi, T., & Chang, W. (2022). DAVID: A web server for functional enrichment analysis and functional annotation of gene lists (2021 update). *Nucleic Acids Res*, 50(W1), W216–W221. https://doi.org/10.1093/nar/gkac194.

Singhvi, N., Talwar, C., Mahanta, U., Kaur, J., Mondal, K., Ahmad, N., Tyagi, I., Sharma, G., & Gupta, V. (2023). Comparative genomics and integrated system biology approach unveiled undirected phylogeny patterns, mutational hotspots, functional patterns, and molecule repurposing for monkeypox virus. *Funct Integr Genomics*, 23(3), 231. https://doi.org/10.1007/s10142-023-01168-z.

Song, G., Lee, E. M., Pan, J., Xu, M., Rho, H. S., Cheng, Y., Whitt, N., Yang, S., Kouznetsova, J., Klumpp-Thomas, C., Michael, S. G., Moore, C., Yoon, K. J., Christian, K. M., Simeonov, A., Huang, W., Xia, M., Huang, R., Lal-Nag, M., Tang, H., Zheng, W., Qian, J., Song, H., Ming, G. L., & Zhu, H. (2021). An integrated systems biology approach identifies the proteasome as a critical host machinery for ZIKV and DENV replication. *Genomics Proteomics Bioinformatics*, 19(1), 108–122. https://doi.org/10.1016/j.gpb.2020.06.016.

Stark, C., Breitkreutz, B. J., Reguly, T., Boucher, L., Breitkreutz, A., & Tyers, M. (2006). BioGRID: A general repository for interaction datasets. *Nucleic Acids Res*, 34(Database issue), D535–539. https://doi.org/10.1093/nar/gkj109.

Stelzer, G., Rosen, N., Plaschkes, I., Zimmerman, S., Twik, M., Fishilevich, S., Stein, T. I., Nudel, R., Lieder, I., Mazor, Y., Kaplan, S., Dahary, D., Warshawsky, D., Guan-Golan, Y., Kohn, A., Rappaport, N., Safran, M., & Lancet, D. (2016). The GeneCards suite: From gene data mining to disease genome sequence analyses. *Curr Protoc Bioinformatics*, 54, 1–30. https://doi.org/10.1002/cpbi.5.

Szklarczyk, D., Kirsch, R., Koutrouli, M., Nastou, K., Mehryary, F., Hachilif, R., Gable, A. L., Fang, T., Doncheva, N. T., Pyysalo, S., Bork, P., Jensen, L. J., & von Mering, C. (2023). The STRING database in 2023: Protein-protein association networks and functional enrichment analyses for any sequenced genome of interest. *Nucleic Acids Res*, 51(D1), D638–D646. https://doi.org/10.1093/nar/gkac1000.

Szklarczyk, D., Santos, A., von Mering, C., Jensen, L. J., Bork, P., & Kuhn, M. (2016). STITCH 5: Augmenting protein-chemical interaction networks with tissue and affinity data. *Nucleic Acids Res*, 44(D1), D380–384. https://doi.org/10.1093/nar/gkv1277.

Tariq, H., Batool, S., Asif, S., Ali, M., & Abbasi, B. H. (2021). Virus-like particles: Revolutionary platforms for developing vaccines against emerging infectious diseases. *Frontiers in Microbiology*, 12, 790121. https://doi.org/10.3389/fmicb.2021.790121.

Tavassoly, I., Goldfarb, J., & Iyengar, R. (2018). Systems biology primer: The basic methods and approaches. *Essays Biochem*, 62(4), 487–500. https://doi.org/10.1042/EBC20180003.

Vaghasia, V. V., Sharma, K., Mishra, S. K., & Georrge, J. J. (2024). In silico identification of natural product inhibitor for multidrug resistance proteins from selected gram-positive bacteria. In *Nanotechnology and In Silico Tools* (pp. 309–317). Elsevier.

Vakhariya Sakina, S., Mishra, S. K., Sharma, K., & Georrge, J. J. (2023). Designing of a novel curcumin analogue to inhibit mitogen-activated protein kinase: A cheminformatics approach. *Journal of Phytonanotechnology and Pharmaceutical Sciences*, 3(1), 37–47.

Vinjoda, P., Mishra, S. K., Sharma, K., & Georrge, J. J. (2024). In silico identification of novel drug target and its natural product inhibitors for herpes simplex virus. In *Nanotechnology and In Silico Tools* (pp. 377–383). Elsevier.

Zhang, H., Hu, H., Deng, C., Chun, Y., Zhou, S., Huang, F., & Zhou, Q. (2012). Integrative system biology strategies for disease biomarker discovery. *Comb Chem High Throughput Screen*, 15(4), 286–298. https://doi.org/10.2174/138620712799361852.

Zhu, L., Wang, X., Asa, T. A., & Hossain, M. A. (2021). Identification of key genes, pathways, and associated comorbidities in chikungunya infection: Insights from system biology analysis. *Network Modeling Analysis in Health Informatics and Bioinformatics*, 10, 1–12.

Zou, Y., & Laubichler, M. D. (2018). From systems to biology: A computational analysis of the research articles on systems biology from 1992 to 2013. *PLoS One*, 13(7), e0200929. https://doi.org/10.1371/journal.pone.0200929.

17 SARS, MERS, and COVID-19
Knowledge Mining through Bioinformatics and Computational Biology

Anusree S. Nair, Saurav Kumar Mishra, Anoop Johny, Krishitha Prabhu Chandran, and John J. Georrge

17.1 INTRODUCTION

SARS-CoV-2 (severe acute respiratory syndrome 2) is a recently emerged virus worldwide; however, infection by SARS is not new. The virus belongs to the Coronaviridae family, encompassing *Alphacoronavirus*, *Betacoronavirus*, *Gammacoronavirus*, and *Deltacoronavirus* genera. This virus has a long history related to the zoonotic transfer of infection and has a wide range of animal host. It is well known for its ability to infect humans and animals. The SRAS infection is a viral respiratory-related infection; its first case was noticed in 2002 in southern China (Guangdong Province). After that, it spread and transited to several countries, which led to global outbreaks in 2003 due to this infection (Hofmann & Pohlmann, 2004; Holmes & Rambaut, 2004). Subsequently, in 2012, the Middle East respiratory syndrome coronavirus (MERS-CoV) emerged; it comes from the same family and is also responsible for respiratory illness; however, the origin of this virus was reported in Saudi Arabia (Who Mers-Cov Research, 2013). In the case of MARS, it is suspected that the virus originated from bats, and their transmission to human hosts was accomplished by camels, which serve as intermediate hosts. Being that this virus is a member of an identical family to SARS-CoV, its clinical symptoms are also similar, such as fever, cough, and breathing problems; apart from this, the infection in its severe stage leads to pneumonia. However, this virus has a high fatality rate compared to SARS-CoV, approximately 35% (Chafekar & Fielding, 2018; Hajjar et al., 2013). At the end of 2019, a pneumonia cluster affected Wuhan's large population, a city in China's Hubei Province. Upon being identified as a novel coronavirus, it subsequently disseminated globally. SARS-CoV-2 is the infectious culprit triggering the spread of COVID-19 (Rabaan et al., 2020). In 2019, the WHO classified COVID-19 as an international outbreak. According to the WHO Dashboard for COVID-19 (https://covid19.who.int/), as of December 14, 2023, confirmed COVID-19 incidents are 772,138,818, including 6,985,964 deaths worldwide. Recently, many variants of SARS-CoV-2 have occurred. As the history of this infection is long, several approaches were applied in this to get a better understanding. Among others, bioinformatics and computational biology are a new hope as they can help knowledge mining like data collection, data processing, data analysis, knowledge representation, natural language processing, data visualization, and knowledge discovery (Ray et al., 2021). Knowledge mining mainly deals with the initial step to the end of the investigation, which is composed of various steps of collection, screening, and validation that are useful for the advancement of the healthcare system along with the medical science as it also sets the path or pipeline that can be utilized for other infectious diseases as well. Therefore, as the involvement of bioinformatics and computational biology in medical science is occurring, it is essential to get better insight through knowledge mining and other associated steps that can be helpful for better tumors.

DOI: 10.1201/9781032721811-17

17.2 MERS, SARS, COVID-19: KNOWLEDGE MINING USING BIOINFORMATICS

Bioinformatics-based knowledge mining is the process of using computers to pull out useful trends, information, or knowledge from biological data. This method analyzes different kinds of biological data, like DNA sequences, protein structures, gene expression patterns, and biochemical pathways, to find secret connections biomarkers, better understand biological processes, and find new medicines or treatments. Bioinformatics tools and programs make the ability to combine, analyze, and make sense of big and complicated biological information possible (Kumar et al., 2022; Pereira et al., 2020). Phylogenetic analysis, molecular modeling, network analysis, and machine learning are some of the most popular bioinformatics methods used for information mining. Bioinformatics knowledge mining helps scientists understand biological systems better, find new things faster, and progress in personalized medicine, drug development, and farming technology (Cao et al., 2020; Iuchi et al., 2023). As bioinformatics covered various stages and found to be promising, worldwide researchers also utilized the concept in the MERS, SARS, and COVID-19 data handling, followed by the identification of the potential novel target, screening of large legend library for the prediction of potential inhibitors and their stability through the investigation of the dynamics (Ma et al., 2021; Ray et al., 2021). Not only the drug design, host interaction, and vaccine development, the researcher also utilizes the concept of biological data investigation in the tools and disease-specific databases such as COVID1–19. A server (https://ncov.schanglab.org.cn/) named "COVID-19 Docking Server" was developed by Kong et al. and can be utilized for the docking analysis of COVID-19 target with the compound, peptide, and small molecules (Kong et al., 2020). Similarly, another server (https://bis.zju.edu.cn/overcovid/) named "OverCOVID" was developed by Ahsan et al., which is based on the complete collection along with the other application such as network analyses which will be helpful to gather all the information of COVID-19 on a single server (Ahsan et al., 2021). Another server (https://cov3d.ibbr.umd.edu/) named "CoV3D," developed by Gowthaman et al., mainly works on the COVID-19 target structure design in high resolution (Gowthaman et al., 2021). Similarly, it is essential to understand the host–pathogen interaction and therefore, a server (https://interactome.ku.edu.tr/sars/) named "SARS-CoV-2 Interactome 3D" was developed by Ovek et al., which is mainly utilized for the interaction analysis along with the visualization (Ovek et al., 2022). Moreover, apart from COVID-19, the researcher also utilized bioinformatics in various pathogenesis as well (Vaghasia et al., 2024; Vakhariya et al., 2023; Vakhariya Sakina et al., 2023; Vinjoda et al., 2024). In the initial step of data collection in infectious disease, it is essential to check the similarity between the available infectious disease information and collected sample data and for bioinformatics with the help of sequence alignment involved as a crucial one followed by phylogenetics analysis can help to correlate the collected data and their earlier evidence (Junejo et al., 2020). Similarly, many approved drugs are available and their information is also available on DrugBank (https://go.drugbank.com/). In the case of coronavirus, the information from MERS, SARS, and COVID-19 was successfully explored. Various algorithms were applied for the drug discovery process, such as drug repurposing, novel inhibitor, and advanced bioinformatics approaches such as structure-based, ligand-based drug design, pharmacophore modeling, and many more (Gurung et al., 2021). Moreover, severe mutations have occurred in COVID-19, and it is essential to keep the resistance-related information to develop impactful therapeutics, which gives the mutational information and effects (Feng et al., 2024). Moreover, different vaccines were successfully designed to combat the COVID-19 infection. Utilizing the concept of bioinformatics in another area is a new hope, that is, immunoinformatics composed of *in silico* and computational investigation. A study performed by Mishra et al. utilized the concept of immunoinformatics and successfully designed an epitope-based peptide vaccine against the COVID-19 infection (Mishra et al., 2023). Apart from that, bioinformatics is involved in many other essential aspects as well, such as host–pathogen investigation to find the most prominent target, structure analysis to get an understanding of the target and their structure representation, surveillance, and also the omics data analysis, which comprises the genomics, transcriptomics, and many more, will give the complete insightful information

(Iuchi et al., 2023). Moreover, Bioinformatics tools have been used to make models of how virus proteins are put together and how they might react with possible drugs or antibodies. This helps us determine how viruses attack cells at the molecular level and make better medicines. Bioinformatics has been used to model how COVID-19 spread and guess where it will go. To stop the virus from spreading, this information has helped guide public health actions and plans. Using bioinformatics, it is possible to look at genomic, proteomic, and clinical data from COVID-19 cases. Thus, finding biomarkers for disease severity, predicting patient outcomes, and directing personalized treatment plans all have become easier. Overall, bioinformatics, along with the computational biology investigation, was found to be a prominent approach in dealing with MERS, SARS, and COVID-19 (Fahmi et al., 2021; Ishack & Lipner, 2021; Li et al., 2021).

17.3 SARS, MERS, COVID-19: KNOWLEDGE MINING USING COMPUTATIONAL BIOLOGY

The variation in different strains and the evolutionary changes throughout the journey of SARS, MERS, and COVID-19 show a complex mechanism as it also varied from geographical location. Computational biology is involved throughout the data collection and therapeutics development, followed by different predictions, analysis, screening, and consensus on the novelty. The computational biology in the SARS, MERS, and COVID-19 is crucial in many areas that can be helpful for better understanding of SARS, MERS, and COVID-19 (Alharbi & Alrefaei, 2021; Ray et al., 2021). Moreover, their evolution and strains vary time to time, and to get insightful information, the genomics analyses computed through computational biology are helpful as they can help examine the phylogenetic connection, similarity, variation, mutation, and many more aspects that are related to genomics interpretation (Gurung et al., 2021; Mishra et al., 2023). Furthermore, through the genomics analysis, the strain and the variation can be identified, and also, among them, more potential targets can be identified and further utilized for the drug discovery process implementing the drug design approach, which is also one of the aspects of computational biology. The drug discovery through computational biology in SARS, MERS, and COVID-19 was successfully performed by worldwide researchers, followed by high-throughput virtual screening and many more (Chen et al., 2023; Hossain et al., 2023; Mishra et al., 2023). For vaccine development, it is essential that the selected target has a prominent role in the pathogenesis, and the computation biology can easily identify the most prominent one, such as spike protein, with the main one in the COVID-19 for the host–pathogen interaction (Mishra et al., 2023; Shah & Coiado, 2022). Therapeutic development mainly comes when a number of high infection transmissions occur; therefore, data related to infection is required. In this case, the computation approaches were used in the epidemiological data handling (computerized models are used to predict how the virus will spread and guess where it will go next. Officials in charge of public health can use these models to help them make smart choices about treatments and control measures), which shows the number of cases and the fatalities rate (Aborode et al., 2023; Tan et al., 2023). Further, in therapeutic development, the availability of the virus structure is also important as it can be helpful for the drug design approach. Several tools and databases are available for structure modeling, refining, and validation to overcome this. Moreover, the collected data and modeled structure need to be interpreted exactly to help achieve a better outcome. Overall, computational biology in SARS, MERS, and COVID-19 was applied in various forms, such as host–pathogen interaction, drug design, vaccine development, transcriptomics analysis, and genomics investigations (Alharbi & Alrefaei, 2021; Fahmi et al., 2021; Gurung et al., 2021; Pereira et al., 2020; Ray et al., 2021).

17.4 CHALLENGES AND FUTURE PROSPECTS

Emerging and re-emerging infectious diseases and infections occur daily, and their transmission rate is also high due to diverse replication mechanisms. As the pathogen has a complex mechanism and their interaction with the host cell is also fizzy, it cannot be easily understandable due to the high mutation and the frequent changes in their interaction. As MERS, SARS, and COVID-19 have been associated with each other, the transmission rate from each other is different due to the interaction and replication mechanisms with the host cell. As advancements in bioinformatics and computational biology are increasing daily, employing artificial intelligence, machine learning, and deep learning to obtain basic and initial information is the only hope and pivotal for novel therapeutic designs such as drugs and vaccines. Similarly, it is also essential that the sample collected before the process has crucial information and is collected precisely. In the computational aspects, the screening and retrieval of bulk datasets is simple; however, the redundancy in the data and retrieved information was one of the leading challenges. Furthermore, especially in COVID-19, several mutations were frequently occurring in the spike protein, and different strains were reported, which require precise data collection along with every essential information without any redundancy, which will be helpful for therapeutic development and also to get a better understanding of the evolutionary mechanisms. As computational aspects are one of the leading pillars in the medical science and healthcare system, it requires the correct data for novel discoveries. Apart from that, the advancement in bioinformatics and computational biology need to be updated as the emerging and re-emerging disease investigation generates a huge amount of data to get a better understanding, and it is essential to manage and find the most crucial information among the datasets.

REFERENCES

Aborode, A. T., Huang, H., Wireko, A. A., Mehta, A., Kalmanovich, J., Abdul-Rahman, T., Sikora, V., & Awaji, A. A. (2023). Approaching COVID-19 with epidemiological genomic surveillance and the sustainability of biodiversity informatics in Africa. *J Med Virol*, 95(1), e28308. https://doi.org/10.1002/jmv.28308.

Ahsan, M. A., Liu, Y., Feng, C., Hofestadt, R., & Chen, M. (2021). OverCOVID: An integrative web portal for SARS-CoV-2 bioinformatics resources. *J Integr Bioinform*, 18(1), 9–17. https://doi.org/10.1515/jib-2020–0046.

Alharbi, S. N., & Alrefaei, A. F. (2021). Comparison of the SARS-CoV-2 (2019-nCoV) M protein with its counterparts of SARS-CoV and MERS-CoV species. *J King Saud Univ Sci*, 33(2), 101335. https://doi.org/10.1016/j.jksus.2020.101335.

Cao, Y., Geddes, T. A., Yang, J. Y. H., & Yang, P. (2020). Ensemble deep learning in bioinformatics. *Nature Machine Intelligence*, 2(9), 500–508.

Chafekar, A., & Fielding, B. C. (2018). MERS-CoV: Understanding the latest human coronavirus threat. *Viruses*, 10(2). https://doi.org/10.3390/v10020093.

Chen, Z., Wang, W., Jue, H., & Hua, Y. (2023). Bioinformatics and system biology approach to identify potential common pathogenesis for COVID-19 infection and osteoarthritis. *Sci Rep*, 13(1), 9330. https://doi.org/10.1038/s41598-023-32555-y.

Fahmi, M., Kharisma, V. D., Ansori, A. N. M., & Ito, M. (2021). Retrieval and investigation of data on SARS-CoV-2 and COVID-19 using bioinformatics approach. *Adv Exp Med Biol*, 1318, 839–857. https://doi.org/10.1007/978-3-030-63761-3_47.

Feng, Y., Yi, J., Yang, L., Wang, Y., Wen, J., Zhao, W., Kim, P., & Zhou, X. (2024). COV2Var, a function annotation database of SARS-CoV-2 genetic variation. *Nucleic Acids Res*, 52(D1), D701–D713. https://doi.org/10.1093/nar/gkad958.

Gowthaman, R., Guest, J. D., Yin, R., Adolf-Bryfogle, J., Schief, W. R., & Pierce, B. G. (2021). CoV3D: A database of high resolution coronavirus protein structures. *Nucleic Acids Res*, 49(D1), D282–D287. https://doi.org/10.1093/nar/gkaa731.

Gurung, A. B., Ali, M. A., Lee, J., Farah, M. A., & Al-Anazi, K. M. (2021). An updated review of computer-aided drug design and its application to COVID-19. *Biomed Res Int*, 2021, 8853056. https://doi.org/10.1155/2021/8853056.

Hajjar, S. A., Memish, Z. A., & McIntosh, K. (2013). Middle east respiratory syndrome coronavirus (MERS-CoV): A perpetual challenge. *Ann Saudi Med*, 33(5), 427–436. https://doi.org/10.5144/0256–4947.2013.427.

Hofmann, H., & Pohlmann, S. (2004). Cellular entry of the SARS coronavirus. *Trends Microbiol*, 12(10), 466–472. https://doi.org/10.1016/j.tim.2004.08.008.

Holmes, E. C., & Rambaut, A. (2004). Viral evolution and the emergence of SARS coronavirus. *Philos Trans R Soc Lond B Biol Sci*, 359(1447), 1059–1065. https://doi.org/10.1098/rstb.2004.1478.

Hossain, M. A., Rahman, M. H., Sultana, H., Ahsan, A., Rayhan, S. I., Hasan, M. I., Sohel, M., Somadder, P. D., & Moni, M. A. (2023). An integrated in-silico pharmaco-BioInformatics approaches to identify synergistic effects of COVID-19 to HIV patients. *Comput Biol Med*, 155, 106656. https://doi.org/10.1016/j.compbiomed.2023.106656.

Ishack, S., & Lipner, S. R. (2021). Bioinformatics and immunoinformatics to support COVID-19 vaccine development. *J Med Virol*, 93(9), 5209–5211. https://doi.org/10.1002/jmv.27017.

Iuchi, H., Kawasaki, J., Kubo, K., Fukunaga, T., Hokao, K., Yokoyama, G., Ichinose, A., Suga, K., & Hamada, M. (2023). Bioinformatics approaches for unveiling virus-host interactions. *Comput Struct Biotechnol J*, 21, 1774–1784. https://doi.org/10.1016/j.csbj.2023.02.044.

Junejo, Y., Ozaslan, M., Safdar, M., Khailany, R. A., Rehman, S., Yousaf, W., & Khan, M. A. (2020). Novel SARS-CoV-2/COVID-19: Origin, pathogenesis, genes and genetic variations, immune responses and phylogenetic analysis. *Gene Reports*, 20, 100752. https://doi.org/10.1016/j.genrep.2020.100752.

Kong, R., Yang, G., Xue, R., Liu, M., Wang, F., Hu, J., Guo, X., & Chang, S. (2020). COVID-19 docking server: A meta server for docking small molecules, peptides and antibodies against potential targets of COVID-19. *Bioinformatics*, 36(20), 5109–5111. https://doi.org/10.1093/bioinformatics/btaa645.

Kumar, S., Kumar, G. S., Maitra, S. S., Maly, P., Bharadwaj, S., Sharma, P., & Dwivedi, V. D. (2022). Viral informatics: Bioinformatics-based solution for managing viral infections. *Brief Bioinform*, 23(5). https://doi.org/10.1093/bib/bbac326.

Li, X., Yu, J., Zhang, Z., Ren, J., Peluffo, A. E., Zhang, W., Zhao, Y., Wu, J., Yan, K., Cohen, D., & Wang, W. (2021). Network bioinformatics analysis provides insight into drug repurposing for COVID-19. *Med Drug Discov*, 10, 100090. https://doi.org/10.1016/j.medidd.2021.100090.

Ma, L., Li, H., Lan, J., Hao, X., Liu, H., Wang, X., & Huang, Y. (2021). Comprehensive analyses of bioinformatics applications in the fight against COVID-19 pandemic. *Comput Biol Chem*, 95, 107599. https://doi.org/10.1016/j.compbiolchem.2021.107599.

Mishra, S. K., Priya, P., Rai, G. P., Haque, R., & Shanker, A. (2023). Coevolution based immunoinformatics approach considering variability of epitopes to combat different strains: A case study using spike protein of SARS-CoV-2. *Comput Biol Med*, 163, 107233. https://doi.org/10.1016/j.compbiomed.2023.107233.

Ovek, D., Taweel, A., Abali, Z., Tezsezen, E., Koroglu, Y. E., Tsai, C. J., Nussinov, R., Keskin, O., & Gursoy, A. (2022). SARS-CoV-2 interactome 3D: A web interface for 3D visualization and analysis of SARS-CoV-2-human mimicry and interactions. *Bioinformatics*, 38(5), 1455–1457. https://doi.org/10.1093/bioinformatics/btab799.

Pereira, R., Oliveira, J., & Sousa, M. (2020). Bioinformatics and computational tools for next-generation sequencing analysis in clinical genetics. *J Clin Med*, 9(1). https://doi.org/10.3390/jcm9010132.

Rabaan, A. A., Al-Ahmed, S. H., Haque, S., Sah, R., Tiwari, R., Malik, Y. S., Dhama, K., Yatoo, M. I., Bonilla-Aldana, D. K., & Rodriguez-Morales, A. J. (2020). SARS-CoV-2, SARS-CoV, and MERS-COV: A comparative overview. *Infez Med*, 28(2), 174–184. https://www.ncbi.nlm.nih.gov/pubmed/32275259.

Ray, M., Sable, M. N., Sarkar, S., & Hallur, V. (2021). Essential interpretations of bioinformatics in COVID-19 pandemic. *Meta Gene*, 27, 100844. https://doi.org/10.1016/j.mgene.2020.100844.

Shah, A., & Coiado, O. C. (2022). COVID-19 vaccine and booster hesitation around the world: A literature review. *Front Med (Lausanne)*, 9, 1054557. https://doi.org/10.3389/fmed.2022.1054557.

Tan, M., Xia, J., Luo, H., Meng, G., & Zhu, Z. (2023). Applying the digital data and the bioinformatics tools in SARS-CoV-2 research. *Comput Struct Biotechnol J*, 21, 4697–4705. https://doi.org/10.1016/j.csbj.2023.09.044.

Vaghasia, V. V., Sharma, K., Mishra, S. K., & Georrge, J. J. (2024). In silico identification of natural product inhibitor for multidrug resistance proteins from selected gram-positive bacteria. In *Nanotechnology and In Silico Tools* (pp. 309–317). Elsevier.

Vakhariya, S., Kumar Mishra, S., & Georrge, J. J. (2023). Identification of novel curcumin analogues to inhibit mitogen-activated protein kinase 1 using in silico combinatorial library design and molecular docking approach. In *Proceedings of XXIX Symposium on Bioinformatics and Computer-Aided Drug Discovery*. September 18–20, Institute of Biomedical Chemistry, Moscow, Russia, pp. 106. ISBN: 978-5-900760-19-3.

Vakhariya Sakina, S., Mishra, S. K., Sharma, K., & Georrge, J. J. (2023). Designing of a novel curcumin analogue to inhibit mitogen-activated protein kinase: A cheminformatics approach. *Journal of Phytonanotechnology and Pharmaceutical Sciences*, 3(1), 37–47.

Vinjoda, P., Mishra, S. K., Sharma, K., & Georrge, J. J. (2024). In silico identification of novel drug target and its natural product inhibitors for herpes simplex virus. In *Nanotechnology and In Silico Tools* (pp. 377–383). Elsevier.

Who Mers-Cov Research Group. (2013). State of Knowledge and Data Gaps of Middle East Respiratory Syndrome Coronavirus (MERS-CoV) in Humans. *PLoS Curr*, 5. https://doi.org/10.1371/currents.outbreaks.0bf719e352e7478f8ad85fa30127ddb8.

18 Bioinformatics and Computational Biology–Based Approaches in Research and Clinical Management of Dengue Fever

Saurav Kumar Mishra, Sneha Roy, Jibin Varghese, and John J. Georrge

18.1 INTRODUCTION

Over the past few decades, dengue fever has been primarily caused by the dengue virus, becoming a significant global healthcare worry. The prevalence of viral infection in both endemic and epidemic transmission cycles across the globe is frequently unnoticed by affected individuals. Hemorrhagic fever, dengue infection, and shock syndrome caused by dengue are all associated with this virus, threatening public health and contributing to substantial healthcare burdens in affected regions (Roy & Bhattacharjee, 2021). The dengue virus' impact in endemic areas affects medical and healthcare systems and causes economic strain. Dengue results in millions of cases annually, leading to thousands of deaths worldwide (Guzman et al., 2016; Halstead et al., 2001; Prompetchara et al., 2020; Wilson & Chen, 2015). World Health Organization reports a significant surge in dengue cases, which has exposed nearly half of the global population to the possibility of contracting dengue fever. Estimates suggest that between 100 million and 400 million people contract dengue each year, but around 80% experience mild or no symptoms. However, about 20% of those infected with severe dengue fever result in roughly22,000 deaths annually (https://www.who.int). It is a member of the Flaviviridae virus family and is composed of single-stranded RNA and split into several serotypes globally (Bifani et al., 2022; Prompetchara et al., 2020). These serotypes 1,2,3,4,5 have variations in how they trigger immune responses (Guzman et al., 2016). When someone gets infected with one of these types, their body learns to defend against that specific type and has a bit of protection against the others. This is helpful for the scientists trying to create medicines and vaccines against dengue. How dengue appears in people can differ depending on which type they've caught. Mild fever can coexist with severe conditions such as DHF and DSS. The symptoms are pretty diverse, too, like headaches, stomach pain, rashes, feeling worn out, joint pain, no appetite, throwing up diarrhea, and having a weird taste in the mouth (Figure 18.1). But it is not just the physical symptoms; sometimes, the virus can also affect how someone grows physically and mentally (Li & Kang, 2022; Prompetchara et al., 2020; Teo et al., 2023). Understanding these different types of dengue and how they affect us is a big deal for scientists. They are trying to find ways to fight the virus and keep people safe and healthy. Dengue, which used to be more contained in certain places, is spreading wider and becoming epidemic. The dengue virus comprises NS1, NS2A, NS2B, NS3, NS4A, NS4B, and NS5 which are the seven non-structural proteins. It primarily consists of three structural proteins, namely, the envelope (E), capsid (C), and membrane (M). These proteins come together to form tiny round

DOI: 10.1201/9781032721811-18

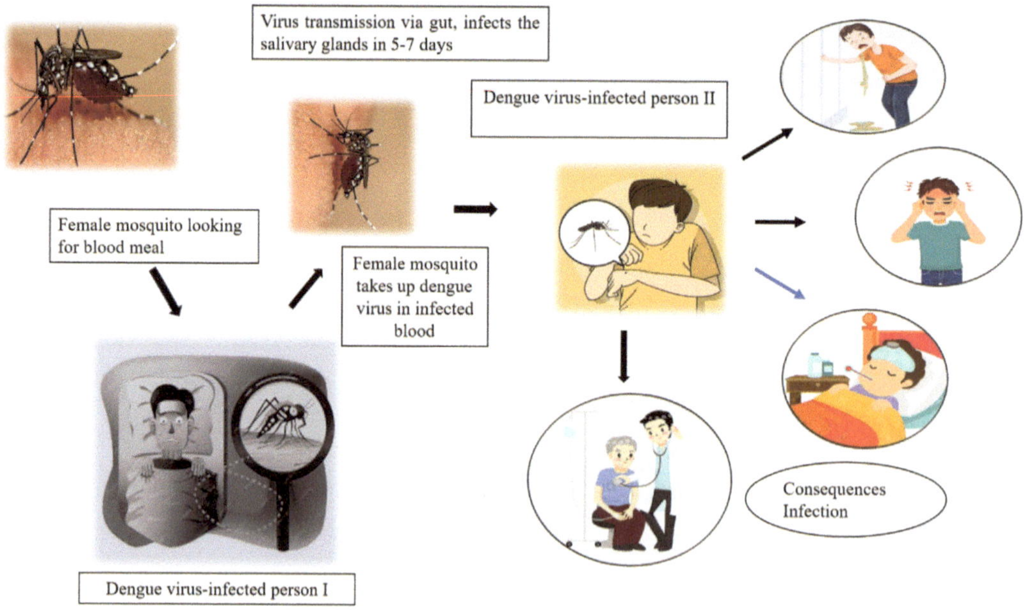

FIGURE 18.1 The basic representation of the dengue virus from transmission to symptoms.

particles about 50 nm in size (Li & Kang, 2022; Perera & Kuhn, 2008). These particles have a smooth surface and are surrounded by an endoplasmic reticulum-derived membrane. Dengue virions evolve from a spiky, embryonic state to a smooth, mature state as they pass through different cell parts, including the Golgi body and endoplasmic reticulum. During the process, the virus gets help from the host cell, picking up specific proteins it needs on its surface to infect more cells (Roy & Bhattacharjee, 2021). Scientists are still working hard to understand these proteins better. They are researching more to determine what these proteins do and how they affect the virus. Unfortunately, there are no particular treatment strategies for dengue infection. Also, no specific drugs are given the thumbs up specifically for treating the dengue virus. However, some ideas exist, and a few treatments are being tried to help against the virus. However, the lasting impact of the dengue virus has put a strain on healthcare systems. This pushes the need for vaccines to control mosquitoes that carry the virus, and taking preventive steps in communities is also really important. In addition, diseases have attracted much attention from scientists in recent years, especially in the current pandemic. However, much research is exploring new ideas, such as using the PLA compounds to develop even better vaccines and treatments (Bifani et al., 2022; Halstead et al., 2001). Nowadays, bioinformatics and computational biology approach contribute to improve the clinical management strategies. In addition, computational analysis of genetic data in present time provides insights into the evolution, population dynamics, and transmission patterns of dengue virus. By analyzing genetic sequences, researchers can track viral lineages, understand genetic diversity, and predict epidemic trends, which are essential for designing effective control measures. However, bioinformatics tools facilitate structural analysis of dengue virus proteins, aiding in understanding viral architecture, protein functions, and potential drug-binding sites, which is valuable for designing therapeutic interventions targeting specific viral components through diagnostics tools development, drug design, structural analysis, and so on (Jagtap et al., 2023; Kabir et al., 2021; Nncube et al., 2018). This chapter will give a quick rundown on the general concept of the dengue virus and focus on the computational approach to easy understanding and effective treatment.

18.2 GENOME OF DENGUE VIRUS

The DENV genome is composed of positive-strand RNA with a single strand of 11 kb length. However, this positive-sense RNA genome generates three structural proteins (C, prM, and E) from a large polyprotein precursor encoded. The nucleocapsid of the virus is composed of the core protein C that surrounds the viral genome, and the prM protein serves as the precursor to the membrane protein, which is essential in both the maturation and assembly processes of the virus. Lastly, the envelope protein or E protein infects the host cells (Vivek Dhar Dwivedi et al., 2017). However, the viral genome and capsid proteins encircled by envelope proteins make up the virus. When a host cell becomes infected, its replicating machinery is taken over. Being a part of Flaviviridae family, the virus split into several serotypes globally, which trigger immune responses in the body. There are four antigenically unique serotypes of the dengue virus (DENV-1,2,3,4) with 65% genome similarity, and there are even recent discoveries of a fifth type spotted in Malaysia in 2007 (Mustafa et al., 2015). It encodes for a single ORF, followed by the highly structured untranslated regions, which are pivotal in replication. The presence of inverted sequences that complement each other in the DENV genome is an essential feature that allows the genome to carry out alterations such as base pairing and various RNA adherence (Kok et al., 2023).

Additionally, the analysis of the DENV genome gives valuable insights into its genetic makeup, evolution and, most importantly, pathogenesis because it is a very complex process that remains incompletely misunderstood, and the mechanisms remain unclear (Bifani et al., 2022; Guzman et al., 2016; Roy & Bhattacharjee, 2021). Nowadays, there is so much research to offer to unravel the complexity computationally. Bioinformatics and computational biology are instrumental in advancing the analysis part and leveraging sophisticated tools, which enable a comprehensive understanding of its complexity with structure and function. One of the most crucial stages of analysis is genome sequencing, which plays a crucial role in uncovering the genetic variations, evolutionary patterns, and structural features that influence viral behavior and pathogenesis. A comprehensive analysis of the dengue virus genome can be conducted by utilizing advanced sequencing technologies and bioinformatics tools. For tracking illnesses, epidemiology, and vaccine development, knowledge of the entire genome's sequence and characteristics is essential (Xie et al., 2021). Furthermore, sequencing efforts can enable phylogenetic analysis of DENV strains from severe and non-severe cases and comparing the complete genome sequences with the phylogenetic analysis (Rahim et al., 2023) can also trace the evolutionary relationships between different viral strains, which can also help to understand the variations in the disease impact. In conjunction with conventional sequencing technologies, whole genome sequences of data, gene variants, and genetic characteristics of dengue patients are examined using sequenced RNA samples. It also aids in identifying the biomarkers associated with the potential pathogenic genes, enabling the identification of potential therapeutic targets or diagnostic markers, which can help to find the drug target for future therapeutic approaches (Anez et al., 2016). Moreover, the comparative genomics approach and its tools are used to develop the pipelines for the analysis of RNA viruses like dengue (Goz et al., 2018). Despite the advances in sequencing technologies, some limitations exist in using bioinformatics tools for analysis, including data quality, biological complexity, data interpretation and validation, and many more. Besides, all these sequencing technologies and other algorithms and integration strategies enhance the accuracy and reliability of bioinformatics analysis in the virology area (Nncube et al., 2018).

18.3 PROTEOMICS AND STRUCTURAL ANALYSIS OF DENGUE VIRUS

The investigation of proteins is called proteomics research, that is, it involves the application of technologies for the identification, interaction, and function of all the proteins. Proteins are crucial elements for any organism, including viruses, and they can help in the survival and growth of the virus. Proteomics is vital in studying dengue viral proteins, their interactions with host cell machinery, and their role in disease severity. The primary goal of structural proteomics has been to comprehend the functions and interactions of structural and non-structural proteins involving capsid, membrane, and envelope proteins (Perera & Kuhn, 2008). Analyzing the three-dimensional structures of these proteins also provides further insights into their functions and interactions. As in the focus of the structural proteins, capsid proteins of DENV are essential for packaging the viral RNA genome within the lipid bilayer. However, the membrane and envelope protein have different structural configurations in mature and immature DENV particles. The membrane protein is used for viral attachment, fusion, assembly, and maturation. The envelope protein, nonetheless, serves as the initial interface between the viral particle and the host cell. These structural proteins are important in identifying drug targets due to their significance in the virus life cycle and interaction with the host. Further, envelope protein is the first protein to show its function and is also a potential site for therapeutic intervention (V. D. Dwivedi et al., 2017; Miao et al., 2019). These can also be a promising target for epitope-based drug design as they can be targeted through antigenic sites. The structural changes in the E protein help uncover new DENV particles and strains for specialized anti-dengue medicines. Membrane protein appears as the transmembrane protein beneath E protein. The 3D structure of the membrane protein is solved at its pH. Studies have linked resistance to medications to the dengue virus genome's capsid coding region, underlining this protein's significance in antiviral drug response (V. D. Dwivedi et al., 2017; S et al., 2021).

In addition to these structural proteins, antiviral pathogens also need non-structural proteins because they do more than one thing. Many individuals use non-structural objectives like NS3 and NS5 as desired outcomes. Since NS3 helicase is essential for the virus to function properly, it can be deployed as a therapeutic target. It also disrupts viral replication, which can also be used in computer-aided drug design, which can help treat dengue (V. D. Dwivedi et al., 2017; Norazharuddin & Lai, 2018). Besides NS3, targeting NS5 helicase of the dengue virus has emerged as a promising strategy for antiviral drug development because inhibiting the helicase activity of NS5 can disrupt viral replication; this makes it a possible candidate for making drugs that fight dengue viruses (V. D. Dwivedi et al., 2017). However, there are only a few reports on targeting the capsid alpha helix of capsid protein (PDB ID:1R6R) (V. D. Dwivedi et al., 2017; Shankaranarayana et al., 2021). Additionally, targeting the Noctyl-β-D-glucoside binding site of the envelope (PDB ID: 1OKE) is also a new study in drug development for the DENV (Shankaranarayana et al., 2021).

18.4 NOVEL DRUG DEVELOPMENT

Research and development efforts are ongoing to address the need for novel drugs to combat dengue virus infection. Even though dengue vaccines are available, individuals continue to seek better safeguarding drugs because the vaccines don't always work against all types of dengue virus. Developing novel drugs for dengue fever through bioinformatics and computational biology is a promising area of research. Different screening methods, such as virtual screening, are employed to find new possible DAA and HAD for treating DENV diseases (Troost & Smit, 2020). Virtual screening methods involve the computational screening of millions of compounds against dengue virus targets, which can lead to the discovery of novel antiviral drugs. A computational drug approach can identify potential inhibitors of viral proteins in dengue-like NS5 methyltransferase. This method also identifies compounds that interact with essential residues in the viral proteins, blocking their catalytic activity and inhibiting their viral replication. Several studies have performed virtual screening against the most promising viral proteins, which can identify several compounds and docked with the targeted viral proteins (Lim et al., 2011; Vakhariya et al., 2023). Moreover, some open-access databases provide information on the interactions of dengue proteins with the host. However, several therapeutic strategies are being explored, including developing antiviral peptides targeting different dengue virus serotypes (Vinjoda et al., 2024). Nevertheless, bioinformatics tools are used to make it easier to find dengue RNA-dependent RNA polymerase inhibiting agents, which provide a comprehensive framework for understanding the viral enzymes involved in viral infection (Shimizu et al., 2019). The use of old drug compounds available in the database also accelerates the discovery of anti-dengue inhibitors. Compounds like prochlorperazine, quinine, and minocycline have shown promise in inhibiting dengue virus replication, which also offers new pathways for therapeutic interventions. This method can also offer a cost-effective and efficient strategy for discovering innovative treatments for dengue fever. In addition, epitopes' role in drug designing for the dengue virus particularly targets the specific viral components to develop effective drugs. This method also allowed for predicting potential drug candidates that can specifically target the epitopes, disrupting viral replication. Nevertheless, it isn't currently a specialized antiviral drug on the market. Many sulphated polysaccharides have been found as a potential feature against DENV (Botta et al., 2018). Some polysaccharide components originating from algae, such as *Caulerpa cupressoides*, suppress the virulence of DENV. Several compounds like curdlan, carrageenan G33d, DL-galactan, and C2S–3 hybrids have demonstrated antiviral activity against all strains of the dengue virus. In modern medicine, ribavirin is a ribonucleic analog also used for DENV treatment to stop the synthesis of viral RNA into the host cell. Glycyrrhizin and their derivatives,NITD008 and uridine analogs, inhibit the posttranslational modifications; on the other hand, certain small-molecule viral inhibitors, including CDDO-me, CP26, SDM25N and UV-4B, reached phase I clinical trials which specifically target DENV replication (Low et al., 2018). Despite all these advancements in drug development, limited treatment options, the complexity of the host–virus interactions, drugs for specific targets, and many more significant challenges are part of the pathway of novel drug development. Recent advances and technology with innovative research approaches improve the treatment outcomes and critical aspects of drug development (Vaghasia et al., 2024; Vakhariya Sakina et al., 2023).

18.5 CLINICAL DIAGNOSTICS OF DENGUE FEVER

As dengue infection is not new, several detection methods are valid, and the detection methods are still under clinical trials. Furthermore, two methods can be utilized for detecting dengue: diagnostic laboratories and anti-dengue antibody testing (Roy & Bhattacharjee, 2021). However, serological examinations, predominantly serum and cerebrospinal fluids, are extensively employed to identify the infection (Malik et al., 2023). These tests are low-cost and most frequently used to test the presence of immunoglobin M (IgM). Other tests besides the serological tests include IgM-based, IgG, hemagglutination inhibition, NS1-based, and polymerase chain reaction. During the infection, IgM is produced within approximately seven days. Primary dengue infection has a higher IgM response, and capturing this MAC ELISA test is most commonly used (Kabir et al., 2021). Conversely, subsequent infections result in the overproduction of IgG. IgG-based ELISA can be useful for standard diagnostics and can be done approximately 10–15 days after symptoms (Blacksell et al., 2012). ELISA tools are developed within software environments like R to address variability between ELISA kits, treating it as a batch effect. This correction helps improve biomarker reproducibility and ensures more reliable results across different batches or lots of ELISA kits. Computational biology is pivotal for enhancing ELISA kits' performance, reliability, and reproducibility for detecting biomarkers related to dengue virus infection. By utilizing computational tools for primer design, batch effect correction, data analysis, and assay optimization, researchers can develop highly sensitive and specific ELISA assays essential for effectively diagnosing and monitoring dengue infections (Gandikota et al., 2021). However, these IgM and IgG diagnostics are not as dependable for detecting primary infection as a consequence of cross-reactivity with alternative *flavivirus* strains (Wellekens et al., 2022). However, a standard test called hemagglutinin inhibition can be utilized to differentiate between primary and secondary illnesses. However, IgG-based tests are better and easier to perform for secondary infection than hemagglutinin inhibition. Also, HI fails to differentiate from other flaviviruses. Nevertheless, the NS1 antigen is crucial for DENV replication within the recipient cell. NS1 tests are mainly essential to detect the particular phase of dengue. However, antigen capture ELISA is primarily used, and some commercial types of immunoassays and ELISA kits are also used to detect NS1. In addition, NS1 can be done within approximately seven days of infection (Kok et al., 2023). The most common test for the dengue virus, though, is RT-PCR.

This is the best way to find the virus in blood, serum, plasma, and urine samples. It typically targets all dengue virus strains and can distinguish between them. The RT-PCR approach can detect infection early but is time-consuming and requires complex procedures. In certain situations, NS1 recognition is preferable to viral RNA undetectable by RT-PCR. RT-PCR techniques primarily recognize dengue viruses via the utilization of a computational approach. This entails the development of specific primers and probes, as well as the identification of the targeted sequence. It is susceptible and specific, which allows the detection of low levels of viral clinical RNA samples. It detects the amplicons from the whole genome sequence data. However, sequencing plays the most crucial role in PCR and RT-PCR techniques. Advances in PCR technology have led to the development of novel assays that automate sample extraction to amplification processes, improving efficiency and accuracy in DENV detection (Kabir et al., 2021). Moreover, commercial kits are also available on the market that can be primarily employed to detect NS1 as a possible early-stage biomarker for DENV (Kok et al., 2023).

18.6 DEVELOPMENT OF VACCINES FOR DENGUE FEVER

The development of DENV vaccines that are effective and safe has encountered significant challenges as a result of the virus' intricacy and its demand for full protection against all four serotypes. The sustained nature of dengue fever has posed a significant healthcare cost over the years, implicating the anti-dengue vaccine against them. Different vaccine candidates, namely, Dengvaxia, TV005, and DENVax, are undergoing trials. However, TAK-003 is getting licensed in Japan and actively works against four serotypes (Torresi et al., 2017; Wellekens et al., 2022). Also, WHO advocates using Dengvaxia in people aged 9–45 who live in endemic areas, but other formulations against DENV are in the experimental phases (Roy & Bhattacharjee, 2021; Wellekens et al., 2022). Unfortunately, even when all vaccine criteria are met, complete protection is not assured. Developing a tetravalent vaccine that offers sustained immunity to all four serotypes of DENV is critical. Several studies have focused on developing epitope-based vaccines, and studies have utilized a multi-epitope approach to design vaccines that target specific epitopes from different DENV proteins. By combining B-cell and T-cell epitopes, these vaccines aim to elicit a robust response to dengue virus (DENV) using immunoinformatics and computational approaches (Mishra et al., 2023). Epitope-based vaccine candidates are evaluated for their antigenicity, solubility, allergenicity, toxicity, flexibility, and interaction with immune receptors like TLR-2, TLR-4, MHC-1, and MHC-2 (Hou et al., 2020; Park et al., 2022). However, by targeting specific epitopes from DENV proteins and utilizing computational modeling techniques, researchers aim to create vaccines that can elicit strong and protective immune responses while minimizing the risk of adverse reactions. While traditional vaccines have shown limited efficacy in managing dengue, epitope-based vaccines represent a novel and potentially more targeted approach to vaccine development. Further research and experimental validation are needed to assess the comparative efficacy of epitope-based dengue vaccines against traditional vaccines and to optimize their design for enhanced protection against all DENV serotypes (Dixit, 2021; Fadaka et al., 2021). The antibody-dependent enhancement in dengue is a well-known feature. In this context, the development of dengue vaccines for all serotypes is a challenging task. In conclusion, understanding ADE against the dengue virus is a significant concern for vaccine development and treating dengue infection, where the computational vaccine approach is essential for vaccine design (Morrone & Lok, 2019; Xu et al., 2021).

18.7 EPIDEMIOLOGICAL MODELING OF DENGUE

Dengue is presently a worldwide infectious disease that spreads via mosquitoes. Dengue has its historical origins in the Chin and Jin dynasties. More than 3.6 billion people are infected by DENV in more than 100 countries yearly. The first documented dengue waves occurred in 1779 in Egypt, Indonesia, Jakarta, and coastal and temperate regions of Asia and Europe. The first verified dengue epidemic in Philadelphia was documented in 1780, spreading throughout North America (Wu et al., 2011). In 2016, the United States experienced its most severe epidemic, with over 2.38 million reported cases and 1.5 million occurrences in Brazil, with the highest contribution (Bifani et al., 2022; Prompetchara et al., 2020; Roy & Bhattacharjee, 2021; Wilson & Chen, 2015). In India, Madras was first affected by a disease similar to dengue in 1780, whereas the first dengue epidemic struck Kolkata and the Eastern coast in 1963–1964. Dengue became a burden after World War II following two outbreaks in 1953 and 1956 due to rapid urbanization, population growth, and global warming. Dengue fever cases have exhibited an increasing pattern in the Americas, Southeast Asia, the eastern Mediterranean, and the West Pacific regions since 1970 (Messina et al., 2014). About 50–100 million individuals are believed to be infected with dengue fever. There is a worldwide fever outbreak in a year, with a mortality rate of 1–5% without treatment and less than 1% with treatment. The prevalence of dengue globally has experienced an enormous increase in recent decades, as the WHO report indicates that the number of incidents has escalated from 505 to 430 in 2000 to 5.2 million in 2019. The latest dengue outbreak, mainly from 2021, created a desolation on the health system, which was already suffering from the COVID-19 pandemic. A recent research report assessing the potential exposure to dengue viruses suggests that dengue infection poses a risk to an estimated 3.9 billion individuals worldwide. The largest dengue reported globally was in 2019 (Bifani et al., 2022; Redoni et al., 2020). An estimated 3.1 million cases are documented in the American area, with a significant proportion of infections reported (Bangladesh, Malaysia, Philippines, and Vietnam) in Asia (https://www.who.int/). However, epidemiological modeling of dengue fever plays a crucial role in understanding the dynamics of the disease, predicting outbreaks, and evaluating control strategies. In addition, by integrating immunological aspects, vector populations, and serotype interactions into these models, researchers can enhance our understanding of dengue epidemiology and inform evidence-based disease prevention and control interventions. Artificial intelligence (AI) plays an important role in epidemiological modeling, particularly concerning infectious diseases like dengue fever. AI applications in epidemiology primarily focus on predictive analytics, utilizing historical data to forecast disease origins and spread patterns. Ongoing advancements in modeling techniques continue to contribute significantly to our ability to combat dengue fever effectively (Aguiar et al., 2022; Hu et al., 2013). About the presence of various flavivirus strains in circulation, which often results in a situation known as hyperendemicity due to the constant reshuffling of viral species, the scientific community must gain a comprehensive understanding and work toward the development of an effective vaccine strategy to combat dengue virus (DENV) infections (Malik et al., 2023; Redoni et al., 2020).

18.8 CHALLENGES AND FUTURE PROSPECTS

Dengue has become an increasingly severe threat to human health, lacking a highly effective vaccine that covers all virus strains. How individuals gain protection from the dengue virus remains uncertain. This chapter provides readers with a thorough and all-encompassing perspective of the dengue virus, which helps them understand the correct ongoing scenario of this infection. Due to the complex mechanisms of dengue virus infection, worldwide researchers are still facing the design and development of a potential treatment to combat this infection. However, as cases of dengue are increasing daily, worldwide researchers need to utilize advanced approaches and precise methods that can be helpful for the successful development of therapeutics against this infection. Bioinformatics and computational biology offer valuable tools and algorithms for understanding and combating dengue fever, which can also help understand DENV evolution, population dynamics, and transmission patterns. In conclusion, leveraging bioinformatics and computational biology approaches presents promising avenues for addressing the challenges associated with dengue fever. By adopting advanced computational tools, existing limitations can be overcome, and new and more effective ways can be opened to control programs, vaccination strategies, and therapeutic interventions against the dengue virus.

REFERENCES

Aguiar, M., Anam, V., Blyuss, K. B., Estadilla, C. D. S., Guerrero, B. V., Knopoff, D., Kooi, B. W., Srivastav, A. K., Steindorf, V., & Stollenwerk, N. (2022). Mathematical models for dengue fever epidemiology: A 10-year systematic review. *Phys Life Rev*, 40, 65–92. https://doi.org/10.1016/j.plrev.2022.02.001.

Anez, G., Heisey, D. A., Volkova, E., & Rios, M. (2016). Complete genome sequences of dengue virus type 1 to 4 strains used for the development of CBER/FDA RNA reference reagents and WHO international standard candidates for nucleic acid testing. *Genome Announc*, 4(1). https://doi.org/10.1128/genomeA.01583–15.

Bifani, A. M., Siriphanitchakorn, T., & Choy, M. M. (2022). Intra-host diversity of dengue virus in Mosquito vectors. *Frontiers in Cellular and Infection Microbiology*, 12, 888804. https://doi.org/10.3389/fcimb.2022.888804.

Blacksell, S. D., Jarman, R. G., Gibbons, R. V., Tanganuchitcharnchai, A., Mammen, M. P., Jr., Nisalak, A., Kalayanarooj, S., Bailey, M. S., Premaratna, R., de Silva, H. J., Day, N. P., & Lalloo, D. G. (2012). Comparison of seven commercial antigen and antibody enzyme-linked immunosorbent assays for detection of acute dengue infection. *Clin Vaccine Immunol*, 19(5), 804–810. https://doi.org/10.1128/CVI.05717–11.

Botta, L., Rivara, M., Zuliani, V., & Radi, M. (2018). Drug repurposing approaches to fight dengue virus infection and related diseases. *Front Biosci (Landmark Ed)*, 23(6), 997–1019. https://doi.org/10.2741/4630.

Dixit, N. K. (2021). Design of monovalent and chimeric tetravalent dengue vaccine using an immunoinformatics approach. *Int J Pept Res Ther*, 27(4), 2607–2624. https://doi.org/10.1007/s10989-021-10277-x.

Dwivedi, V. D., Tripathi, I. P., Tripathi, R. C., Bharadwaj, S., & Mishra, S. K. (2017). Genomics, proteomics and evolution of dengue virus. *Briefings in Functional Genomics*, 16(4), 217–227. https://doi.org/10.1093/bfgp/elw040.

Fadaka, A. O., Sibuyi, N. R. S., Martin, D. R., Goboza, M., Klein, A., Madiehe, A. M., & Meyer, M. (2021). Immunoinformatics design of a novel epitope-based vaccine candidate against dengue virus. *Sci Rep*, 11(1), 19707. https://doi.org/10.1038/s41598-021-99227-7.

Gandikota, C., Gandhi, L., Maisnam, D., Kesavulu, M. M., Billoria, A., Prasad, V. S. V., & Venkataramana, M. (2021). A novel anti-NS2BNS3pro antibody-based indirect ELISA test for the diagnosis of dengue virus infections. *J Med Virol*, 93(6), 3312–3321. https://doi.org/10.1002/jmv.26024.

Goz, E., Tsalenchuck, Y., Benaroya, R. O., Zafrir, Z., Atar, S., Altman, T., Julander, J., & Tuller, T. (2018). Generation and comparative genomics of synthetic dengue viruses. *BMC Bioinformatics*, 19(Suppl. 6), 140. https://doi.org/10.1186/s12859-018-2132-3.

Guzman, M. G., Gubler, D. J., Izquierdo, A., Martinez, E., & Halstead, S. B. (2016). Dengue infection. *Nat Rev Dis Primers*, 2, 16055. https://doi.org/10.1038/nrdp.2016.55.

Halstead, S. B., Streit, T. G., Lafontant, J. G., Putvatana, R., Russell, K., Sun, W., Kanesa-Thasan, N., Hayes, C. G., & Watts, D. M. (2001). Haiti: Absence of dengue hemorrhagic fever despite hyperendemic dengue virus transmission. *Am J Trop Med Hyg*, 65(3), 180–183. https://doi.org/10.4269/ajtmh.2001.65.180.

Hou, J., Ye, W., Loo, H. L., Wong, L. H., & Chen, J. (2020). Successive immunization with epitope-decreasing dengue antigens induced conservative anti-dengue immune responses. *Frontiers in Immunology*, 11, 585133. https://doi.org/10.3389/fimmu.2020.585133.

Hu, K., Thoens, C., Bianco, S., Edlund, S., Davis, M., Douglas, J., & Kaufman, J. (2013). Modeling the dynamics of dengue fever. In A. M. Greenberg, W. G. Kennedy, & N. D. Bos (Eds.), *Social Computing, Behavioral-Cultural Modeling and Prediction*. Berlin, Heidelberg: Springer.

Jagtap, S., Pattabiraman, C., Sankaradoss, A., Krishna, S., & Roy, R. (2023). Evolutionary dynamics of dengue virus in India. *PLOS Pathogens*, 19(4), e1010862. https://doi.org/10.1371/journal.ppat.1010862.

Kabir, M. A., Zilouchian, H., Younas, M. A., & Asghar, W. (2021). Dengue detection: Advances in diagnostic tools from conventional technology to point of care. *Biosensors (Basel)*, 11(7). https://doi.org/10.3390/bios11070206.

Kok, B. H., Lim, H. T., Lim, C. P., Lai, N. S., Leow, C. Y., & Leow, C. H. (2023). Dengue virus infection—a review of pathogenesis, vaccines, diagnosis and therapy. *Virus Res*, 324, 199018. https://doi.org/10.1016/j.virusres.2022.199018.

Li, Q., & Kang, C. (2022). Structures and dynamics of dengue virus non-structural membrane proteins. *Membranes (Basel)*, 12(2). https://doi.org/10.3390/membranes12020231.

Lim, S. V., Rahman, M. B., & Tejo, B. A. (2011). Structure-based and ligand-based virtual screening of novel methyltransferase inhibitors of the dengue virus. *BMC Bioinformatics*, 12(Suppl. 13), S24. https://doi.org/10.1186/1471-2105-12-S13-S24.

Low, J. G., Gatsinga, R., Vasudevan, S. G., & Sampath, A. (2018). Dengue antiviral development: A continuing journey. *Adv Exp Med Biol*, 1062, 319-332. https://doi.org/10.1007/978-981-10-8727-1_22.

Malik, S., Ahsan, O., Mumtaz, H., Tahir Khan, M., Sah, R., & Waheed, Y. (2023). Tracing down the updates on dengue virus-molecular biology, antivirals, and vaccine strategies. *Vaccines (Basel)*, 11(8). https://doi.org/10.3390/vaccines11081328.

Messina, J. P., Brady, O. J., Scott, T. W., Zou, C., Pigott, D. M., Duda, K. A., Bhatt, S., Katzelnick, L., Howes, R. E., Battle, K. E., Simmons, C. P., & Hay, S. I. (2014). Global spread of dengue virus types: Mapping the 70 year history. *Trends Microbiol*, 22(3), 138–146. https://doi.org/10.1016/j.tim.2013.12.011.

Miao, M., Yu, F., Wang, D., Tong, Y., Yang, L., Xu, J., Qiu, Y., Zhou, X., & Zhao, X. (2019). Proteomics profiling of host cell response via protein expression and phosphorylation upon dengue virus infection. *Virol Sin*, 34(5), 549–562. https://doi.org/10.1007/s12250-019-00131-2.

Mishra, S. K., Priya, P., Rai, G. P., Haque, R., & Shanker, A. (2023). Coevolution based immunoinformatics approach considering variability of epitopes to combat different strains: A case study using spike protein of SARS-CoV-2. *Comput Biol Med*, 163, 107233. https://doi.org/10.1016/j.compbiomed.2023.107233.

Morrone, S. R., & Lok, S. M. (2019). Structural perspectives of antibody-dependent enhancement of infection of dengue virus. *Curr Opin Virol*, 36, 1–8. https://doi.org/10.1016/j.coviro.2019.02.002.

Mustafa, M. S., Rasotgi, V., Jain, S., & Gupta, V. (2015). Discovery of fifth serotype of dengue virus (DENV-5): A new public health dilemma in dengue control. *Med J Armed Forces India*, 71(1), 67–70. https://doi.org/10.1016/j.mjafi.2014.09.011.

Nncube, N. B., Ramharack, P., & Soliman, M. E. S. (2018). Using bioinformatics tools for the discovery of dengue RNA-dependent RNA polymerase inhibitors. *PeerJ*, 6, e5068. https://doi.org/10.7717/peerj.5068.

Norazharuddin, H., & Lai, N. S. (2018). Roles and prospects of dengue virus non-structural proteins as antiviral targets: An easy digest. *Malays J Med Sci*, 25(5), 6–15. https://doi.org/10.21315/mjms2018.25.5.2.

Park, J., Kim, J., & Jang, Y. S. (2022). Current status and perspectives on vaccine development against dengue virus infection. *J Microbiol*, 60(3), 247–254. https://doi.org/10.1007/s12275-022-1625-y.

Perera, R., & Kuhn, R. J. (2008). Structural proteomics of dengue virus. *Curr Opin Microbiol*, 11(4), 369–377. https://doi.org/10.1016/j.mib.2008.06.004.

Prompetchara, E., Ketloy, C., Thomas, S. J., & Ruxrungtham, K. (2020). Dengue vaccine: Global development update. *Asian Pac J Allergy Immunol*, 38(3), 178–185. https://doi.org/10.12932/AP-100518–0309.

Rahim, R., Hasan, A., Phadungsombat, J., Hasan, N., Ara, N., Biswas, S. M., Nakayama, E. E., Rahman, M., & Shioda, T. (2023). Genetic analysis of dengue virus in severe and non-severe cases in Dhaka, Bangladesh, in 2018–2022. *Viruses*, 15(5). https://doi.org/10.3390/v15051144.

Redoni, M., Yacoub, S., Rivino, L., Giacobbe, D. R., Luzzati, R., & Di Bella, S. (2020). Dengue: Status of current and under-development vaccines. *Rev Med Virol*, 30(4), e2101. https://doi.org/10.1002/rmv.2101.

Roy, S. K., & Bhattacharjee, S. (2021). Dengue virus: Epidemiology, biology, and disease aetiology. *Canadian Journal of Microbiology*, 67(10), 687–702.

Shankaranarayana, A. H., Pujar, G. V., Sethu, A. K., Bhagyalalitha, M., & Singh, M. (2021). Dengue structural proteins as antiviral drug targets: Current status in the drug discovery & development. *Eur J Med Chem*, 221, 113527. https://doi.org/10.1016/j.ejmech.2021.113527.

Shimizu, H., Saito, A., Mikuni, J., Nakayama, E. E., Koyama, H., Honma, T., Shirouzu, M., Sekine, S. I., & Shioda, T. (2019). Discovery of a small molecule inhibitor targeting dengue virus NS5 RNA-dependent RNA polymerase. *PLoS Negl Trop Dis*, 13(11), e0007894. https://doi.org/10.1371/journal.pntd.0007894.

Teo, A., Tan, H. D., Loy, T., Chia, P. Y., & Chua, C. L. L. (2023). Understanding antibody-dependent enhancement in dengue: Are afucosylated IgG1s a concern? *PLOS Pathogens*, 19(3), e1011223. https://doi.org/10.1371/journal.ppat.1011223.

Torresi, J., Ebert, G., & Pellegrini, M. (2017). Vaccines licensed and in clinical trials for the prevention of dengue. *Hum Vaccin Immunother*, 13(5), 1059–1072. https://doi.org/10.1080/21645515.2016.1261770.

Troost, B., & Smit, J. M. (2020). Recent advances in antiviral drug development towards dengue virus. *Curr Opin Virol*, 43, 9–21. https://doi.org/10.1016/j.coviro.2020.07.009.

Vaghasia, V. V., Sharma, K., Mishra, S. K., & Georrge, J. J. (2024). In silico identification of natural product inhibitor for multidrug resistance proteins from selected gram-positive bacteria. In *Nanotechnology and In Silico Tools* (pp. 309–317). Elsevier.

Vakhariya, S., Kumar Mishra, S., & Georrge, J. J. (2023). Identification of novel curcumin analogues to inhibit mitogen-activated protein kinase 1 using in silico combinatorial library design and molecular docking approach. In *Proceedings of XXIX Symposium on Bioinformatics and Computer-Aided Drug Discovery*. September 18–20, Institute of Biomedical Chemistry, Moscow, Russia, pp. 106. ISBN: 978-5-900760-19-3.

Vakhariya Sakina, S., Mishra, S. K., Sharma, K., & Georrge, J. J. (2023). Designing of a novel curcumin analogue to inhibit mitogen-activated protein kinase: A cheminformatics approach. *Journal of Phytonanotechnology and Pharmaceutical Sciences*, 3(1), 37–47.

Vinjoda, P., Mishra, S. K., Sharma, K., & Georrge, J. J. (2024). In silico identification of novel drug target and its natural product inhibitors for herpes simplex virus. In *Nanotechnology and In Silico Tools* (pp. 377–383). Elsevier.

Wellekens, K., Betrains, A., De Munter, P., & Peetermans, W. (2022). Dengue: Current state one year before WHO 2010–2020 goals. *Acta Clin Belg*, 77(2), 436–444. https://doi.org/10.1080/17843286.2020.1837576.

Wilson, M. E., & Chen, L. H. (2015). Dengue: Update on epidemiology. *Curr Infect Dis Rep*, 17(1), 457. https://doi.org/10.1007/s11908-014-0457-2.

Wu, W., Bai, Z., Zhou, H., Tu, Z., Fang, M., Tang, B., Liu, J., Liu, L., Liu, J., & Chen, W. (2011). Molecular epidemiology of dengue viruses in southern China from 1978 to 2006. *Virol Journal*, 8, 322. https://doi.org/10.1186/1743–422X-8–322.

Xie, L. M., Yin, X., Bi, J., Luo, H. M., Cao, X. J., Ma, Y. W., Liu, Y. L., Su, J. W., Lin, G. L., & Guo, X. G. (2021). Identification of potential biomarkers in dengue via integrated bioinformatic analysis. *PLoS Negl Trop Dis*, 15(8), e0009633. https://doi.org/10.1371/journal.pntd.0009633.

Xu, L., Ma, Z., Li, Y., Pang, Z., & Xiao, S. (2021). Antibody dependent enhancement: Unavoidable problems in vaccine development. *Adv Immunol*, 151, 99–133. https://doi.org/10.1016/bs.ai.2021.08.003.

19 Bioinformatics and Computational Biology–Based Approaches in Research and Clinical Management of Emerging Viral Zoonoses (LHF–NiVF)

Saurav Kumar Mishra, Tabsum Chhetri, Ani Johnson, and John J. Georrge

19.1 INTRODUCTION

The term "zoonosis" was first coined by German physician and researcher Rudolph Virchow (1821–1902). Zoonotic diseases, which stem from animals but can be transferred to people, are an enormous threat to global public health (Vourc'h et al., 2022). Lassa fever and Nipah virus infection are two well-known zoonoses that have received recent media interest. These illnesses highlight the complex interactions between people, animals, and the environment and how important it is to comprehend the mechanics of zoonotic transmission to create preventative and control strategies that work. Lassa fever, caused by the Lassa virus primarily carried by rodents, is endemic to West Africa and significantly burdens affected communities. Similarly, the Nipah virus, transmitted from fruit bats to humans through intermediate hosts like pigs, has caused outbreaks in South and Southeast Asia with high fatality rates. The origin and transmission of zoonotic pathogens are facilitated by the interdependence of ecosystems, animals, livestock, and human populations, as both illnesses highlight (Richmond, 2003; Singh et al., 2019). Various factors influence the appearance of zoonotic virus outbreaks, encompassing animal-to-human contact and significant environmental shifts. Important contributors to viral disease epidemics include forest destruction, natural environment modifications, warming temperatures, manufacturing, and depletion of natural resources (Soman Pillai et al., 2020).

19.2 GENOMIC ANALYSIS

With the recent development in computational biology, it is easier to perform genomic surveillance of zoonotic diseases. This section focuses on genomics and its advancements. Predictive modeling and genomic surveillance are crucial instruments for tracking and forecasting pathogen behavior, particularly in relation to zoonotic illnesses. This real-time surveillance makes the identification of novel strains with elevated transmissibility or potential for virulence possible (Becker et al., 2022a; Mollentze & Streicker, 2023a). Finding the pathogen causing the illness is the first step in diagnosing and treating any disease. This can be done quickly using genomic technologies like sequencing, array-based single nucleotide polymorphism, and m(RNA) expression profiling (Haagmans et al., 2009;

DOI: 10.1201/9781032721811-19

Happi et al., 2024). NiVF is a paramyxovirus with a high fatality rate that causes zoonotic diseases spread mostly by bats. qRT-PCR is a sensitive approach for diagnosing acute NiVF infections by detecting the N gene in materials such as nasal swabs, cerebrospinal fluid, urine, and blood. The RNA genome of LHF, a negatively polarized single-stranded arenavirus, is divided into two segments and encodes various proteins, including Z, L, NP, and GPC. The phylogenetic study contributes to understanding LHF genetic variation across lineages and locales (Becker et al., 2022a; Mollentze & Streicker, 2023a).

19.3 HOST–PATHOGEN INTERACTION

Zoonotic diseases pass from animals to humans (Vourc'h et al., 2022). Computational biology and bioinformatics have completely transformed the investigation of host–pathogen interactions in zoonotic illnesses. Predicting how host components interact with pathogen proteins, locating possible therapeutic targets, and comprehending the mechanisms behind disease and possible intervention sites have all grown dependent on these methods. Applying systems biology models is a fundamental component of computational biology in host–pathogen interactions. These models aid in illuminating the molecular mechanisms underlying pathogen adaptability to their hosts, pathogen immune evasion, and host immunological response (Durmuş et al., 2015). Computational analysis of pathogen–host interaction networks (PHI networks) offers insights into the mechanics of host–pathogen interactions by examining the characteristics of pathogen-targeted host proteins (Cesur & Durmuş, 2018). System-level analysis of host–pathogen interactions is made possible by various computational techniques. This facilitates the mining and utilization of network data to obtain a more profound comprehension of the complex interactions which transpire between pathogens and host cells. Systems biology has been used to unravel the networks of interactions between hosts and pathogens, resulting in medicinal applications that enhance treatment and diagnosis, therefore raising the likelihood of survival (Cesur & Durmuş, 2018; Yeung et al., 2019). Moreover, as the zoonotic virus emerges and re-emerges frequently, computational approaches will help understand the evolutionary relationship and the interactive target with the host for therapeutic development networks.

19.4 EPIDEMIOLOGICAL MODELING

The introduction and transmission of zoonoses cause serious threats to public health worldwide. Effective control methods can be developed using epidemiological modeling, which is essential for comprehending these diseases' geographic distribution and transmission dynamics. The use of computational models in modeling transmission dynamics and spatial analysis of newly emerging zoonotic illnesses is covered in this section. Computer models, such as spatial regression models and cluster detection approaches, aid in the identification of environmental, socioeconomic, and ecological risk factors for disease transmission. These models can shed light on the intricate interconnections that drive zoonotic disease dynamics by combining data from animal reservoirs, vector habitats, and human populations. For example, research on LHV has used spatial analysis to pinpoint high-risk locations in West Africa, leading surveillance efforts and public health measures (Jato-Espino et al., 2023; Nwafor et al., 2021). The epidemic of Lassa fever especially prevalent in West Africa, resulting in a considerable number of illnesses and deaths each year. The disease has been detected in various West African countries and provides complications due to its variable symptoms and challenging clinical diagnosis (https://www.who.int/health-topics/lassa-fever#tab=tab_1). An arenavirus causes Lassa fever, and individuals usually get it by contacting food or household goods that have been tainted by the urine or feces of *Mastomy* rats, which have become infectious. Infections between individuals can also happen, particularly in hospital environments with insufficient infection control protocols (Richmond, 2003). Nipah infection is mostly caused by *Pteropus* spp. Fruit bats were initially found in Malaysia in 1998. NiVF is a threat to all individuals around the world because it has a high death rate, no effective treatment, and can give people dangerous encephalitis and serious respiratory illnesses. Bats of Pteropus Nipah virus epidemics have been connected to *Pteropus vampyrus*, *hypomelanus*, *lylei*, and *gigantea* in the countries of Bangladesh, Cambodia, East Timor, Indonesia, India, Malaysia, Papua New Guinea, Vietnam, and Thailand. The first instance, pigs served as intermediary hosts (Aditi & Shariff, 2019; Singh et al., 2019). In 2004, humans in Bangladesh contracted NiVF by consuming date palm juice contaminated by sick fruit bats. Additionally, transmission from person to person has been recorded, most notably in the Indian region, through clinical testing (Verma et al., 2018).

19.5 VACCINE AND DRUG DEVELOPMENT

Computational techniques are employed in the drug development process to search chemical libraries for potential antiviral medicines. These methods use virtual screening algorithms and molecular modeling techniques to estimate chemical binding affinity to viral targets. Computational methods greatly speed up drug development by reducing the amount of substances required to be subjected to tested in investigations. This is done by simulating how these chemicals interact with viral proteins that are being studied (Cerqueira et al., 2010; Gentile et al., 2022; Murgueitio et al., 2012; S et al., 2023). For the purpose of predicting zoonotic viral proteins' epitope-specific regions that have the ability to trigger an immune response, computational methods are crucial. These methods detect possible immune system targets by evaluating the sequencing and structure of viral proteins. This information is critical for developing vaccines that can elicit a strong immune response to zoonotic viruses. Here are some important tools and methodologies for epitope prediction: DiscoTope predicts discontinuous B-cell epitopes by examining 3D protein structure and is also useful for developing medications, peptide-based vaccinations, and diagnostic tools (http://tools.iedb.org/discotope/). ElliPro utilized 3D model of a target antigen to predict both discontinuous and linear epitopes (http://tools.iedb.org/ellipro/). Machine learning and AI are the foundation of HLAP. MHC-I binding peptides are predicted using ANN and quantitative matrices (QM). Support vector machines (SVMs) are used to separate data into groups and speed up the discovery of T-cell epitopes. These computational techniques use a variety of ways to study peptide interactions and map B and T-cell epitopes in pathogens, including motif-based systems, QSAR analysis, structure-based methods, and different advance algorithms (S. K. Mishra et al., 2023; Soria-Guerra et al., 2015; Sunita et al., 2020). The unavailability of a licensed medication or vaccine, especially for NiVF illness, is a severe problem. Although it is not a specific treatment for NiVF, ribavirin has been utilized as a management tactic in certain situations. Amidst the numerous RNA viruses is the respiratory syncytial virus (RSV). Wide-ranging medication known as ribavirin has proven effective against this. Nevertheless, its effectiveness against NiVF is still under investigation, and its use is still experimental (Chong et al., 2001; Johnson et al., 2021a; Vaghasia et al., 2024). Supportive care is an essential component of treatment for NiVF. This includes treating the resulting symptoms and issues, like respiratory distress. In addition to additional steps to stabilize the patient, intensive supportive care may involve IV infusions to maintain hydration and mechanical ventilation for respiratory support. Ongoing research and development is being done to find certain antiviral medications and create NiVF vaccines. The difficulties stem from the virus' complexity and our incomplete knowledge of its pathophysiology. In the interim, medical professionals are forced to depend on ribavirin, supportive care, and investigative therapies to help NiVF patients live better lives (Chong et al., 2001; Vinjoda et al., 2024). Amid the NiVF epidemic that swept Malaysia, some infected people were administered ribavirin, which reduced their mortality rate by 36%. In the recent 2018 outbreak in Kerala, two patients who got ribavirin orally showed signs of recovery. However, in an animal model, this medication had no impact. Remdesivir and favipiravir are the other two antivirals being tested in animal models, with favipiravir showing promising results in the Syrian golden hamster model (Freiberg et al., 2010; Johnson et al., 2021b). Griffithsin, also known as GRFT, is a red algae-derived protein that inhibits NiVF's ability to enter and fuse with the host cell, exhibiting antiviral properties. In the case of passive immunization, human monoclonal antibody m102.4 is isolated, which is specific for G protein binding epitope in the ephrin B2/B3 and has shown neutralizing activity against NiVF (G. Mishra et al., 2023; Thakur & Bailey, 2019). For the purpose of treating Lassa fever, antiviral drugs like ribavirin have been used; studies have shown that ribavirin works best when used early in the course of an illness. It is essential for reducing mortality and enhancing patient results. Patients with Lassa fever need to take antiviral drugs based on ribavirin and supportive therapy. This includes treating complex infections, ensuring enough oxygenation and blood pressure, and keeping the right balance of electrolytes and water. Treating the symptoms and aftereffects of Lassa fever requires supportive treatment (Hansen et al., 2021; Salam et al., 2022).

19.6 CLINICAL DIAGNOSTICS AND SURVEILLANCE

Surveillance and early diagnosis of freshly emerging zoonotic diseases are crucial for averting outbreaks and reducing public health consequences (Figure 19.1). Genomic surveillance and predictive modeling are critical techniques for monitoring and forecasting pathogen behavior, particularly zoonotic illnesses. Researchers can use computational biology methods to continuously examine genomic sequences for alterations that may affect pathogen activity. This real-time monitoring enables the detection of developing strains with greater transmissibility or virulence potential. Predictive modeling combines genetic data with epidemiological and environmental parameters to predict spillover risk and prioritize surveillance efforts in high-risk areas or host populations. Models predict zoonotic risk by considering virus characteristics, host range, and genome composition. These models assist researchers in understanding the possible influence of genetic sequence changes on the spread and severity of zoonotic illnesses, allowing for the creation of tailored surveillance and control strategies (Becker et al., 2022b; Mollentze & Streicker, 2023b). NiVF poses a significant healthcare concern owing to its exceptionally pathogenic characteristics and inclination for transmission between individuals through body fluids. The difficulty with NiVF is its symptoms might be vague and similar to other common ailments such as fever, nausea, and respiratory issues. Some people may even carry the virus without showing symptoms (asymptomatic instances), making early detection even more difficult. To address these issues, a range of laboratory procedures for diagnosing NiVF are used; one of the most common tests used to diagnose NiVF is ELISA (enzyme-linked immunosorbent assay); apart from this, immunohistochemistry, histopathology, and PCR (polymerase chain reaction) are some of the techniques used for diagnosis (Ambat et al., 2019; Thakur & Bailey, 2019) Early detection is critical for regulating and limiting NiVF transmission. These several laboratory tests allow for a more thorough approach to diagnosis, considering the virus' various symptoms and probable asymptomatic instances. It emphasizes the significance of a multifaceted diagnostic technique in dealing with the issues faced by this hazardous virus (Mazzola & Kelly-Cirino, 2019). Nipah is a zoonotic virus, and understanding how the virus spreads is essential for monitoring and controlling transmission. Most infections in Malaysia spread from swine to individuals via bodily secretions (blood, sputum). Because the afflicted individuals were primarily those with intimate proximity to the swine, the pig farmer and male abattoirs were the primary individuals exposed to the disease (Sahani et al., 2001). Although contact between individuals was the predominant mode of spread in Bangladesh and India, bats initially transmitted the virus to humans. NiVF originated from bats to individuals via excretory secretions and infected produce, while NiVF was passed from humans to humans via body fluids. Most of the cases of NiVF were fatal, with the risk of encephalitis (Hassan et al., 2018; Plowright et al., 2019). Malaysia, one of the countries where pig farming is an important sector, experienced losses during the outbreak. Various control measures were implemented, including culling millions of pigs and their disposal by deep burial and fast lime decontamination. A swine imported from an infected

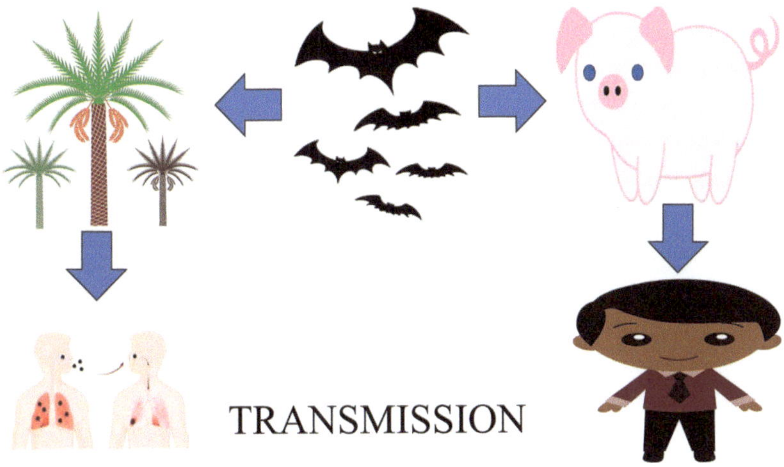

FIGURE 19.1 Representation of the basic transmission of the NiV.

region of Malaysia was responsible for the pandemic in Singapore. Singapore's government also implemented control measures, including culling pigs and prohibiting imports (Aditi & Shariff, 2019; Sharma et al., 2019; Singh et al., 2019). The entire region was quarantined during the outbreak in India and Bangladesh to prevent further transmission. Attempts were made to avoid bat–human transmission by keeping bats away from date palm trees and fruits (https://www.who.int/). The vague symptoms of Lassa fever make clinical identification difficult since they can mimic other diseases such as Ebola, malaria, and typhoid fever. Early detection is important yet difficult, especially without precise diagnostic testing (https://www.who.int/news-room/fact-sheets/detail/lassa-fever). A definitive diagnosis of Lassa fever can only be made through laboratory tests, which can only be done in reference labs. Testing involves detecting antigens, RT-PCR, antibody enzyme-linked immunosorbent assays (ELISA), and extracting viral genomes using cell culture (Emperador et al., 2019; Richmond, 2003).

19.7 CHALLENGES AND FUTURE PROSPECTS

This chapter highlights the complex issues raised by NiVF and Lassa fever. The lack of vaccines and difficulty in controlling the transmission are other challenges. Although several antiviral drugs, including ribavirin, have been studied, the lack of conclusive therapeutic sheds light on the gap in our understanding. More research and clinical trials are necessary to fully understand the complexities of NiVF–host interactions and find focused therapies to lessen the severity of NiVF infections. Many advancements have been made with the help of bioinformatics and computational biology, and the challenges of bulk data integration and standardization are to be noticed. Apart from these, being well-equipped with the technologies and their functions is also essential. Finally, by facilitating the identification of important viral proteins, improving the development of prospective vaccines, and boosting drug discovery initiatives, bioinformatics and computational biology have completely transformed the research of zoonotic illnesses like Lassa fever and Nipah virus. These methods show potential for resolving the issues these fatal viruses present and open the door for additional study and treatments in the future.

REFERENCES

Aditi, & Shariff, M. (2019). Nipah virus infection: A review. *Epidemiology and Infection*, 147, e95. https://doi.org/10.1017/S0950268819000086.

Ambat, A. S., Zubair, S. M., Prasad, N., Pundir, P., Rajwar, E., Patil, D. S., & Mangad, P. (2019). Nipah virus: A review on epidemiological characteristics and outbreaks to inform public health decision making. *Journal of Infection and Public Health*, 12(5), 634–639. https://doi.org/10.1016/j.jiph.2019.02.013.

Becker, D. J., Albery, G. F., Sjodin, A. R., Poisot, T., Bergner, L. M., Chen, B., Cohen, L. E., Dallas, T. A., Eskew, E. A., Fagre, A. C., Farrell, M. J., Guth, S., Han, B. A., Simmons, N. B., Stock, M., Teeling, E. C., & Carlson, C. J. (2022a). Optimising predictive models to prioritise viral discovery in zoonotic reservoirs. *The Lancet Microbe*, 3(8), e625–e637. https://doi.org/10.1016/S2666–5247(21)00245-7.

Cerqueira, N. M. F. S. A., Sousa, S. F., Fernandes, P. A., & Ramos, M. J. (2010). *Virtual Screening of Compound Libraries* (pp. 57–70). https://doi.org/10.1007/978-1-60761-244-5_4.

Cesur, M. F., & Durmuş, S. (2018). *Systems Biology Modeling to Study Pathogen–Host Interactions* (pp. 97–112). https://doi.org/10.1007/978-1-4939-7604-1_10.

Chong, H.-T., Kamarulzaman, A., Tan, C.-T., Goh, K.-J., Thayaparan, T., Kunjapan, S. R., Chew, N.-K., Chua, K.-B., & Lam, S.-K. (2001). Treatment of acute Nipah encephalitis with ribavirin. *Annals of Neurology*, 49(6), 810–813. https://doi.org/10.1002/ana.1062.

Durmuş, S., Çakır, T., Özgür, A., & Guthke, R. (2015). A review on computational systems biology of pathogen-host interactions. *Frontiers in Microbiology*, 6. https://doi.org/10.3389/fmicb.2015.00235.

Emperador, D. M., Yimer, S. A., Mazzola, L. T., Norheim, G., & Kelly-Cirino, C. (2019). Diagnostic applications for Lassa fever in limited-resource settings. *BMJ Global Health*, 4(Suppl. 2), e001119. https://doi.org/10.1136/bmjgh-2018–001119.

Freiberg, A. N., Worthy, M. N., Lee, B., & Holbrook, M. R. (2010). Combined chloroquine and ribavirin treatment does not prevent death in a hamster model of Nipah and Hendra virus infection. *Journal of General Virology*, 91(3), 765–772. https://doi.org/10.1099/vir.0.017269–0.

Gentile, F., Yaacoub, J. C., Gleave, J., Fernandez, M., Ton, A.-T., Ban, F., Stern, A., & Cherkasov, A. (2022). Artificial intelligence–enabled virtual screening of ultra-large chemical libraries with deep docking. *Nature Protocols*, 17(3), 672–697. https://doi.org/10.1038/s41596-021-00659-2.

Haagmans, B. L., Andeweg, A. C., & Osterhaus, A. D. M. E. (2009). The application of genomics to emerging zoonotic viral diseases. *PLoS Pathogens*, 5(10), e1000557. https://doi.org/10.1371/journal.ppat.1000557.

Hansen, F., Jarvis, M. A., Feldmann, H., & Rosenke, K. (2021). Lassa virus treatment options. *Microorganisms*, 9(4), 772. https://doi.org/10.3390/microorganisms9040772.

Happi, A. N., et al. (2024). Lassa virus in novel hosts: Insights into the epidemiology of Lassa virus infections in southern Nigeria. *Emerging Microbes & Infections*, 13(1). https://doi.org/10.1080/22221751.2023.2294859.

Hassan, M. Z., Sazzad, H. M. S., Luby, S. P., Sturm-Ramirez, K., Bhuiyan, M. U., Rahman, M. Z., Islam, M. M., Ströher, U., Sultana, S., Kafi, M. A. H., Daszak, P., Rahman, M., & Gurley, E. S. (2018). Nipah virus contamination of hospital surfaces during outbreaks, Bangladesh, 2013–2014. *Emerging Infectious Diseases*, 24(1), 15–21. https://doi.org/10.3201/eid2401.161758.

Jato-Espino, D., Mayor-Vitoria, F., Moscardó, V., Capra-Ribeiro, F., & Bartolomé del Pino, L. E. (2023). Toward one health: A spatial indicator system to model the facilitation of the spread of zoonotic diseases. *Frontiers in Public Health*, 11. https://doi.org/10.3389/fpubh.2023.1215574.

Johnson, K., Vu, M., & Freiberg, A. N. (2021a). Recent advances in combating Nipah virus. *Faculty Reviews*, 10. https://doi.org/10.12703/r/10–74.

Mazzola, L. T., & Kelly-Cirino, C. (2019). Diagnostics for Nipah virus: A zoonotic pathogen endemic to Southeast Asia. *BMJ Global Health*, 4(Suppl. 2), e001118. https://doi.org/10.1136/bmjgh-2018–001118.

Mishra, G., Prajapat, V., & Nayak, D. (2023). Advancements in nipahvirus treatment: Analysis of current progress in vaccines, antivirals, and therapeutics. *Immunology*. https://doi.org/10.1111/imm.13695.

Mishra, S. K., Priya, P., Rai, G. P., Haque, R., & Shanker, A. (2023). Coevolution based immunoinformatics approach considering variability of epitopes to combat different strains: A case study using spike protein of SARS-CoV-2. *Comput Biol Med*, 163, 107233. https://doi.org/10.1016/j.compbiomed.2023.107233.

Mollentze, N., & Streicker, D. G. (2023). Predicting zoonotic potential of viruses: Where are we? *Current Opinion in Virology*, 61, 101346. https://doi.org/10.1016/j.coviro.2023.101346.

Murgueitio, M. S., Bermudez, M., Mortier, J., & Wolber, G. (2012). In silico virtual screening approaches for antiviral drug discovery. *Drug Discovery Today: Technologies*, 9(3), e219–e225. https://doi.org/10.1016/j.ddtec.2012.07.009.

Nwafor, C. D., et al. (2021). The one health approach to incident management of the 2019 Lassa fever outbreak response in Nigeria. *One Health*, 13, 100346. https://doi.org/10.1016/j.onehlt.2021.100346.

Plowright, R. K., Becker, D. J., Crowley, D. E., Washburne, A. D., Huang, T., Nameer, P. O., Gurley, E. S., & Han, B. A. (2019). Prioritising surveillance of Nipah virus in India. *PLOS Neglected Tropical Diseases*, 13(6), e0007393. https://doi.org/10.1371/journal.pntd.0007393.

Richmond, J. K. (2003). Lassa fever: Epidemiology, clinical features, and social consequences. *BMJ*, 327(7426), 1271–1275. https://doi.org/10.1136/bmj.327.7426.1271.

Sahani, M., Parashar, U. D., Ali, R., Das, P., Lye, M., Isa, M. M., Arif, M. T., Ksiazek, T. G., & Sivamoorthy, M. (2001). Nipah virus infection among abattoir workers in Malaysia, 1998–1999. *International Journal of Epidemiology*, 30(5), 1017–1020. https://doi.org/10.1093/ije/30.5.1017.

Salam, A. P., Duvignaud, A., Jaspard, M., Malvy, D., Carroll, M., Tarning, J., Olliaro, P. L., & Horby, P. W. (2022). Ribavirin for treating Lassa fever: A systematic review of pre-clinical studies and implications for human dosing. *PLOS Neglected Tropical Diseases*, 16(3), e0010289. https://doi.org/10.1371/journal.pntd.0010289.

Shankaranarayana, V. S., Mishra, S. K., Sharma, K., & Georrge, J. J. (2023). Designing of a novel curcumin analogue to inhibit mitogen-activated protein kinase: A cheminformatics approach. *Journal of Phytonanotechnology and Pharmaceutical Sciences*, 3(1). https://doi.org/10.54085/jpps.2023.3.1.5.

Sharma, V., Kaushik, S., Kumar, R., Yadav, J. P., & Kaushik, S. (2019). Emerging trends of Nipah virus: A review. *Reviews in Medical Virology*, 29(1). https://doi.org/10.1002/rmv.2010.

Singh, R. K., Dhama, K., Chakraborty, S., Tiwari, R., Natesan, S., Khandia, R., Munjal, A., Vora, K. S., Latheef, S. K., Karthik, K., Singh Malik, Y., Singh, R., Chaicumpa, W., & Mourya, D. T. (2019). Nipah virus: Epidemiology, pathology, immunobiology and advances in diagnosis, vaccine designing and control strategies – a comprehensive review. *Veterinary Quarterly*, 39(1), 26–55. https://doi.org/10.1080/01652176.2019.1580827.

Soman Pillai, V., Krishna, G., & Valiya Veettil, M. (2020). Nipah virus: Past outbreaks and future containment. *Viruses*, 12(4), 465. https://doi.org/10.3390/v12040465.

Soria-Guerra, R. E., Nieto-Gomez, R., Govea-Alonso, D. O., & Rosales-Mendoza, S. (2015). An overview of bioinformatics tools for epitope prediction: Implications on vaccine development. *Journal of Biomedical Informatics*, 53, 405–414. https://doi.org/10.1016/j.jbi.2014.11.003.

Sunita, S. A., Singh, Y., & Shukla, P. (2020). Computational tools for modern vaccine development. *Human Vaccines & Immunotherapeutics*, 16(3), 723–735. https://doi.org/10.1080/21645515.2019.1670035.

Thakur, N., & Bailey, D. (2019). Advances in diagnostics, vaccines and therapeutics for Nipah virus. *Microbes and Infection*, 21(7), 278–286. https://doi.org/10.1016/j.micinf.2019.02.002.

Vaghasia, V. V., Sharma, K., Mishra, S. K., & Georrge, J. J. (2024). In silico identification of natural product inhibitor for multidrug resistance proteins from selected gram-positive bacteria. In *Nanotechnology and In Silico Tools* (pp. 309–317). Elsevier. https://doi.org/10.1016/B978-0-443-15457-7.00015-0.

Verma, M. K., Verma, P., Singh, S., Gaur, P., Siddiqui, A. H., & Pandey, S. (2018). Nipah virus-infectious agent: An overview. *International Journal of Life-Sciences Scientific Research*, 4(3), 1844–1850. https://doi.org/10.21276/ijlssr.2018.4.3.16.

Vinjoda, P., Mishra, S. K., Sharma, K., & Georrge, J. J. (2024). In silico identification of novel drug target and its natural product inhibitors for herpes simplex virus. In *Nanotechnology and In Silico Tools* (pp. 377–383). Elsevier. https://doi.org/10.1016/B978-0-443-15457-7.00007-1.

Vourc'h, G., Jourdain, E., Morand, S., & Moutou, F. (2022). *Zoonoses the Ties that Bind Humans to Animals*. éditions Quae.

Yeung, A., Hale, C., Clare, S., Palmer, S., Bartholdson Scott, J., Baker, S., & Dougan, G. (2019). Using a systems biology approach to study host-pathogen interactions. *Microbiology Spectrum*, 7(2). https://doi.org/10.1128/microbiolspec.BAI-0021-2019.

20 Role of Bioinformatics in Emerging and Re-Emerging Infectious Diseases

A Drug and Vaccine Development Prospective

Saurav Kumar Mishra, Umesh Kalathiya, Monikaben Padariya, and John J. Georrge

20.1 INTRODUCTION

As the emerging and re-emerging infectious disease (EID-REID) is creating an endemic situation worldwide, the bioinformatics-assisted application and their advanced approaches, applicable tools, and databases are being utilized by the researchers to analyze and understand all crucial data generated through biological experiments (Bah et al., 2018). The involvement of bioinformatics in biological experiments is not limited to various biological fields that are associated with it, such as proteomics, transcriptomics, genomics, metabolomics, chemoinformatics, immunoinformatics, and many more that are the base of the bioinformatic bases study (Arasu et al., 2023; Gauthier et al., 2019; Guedes et al., 2021; Raslan et al., 2023). In general, with every EID-REID, a bulk amount of data was also generated with a different entity, and handling these data and finding the essential information from these data is also challenging. To answer and handle this situation, several approaches and tools were developed to overcome this situation and also utilized for drug design and vaccine development (Akash, Baeza, et al., 2023; Shahab et al., 2023; Ullah et al., 2023; Vaghasia et al., 2024; Vakhariya Sakina et al., 2023). Also, various samples were collected every day and had different uniqueness and similarity; for this, the bioinformatics-assisted, next-generation sequencing (NGS) data analysis emerged as a preeminent discipline fields capable of finding the collected samples' similarities, uniqueness, and variations (Maljkovic Berry et al., 2020). Apart from that, with every EID-REID endemic situation, the surveillance of the infectious disease is essential, and the researcher is utilizing the bioinformatics-assisted application to monitor the endemic situation (One Health High-Level Expert et al., 2023). Further, websites (https://www.who.int/) are available to get the exact information of the current ongoing EID and REID; the precaution (https://www.fda.gov/) is being utilized. Moreover, the interdisciplinary knowledge of bioinformatics employs statistical (for screening the most promising ones) and computer methods (to handle the bulk data) to examine and comprehend biological information. It functions as an interface between biological and computer science and technology, increasing multiple areas of research and implementation while becoming fundamental in helping to comprehend complicated biological mechanisms and analyze genetic information (to get information about genetic similarities) (Gauthier et al., 2019; Kumar et al., 2022; Vinjoda et al., 2024).

Nearly always, the bioinformatics-assisted approach begins from the primary sequence information based on protein and nucleotide; based on the primary information, the researcher can easily identify the target genes, essential regulatory elements, and many more. Interestingly, in the case

 DOI: 10.1201/9781032721811-20

of host–pathogen interaction analysis, the approaches assisted through bioinformatics, that is, protein–protein interaction, and for the identification of essential, the generation of the network is one of the leading interests in the research community, and these interests come under the system biology–assisted approaches (Gauthier et al., 2019; Iuchi et al., 2023; Kumar et al., 2022). Moreover, several biological fields use bioinformatics in daily research life to overcome the EID and REID. So, this study aims to summarize the concept of bioinformatics, available tools and databases, and how the research communities are utilizing bioinformatics concepts in the case of drug design and vaccine development.

20.2 APPLICABLE TOOLS AND DATABASES IN BIOINFORMATICS

In the case of EID-REID, one of the challenges is to develop a potential drug and vaccine in less time with high accuracy. The conventional methods are time-consuming and require a high cost for successful therapeutic development to combat EID and REID. Moreover, bioinformatics is one of the leading approaches researchers worldwide utilize to develop and design therapeutics against infection. In the case of drug and vaccine development, bioinformatics plays a crucial role, and utilizing bioinformatics in research not only reduces the time but is also cost-effective and highly accurate. Further, bioinformatics concepts lie on computer-assisted technology, so there are several tools and databases available that are the helping hand of bioinformatics-based assisted research and are utilized mainly by worldwide researchers for successful inhibitor identification and vaccine development. This section will be helpful for the reader to get to know the available tools and database for the bioinformatics study. (*Note:* For more details on the mentioned tools and database, please go through the mentioned URLs) as shown in Table 20.1.

TABLE 20.1
Selective Applicable Servers in Bioinformatics for Drug or Vaccine Developments

Sl. No	Name	URLs	Remarks	Reference
1.	UniProt	https://www.uniprot.org/	It can help retrieve the protein sequence of the query ID	(UniProt, 2023)
2.	NCBI	https://www.ncbi.nlm.nih.gov/	It can help to retrieve the sequence in different forms of different targets	(Sayers et al., 2022)
3.	PDB	https://www.rcsb.org/	It can help retrieve the 3D structure and other properties	(Berman et al., 2000)
4.	Expasy	https://www.expasy.org/	It can help predict the query input's physiochemical properties and many more	(Gasteiger et al., 2003)
5.	SAVES	https://saves.mbi.ucla.edu/	It can help to validate the structure based on the Ramachandran plot and many more	(Laskowski et al., 1996)
6.	ProSA	https://prosa.services.came.sbg.ac.at/prosa.php	It can help predict the structure's general state based on Z-score	(Wiederstein & Sippl, 2007)
7.	IEDB	https://www.iedb.org/	It can help to identify B-cell and T-cell epitopes along with other essential immunogen attributes	(Vita et al., 2019)
8.	ABCpred	http://crdd.osdd.net/raghava/abcpred/	It can help to identify LBL epitope prediction	(Saha & Raghava, 2006)
9.	VaxiJen	https://www.ddg-pharmfac.net/vaxijen/	It can help to evaluate antigenicity and non-antigenicity	(Doytchinova & Flower, 2007)
10.	AllerTOP v.2	https://www.ddg-pharmfac.net/AllerTOP/	It can help predict the allergenicity and non-allergenicity of the query input	(Dimitrov et al., 2014)

(*Continued*)

TABLE 20.1
Selective Applicable Servers in Bioinformatics for Drug or Vaccine Developments

Sl. No	Name	URLs	Remarks	Reference
11.	ToxinPred	http://crdd.osdd.net/raghava/toxinpred/	It can help to evaluate toxicity of the query input	(Gupta et al., 2013)
12.	I-TASSER	https://zhanggroup.org/I-TASSER/	It can help to identify the model of the target sequence	(Zhang, 2008)
13.	SWISS-MODEL	https://swissmodel.expasy.org/	It can help to model the structure based on the template	(Schwede et al., 2003)
14.	AlphaFold	https://alphafold.ebi.ac.uk/	It can help to model the 3D structure through deep learning	(Jumper et al., 2021)
15.	Robetta	https://robetta.bakerlab.org/submit.php	It can help to predict 3D structures of the query sequence	(Kim et al., 2004)
16.	ZDOCK	https://zdock.umassmed.edu/	It can help to perform receptor-ligand docking	(Pierce et al., 2014)
17.	SwissDock	http://www.swissdock.ch/docking	It can help to perform molecular docking.	(Grosdidier et al., 2011)
18.	Cluspro	https://cluspro.bu.edu/login.php	It can help to perform protein–peptide docking	(Kozakov et al., 2017)
19.	HADDOCK	https://wenmr.science.uu.nl/haddock2.4/	It can help to perform protein–protein docking.	(Dominguez et al., 2003)
20.	PRODIGY	https://wenmr.science.uu.nl/prodigy/	It can help calculate the binding energy of the docked cluster	(Xue et al., 2016)
21.	PDBsum	https://www.ebi.ac.uk/thornton-srv/databases/pdbsum/	It can help identify the molecular interaction in the docked complex	(Laskowski et al., 2018)
22.	DockThor	https://dockthor.lncc.br/v2/	It can help to perform molecular docking	(Guedes et al., 2021)
23.	STRING	https://string-db.org/	It can help to predict the network between proteins	(Szklarczyk et al., 2019)
24.	ChEMBL	https://www.ebi.ac.uk/chembl/	It can help to get information on biologically active compounds and other properties	(Gaulton et al., 2012)
25.	DrugBank	https://go.drugbank.com/	It can help get gathering data on drugs and target substances	(Wishart et al., 2008)
26.	IMPPAT 2.0	https://cb.imsc.res.in/imppat/	It can help to get an extensive collection of Indian-origin-based phytochemical compounds	(Vivek-Ananth et al., 2023)
27.	LOTUS	https://lotus.naturalproducts.net/	It can help to get a large compilation of annotated natural substance	(Rutz et al., 2022)
28.	COCONUT	https://coconut.naturalproducts.net/	It can help to get a large compilation natural substance	(Sorokina et al., 2021)
29.	ZINC	https://zinc.docking.org/	It can help to get large commercially available compounds	(Irwin & Shoichet, 2005)
30.	PubChem	https://pubchem.ncbi.nlm.nih.gov/	It can help to get large chemical compound information	(Kim et al., 2016)

20.3 BIOINFORMATICS IN DRUG DESIGN

In the present situation, drug design through bioinformatics and computational biology is one of the leading interests in the researchers worldwide, as through bioinformatics, it is easy to identify the most potential inhibitor among the bulk dataset, and also helpful to understand the compound and

TABLE 20.2
Bioinformatics-Assisted Drug Design Studies

Sl. No	Target	Remarks	Reference
1.	Hepatitis C virus (HCV)	Identification of potential inhibitors targeting Nonadec-7-ene-4-carboxylic acid derivatives	(Ejeh et al., 2023)
2.	SARS-CoV-2	Finding of new inhibitors that specifically inhibit the main protease of SARS-CoV	(Mukherjee et al., 2023)
3.	Lassa virus	Identification of potential evodiamine derivatives inhibitor for Lassa virus	(Akash, Baeza, et al., 2023)
4.	Black fungus, MPXV, and Marburg	Identification of putative inhibitory coptisine derivatives	(Akash, Hossain, et al., 2023)
5.	Monkeypox virus	Identification of molecular interaction between monkeypox and approved drugs	(Arasu et al., 2023)

the target. As the EID and REID become a global challenge, exploring the concept of bioinformatics and utilizing their approach can be helpful for novel inhibitor identification to control and reduce the infection. The bioinformatics revolution also has a particular field that is mostly related to drug design chemoinformatics. The concept of bioinformatics in drug design is not new; however, the advances in this approach make it an ideal approach worldwide. Interestingly, it has also been successfully designed to find a novel approach that can be helpful for drug development. Apart from that, several studies are available that utilize the concepts of bioinformatics and cheminformatics to design a study successfully. This section will help readers understand the basic concepts of drug design through the bioinformatics approach (Table 20.2).

Hepatitis C virus (HCV) is one of the common viruses, and infection with it leads to the progression of chronic liver complications. To overcome this, a study performed by Ejeh et al. employed the bioinformatics approach to study the interaction of derivatives of Nonadec-7-ene-4-carboxylic on the HCV target. In this study, the authors performed molecular docking simulations to identify novel inhibitors targeting the HCV protease (NS3/4a). The overall finding shows that Molecule 45 (PubChem68792528) is more suitable than the reference drug molecule, which has more stability and can lead to therapeutic developments (Ejeh et al., 2023).

The COVID-19 infection is not over yet, which requires the identification of novel inhibitors that can be helpful to reduce the infection. An *in silico* approach was utilized in a study conducted by Mukherjee et al. to identify potential inhibitors for SARS-CoV-2 transmission. In this study, authors targeted Mpro, a collection of natural substance and approved drugs considered a ligand library for molecular docking analysis. Based on the docking analysis and stability analysis, ZINC70699832 was found promising among the retrieved ligands based on its molecular interaction stability and physiological quality, which can be used for drug development (Mukherjee et al., 2023).

Lassa virus is a virus responsible for hemorrhagic fever and transfusions from human to human. A study performed by Akash et al. utilized a computational approach to identify evodiamine derivative inhibitors to combat the Lassa virus infection by targeting its glycoprotein and nucleoprotein. In this study, the authors utilized glycoprotein and nucleoprotein (PDB ID; 5FT2 and 3MX5) as the target, and the evodiamine derivative was used as a ligand and performed the molecular docking analysis further; the stability was analyzed through MD simulation. The overall finding shows that the evodiamine derivative can serve as a potential inhibitor based on the molecular interaction stability and rug likeliness to combat the Lassa virus infection (Akash, Baeza, et al., 2023).

Another study by Akash et al. employed an advanced drug design approach for identifying coptisine derivatives based on inhibitors targeting Black fungus, MPXV, and Marburg infection. In this study, authors used PDB ID; 7D6X, 4WTP, 4QWO, 4OR8 of the target virus as a target receptor and the coptisine derivatives as the ligand and performed the molecular docking, molecular dynamics simulation, QSAR, and pIC50 calculation and pharmacokinetics analysis. The finding shows that bioactive coptisine derivatives have promising molecular interaction, stability, and pharmacokinetics properties and can be utilized to reduce the infection rate of the selected virus (Akash, Hossain, et al., 2023).

As numerous EIDs and REIDs are going on worldwide, the research community is also interested in utilizing the earlier approved drug to identify new inhibitors, utilizing the concept of a drug repurposing approach. A study was performed by Arasu et al. on the recently re-emerged virus (Monkeypox) for the identification of suitable inhibitors utilizing FDA-approved drugs. In this study, the authors utilized cysteine proteinase as a target; however, the three-dimensional structure was unavailable, so the author modeled the target structure and performed molecular docking with the FDA-approved drugs. The overall finding shows that selected FDA-approved drugs can be potential inhibitors to reduce monkeypox infection (Arasu et al., 2023).

20.4 BIOINFORMATICS IN VACCINE DEVELOPMENT

Bioinformatics plays the most crucial role in the current situation; as we have seen in the recent outbreaks (COVID-19), worldwide researchers are applying different bioinformatics applications and algorithms to handle this endemic situation. Notably, the EID and REID researchers are more interested in vaccine development aspects. Immunoinformatics is an essential part of bioinformatics, which came as a discipline that can help design a peptide-based vaccine and easily help identify the most antigenic peptide that can lead to vaccine development. Most of the studies in the past year successfully applied the immunoinformatics approach in their study against the EID and REID. This section will summarize some selective studies targeting an infectious disease that can help the reader understand the role of immunoinformatics and other essential aspects (Table 20.3). Apart from that, the immunoinformatics study mainly lies in predicting B cells and T cells as both receptors are the major in vaccine development in the context of immune response.

Due to viral infection, inflammation of the myocardium leads to heart failure, so the goal of this study is to develop an immunodominant muti-epitope vaccine targeting coxsackievirus B3. Twenty-one immunodominant epitopes were screened based on their immunogenic properties and utilized for vaccine construction considering adjuvants for more effective profile. The designed vaccine successfully binds with the TLR-4 and is found steady through the simulation. Additionally, authors performed the immune simulation and expression analysis and found that designed vaccine can evoke promising response and have a high expression level in cloned vectors (Ullah et al., 2023).

TABLE 20.3
Different Immunoinformatics and Bioinformatics-Assisted Vaccine Development Studies

Sl. No	Target	Remarks	Reference
1	Coxsackievirus B3	Prediction of promising B-cell, CD8+ and CD4+ T-cell antigen for multi-peptide vaccine design	Ullah et al., (2023)
2	SARS-CoV-2	Development of coevolution-based peptide vaccine targeting spike of COVID-19	Mishra et al., (2023)
3	Monkeypox	Designing of a novel epitope vaccine using its envelop protein	Lahimchi et al., (2023)
4	Jamestown Canyon virus (JCV)	Identification of highly immunogenic epitope for the multi-peptide vaccine development	Shahab et al., (2023)
5	Lymphocytic choriomeningitis virus (LCMV)	Identification of B-cell, CTL and HTL epitope for multi-epitope-based subunit vaccine design	Waqas et al., (2023)

The mutation in the SARS-CoV is one of the challenges research communities face in designing and developing a vaccine. A study performed by Mishra et al. employed a coevolution-based strategy to develop vaccine candidate targeting spike protein. In this study, the authors designed a vaccine based on the wild type, and based on the coevolution, the mutated vaccine was also designed further; the designed vaccine was validated based on the molecular docking with the TLR-5, and their stability was analyzed through the MD simulation. Additionally, the authors performed *in silico* cloning and found that both designed vaccines have high expression in the vector, and the mutation in the designed vaccine affects its efficacy and can lead to combating the SARS-CoV-2 viral infection (Mishra et al., 2023). Monkeypox is a re-emerging zoonotic viral infection, and their cases were increasing daily to overcome the study performed by Lahimchi et al., which employed immunoinformatics and bioinformatics strategies to construct a multi-epitope-based vaccine. A total of three immunogenic proteins were targeted for the epitope prediction and based on the immunogen-assisted screening, the potential epitope was selected and used for vaccine construction, considering PADRE and RS01 as adjuvants to boost the vaccine response. Further, the designed vaccine was successfully docked with the TLR-4. Further, the vaccine was successfully inserted in pET-21b (+) plasmid, and the immune simulation analysis revealed that the developed candidate has a promising response followed by the generation of IgM, IgG1, and IgG2 antibodies (Lahimchi et al., 2023). Likewise, a study performed by Shahab et al. employed immunoinformatics to develop a multi-peptide vaccine targeting the Jamestown canyon virus (JCV), one of the major viruses in the case of encephalitis. In this study, through immunogen-assisted screening, such as antigenicity, allergenic, toxicity, and many more, a total of 15 epitopes comprised of CTL, HTL, and LBL were found suitable and used for vaccine construction and to strengthen the immune response of this vaccine, 50s ribosomal protein sequence were also integrated into the vaccine formulation. Further, the molecular docking with the TLR-2 and TLR-4 shows molecular interaction and immune reaction, and expression analysis revealed the promising immune profile of the designed vaccine (Shahab et al., 2023). Another study performed by Waqas et al. employed an immunoinformatics strategy for developing an effective multi-epitope-based subunit vaccine as opposed to the lymphocytic choriomeningitis virus. In this study, 26 immunodominant epitopes were found to be more suitable bases on the screening and used for vaccine construction and along with the epitope, the adjuvant was also fused for robust immune response. Further, the molecular docking revealed that the vaccine is successfully bound with the Toll-like receptors TLR-2 and TLR-4. The immune reaction and expression analysis reveal the potential immunogenic profile of the designed vaccine. Several successful studies were also performed using bioinformatics and immunoinformatics approaches for the subunit vaccine developments (Waqas et al., 2023).

20.5 CONCLUSION

Bioinformatics is one of the leading approaches in the different biological fields, as the EID and REID create global endemic situations; the researchers utilized bioinformatics for drug and vaccine development. The advanced bioinformatics approach can help understand the complex biological system and further lead to therapeutic development. Most of the studies utilized the concept of bioinformatics; however, as the infection is increasing, the advancement in the bioinformatics concepts and more research related to that are required that can be helpful for a better tomorrow.

REFERENCES

Akash, S., Baeza, J., Mahmood, S., Mukerjee, N., Subramaniyan, V., Islam, M. R., Gupta, G., Rajakumari, V., Chinni, S. V., Ramachawolran, G., Saleh, F. M., Albadrani, G. M., Sayed, A. A., & Abdel-Daim, M. M. (2023). Development of a new drug candidate for the inhibition of Lassa virus glycoprotein and nucleoprotein by modification of evodiamine as promising therapeutic agents. *Frontiers in Microbiology*, 14, 1206872. https://doi.org/10.3389/fmicb.2023.1206872.

Akash, S., Hossain, A., Mukerjee, N., Sarker, M. M. R., Khan, M. F., Hossain, M. J., Rashid, M. A., Kumer, A., Ghosh, A., Leon-Figueroa, D. A., Barboza, J. J., Padhi, B. K., & Sah, R. (2023). Modified coptisine derivatives as an inhibitor against pathogenic Rhizomucor miehei, Mycolicibacterium smegmatis (Black Fungus), Monkeypox, and Marburg virus by molecular docking and molecular dynamics simulation-based drug design approach. *Front Pharmacol*, 14, 1140494. https://doi.org/10.3389/fphar.2023.1140494.

Arasu, M. V., Vijayaragavan, P., Purushothaman, S., Rathi, M. A., Al-Dhabi, N. A., Gopalakrishnan, V. K., Choi, K. C., & Ilavenil, S. (2023). Molecular docking of monkeypox (mpox) virus proteinase with FDA approved lead molecules. *J Infect Public Health*, 16(5), 784–791. https://doi.org/10.1016/j.jiph.2023.03.004.

Bah, S. Y., Morang'a, C. M., Kengne-Ouafo, J. A., Amenga-Etego, L., & Awandare, G. A. (2018). Highlights on the application of genomics and bioinformatics in the fight against infectious diseases: Challenges and opportunities in Africa. *Front Genet*, 9, 575. https://doi.org/10.3389/fgene.2018.00575.

Berman, H. M., Westbrook, J., Feng, Z., Gilliland, G., Bhat, T. N., Weissig, H., Shindyalov, I. N., & Bourne, P. E. (2000). The protein data bank. *Nucleic Acids Res*, 28(1), 235–242. https://doi.org/10.1093/nar/28.1.235.

Dimitrov, I., Bangov, I., Flower, D. R., & Doytchinova, I. (2014). AllerTOP v.2-a server for in silico prediction of allergens. *J Mol Model*, 20(6), 2278. https://doi.org/10.1007/s00894-014-2278-5.

Dominguez, C., Boelens, R., & Bonvin, A. M. (2003). HADDOCK: A protein-protein docking approach based on biochemical or biophysical information. *J Am Chem Soc*, 125(7), 1731–1737. https://doi.org/10.1021/ja026939x.

Doytchinova, I. A., & Flower, D. R. (2007). VaxiJen: A server for prediction of protective antigens, tumour antigens and subunit vaccines. *BMC Bioinformatics*, 8, 4. https://doi.org/10.1186/1471-2105-8-4.

Ejeh, S., Uzairu, A., Shallangwa, G. A., Abechi, S. E., Ibrahim, M. T., Ramu, R., & Al-Ghorbani, M. (2023). Chemical bioinformatics study of Nonadec-7-ene-4-carboxylic acid derivatives via molecular docking, and molecular dynamic simulations to identify novel lead inhibitors of hepatitis c virus NS3/4a protease. *Scientific African*, 20, e01591.

Gasteiger, E., Gattiker, A., Hoogland, C., Ivanyi, I., Appel, R. D., & Bairoch, A. (2003). ExPASy: The proteomics server for in-depth protein knowledge and analysis. *Nucleic Acids Res*, 31(13), 3784–3788. https://doi.org/10.1093/nar/gkg563.

Gaulton, A., Bellis, L. J., Bento, A. P., Chambers, J., Davies, M., Hersey, A., Light, Y., McGlinchey, S., Michalovich, D., Al-Lazikani, B., & Overington, J. P. (2012). ChEMBL: A large-scale bioactivity database for drug discovery. *Nucleic Acids Res*, 40(Database issue), D1100–1107. https://doi.org/10.1093/nar/gkr777.

Gauthier, J., Vincent, A. T., Charette, S. J., & Derome, N. (2019). A brief history of bioinformatics. *Brief Bioinform*, 20(6), 1981–1996. https://doi.org/10.1093/bib/bby063.

Grosdidier, A., Zoete, V., & Michielin, O. (2011). SwissDock, a protein-small molecule docking web service based on EADock DSS. *Nucleic Acids Res*, 39(Web Server issue), W270–277. https://doi.org/10.1093/nar/gkr366.

Guedes, I. A., Costa, L. S. C., Dos Santos, K. B., Karl, A. L. M., Rocha, G. K., Teixeira, I. M., Galheigo, M. M., Medeiros, V., Krempser, E., Custodio, F. L., Barbosa, H. J. C., Nicolas, M. F., & Dardenne, L. E. (2021). Drug design and repurposing with DockThor-VS web server focusing on SARS-CoV-2 therapeutic targets and their non-synonym variants. *Sci Rep*, 11(1), 5543. https://doi.org/10.1038/s41598-021-84700-0.

Gupta, S., Kapoor, P., Chaudhary, K., Gautam, A., Kumar, R., Open Source Drug Discovery, C., & Raghava, G. P. (2013). In silico approach for predicting toxicity of peptides and proteins. *PLoS One*, 8(9), e73957. https://doi.org/10.1371/journal.pone.0073957.

Irwin, J. J., & Shoichet, B. K. (2005). ZINC-a free database of commercially available compounds for virtual screening. *J Chem Inf Model*, 45(1), 177–182. https://doi.org/10.1021/ci049714.

Iuchi, H., Kawasaki, J., Kubo, K., Fukunaga, T., Hokao, K., Yokoyama, G., Ichinose, A., Suga, K., & Hamada, M. (2023). Bioinformatics approaches for unveiling virus-host interactions. *Comput Struct Biotechnol J*, 21, 1774–1784. https://doi.org/10.1016/j.csbj.2023.02.044.

Jumper, J., Evans, R., Pritzel, A., Green, T., Figurnov, M., Ronneberger, O., Tunyasuvunakool, K., Bates, R., Zidek, A., Potapenko, A., Bridgland, A., Meyer, C., Kohl, S. A. A., Ballard, A. J., Cowie, A., Romera-Paredes, B., Nikolov, S., Jain, R., Adler, J., Back, T., Petersen, S., Reiman, D., Clancy, E., Zielinski, M., Steinegger, M., Pacholska, M., Berghammer, T., Bodenstein, S., Silver, D., Vinyals, O., Senior, A. W., Kavukcuoglu, K., Kohli, P., & Hassabis, D. (2021). Highly accurate protein structure prediction with AlphaFold. *Nature*, 596(7873), 583–589. https://doi.org/10.1038/s41586-021-03819-2.

Kim, D. E., Chivian, D., & Baker, D. (2004). Protein structure prediction and analysis using the Robetta server. *Nucleic Acids Res*, 32(Web Server issue), W526–531. https://doi.org/10.1093/nar/gkh468.

Kim, S., Thiessen, P. A., Bolton, E. E., Chen, J., Fu, G., Gindulyte, A., Han, L., He, J., He, S., Shoemaker, B. A., Wang, J., Yu, B., Zhang, J., & Bryant, S. H. (2016). PubChem Substance and Compound databases. *Nucleic Acids Res*, 44(D1), D1202–1213. https://doi.org/10.1093/nar/gkv951.

Kozakov, D., Hall, D. R., Xia, B., Porter, K. A., Padhorny, D., Yueh, C., Beglov, D., & Vajda, S. (2017). The ClusPro web server for protein-protein docking. *Nat Protoc*, 12(2), 255–278. https://doi.org/10.1038/nprot.2016.169.

Kumar, S., Kumar, G. S., Maitra, S. S., Malý, P., Bharadwaj, S., Sharma, P., & Dwivedi, V. D. (2022). Viral informatics: Bioinformatics-based solution for managing viral infections. *Briefings in Bioinformatics*, 23(5), bbac326.

Lahimchi, M. R., Madanchi, H., Ahmadi, K., Shahbazi, B., & Yousefi, B. (2023). In silico designing a novel TLR4-mediating multi-epitope vaccine against monkeypox via advanced immunoinformatics and bioinformatics approaches. *J Biomol Struct Dyn*, 1–17. https://doi.org/10.1080/07391102.2023.2203253.

Laskowski, R. A., Jablonska, J., Pravda, L., Varekova, R. S., & Thornton, J. M. (2018). PDBsum: Structural summaries of PDB entries. *Protein Sci*, 27(1), 129–134. https://doi.org/10.1002/pro.3289.

Laskowski, R. A., Rullmannn, J. A., MacArthur, M. W., Kaptein, R., & Thornton, J. M. (1996). AQUA and PROCHECK-NMR: Programs for checking the quality of protein structures solved by NMR. *J Biomol NMR*, 8(4), 477–486. https://doi.org/10.1007/BF00228148.

Maljkovic Berry, I., Melendrez, M. C., Bishop-Lilly, K. A., Rutvisuttinunt, W., Pollett, S., Talundzic, E., Morton, L., & Jarman, R. G. (2020). Next generation sequencing and bioinformatics methodologies for infectious disease research and public health: Approaches, applications, and considerations for development of laboratory capacity. *J Infect Dis*, 221(Suppl. 3), S292–S307. https://doi.org/10.1093/infdis/jiz286.

Mishra, S. K., Priya, P., Rai, G. P., Haque, R., & Shanker, A. (2023). Coevolution based immunoinformatics approach considering variability of epitopes to combat different strains: A case study using spike protein of SARS-CoV-2. *Comput Biol Med*, 163, 107233. https://doi.org/10.1016/j.compbiomed.2023.107233.

Mukherjee, A., Pandey, K. M., Ojha, K. K., & Sahu, S. K. (2023). Identification of possible SARS-CoV-2 main protease inhibitors: In silico molecular docking and dynamic simulation studies. *Beni-Suef University Journal of Basic and Applied Sciences*, 12(1), 69.

One Health High-Level Expert Panel, Hayman, D. T. S., Adisasmito, W. B., Almuhairi, S., Behravesh, C. B., Bilivogui, P., Bukachi, S. A., Casas, N., Becerra, N. C., Charron, D. F., Chaudhary, A., Ciacci Zanella, J. R., Cunningham, A. A., Dar, O., Debnath, N., Dungu, B., Farag, E., Gao, G. F., Khaitsa, M., Machalaba, C., Mackenzie, J. S., Markotter, W., Mettenleiter, T. C., Morand, S., Smolenskiy, V., Zhou, L., & Koopmans, M. (2023). Developing one health surveillance systems. *One Health*, 17, 100617. https://doi.org/10.1016/j.onehlt.2023.100617.

Pierce, B. G., Wiehe, K., Hwang, H., Kim, B. H., Vreven, T., & Weng, Z. (2014). ZDOCK server: Interactive docking prediction of protein-protein complexes and symmetric multimers. *Bioinformatics*, 30(12), 1771–1773. https://doi.org/10.1093/bioinformatics/btu097.

Raslan, M. A., Raslan, S. A., Shehata, E. M., Mahmoud, A. S., & Sabri, N. A. (2023). Advances in the applications of bioinformatics and chemoinformatics. *Pharmaceuticals (Basel)*, 16(7). https://doi.org/10.3390/ph16071050.

Rutz, A., Sorokina, M., Galgonek, J., Mietchen, D., Willighagen, E., Gaudry, A., Graham, J. G., Stephan, R., Page, R., Vondrasek, J., Steinbeck, C., Pauli, G. F., Wolfender, J. L., Bisson, J., & Allard, P. M. (2022). The LOTUS initiative for open knowledge management in natural products research. *Elife*, 11. https://doi.org/10.7554/eLife.70780.

Saha, S., & Raghava, G. P. (2006). Prediction of continuous B-cell epitopes in an antigen using recurrent neural network. *Proteins*, 65(1), 40–48. https://doi.org/10.1002/prot.21078.

Sayers, E. W., Bolton, E. E., Brister, J. R., Canese, K., Chan, J., Comeau, D. C., Connor, R., Funk, K., Kelly, C., Kim, S., Madej, T., Marchler-Bauer, A., Lanczycki, C., Lathrop, S., Lu, Z., Thibaud-Nissen, F., Murphy, T., Phan, L., Skripchenko, Y., Tse, T., Wang, J., Williams, R., Trawick, B. W., Pruitt, K. D., & Sherry, S. T. (2022). Database resources of the national center for biotechnology information. *Nucleic Acids Res*, 50(D1), D20–D26. https://doi.org/10.1093/nar/gkab1112.

Schwede, T., Kopp, J., Guex, N., & Peitsch, M. C. (2003). SWISS-MODEL: An automated protein homology-modeling server. *Nucleic Acids Res*, 31(13), 3381–3385. https://doi.org/10.1093/nar/gkg520.

Shahab, M., Aiman, S., Alshammari, A., Alasmari, A. F., Alharbi, M., Khan, A., Wei, D. Q., & Zheng, G. (2023). Immunoinformatics-based potential multi-peptide vaccine designing against Jamestown Canyon Virus (JCV) capable of eliciting cellular and humoral immune responses. *Int J Biol Macromol*, 253(Pt 2), 126678. https://doi.org/10.1016/j.ijbiomac.2023.126678.

Sorokina, M., Merseburger, P., Rajan, K., Yirik, M. A., & Steinbeck, C. (2021). COCONUT online: Collection of open natural products database. *J Cheminform*, 13(1), 2. https://doi.org/10.1186/s13321-020-00478-9.

Szklarczyk, D., Gable, A. L., Lyon, D., Junge, A., Wyder, S., Huerta-Cepas, J., Simonovic, M., Doncheva, N. T., Morris, J. H., Bork, P., Jensen, L. J., & Mering, C. V. (2019). STRING v11: Protein-protein association networks with increased coverage, supporting functional discovery in genome-wide experimental datasets. *Nucleic Acids Res*, 47(D1), D607–D613. https://doi.org/10.1093/nar/gky1131.

Ullah, A., Waqas, M., Aziz, S., Rahman, S. U., Khan, S., Khalid, A., Abdalla, A. N., Uddin, J., Halim, S. A., Khan, A., & Al-Harrasi, A. (2023). Bioinformatics and immunoinformatics approach to develop potent multi-peptide vaccine for coxsackievirus B3 capable of eliciting cellular and humoral immune response. *Int J Biol Macromol*, 239, 124320. https://doi.org/10.1016/j.ijbiomac.2023.124320.

UniProt, C. (2023). UniProt: The universal protein knowledgebase in 2023. *Nucleic Acids Res*, 51(D1), D523–D531. https://doi.org/10.1093/nar/gkac1052.

Vaghasia, V. V., Sharma, K., Mishra, S. K., & Georrge, J. J. (2024). In silico identification of natural product inhibitor for multidrug resistance proteins from selected gram-positive bacteria. In *Nanotechnology and In Silico Tools* (pp. 309–317). Elsevier.

Vakhariya Sakina, S., Mishra, S. K., Sharma, K., & Georrge, J. J. (2023). Designing of a novel curcumin analogue to inhibit mitogen-activated protein kinase: A cheminformatics approach. *Journal of Phytonanotechnology and Pharmaceutical Sciences*, 3(1), 37–47.

Vinjoda, P., Mishra, S. K., Sharma, K., & Georrge, J. J. (2024). In silico identification of novel drug target and its natural product inhibitors for herpes simplex virus. In *Nanotechnology and In Silico Tools* (pp. 377–383). Elsevier.

Vita, R., Mahajan, S., Overton, J. A., Dhanda, S. K., Martini, S., Cantrell, J. R., Wheeler, D. K., Sette, A., & Peters, B. (2019). The immune epitope database (IEDB): 2018 update. *Nucleic Acids Res*, 47(D1), D339–D343. https://doi.org/10.1093/nar/gky1006.

Vivek-Ananth, R. P., Mohanraj, K., Sahoo, A. K., & Samal, A. (2023). IMPPAT 2.0: An enhanced and expanded phytochemical atlas of Indian medicinal plants. *ACS Omega*, 8(9), 8827–8845. https://doi.org/10.1021/acsomega.3c00156.

Waqas, M., Aziz, S., Bushra, A., Halim, S. A., Ali, A., Ullah, S., Khalid, A., Abdalla, A. N., Khan, A., & Al-Harrasi, A. (2023). Employing an immunoinformatics approach revealed potent multi-epitope based subunit vaccine for lymphocytic choriomeningitis virus. *J Infect Public Health*, 16(2), 214–232. https://doi.org/10.1016/j.jiph.2022.12.023.

Wiederstein, M., & Sippl, M. J. (2007). ProSA-web: Interactive web service for the recognition of errors in three-dimensional structures of proteins. *Nucleic Acids Res*, 35(Web Server issue), W407–410. https://doi.org/10.1093/nar/gkm290.

Wishart, D. S., Knox, C., Guo, A. C., Cheng, D., Shrivastava, S., Tzur, D., Gautam, B., & Hassanali, M. (2008). DrugBank: A knowledgebase for drugs, drug actions and drug targets. *Nucleic Acids Res*, 36(Database issue), D901–906. https://doi.org/10.1093/nar/gkm958.

Xue, L. C., Rodrigues, J. P., Kastritis, P. L., Bonvin, A. M., & Vangone, A. (2016). PRODIGY: A web server for predicting the binding affinity of protein-protein complexes. *Bioinformatics*, 32(23), 3676–3678. https://doi.org/10.1093/bioinformatics/btw514.

Zhang, Y. (2008). I-TASSER server for protein 3D structure prediction. *BMC Bioinformatics*, 9, 40. https://doi.org/10.1186/1471-2105-9-40.

21 One Health Approach in EID-REID Mitigation

A Comprehensive Review of Interactions

Aniruddha Sen, Ayan Chatterjee, Ankita Gon, and Jewelson Santos

21.1 INTRODUCTION

The fast changes in climate, environment, and population are causing an upsurge in the emergence and re-emergence of infectious illnesses. The abrupt onset and swift dissemination of these illnesses make them serious hazards to the health of the entire world (Aggarwal & Ramachandran, 2020). Understanding infectious disease epidemiology and transmission dynamics is essential for improving public health decision-making for integrated, real-time community-based control methods at the individual and community levels (Ryu et al., 2022). Periodically, epidemics of both newly discovered and previously identified infectious illnesses arise against a background of known infections, significantly increasing the burden of these diseases worldwide (Kapoor, 2022). Zoonoses, contagious diseases that can be transferred between animals and humans, account for approximately 75% of new emerging and re-emerging disease pathogens (Aggarwal & Ramachandran, 2020). A zoonotic illness arises when human infections from animal populations spread to human populations due to the disturbance of global ecosystems. The most successful treatments for infectious illnesses, antimicrobial medications, are becoming less effective as a result of rising antimicrobial resistance (AMR), which is mostly caused by improper use of antimicrobials in people, animals, and plants (Dafale et al., 2020).

The One Health concept acknowledges the relationships and mutual reliance on health that exist between people, other species, and shared surroundings (Kapoor, 2022). This collaborative, multidisciplinary approach is designed to address the risks arising from human–animal–ecosystem interactions (Kapoor, 2022). The One Health concept places a strong emphasis on the environment's role in the formation and danger of illnesses that people and animals may get from one another (Li et al., 2022). Considering the diverse levels of direct and indirect mitigating effects on distinct infectious illnesses and their loads, it is crucial to take into account the "One Health" integrated assessment method in order to comprehend the dynamics and management of infectious illnesses (Shingankar & Ugemuge, 2022).

21.2 INTERCONNECTED HEALTH REALMS

There is a close relationship between the environment, animal health, and human health; one domain affects the others. The transmission of zoonotic illnesses is one area in which this interconnection is most clear, which often happens at the intersection of household animals, people, and wildlife (Suu-Ire et al., 2021). The chances of the spread of zoonotic diseases have grown due to many factors. including habitat degradation, increased contacts between humans and wildlife, and the wildlife trade (Suu-Ire et al., 2021). Southeast Asian regions, in particular, are extremely prone

DOI: 10.1201/9781032721811-21

to the introduction and resurgence of zoonotic viral infections due to the continuous substantial environmental and societal changes in these areas (Saba Villarroel et al., 2023).

Zoonotic illnesses, or infections and diseases that are naturally passed from vertebrate animals to people, have grown to be serious hazards to world health (Tomori & Oluwayelu, 2023). Currently making up 73% of emerging infectious diseases (EID), they represent the major category of illnesses within EID (Tomori & Oluwayelu, 2023). Many pathogens are the etiological agents of zoonoses, with viruses making up around 30% of all zoonotic illnesses (Tomori & Oluwayelu, 2023). Direct or indirect transmission of zoonotic illnesses can occur by contact, aerosols, vectors, or vertical transmission in utero (Tomori & Oluwayelu, 2023). Human activities, environmental factors, and pathogen characteristics play important roles in the development and spread of many illnesses (Tomori & Oluwayelu, 2023). Over the past century, regions like Southeast Asia have faced major viral outbreaks, including arboviruses, severe acute respiratory syndrome coronavirus 2 (SARS-CoV-2), highly pathogenic avian influenza (H5N1), and severe acute respiratory syndrome (SARS-CoV) (9). The necessity of strengthening the interaction between humans, animals, plants, and the environment in order to effectively prevent, detect, and respond to health hazards while encouraging sustainable development highlights the significance of One Health (Saba Villarroel et al., 2023).

21.3 CASE STUDIES AND SUCCESS STORIES

- *RSV vaccines development in Europe:* One Health (OH) concept places emphasis on the interdependence of environmental, animal, and human health. However, coordination among the veterinary and human domains is not always apparent, particularly when it comes to non-zoonotic illnesses. Research explored the factors that support and hinder cooperative efforts to produce vaccines against the non-zoonotic RSV virus in the European context. The research highlighted that majority of experts working on the RSV vaccine saw themselves as belonging to two different groups, which suggests that there is no shared objective for cooperation. The study sheds light on how cooperation between veterinary and human health experts influenced the creation of vaccinations, highlighting the significance of respect for one another and common objectives. Awareness, admiration for one another, and goal-setting in unison are necessary for collaboration. Even for non-zoonotic situations, One Health approach can provide valuable insights and foster collaboration (Janse et al., 2023).
- *One Health in Belgium:* Belgium has been exploring the potential of population health management, especially concerning atherosclerotic cardiovascular disease. Through the elicitation of implementation challenges and suggestions, research sought to increase knowledge of community health management in Belgium. The research found seven interconnected obstacles and suggestions for the advancement of population health management in Belgium. These have to do with community engagement, shared accountability for population health, and the duties of various governmental levels. Belgium has to develop a shared population-oriented vision, and this requires the active participation and support of all stakeholders at the national and regional levels (Steenkamer et al., 2023).
- *One Health initiatives in Ethiopia and Pakistan:* The roles and contributions of animal health in One Health programs were examined in a transdisciplinary research study, focusing on Ethiopia and Pakistan. According to the report, operationalizing One Health at the national and subnational levels continues to present substantial problems, despite the fact that the approach has grown more inclusive and multidisciplinary on a global scale. The integration of animal health into One Health frameworks is threatened by deficiencies in competence, political will, and governance. Stronger multilevel linkages and collaboration between the animal health and other sectors is essential to enable the adoption of One Health initiatives that are successful, inclusive, and well-resourced (Braam et al., 2023).
- *One Health tools integration in Pakistan:* A workshop in Pakistan integrated two One Health tools—the resource toolkit (OH-SMART) for One Health Systems Mapping and

Analysis and One Health Zoonotic Disease Prioritization (OHZDP). The workshop employed OH-SMARTTM for a partner evaluation and disease-specific gap analysis, and the OHZDP to create a list of priority zoonotic illnesses for Pakistan. Eleven in-country facilitators representing the human and animal health sectors were educated in both methods during the integrated workshop, which also highlighted opportunities for One Health partnership for the priority zoonotic illnesses. In-country facilitators were given the chance to further their professional development while maintaining consistency between the two processes through the combined use of the OHZDP Process and OH-SMARTTM (Varela et al., 2023).

21.4 STRENGTHENING EARLY DETECTION AND SURVEILLANCE

One Health approach in Uganda: Uganda has been actively implementing One Health strategy to improve health security. National Action Plan for Health Security (NAPHS) in 2019–2023 was developed based on the results of a Joint External Evaluation (JEE) that the nation undertook in 2017. Workforce development, real-time surveillance, combating zoonotic illnesses, strengthening national laboratory systems, and expanding emergency response operations all have benefited from the One Health strategy. By using an electronic integrated disease surveillance and response platform, the strategy has facilitated the reporting of animal health incidents and promoted multisectoral trainings. The study underscores the importance of full funding for activities, with 36.5% of activities fully funded. It also emphasizes the need for continuous progress, with 65% of activities still in progress as of 2020 (Bakiika et al., 2023).

A single system for health intelligence surveillance: The rise in illnesses and health risks throughout the world has brought attention to how critical it is to improve surveillance systems for early diagnosis and monitoring. Countries are now more vulnerable to risks to global health as a result of the recent COVID-19 pandemic, which calls for coordinated efforts beyond the scope of individual organizations and nations. There has been a widespread appeal to switch from the conventional reactive methods of surveillance to proactive ones that involve system integration. Responses to new health risks may be made more thorough and prompter with the inclusion of One Health Intelligence surveillance systems.

Tracking variants of SARS-CoV-2 in wastewater: It is well known that one of the most effective methods for detecting and tracking the dynamics of SARS-CoV-2 and its circulating lineages in communities is wastewater-based epidemiological surveillance. The genetic diversity of SARS-CoV-2 in wastewater samples was found in a study conducted in Dhaka, Bangladesh. Interestingly, as early as September 2020, wastewater samples were found to contain the Delta variant (B.1.617.2), which was discovered in clinical samples in May 2021. Surveillance of the environment such as wastewater-based epidemiology can yield important information on the dynamics of virus variations and support evidence-based public health measures (Haque et al., 2023).

A single system for health early warning and response (OH-EWRS): Through the reinforcement of "bottom-up" techniques, the OH-EWRS seeks to enhance zoonotic disease surveillance and notification, especially in hotspot areas where diseases emerge. In order to create an integrated One Health preventive and control system, the system encourages cooperation among stakeholders, including NGOs, governmental organizations, research institutions, the commercial sector, and local communities. It should be the duty of governmental entities to operationalize, manage, and institutionalize the OH-EWRS. However, it is crucial to get input from and feedback to pertinent stakeholders using a top-down and bottom-up strategy (Hassan et al., 2023).

21.5 METHODS OF PREVENTION AND CONTROL

One Health WASH in farming communities: One Health approach emphasizes the importance of various preventative measures across the human, animal, and environmental spectrum. The World Antimicrobial Awareness Week in 2022 centered on the idea of "Preventing AMR together" in order to advance best practices and raise public knowledge of antimicrobial resistance (AMR). Infections are often prevented and controlled by using strategies like good cleanliness, biosecurity, immunizations, and other ways to boost immunity. Terms like "AMR-specific" and "AMR-sensitive" were coined by the World Bank to characterize actions that either directly or indirectly aid in the reduction of AMR. Since they have the ability to tangentially address AMR, for example, policies like as water, sanitation, and hygiene (WASH) are acknowledged as critical to supporting AMR initiatives. Investments in interventions like WASH are "AMR-Smart" and essential for preventing and controlling infections, especially in healthcare settings (Jimenez et al., 2023).

Hand Hygiene for All: More emphasis was placed on hand cleanliness by the World Health Organization (WHO) during COVID-19 pandemic. The "Hand Hygiene for All" campaign was started by the WHO to continuously enhance hand hygiene practices as a way to limit the spread of SARS-CoV-2 and maintain excellent practices after the epidemic. To guarantee proper hand hygiene practices, proper infrastructures should be given in healthcare facilities and communal areas, such as schools and public transportation.

The WHO's mission is to save lives: achieving proper hand hygiene at the point of care is the goal of the Clean Your Hands initiative. The proper infrastructure and materials must be available for the effective execution of hand hygiene actions, which calls for system reform (Allegranzi et al., 2021).

21.6 COVID-19: ONE YEAR AND SOMEWHAT FURTHER?

China quickly launched a nationwide push to combat COVID-19 after discovering the virus. In order to ensure that all control measures are effective, the nation stressed the importance of the "Four Earlies": early identification, early reporting, early isolation/quarantine, and early treatment. The implementation of One Health strategy has proven beneficial in several areas such as workforce development, real-time surveillance, zoonotic disease management, national laboratory system enhancement, and emergency response operations. Continuous progress is essential, and the overarching strategy was early containment. Every community, organization, and individual were encouraged to take part in the prevention and control of the pandemic (Gao, 2021).

Mosquito control in Western Australia: In Western Australia, vector-borne illnesses pose a serious threat to public health. Numerous diseases may be spread by mosquitoes, which can also be a major annoyance. Statewide initiatives for prevention include physical, chemical, and cultural control methods for restricting mosquito breeding. However, in developed metropolitan settings, individuals have the primary duty for mosquito protection as breeding places are frequently situated within private residential landholdings. The research evaluated locals' knowledge, attitudes, and practices about mosquito management in their homes as well as the extent to which "backyard" mosquito breeding contributes to mosquito issues in the area (Mainali et al., 2017).

21.7 CHALLENGES AND HURDLES

Unsustainable funding models: Programs for the community-based management of acute malnutrition (CMAM), which are crucial for treating malnutrition in communities that are susceptible, sometimes depend on limited funding. These initiatives are carried out in nations with inadequate health systems and policies and are mostly supported by donors. Unsustainable financing models, high operating expenses, cultural preconceptions, and obstacles pertaining to access are the primary issues facing CMAM programs. Government funding for nutrition programs is either non-existent or allocated in small amounts in national budgets and fiscal policies; nutrition supplies are not included on the list of necessary drugs. For effective implementation and sustainability of such programs, there is a need for long-term funding commitments and integration into national health and nutrition strategies (Renzaho et al., 2022).

Combining adaptation to climate change and disaster risk reduction: The frequency and magnitude of natural disasters are increasing due to climate change. While both disaster risk reduction (DRR) and climate change adaptation (CCA) aim to reduce vulnerability and increase resilience, integrating interventions for DRR and CCA remains a challenge. DRR and CCA efforts often operate in silos, leading to overlapping efforts and inefficient use of resources. This is particularly problematic for resource-constrained countries, which face challenges related to scarce resources and inconsistent policies surrounding DRR and CCA. A political economy perspective can provide insights into the actions and funding allocation decisions of DRR and CCA by different stakeholders. Effective integration requires understanding the influence of powerful decision-makers, resource distribution mechanisms, and government interactions with stakeholders (Wen et al., 2023).

Challenges in preventing Clostridioides difficile infection: One of the main causes of illnesses related to healthcare in the United States is *Clostridium difficile*. Despite efforts to improve prevention, many hospitals continue to have elevated infection rates. Challenges include the need for standardized surveillance definitions, consistent prevention efforts, and adequate infrastructure for hand hygiene practices. Acute care hospitals are required by the Centers for Medicare & Medicaid Services (CMS) to report to the CDC's National Healthcare Safety Network (NHSN), but *C. difficile* remains a leading cause of healthcare-associated infections. Continuous efforts, including standardized surveillance, consistent prevention measures, and public reporting, are essential to address the challenges of preventing *C. difficile* infections (Dubberke et al., 2019).

Payment mechanisms for patient-centered care integration: The treatment given may vary depending on how health services are reimbursed. Policies to improve provider collaboration and promote integrated care are being introduced in several nations. However, integrating interventions remains a challenge due to distinct origins and articulations by different researchers and practitioners. Aligning budgets and goals, distributing financial risk and rewards across providers, and fully integrating finance and administration are among the difficulties. Effective integration requires understanding the interactions between different stakeholders, resource distribution mechanisms, and government interactions with stakeholders (Kasteng et al., 2016).

21.8 CHARTING THE PATH FORWARD

The future of One Health research and practice: In order to combat newly developing and re-emerging infectious illnesses, the future of One Health research and practice offers considerable potential (EID-REID). As our knowledge of the relationships between animal, human, and environmental health expands, we may anticipate more focused and efficient approaches to illness prevention and management (Friedson-Ridenour et al., 2019). Advances in technology, including genomics and data analytics, will enable quicker detection and characterization of pathogens, facilitating rapid response. Additionally, One Health approach will expand beyond infectious diseases to tackle broader health challenges, such as the impact of environmental degradation on chronic diseases and the intersection of mental health and ecosystems. Research will increasingly focus on holistic health assessments, recognizing that human well-being is intricately linked to the health of the planet (Myers et al., 2017).

Recommendations for policymakers: Policymakers play a pivotal role in advancing One Health agenda. They should prioritize legislation and policies that support interdisciplinary collaboration, allocate resources for integrated surveillance systems, and promote research into zoonotic diseases and environmental health. This includes establishing clear guidelines for responsible antimicrobial use and supporting vaccination campaigns (Asim Anwar et al., 2019). Additionally, governments should invest in public education initiatives to raise awareness about One Health approach and its importance in safeguarding global health security. They should foster international collaboration, recognizing that infectious diseases have no borders (Karesh et al., 2012).

Recommendations for researchers: Researchers should continue to explore the complexities of the interactions between human, animal, and environmental health. This involves studying emerging zoonotic diseases, tracking the evolution of pathogens, and identifying potential spillover events.

Advanced molecular diagnostic techniques and data analysis will be invaluable tools in this pursuit (Woolhouse et al., 2012). Moreover, interdisciplinary research teams should collaborate to bridge gaps between sectors and generate holistic solutions. It will be crucial to do research on how diseases are regulated by biodiversity, how urbanization affects disease transmission, and how climate change affects human health (Ostfeld, 2017).

Recommendations for healthcare professionals: Healthcare professionals should embrace One Health mindset in their daily practice. This means considering not only the individual patient but also their broader context, including potential environmental exposures and animal interactions (Okello et al., 2014). They should stay updated on vaccination guidelines, antimicrobial stewardship, and vector-borne disease risks. Public health practitioners should be prepared to respond swiftly to outbreaks, leveraging integrated surveillance systems and collaborating with veterinarians and environmental experts (Morse et al., 2012).

Continued interdisciplinary collaboration: Interdisciplinary collaboration remains the linchpin of One Health approach. It is essential to convene experts from diverse fields, including medicine, veterinary science, ecology, epidemiology, and social sciences, to jointly address health challenges (Degeling et al., 2015). Public–private partnerships, NGOs, and international organizations should facilitate knowledge exchange and resource sharing. Academic institutions should promote interdisciplinary education and research, nurturing the next generation of One Health leaders (Sinclair, 2019).

21.9 CONCLUSION

In light of the constantly changing risks to global health presented by virally-based emerging and re-emerging infectious diseases (REID) and infections (EID), the One Health approach stands as an indispensable beacon of hope. This comprehensive review has illuminated the path toward a more resilient and prepared world by uncovering key findings and their far-reaching implications. The One Health approach's fundamental component is its understanding of the complex relationships that exist between the health of people, animals, and the environment. Through this holistic framework, we have learned that addressing viral EID-REID is not a one-dimensional task; rather, it requires multifaceted, interconnected strategies. The interdependencies between these health realms have been underscored, emphasizing how intricate interactions can either facilitate or impede the emergence and transmission of viral diseases. Zoonotic diseases, with their ability to jump between species, serve as poignant examples of these complex relationships. This chapter has also shone a spotlight on real-world success stories, showcasing the transformative impact of collaborative efforts among healthcare experts, veterinarians, ecologists, and other stakeholders. From RSV vaccine development in Europe to One Health initiatives in Ethiopia and Pakistan, these case studies underscore the profound influence of interdisciplinary coordination on illness prevention and management. Moreover, One Health approach's pivotal role in early detection and surveillance of viral outbreaks has been highlighted. Integrated surveillance systems have proven to be vital, exemplified by Uganda's Joint External Evaluation and the use of wastewater-based epidemiology to monitor SARS-CoV-2 variants. Timely detection has made a profound difference in our ability to respond effectively. Strategies for prevention and control discussed in this chapter emphasize the importance of vaccination campaigns, vector control measures, responsible antimicrobial use, and public education initiatives. These approaches are essential components of our arsenal against viral EID-REID, and their successful implementation can significantly reduce the burden of disease. Yet, this journey has not been without its challenges and hurdles. The road to implementing One Health effectively has been paved with challenges, ranging from unsustainable finance methods to problems with coordination among many stakeholders, policy alignment, and regulatory barriers. It is critical that we take these issues head-on in order to strengthen our reaction as a group to risks to world health.

REFERENCES

Aggarwal, D., & Ramachandran, A. (2020). One health approach to address zoonotic diseases. *Indian Journal of Community Medicine,* 45(Suppl. 1), S6–S8.

Allegranzi, B., Tartari, E., & Pittet, D. (2021). An important announcement from the World Health Organization:"Seconds save lives—clean your hands": The May 5, 2021, World Health Organization SAVE LIVES: Clean your hands campaign. *Infection Control & Hospital Epidemiology*, 42(5), 649–651.

Asim Anwar, A. A., Sajid Anwar, S. A., Muhammad Ayub, M. A., Faisal Nawaz, F. N., Shabir Hyder, S. H., Noman Khan, N. K., & Imran Malik, I. M. (2019). Climate change and infectious diseases: Evidence from highly vulnerable countries. *Iran Journal Public Health*, 48(12), 2187–2195.

Bakiika, H., et al. (2023). Contribution of the one health approach to strengthening health security in Uganda: A case study. *BMC Public Health*, 23(1), 1498.

Braam, D. H., Tasker, A., Parravani, E., & Saville, K. (2023). A fair share? Animal health actors and resources in one health initiatives: A multisite case study in Ethiopia and Pakistan. *CABI One Health*, 2023, ohcs202300015.

Dafale, N. A., Srivastava, S., & Purohit, H. J. (2020). Zoonosis: An emerging link to antibiotic resistance under "one health approach". *Indian Journal of Microbiology,* 60(2), 139–152.

Degeling, C., Johnson, J., Kerridge, I., Wilson, A., Ward, M., Stewart, C., & Gilbert, G. (2015). Implementing a one health approach to emerging infectious disease: Reflections on the socio-political, ethical and legal dimensions. *BMC Public Health*, 15, 1–11.

Dubberke, E. R., Rohde, J. M., Saint, S., Jones, K., Snyder, A., Rolle, A. J., & Chopra, V. (2019). Quantitative results of a national intervention to prevent Clostridioides difficile infection: A pre–post observational study. *Annals of Internal Medicine*, 171(Suppl. 7), S52–S58.

Friedson-Ridenour, S., Dutcher, T. V., Calderon, C., Brown, L. D., & Olsen, C. W. (2019). Gender analysis for one health: Theoretical perspectives and recommendations for practice. *Ecohealth*, 16, 306–316.

Gao, G. F. (2021). COVID-19: A year long and beyond? *China CDC Weekly*, 3(8), 157.

Haque, R., et al. (2023). Monitoring SARS-CoV-2 variants in wastewater of Dhaka City, Bangladesh: Approach to complement public health surveillance systems. *Human Genomics*, 17(1), 58.

Hassan, O. A., de Balogh, K., & Winkler, A. S. (2023). One health early warning and response system for zoonotic diseases outbreaks: Emphasis on the involvement of grassroots actors. *Veterinary Medicine and Science*, 9(4), 1881–1889.

Janse, M., Sesa, G., & van de Burgwal, L. (2023). A case study of European collaboration between the veterinary and human field for the development of RSV vaccines. *Vaccines*, 11(7), 1137.

Jimenez, C. E. P., Keestra, S. M., Tandon, P., Pickering, A. J., Moodley, A., Cumming, O., & Chandler, C. I. (2023). One health WASH: An AMR-smart integrative approach to preventing and controlling infection in farming communities. *BMJ Global Health*, 8(3), e011263.

Kapoor, S. (2022). Developing framework toward one health approach: Relevance and future implications. *Journal of Public Health and Primary Care,* 4.

Karesh, W. B., et al. (2012). Ecology of zoonoses: Natural and unnatural histories. *The Lancet*, 380(9857), 1936–1945.

Kasteng, F., Magnusson, J., Borgermans, L., & Kalseth, J. (2016). An overview of payment mechanisms and contractual models to support patient-centred care integration. *International Journal of Integrated Care*, 16(6), A337.

Li, Q., Bergquist, R., Grant, L., Song, J. X., Feng, X. Y., & Zhou, X. N. (2022). Consideration of COVID-19 beyond the human-centred approach of prevention and control: The ONE-HEALTH perspective. *Emerging Microbes & Infections*, 11(1), 2520–2528.

Mainali, S., Lamichhane, R. S., Clark, K., Beatty, S., Fatouros, M., Neville, P., & Oosthuizen, J. (2017). "Looking over the backyard fence": Householders and mosquito control. *International Journal of Environmental Research and Public Health*, 14(3), 246.

Morse, S. S., et al. (2012). Prediction and prevention of the next pandemic zoonosis. *The Lancet*, 380(9857), 1956–1965.

Myers, S. S., et al. (2017). Climate change and global food systems: Potential impacts on food security and undernutrition. *Annual Review of Public Health*, 38(1), 259–277.

Okello, A. L., Bardosh, K., Smith, J., & Welburn, S. C. (2014). One health: Past successes and future challenges in three African contexts. *PLoS Neglected Tropical Diseases*, 8(5), e2884.

Ostfeld, R. S. (2017). Biodiversity loss and the ecology of infectious disease. *The Lancet Planetary Health*, 1(1), e2–e3.

Renzaho, A. M., Dachi, G., Ategbo, E., Chitekwe, S., & Doh, D. (2022). Pathways and approaches for scaling-up of community-based management of acute malnutrition programs through the lens of complex adaptive systems in South Sudan. *Archives of Public Health*, 80(1), 203.
Ryu, S., et al. (2022). Epidemiology and transmission dynamics of infectious diseases and control measures. *Viruses,* 14(11), 2510.
Saba Villarroel, P. M., et al. (2023). Emerging and re-emerging zoonotic viral diseases in Southeast Asia: One health challenge. *Frontiers in Public Health*, 11, 1141483.
Shingankar, S., & Ugemuge, P. (2022). One health approach prevents future pandemics: A review article. *Journal of Datta Meghe Institute of Medical Sciences University*, 17(Suppl. 1), S88–S93.
Sinclair, J. R. (2019). Importance of a one health approach in advancing global health security and the sustainable development goals. *Rev Sci Tech*, 38(1), 145–154.
Steenkamer, B., et al. (2023). Population health management in Belgium: A call-to-action and case study. *BMC Health Services Research*, 23(1), 659.
Suu-Ire, R. D., Obodai, E., Bonney, J. K., Bel-Nono, S. O., Ampofo, W., & Kelly, T. R. (2021). Viral zoonoses of national importance in Ghana: Advancements and opportunities for enhancing capacities for early detection and response. *Journal of Tropical Medicine*, 2021(1), 8938530.
Tomori, O., & Oluwayelu, D. O. (2023). Domestic animals as potential reservoirs of zoonotic viral diseases. *Annual Review of Animal Biosciences*, 11(1), 33–55.
Varela, K., Goryoka, G., Suwandono, A., Mahero, M., Valeri, L., Pelican, K., & Salyer, S. J. (2023). One health zoonotic disease prioritization and systems mapping: An integration of two one health tools. *Zoonoses and Public Health*, 70(2), 146–159.
Wen, J., Wan, C., Ye, Q., Yan, J., & Li, W. (2023). Disaster risk reduction, climate change adaptation and their linkages with sustainable development over the past 30 years: A review. *International Journal of Disaster Risk Science*, 14(1), 1–13.
Woolhouse, M., Scott, F., Hudson, Z., Howey, R., & Chase-Topping, M. (2012). Human viruses: Discovery and emergence. *Philosophical Transactions of the Royal Society B: Biological Sciences*, 367(1604), 2864–2871.

22 The Impact of Balanced Diet in Managing Degenerative Consequences of SARS-CoV-2 Infection

Sangita Agarwal, Soumendra Darbar, and Srimoyee Saha

22.1 INTRODUCTION

Nutrition may have a substantial impact on a person's long-term COVID susceptibility. Natural resources, including plants, micronutrients, and/or macronutrients, can enhance the immune response during COVID-19 in addition to reducing infection (Motti et al., 2022). The protection against neuroinflammation and oxidative stress that nutrients, including vitamins (B1, B6, B9, B12, C, D, and E), fatty acids, minerals (iron, zinc), and oligoelements (selenium), provide is well known. They significantly improve cognitive processes. The maintenance of immunological homeostasis and resilience, as well as resistance to disease, including infections with viral and bacterial pathogens, may be significantly influenced by balanced nutrition (Scarmeas et al., 2018). By affecting gene expression, cell activation, and interfering with signaling molecules that shape and modulate the immune system, malnutrition has long-lasting impacts on both physical and mental health (Curtis et al., 2016). Therefore, a poor diet and inadequate nutrition may dramatically impair the immune system and increase vulnerability to viral diseases, such as SARS-CoV-2. Human body is protected against the pathogens by the immune system. The body's first immunological response to any foreign substance is the formation of physical barriers such as ciliated epithelial cells, skin, mucous membranes, and blankets. When the first line of defense fails, the body's immune system responds to differentiate between non-self and self-particles through innate and/or adaptive immunity (2021; WHO, 2021a). The foreign substance triggers a well synchronized response in the body through various cell-mediated processes utilizing the receptors, various types of immune cells, and cytokines. Immediate response is triggered in innate immunity through the pathogen-associated molecular patterns which is non-specific, and mutable. Adaptive responses are through recognition of specific antigens and generation of antibodies to destroy pathogens (Pecora et al., 2020; Cai et al., 2020). The health and vitality of all living beings is maintained by nutrients generally through diet, which help in energy supply, growth, immune defense, cellular function, and tissue development (Koithan & Devika, 2010). Human diet consists of both macro- and micronutrients along with dietary fiber, all of which have a vital role in biological activities (Walls et al., 2016). Lack of sufficient macro- and micronutrients in the diet can impair the immune system's ability to defend against pathogen attack. Additionally, a number of diseases result in the elimination of crucial nutrients, which causes nutrient deficiencies (Oz, 2017). Nutrients can help in modulating chronic inflammatory and autoimmune diseases, can mediate pro- and anti-inflammatory responses (Walls et al., 2016; Wu et al., 2019). The presence of some micronutrients at every stage is important for immune response. The efficiency of immune system is determined by the person's nutritional condition. Body's ability to support innate immune response is impaired in undernourished persons, who consume inadequate micronutrients. However, all illness are not influenced by nutritional status. Although the clinical progression and

DOI: 10.1201/9781032721811-22

prognosis of number of diseases, including measles, viral diarrhea, tuberculosis, pneumonia, and others, can be predicted, and many ailments can be minimally or moderately influenced by nutritional status (Alpert, 2017). Many researchers have presented experiential evidences of vitamins and minerals in boosting immune responsiveness. It has been demonstrated that minerals, vitamins, and trace elements are essential to maintain immunocompetence in animal models and in some cases nutritionally deficient humans (Suksmasari, 2015). The vitamins (A, B, C, D, and E) micronutrients like zinc, selenium, riboflavin, and carotene are vital in immunity (Widasari et al., 2020). Addition of supplementation and nutrients through changes in diet can be helpful to improve resistivity to infection through improved immune responses (Gombart et al., 2020, Motti et al., 2022b). There is an interaction between nutrients, the microbiota present in the gut, and the immune system. The gastrointestinal tract with its lymphoid tissue constitutes the major component of immune system, which is providing protection to gut from pathogens (WHO, 2021b; WHO, 2021c). The microbiota present in the colon, such as fungi, bacteria, and others, can trigger immunological responses. The nutrients are absorbed in gut-associated lymphoid tissue.

22.2 MACRO- AND MICRONUTRIENTS SUPPLEMENTS FOR SUPPORTING IMMUNE SYSTEM

The host immune responses are suppressed and dysregulated during nutrient deficiency. Since we were children, the concept of balanced diet has been iterated by our parents and doctors which should consist of fruits, nuts, vegetables, legumes, dairy products, and whole grains. The healthy diet is essential for maintaining sufficient immune cells and antibodies to combat infections. Deficiency of Zn escalates the body's vulnerability to diseases and infections. It is important in growth and maintenance of immune cells (Bonaventura et al., 2014). Iron deficiency can debilitate the immunity as it is required in growth and development of immune system. Magnesium is significant in regulating functions like immunoglobulin synthesis, antibody-independent cytolysis, adherence of immune cells, binding of immunoglobulin M lymphocyte, and response of macrophage toward lymphokines (16). According to report (Jayawardena et al., 2020), magnesium is essential for the immune system's defense against viral infections. Calcium is important for maintaining healthy bones as well as for removing infections from our body. One indicator of the severity of a SARS-CoV-2 infection is hypocalcemia.

22.3 ROLE OF VITAMINS IN ADDRESSING INFECTIONS

Vitamins and trace elements are implicated in augmenting the immunity and reducing infections. Infections can impair the immune system, and this is especially true for someone who is nutritionally undernourished. Vitamins are important and essential components of a balanced diet because of their immunomodulatory effects and antioxidant properties (Shakoor et al., 2021). Vitamins E and C can fight free radical species owing to their antioxidant properties (Gombart et al., 2020). Vitamin B complex is essential in immune regulation in colon and plays a supportive role as an intestinal barrier (Lindschinger et al., 2019). Vitamin A facilitates both types of immunity (innate and adaptive) by acting as T-cell effectors (Raverdeau & Mills, 2014). Vitamin D plays an immunomodulatory role in autoimmune and infectious diseases. Vitamin K is crucial in hemostasis because it functions as co-enzyme and co-factor, involved in other physiological functions (Janssen & Walk, 2020).

22.4 ROLE OF VITAMIN A IN PREVENTION OF SARS-COV-2

Significant advantages for the immune system have been associated with the fat-soluble vitamin retinol. Lack of vitamin A (Vit A) can worsen immune system irregularities in the gastrointestinal, respiratory, and ocular systems and raise those systems' vulnerability to typical infections (Alpert, 2017b). Different recognized forms of vitamin A obtained from food include all-trans retinol, retinyl

TABLE 22.1
Recommended Doses of Vitamin A in Different Ages, Including Pregnant and Lactating Women

		Recommended Dose	
Sl. No.	**Age**	**Female**	**Male**
1.	0–6 months	400 mcg (16,000 IU)	400 mcg (16,000 IU)
2.	7–12 months	500 mcg (20,000 IU)	500 mcg (20,000 IU)
3.	1–4 years	300 mcg (12,000 IU)	300 mcg (12,000 IU)
4.	5–8 years	400 mcg (16,000 IU)	400 mcg (16 000 IU)
5.	9–14 years	600 mcg (24.000 IU)	600 mcg (24,000 IU)
6.	15–18 years	700 mcg (28,000 IU)	900 mcg (36,000 IU)
7.	Above 19 years	700 mcg (28,000 IU)	900 mcg (36,000 IU)
8.	Pregnant woman	770 mcg (30,800 IU)	
9.	Lactating woman	1,300 mcg (52,000 IU)	

esters, and beta-carotene (Mora et al., 2008; Blomhoff & Blomhoff, 2006; Moise et al., 2007). Commonly occurring vitamin A metabolites have been found to have significant effects on adaptive immune responses (Mora et al., 2008). Retinoic acid has been strongly linked to increased cytotoxicity (Dennert & Lotan, 1978), T-cell proliferation (Ertesvag et al., 2002), Il-2 secretion stimulation, T-cell signaling (Ertesvag et al., 2002), and T-cell activities. Mice with severe abnormalities in TH-cell activity had also been linked to vitamin A deprivation (Carman et al., 1989). Taking into account the aforementioned points, vitamin A (Table 22.1) can be regarded as a crucial component in immune system enhancement. During viral infections, such as COVID-19, it can strengthen the immunological state and serve as an effective preventive component.

22.5 ROLE OF VITAMIN E AND C IN PREVENTION OF SARS-COV-2

A fat-soluble vitamin, vitamin E alpha tocopherol, can be found in nuts, seeds, green leafy vegetables, and vegetable oils (Carr & Maggini, 2017). Vitamin E is also abundant in green leafy vegetables (Table 22.2). Due to its antioxidant properties, it can neutralize free radicals and is implicated in immune function modulation (Lee & Han, 2018). Although it is rarely deficient, supplementation improves immunological responses, even in elderly people (Galmés et al., 2018). For both men and women, a daily dietary intake of 75 mg and 90 mg of vitamin C, respectively, is advised (Table 22.2). By boosting immunity, vitamin C, a potent antioxidant, aids in the prevention of infections (Mikkelsen & Apostolopoulos, 2019). It helps in the production of the lymphocytes and phagocytes that make up the body's defense system. Additionally linked to defense against reactive oxygen species and free radicals is vitamin C. The skin's protective barrier can be strengthened by vitamin C (Beck et al.,1994). Among the foods high in vitamin C include broccoli, bell peppers, strawberries, gooseberries, and lemons. Vitamin C supplements have been discovered to be crucial in the fight against the corona virus.

22.6 ROLE OF VITAMIN K IN PREVENTION OF SARS-COV-2

Vitamin K is natural content present in some food, it functions as a co-enzyme, synthesizes proteins, and is also involved in various biological functions (Janssen & Walk, 2020). The dietary supplements of vitamin K (Table 22.3) are K1 and K2 (Luo et al., 2012). In case of inflammation in the body due to SARS-CoV-2 infection, synthesis of Matrix Gla protein in the lungs increases to safeguard the extracellular matrix of the lungs from degradation. The synthesis of Matrix Gla

TABLE 22.2

Recommended Doses of Vitamin E and Vitamin C in Different Ages, Including Pregnant and Lactating Women

Vitamin E

		Recommended Dose	
Sl. No.	**Age**	**Female**	**Male**
1.	0–6 months	4 mg (6 IU)	4 mg (6 IU)
2.	7-12 months	5 mg (7.5 IU)	5 mg (7.5 IU)
3.	1–4 years	6 mg (9 IU)	6 mg (9 IU)
4.	5–8 years	7 mg (10.4 IU)	7 mg (10.4 IU)
5.	9–14 years	11 mg (16.4 IU)	11 mg (16.4 IU)
6.	Above 15 years	15 mg (22.4 IU)	15 mg (22.4 IU)
7.	Pregnant woman	15 mg (22.4 IU)	–
8.	Lactating woman	19 mg (28.4 IU)	–

Vitamin C

		Recommended dose	
Sl. No.	**Age**	**Female**	**Male**
1.	0–6 months	40 mg (800 IU)	40 mg (800 IU)
2.	7–12 months	50 mg (1,000 IU)	50 mg (1,000 IU)
3.	1–4 years	15 mg (300 IU)	15 mg (300 IU)
4.	5–8 years	25 mg (500 IU)	25 mg (500 IU)
5.	9–14 years	45 mg (900 IU)	45 mg (900 IU)
6.	15–18 years	65 mg (1,300 IU)	75 mg (1,500 IU)
7.	Above 19 years	75 mg (1,500 IU)	90 mg (1,800 IU)
8.	Pregnant woman	80 mg (1,600 IU)	–
9.	Lactating woman	85 mg (1700, IU)	–
10.	Smokers	35 mg (700 IU)	

protein is dependent on vitamin K, thereby during this type of infection utilization of vitamin K is also increased (McCann & Ames, 2009). Vitamin K1 can help patients with COVID-19 who experience thrombotic problems since it is necessary for the activation of hepatic coagulation factors (Klok et al., 2020).

22.7 ROLE OF VITAMIN D IN PREVENTION OF SARS-COV-2

Vitamin D has an immunomodulatory effect that boosts both innate and adaptive immune responses as well as the production of antiviral peptides. Acute respiratory tract infections (RTI), particularly epidemic influenza, are directly correlated with low serum vitamin D levels (Bikle, 2022). Randomized clinical trials showed that vitamin D supplementation reduced the risk when compared to people who are more likely to experience vitamin D insufficiency during this global pandemic. This is due to the SARS-CoV-2 infection's rapid dissemination among patients. In order to keep the circulating 25(OH)D at the ideal levels (75–125 nmol/L), doctors have advised vitamin D supplements to patients who have insufficient vitamin D levels in their blood during this global pandemic (Malaguarnera, 2020). Researchers from various parts of the world have noted a strong correlation between the global average vitamin D level (Table 22.3) and both the incidence of COVID-19 infections and infections-related deaths (Ilie et al., 2020; Oristrell et al., 2022). Vitamin D supplementation is advised to lower the incidence of COVID-19 infection, criticality, and mortality (Grant et al., 2020). Vitamin D supplementation can

TABLE 22.3
Recommended Doses of Vitamin K and Vitamin D in Different Ages, Including Pregnant and Lactating Women

Vitamin K

		Recommended Dose	
Sl. No.	**Age**	**Female**	**Female**
1.	0–6 months	2.0 mcg (80 IU)	2.0 mcg (80 IU)
2.	7–12 months	2.5 mcg (100 IU)	2.5 mcg (100 IU)
3.	1–3 years	30 mcg (1,200 IU)	30 mcg (1,200 IU)
4.	4–8 years	55 mcg (2,200 IU)	55 mcg (2,200 IU)
5.	9–13 years	60 mcg (2,400 IU)	60 mcg (2,400 IU)
6.	14–18 years	75 mcg (3,000 IU)	75 mcg (3,000 IU)
7.	19–50 years	90 mcg (3,600 IU)	120 mcg (4,800 IU)
	50–70 years	90 mcg (3,600 IU)	120 mcg (4,800 IU)
	Above 70 years	90 mcg (3,600 IU)	120 mcg (4,800 IU)
8.	Pregnant woman	75mcg (3,000 IU)	
9.	Lactating woman	90 mcg (3,600 IU)	

Vitamin D

Sl. No.	**Age**	**Recommended Dose**
1.	0–12 months	10 mcg (400 IU)
2.	1–14 years	15 mcg (600 IU)
3.	15 –18 years	15 mcg (600 IU)
4.	19 –70 years	15 mcg (600 IU)
5.	Above 71 years	20 mcg (800 IU)
6.	Pregnant and lactating women	15 mcg (600 IU)

also be used to treat influenza, coronavirus, and pneumonia infections. According to data from a retrospective cohort study by Demir et al., vitamin D insufficiency was linked to a higher risk of coronavirus infection (Grinstein & Klip, 1989). The patients who tested positive for COVID-19 and had sufficient vitamin D levels had significantly lower D-dimer levels and CRP levels in their blood, as well as less recurrence of ground-glass opacity on chest CT scans and shorter hospital stays.

22.8 ROLE OF VITAMIN B COMPLEXES IN PREVENTION OF SARS-COV-2

The benefits of vitamin B include improved immunological, respiratory, and cell functions. For healthy cellular activity, the following vitamins are necessary (Zhang & Liu, 2020; Michele et al., 2020). Thiamine assists in immune response against infection. It plays an effective role in decreasing the neuro inflammation and spikes antibody response, improving the functioning of the immune system. This effective antibody responses helped combating the COVID-19 infection and also decreasing the stay in hospitals (Özdemir et al., 2013). It is also implicated in inhibiting the enzyme carbonic anhydrase which aids in improving hypoxia associated with COVID-19 infection. More intensive research is required to support the importance of thiamine in infection caused by SARS-CoV-2. Both riboflavin and UV rays have the capacity to prevent the replication of DNA and RNA and have been proven to be effective against MERS-CoV infections. The effectiveness of riboflavin in lowering blood viral titer and, thus, lowering illness severity, has been demonstrated (Ragan et al., 2020). With its immunomodulatory properties, niacin, one of the components of nicotinamide adenine dinucleotide phosphate and nicotinamide adenine dinucleotide, aids in lowering inflammatory mediators such as interleukin-1, interleukin-6, and homo trimer protein, also known as tumor necrosis factor, during an infection (Mikkelsen & Apostolopoulos, 2018; Mikkelsen et al., 2016, 2017). Patients who had contracted the SARS-CoV-2 virus were clearly affected by IL-6. Furthermore, nicotinamide prevents viruses like hepatitis B and human immunodeficiency virus from multiplying. It aids in fortifying the immune system's defense against infection and may be a useful agent in decreasing inflammation (Liu et al., 2020; Mehmel et al., 2020). Vitamin B5 aids in the healing of wounds and the decrease of inflammation. It is yet unclear what role vitamin B5 has in COVID-19. Due to its function in inflammatory pathways, vitamin B6 deficiency or pyridoxal 5′-phosphate deficiency reduces immunological responses (Table 22.4). It helps to relieve COVID-19 symptoms by lowering pro-inflammatory cytokines and stopping hypercoagulation (Nix et al., 2015). The creation of DNA, proteins, and the adaptive immune system all depend on vitamin B9, often known as folic acid. It can prevent the binding of the SARS-CoV-2 spike protein as well as furin, a component of bacterial and viral reproduction. Folic acid plays a crucial role in treating respiratory diseases linked to COVID-19 (Ni et al., 2020). Tetrahydrofolic acid and its derivatives, such as 5-methyl tetrahydrofolic acid, have affinity for SARS-CoV-2 and can attach to the structure by molecular docking (Kumar et al., 2021). Oxidative stress, reactive oxygen species, and inflammation are brought on by a vitamin B12 or cobalamin deficiency; coronavirus infection exhibits very identical symptoms (Table 22.4). Clinical research has demonstrated the ability of magnesium, vitamin D, and vitamin B12 to lessen the severity of SARS-CoV-2 virus-related symptoms (Sabry et al., 2020; Fuhrman, 2020; Calder et al., 2020).

TABLE 22.4

Recommended Daily Intakes and Regular Common Sources of the Fat- and Water-Soluble Vitamins

Sl. No.	Vitamins	Regular Common Sources	Daily Requirement (Adults >19 years)
1.	A (retinol)	Retinol (liver, dairy, fish), carotenoids (sweet potatoes, carrots, spinach), green leafy vegetables, nuts, tomatoes, oranges, ripe yellow fruits, guava, milk, liver, carrots, broccoli and watermelon	650–850 mcg
2.	B1 (thiamine)	Fresh fruits, corn, cashew nuts, potatoes, sweet potatoes, peas, wheat, milk, dates, black beans, etc.	1.1–1.2 mg
3.	B2 (riboflavin)	Bananas, grapes, mangoes, peas, pumpkin, dates, yoghurt, milk, mushrooms, popcorn, beef liver, etc.	1.1–1.3 mg
4.	B3 (niacin)	Meat, eggs, fish, milk products, guava, mushroom, peanuts, cereals, green peas, etc.	14–17 mg
5.	B5 (pantothenic acid)	Meat, kidney, egg yolk, broccoli, peanuts, fish, chicken, milk, yoghurt, legumes, mushrooms, avocado, etc.	4–5 mg
6.	B6 (pyridoxine)	Pork, chicken, fish, bread, wholegrain cereals, eggs, vegetables, soya beans, etc.	1.3 mg
7.	B7 (biotin)	Walnuts, peanuts, cereals, milk, egg yolks, salmon, pork, mushroom, cauliflower, avocados, bananas, raspberries, etc.	30–35 mcg
8.	B9 (folic acid)	Citrus fruits, green leafy vegetables, whole grains, legumes, beets, etc.	400 mcg
9.	B12 (cobalamin)	Fish, meat, poultry, eggs, milk, etc.	2.4 mcg
10.	C (ascorbic acid)	Fresh citrus fruits such as orange and grapefruit, broccoli, goat milk, black currant and chestnuts	75–95 mg
11.	E (tocopherol)	Sunflower seeds, wheat germ, almonds, potatoes, pumpkin, guava, mango, milk, nuts and seeds	14–18 mg
12.	D (calciferol)	Sunlight, fish oil, milk, fish, beef, cod liver oil, egg yolk, liver, chicken breast and cereals	620–860 IU
13.	K (phytonadione)	Leafy greens, soybeans, pumpkin, tomatoes, broccoli, mangoes, grapes, chestnuts, cashew nuts, beef and lamb	85–125 mcg

22.9 CONTRIBUTION OF MACRO- AND MICRONUTRIENTS TO SARS-COV-2 PREVENTION

- *Role of magnesium:* Humans have a high need for the macronutrient magnesium. Low levels of magnesium in the body, particularly in Western nations, are a result of the greater use of processed foods and fertilizers. Since the majority of magnesium (Table 22.5) is found in the inner layer of bones and is stored intracellularly, magnesium insufficiency is difficult to identify in its early stages. Magnesium is important for more than 300 ATP-dependent metabolic processes, including protein synthesis, RNA synthesis, and DNA synthesis. Additionally, it maintained blood pressure and controlled neuromuscular, signal, and blood glucose levels. The cardiovascular system, gastrointestinal tract, central nervous system, metabolism, and mood disorders are just a few of the organs and systems whose functioning are impacted by magnesium deficiency. A small number of researchers (Price et al., 2020) have hypothesized that patients with hypomagnesia and COVID-19 share pathophysiology, such as elevated inflammatory cytokines, decreased T cells, and dysfunctional endothelia. Due to endothelial damage, poor vascular integrity, and decreased CD4+ and CD8+ T lymphocytes, the effect on individuals who have both diabetes and hypertension manifests as lower organ function and a compromised immune system (Theofilis et al. 2021). By inhibiting L-type calcium channels, magnesium plays a critical role in reducing the levels of inflammatory markers in lung tissues, including prostaglandin E2, C-reactive protein, cytokines (IL-6), and cyclooxygenase-2. This, in turn, reduces the smooth muscle contraction of the airways, which is primarily seen in asthmatic patients (Grinstein & Klip, 1989). By increasing plasminogen activator inhibition, which prevents plasmin from developing on vascular endothelial cells, magnesium decreases the formation of blood clots. A lack of magnesium may promote thromboembolism (Lewis 2001), which has been seen in COVID-19 hospital patients. The health of COVID-19 patients is impacted by the delay in diagnosing hypomagnesemia. The dosage of magnesium and its dietary supplement varies depending on the individual, thus speaking with a doctor is advised for the right dosing and consumption.
- *Role of calcium:* Numerous studies have established the connection between calcium and immunological responses. The important immunological responses, such as the creation of antibodies and the killing of pathogens by neutrophils, are mediated by changes in the availability of calcium that is available in the cytoplasm in free form. The increase in free cytoplasmic calcium (Gronski et al., 2009, https://www.who. int/emergencies/

TABLE 22.5
Recommended Daily Intakes and Regular Common Sources of the Essential Major Minerals

Sl. No.	Essential Major Minerals	Regular Common Sources	Daily Requirement (Adults >19 years)
1.	Sodium (Na)	Salt, processed foods, green vegetables, canned soup, and dairy products	2,300 mg
2.	Potassium (K)	Beans, lentils, nuts, acorn squash, bananas, and dairy foods	4,700 mg
3.	Magnesium (Mg)	Green leafy vegetables, such as spinach, legumes, nuts, seeds, and whole grains. Almonds, cashews, and black beans	340–420 mg
4.	Calcium (Ca)	Milk and milk products, green leafy vegetables such as curly kale, okra, spinach, broccoli, and cheese	2,100–2,400 mg
2.	Chloride (Cl^-)	Tomatoes, lettuce, salt, seaweed, rye, celery, and olives	1,800–2,400 mg
5.	Phosphorus (P)	Red meat, seafood, salmon, poultry, yogurt, turkey and nut	600–800 mg

diseases/novel-coronavirus-2019/technical-guidance/naming-the-coronavirus-disease-(covid-2019)-and-the-virus-that-causes-it.) causes the membranes' permeability to cations to change, which in turn triggers the necessary physiological changes such as the production of free radicals, lymphocyte proliferation, and neutrophil degranulation (https://www.who.int/emergencies/diseases/novel-coronavirus-2019/events-as-they-happen). Calcium channel blockers verapamil and nifedipine suppress immunological responses by immobilizing calcium (Table 22.5). Immune response and free cytoplasmic calcium are related in a causal manner. In the signaling of calcium, calcium serves as the second messenger.

- *Role of iron:* Additionally, an important micronutrient called iron has a different recommended dietary daily limit depending on a person's age and gender. Pregnant ladies need more than non-pregnant women do. Due to its changing oxidation state (ferrous and ferric state), it can function as a catalyst in a number of intracellular processes, including DNA replication, as well as take part in the electron transport chain and oxygen transport (Tang et al., 2020). Innate immunity provides defense against germs by locking iron (Table 22.6) intracellularly and reducing intestinal absorption. Pathogens also need iron for survival. Both lactoferrin and hepcidin play this part. The virus needs the cofactor enzymes that need iron in order to survive and reproduce. Limiting metals like iron and chromium, which are also effective against SARS-CoV-2, are known to fight and impede the proliferation of viruses such as the human immunodeficiency virus, influenza, hepatitis, and others (Tang et al., 2020; Iotti et al., 2020). According to certain case studies, patients with hemochromatosis had poor prognoses for infections, but anemia makes them more vulnerable to infections that can be fatal. Hence, having iron is necessary.
- *Role of zinc:* A vital nutrient having a key function in immunity is zinc. Immune cells in innate and acquired immunity require supplemental zinc for cellular development (Table 22.6). For the production of antibodies, zinc is a crucial and necessary nutrient (Maares & Haase, 2016). It is also crucial for the production of proteins and enzymes that are necessary for healthy immunological function. The immunological responses are predominantly brought on by macrophages and immunoglobulins (Maywald et al., 2017). In addition to being essential for the growth and development of macrophages, T and B lymphocytes, and immunoglobulins, zinc also plays a crucial role in preventing the reproduction of RNA viruses, such as SARS-CoV (Fuhrman 2020). It is well known that a zinc deficiency can affect several processes, including phagocytosis, T lymphocyte

TABLE 22.6
Recommended Daily Intakes and Regular Common Sources of the Macro- and Microminerals

Sl. No.	Macro- and Microminerals	Regular Common Sources	Daily Requirement (Adults >19 years)
1.	Iodine (I)	Egg, seaweed, cod, milk, and yogurt	140–185 mcg
2.	Selenium (Se)	Brazil nuts, seafood, sardines, cereals, ham	310–460 mg
3.	Iron (Fe)	Nuts, dried fruit, spinach, wholemeal pasta and bread oysters, white beans	7–21 mg
4.	Copper (Cu)	Cashews, oysters, nuts, seeds, liver, crabs, shitake mushrooms, lobster, liver, leafy greens, and dark chocolate	900–940 mcg
5.	Manganese (Mn)	Mussels, nuts, soybeans, pineapple, coffee, tea, pecans, peanuts	1.8–2.5 mg
6.	Zinc (Zn)	Shellfish, oysters, poultry, crab, beef, lobster, pork, legumes, nuts, seeds, whole grains, fortified breakfast cereals	8.2–12.5 mg
7.	Fluoride (F)	Fruit juice, oatmeal, fluoridated water, Potatoes, crab, brewed black tea and coffee	3.5–4.5 mg

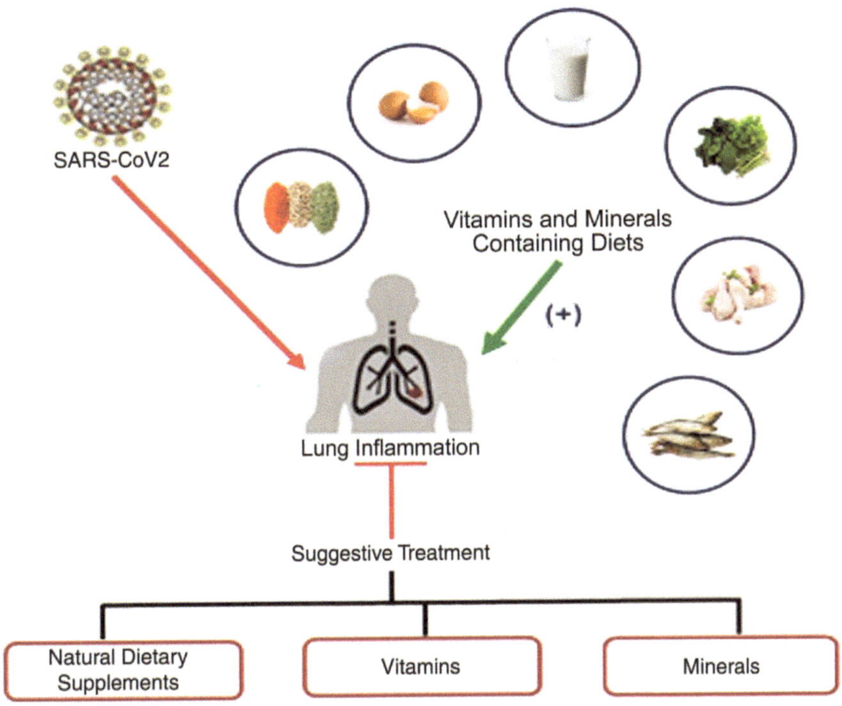

FIGURE 22.1 Role of vitamins and minerals in prevention of SARS-CoV-2.

synthesis, natural killer cell activity, complement activity, and T and B lymphocyte function (Razzaque, 2020; Wessels et al., 2017), and it can also increase the production of inflammatory cytokines (Figure 22.1). The barriers such as the epidermis is affected and also the respiratory and gastrointestinal mucosa is damaged during zinc deficiency (Read et al., 2019). Zinc also functions as an antioxidant, preventing the cells from being harmed by free radicals that are produced when the immune system is activated. As a key ion and cofactor for more than 300 enzymes, zinc plays a crucial role in maintaining the structural stability of the enzyme and regulating enzymatic activity. Zinc deficiency has been linked to lymphoid tissue atrophy in animal models (Te Velthuis et al., 2010), and young zinc-deficient mice displayed decreased immunological responses. In addition to raising the number of regulatory T cells and T17 cells (Figure 22.1), zinc deficiency also results in higher amounts of cytokines such IL-6, IL-8, and TNF-α (Weiss, 2002). Elderly people and young children in poor nations are more vulnerable to reduced respiratory infection. With zinc supplementation, immunological responses can be changed. When moderate doses of zinc were given to elderly people with zinc deficiencies, the overproduction of inflammatory cytokines was reduced, and the risk of respiratory infections was reduced in many persons (Wessling-Resnick, 2018). Zinc supplementation reduced mortality in patients with severe pneumonia, decreased the length of children's colds, and decreased their risk of developing pneumonia.

22.10 DISCUSSION

Humans who contracted SARS-COV-2 during the COVID-19 pandemic saw multiple reductions in body immunity, leaving them helpless to fend off the virus (WHO 2021a; WHO, 2021b; WHO, 2021c). Numerous health data from hospitals and doctors indicate that taking vitamins and key minerals supplements at the proper moment can help fight the illness (Johns Hopkins University and Medicine (Coronavirus Resource Center), 2021, Baden et al., 2021). For the treatment of COVID-19, the use of vitamins and supplements has produced excellent outcomes, including rapid recovery from corona virus infection. Drug treatments such as oral vitamin D, intravenous and oral vitamin C, oral vitamin D/magnesium/vitamin B12, oral zinc, oral combination zinc/ascorbic acid, and intravenous alpha-lipoic acid have shown positive outcomes as supports (Polack et al., 2020; Sadoff et al., 2021; Cavalcanti et al., 2020). A putative advantage of using vitamin D as an adjuvant in treating COVID-19 infections resulted from earlier claims that it had antiviral properties. Vitamin D levels and COVID-19 positive rate have been linked in a number of backward studies (Jee et al., 2013). In a study of patients older than 70, those with positive COVID-19 results had significantly lower median vitamin D levels than those with negative results (9.3 ng/mL vs. 23.1 ng/mL, respectively; $p = 0.037$) (Infusino et al., 2020). Another study that found a connection between positive COVID-19 tests and poor vitamin D status (defined as 20 ng/mL) at the time of testing (relative risk 1.77, 95% CI 1.12–2.81; $p = 0.02$) came to similar conclusions.

Micronutrient deficits have the potential to significantly worsen the clinical course of SARS-CoV-2 and promote the emergence of a number of sequelae. Patients with COVID-19 who receive vitamin C as a therapeutic intervention have a higher survival rate. This is mostly brought about by reducing immune response hyperactivation, which aids in reducing the cytokine storm (Bae & Kim, 2020). Research shows a link between low zinc levels and risk factors for severe COVID-19 issues. A zinc shortage is more common in immunocompromised, obese, or diabetic individuals (Elham et al., 2021). Magnesium plays a significant role in controlling the cytokine storm that occurs during COVID-19 infection. Diabetes mellitus and chronic renal disease both have a clinical effect that might lead to magnesium insufficiency. In order to manage patients with coexisting conditions, magnesium supplementation is highly advised (Shakeri et al., 2022).

Multiple micronutrient deficiencies, including those of vitamin D, vitamin C, vitamin A, zinc, selenium, copper, and magnesium, have a notable impact on the clinical course of COVID-19, where they can change the course and prognosis of the disease (Stipp, 2020; Grangé et al., 2015). There is little doubt that patients with malnutrition, and more especially, low levels of vital micronutrients, which can also have an impact on the innate and humoral immune systems, have a bad prognosis. On the other hand, studies that are now available indicate that a sufficient level of these micronutrients can enhance the outcomes of diseases, reduce complications, and maximize the effectiveness of COVID-19 vaccination.

22.11 CONCLUSION

Reduced immunity has been linked to a higher risk of infection during the COVID-19 pandemic. Numerous researchers from around the world have looked into the possibility of boosting immunity using plant-based foods and supplements. Researchers have also looked into how vitamins E, D, and C may help boost immunity. Although the need of a healthy diet has been taught since childhood, the epidemic made many people more aware of its advantages. In contrast to fruits like guava, kiwi, and oranges, some vegetables like spinach, beetroot, cauliflower, bell peppers, and eggplant are high in vitamin C. The meal should be supplemented with these vitamins and minerals to fight infections because vitamin C is believed to boost immunity. To determine the mechanism of action of supplements, vitamins, and nutrients in relation to the pathogenesis of SARs-CoV-2 infection, extensive research across continents is required.

ACKNOWLEDGMENT

The authors are grateful to the management of Dey's medical & RCCIIT for their valuable support.

REFERENCES

Alpert, P. T. (2017). The role of vitamins and minerals on the immune system. *Home Health Care Management & Practice*, 29(3), 199–202. https://doi.org/10.1177/1084822317713300.

Baden, L. R., et al. (2021). Efficacy and safety of the mRNA-1273 SARS-CoV-2 vaccine. *New England Journal of Medicine*, 384(5), 403–416.

Bae, M., & Kim, H. (2020). The role of vitamin C, vitamin D, and selenium in immune system against COVID–19. *Molecules*, 25(22), 5346.

Beck, M. A., Kolbeck, P. C., Rohr, L. H., Shi, Q., Morris, V. C., & Levander, O. A. (1994). Vitamin E deficiency intensifies the myocardial injury of coxsackievirus B3 infection of mice. *The Journal of Nutrition*, 124(3), 345–358.

Bikle, D. D. (2022). Vitamin D regulation of immune function during covid–19. *Reviews in Endocrine and Metabolic Disorders*, 23(2), 279–285.

Blomhoff, R., & Blomhoff, H. K. (2006). Overview of retinoid metabolism and function. *Journal of Neurobiology*, 66(7), 606–630. https://doi.org/10.1002/neu.20242.

Bonaventura, P., Benedetti, G., Albarède, F., & Miossec, P. (2014). Zinc and its role in immunity and inflammation. *Autoimmunity Reviews*, 14(4), 277–285. https://doi.org/10.1016/j.autrev.2014.11.008.

Cai, N. H., Chen, L., Yin, C., Liao, Y., Meng, X., Lu, C., Tang, S., Li, X., & Wang, X. (2020). The effect of micro-nutrients on malnutrition, immunity and therapeutic effect in patients with pulmonary tuberculosis: A systematic review and meta-analysis of randomised controlled trials. *Tuberculosis*, 125, 101994. https://doi.org/10.1016/j.tube.2020.101994.

Calder, P. C., Carr, A. C., Gombart, A. F., & Eggersdorfer, M. (2020). Optimal nutritional status for a well-functioning immune system is an important factor to protect against viral infections. *Nutrients*, 12(4), 1181.

Carman, J. A., Smith, S. M., & Hayes, C. E. (1989). Characterization of a helper T lymphocyte defect in vitamin A-deficient mice. *Journal of Immunology (Baltimore, Md.: 1950)*, 142(2), 388–393.

Carr, A. C., & Maggini, S. (2017). Vitamin C and immune function. *Nutrients*, 9(11), 1211.

Cavalcanti, A. B., et al. (2020). Hydroxychloroquine with or without Azithromycin in mild-to-moderate Covid–19. *New England Journal of Medicine*, 383(21), 2041–2052.

Curtis, L. J., Bernier, P., Jeejeebhoy, K., Allard, J., Duerksen, D., Gramlich, L., Laporte, M., & Keller, H. H. (2016). Costs of hospital malnutrition. *Clinical Nutrition*, 36(5), 1391–1396. https://doi.org/10.1016/j.clnu.2016.09.009.

Dennert, G., & Lotan, R. (1978). Effects of retinoic acid on the immune system: Stimulation of T killer cell induction. *European Journal of Immunology*, 8(1), 23–29. https://doi.org/10.1002/eji.1830080106.

Elham, A. S., Azam, K., Azam, J., Mostafa, L., Nasrin, B., & Marzieh, N. (2021). Serum vitamin D, calcium, and zinc levels in patients with COVID–19. *Clinical Nutrition ESPEN*, 43, 276–282.

Ertesvag, A., Engedal, N., Naderi, S., & Blomhoff, H. K. (2002). Retinoic acid stimulates the cell cycle machinery in normal T cells: Involvement of retinoic acid receptor-mediated IL-2 secretion. *The Journal of Immunology*, 169(10), 5555–5563. https://doi.org/10.4049/jimmunol.169.10.5555.

Fuhrman, J. (2020). *Immunity Benefits of Zinc as We Age*. verywellhealth.com/surprising-immunity-benefits-of-zinc-4047431 [Google Scholar].

Galmés, S., Serra, F., & Palou, A. (2018). Vitamin E metabolic effects and genetic variants: A challenge for precision nutrition in obesity and associated disturbances. *Nutrients*, 10(12), 1919.

Gombart, A. F., Pierre, A., & Maggini, S. (2020). A review of micronutrients and the immune system–working in harmony to reduce the risk of infection. *Nutrients*, 12(1), 236. https://doi.org/10.3390/nu12010236.

Grangé, S., et al. (2015). Adult-onset renal thrombotic microangiopathy and pulmonary arterial hypertension in cobalamin C deficiency. *The Lancet*, 386(9997), 1011–1012.

Grant, W. B., Lahore, H., & Rockwell, M. S. (2020). The benefits of vitamin D supplementation for athletes: Better performance and reduced risk of COVID–19. *Nutrients*, 12(12), 3741.

Grinstein, S. E. R. G. I. O., & Klip, A. M. I. R. A. (1989). Calcium homeostasis and the activation of calcium channels in cells of the immune system. *Bulletin of the New York Academy of Medicine*, 65(1), 69.

Gronski, M. A., Kinchen, J. M., Juncadella, I. J., Franc, N. C., & Ravichandran, K. S. (2009). An essential role for calcium flux in phagocytes for apoptotic cell engulfment and the anti-inflammatory response. *Cell Death & Differentiation*, 16(10), 1323–1331.

Ilie, P. C., Stefanescu, S., & Smith, L. (2020). The role of vitamin D in the prevention of coronavirus disease 2019 infection and mortality. *Aging Clinical and Experimental Research*, 32(7), 1195–1198.

Infusino, F., et al. (2020). Diet supplementation, probiotics, and nutraceuticals in SARS-CoV-2 infection: A scoping review. *Nutrients*, 12(6), 1718.

Iotti, S., Wolf, F., Mazur, A., & Maier, J. A. (2020). The COVID-19 pandemic: Is there a role for magnesium? Hypotheses and perspectives. *Magnesium Research*, 33(2), 21–27.

Janssen, R., & Walk, J. (2020). Vitamin K epoxide reductase complex subunit 1 (VKORC1) gene polymorphism as determinant of differences in Covid-19-related disease severity. *Medical Hypotheses*, 144, 110218. https://doi.org/10.1016/j.mehy.2020.110218.

Jayawardena, R., Sooriyaarachchi, P., Chourdakis, M., Jeewandara, C., & Ranasinghe, P. (2020). Enhancing immunity in viral infections, with special emphasis on COVID-19: A review. *Diabetes & Metabolic Syndrome Clinical Research & Reviews*, 14(4), 367–382. https://doi.org/10.1016/j.dsx.2020.04.015.

Jee, J., et al. (2013). Effects of dietary vitamin A content on antibody responses of feedlot calves inoculated intramuscularly with an inactivated bovine coronavirus vaccine. *American Journal of Veterinary Research*, 74(10), 1353–1362.

Johns Hopkins University and Medicine (Coronavirus Resource Center). (2021). Retrieved June 3, 2021, from https://coronavirus.jhu.edu/map.html.

Klok, F. A., et al. (2020). Incidence of thrombotic complications in critically ill ICU patients with COVID-19. *Thrombosis Research*, 191, 145–147.

Koithan, M., & Devika, J. (2010). New approaches to nutritional therapy. *The Journal for Nurse Practitioners*, 6(10), 805–806. https://doi.org/10.1016/j.nurpra.2010.07.001.

Kumar, V., Kancharla, S., & Jena, M. K. (2021). In silico virtual screening-based study of nutraceuticals predicts the therapeutic potentials of folic acid and its derivatives against COVID-19. *VirusDisease*, 32(1), 29–37.

Lee, G. Y., & Han, S. N. (2018). The role of vitamin E in immunity. *Nutrients*, 10(11), 1614.

Lewis, R. S. (2001). Calcium signaling mechanisms in T lymphocytes. *Annual Review of Immunology*, 19(1), 497–521.

Lindschinger, M., Tatzber, F., Schimetta, W., Schmid, I., Lindschinger, B., Cvirn, G., Stanger, O., Lamont, E., & Wonisch, W. (2019). A randomized pilot trial to evaluate the bioavailability of natural versus synthetic vitamin B complexes in healthy humans and their effects on homocysteine, oxidative stress, and antioxidant levels. *Oxidative Medicine and Cellular Longevity*, 2019, 1–14. https://doi.org/10.1155/2019/6082613.

Liu, B., Li, M., Zhou, Z., Guan, X., & Xiang, Y. (2020). Can we use interleukin-6 (IL-6) blockade for coronavirus disease 2019 (COVID-19)-induced cytokine release syndrome (CRS)? *Journal of Autoimmunity*, 111, 102452.

Luo, Y., Teng, Z., & Wang, Q. (2012). Development of zein nanoparticles coated with carboxymethyl chitosan for encapsulation and controlled release of vitamin D3. *Journal of Agricultural and Food Chemistry*, 60(3), 836–843.

Maares, M., & Haase, H. (2016). Zinc and immunity: An essential interrelation. *Archives of Biochemistry and Biophysics*, 611, 58–65.

Malaguarnera, L. (2020). Vitamin D3 as potential treatment adjuncts for COVID-19. *Nutrients*, 12(11), 3512.

Maywald, M., Wessels, I., & Rink, L. (2017). Zinc signals and immunity. *International Journal of Molecular Sciences*, 18(10), 2222.

McCann, J. C., & Ames, B. N. (2009). Vitamin K, an example of triage theory: Is micronutrient inadequacy linked to diseases of aging? *The American Journal of Clinical Nutrition*, 90(4), 889–907.

Mehmel, M., Jovanović, N., & Spitz, U. (2020). Nicotinamide riboside—the current state of research and therapeutic uses. *Nutrients*, 12(6), 1616.

Michele, C. A., et al. (2020). Vitamin supplements in the era of SARS-Cov2 pandemic. *GSC Biological and Pharmaceutical Sciences*, 11(2), 007–019.

Mikkelsen, K., & Apostolopoulos, V. (2018). B vitamins and ageing. *Biochemistry and Cell Biology of Ageing: Part I Biomedical Science*, 451–470.

Mikkelsen, K., & Apostolopoulos, V. (2019). Vitamin B1, B2, B3, B5, and B6 and the immune system. *Nutrition and Immunity*, 115–125.

Mikkelsen, K., Stojanovska, L., & Apostolopoulos, V. (2016). The effects of vitamin B in depression. *Current Medicinal Chemistry*, 23(38), 4317–4337.

Mikkelsen, K., Stojanovska, L., Prakash, M., & Apostolopoulos, V. (2017). The effects of vitamin B on the immune/cytokine network and their involvement in depression. *Maturitas*, 96, 58–71.

Moise, A. R., Noy, N., Palczewski, K., & Blaner, W. S. (2007). Delivery of retinoid-based therapies to target tissues. *Biochemistry*, 46(15), 4449–4458. https://doi.org/10.1021/bi7003069.

Mora, J. R., Iwata, M., & Von Andrian, U. H. (2008). Vitamin effects on the immune system: Vitamins A and D take centre stage. *Nature Reviews in Immunology*, 8(9), 685–698. https://doi.org/10.1038/nri2378.

Motti, M. L., Tafuri, D., Donini, L., Masucci, M. T., De Falco, V., & Mazzeo, F. (2022). The role of nutrients in prevention, treatment and post-coronavirus disease-2019 (COVID-19). *Nutrients*, 14(5), 1000. https://doi.org/10.3390/nu14051000.

Ni, W., Yang, X., Yang, D., Bao, J., Li, R., Xiao, Y., Hou, C., Wang, H., Liu, J., Yang, D., Xu, Y., Cao, Z., & Gao, Z. (2020). Role of angiotensin-converting enzyme 2 (ACE2) in COVID–19. *Critical Care*, 24(1). https://doi.org/10.1186/s13054-020-03120-0.

Nix, W. A., Zirwes, R., Bangert, V., Kaiser, R. P., Schilling, M., Hostalek, U., & Obeid, R. (2015). Vitamin B status in patients with type 2 diabetes mellitus with and without incipient nephropathy. *Diabetes Research and Clinical Practice*, 107(1), 157–165.

Oristrell, J., et al. (2022). Vitamin D supplementation and COVID-19 risk: A population-based, cohort study. *Journal of Endocrinological Investigation*, 45, 167–179.

Oz, H. (2017). Nutrients, infectious and inflammatory diseases. *Nutrients*, 9(10), 1085. https://doi.org/10.3390/nu9101085.

Özdemir, Z. Ö., Şentürk, M., & Ekinci, D. (2013). Inhibition of mammalian carbonic anhydrase isoforms I, II and VI with thiamine and thiamine-like molecules. *Journal of Enzyme Inhibition and Medicinal Chemistry*, 28(2), 316–319.

Pecora, F., Persico, F., Argentiero, A., Neglia, C., & Esposito, S. (2020). The role of micronutrients in support of the immune response against viral infections. *Nutrients*, 12(10), 3198. https://doi.org/10.3390/nu12103198.

Polack, F. P., et al. (2020). Safety and efficacy of the BNT162b2 mRNA Covid-19 vaccine. *New England Journal of Medicine*, 383(27), 2603–2615.

Price, L. C., McCabe, C., Garfield, B., & Wort, S. J. (2020). Thrombosis and COVID-19 pneumonia: The clot thickens! *European Respiratory Journal*, 56(1).

Ragan, I., Hartson, L., Pidcoke, H., Bowen, R., & Goodrich, R. (2020). Pathogen reduction of SARS-CoV-2 virus in plasma and whole blood using riboflavin and UV light. *PLoS One*, 15(5), e0233947.

Raverdeau, M., & Mills, K. H. G. (2014). Modulation of T cell and innate immune responses by retinoic acid. *The Journal of Immunology*, 192(7), 2953–2958. https://doi.org/10.4049/jimmunol.1303245.

Razzaque, M. S. (2020). COVID-19 pandemic: Can maintaining optimal zinc balance enhance host resistance? *The Tohoku Journal of Experimental Medicine*, 251(3), 175–181.

Read, S. A., Obeid, S., Ahlenstiel, C., & Ahlenstiel, G. (2019). The role of zinc in antiviral immunity. *Advances in Nutrition*, 10(4), 696–710.

Sabry, W., Elemary, M., Burnouf, T., Seghatchian, J., & Goubran, H. (2020). Vitamin B12 deficiency and metabolism-mediated thrombotic microangiopathy (MM-TMA). *Transfusion and Apheresis Science*, 59(1), 102717.

Sadoff, J., et al. (2021). Interim results of a phase 1–2a trial of Ad26. COV2. S Covid-19 vaccine. *New England Journal of Medicine*, 384(19), 1824–1835.

Scarmeas, N., Anastasiou, C. A., & Yannakoulia, M. (2018). Nutrition and prevention of cognitive impairment. *The Lancet Neurology*, 17(11), 1006–1015. https://doi.org/10.1016/s1474–4422(18)30338–7.

Shakeri, H., Azimian, A., Ghasemzadeh-Moghaddam, H., Safdari, M., Haresabadi, M., Daneshmand, T., & Namdar Ahmadabad, H. (2022). Evaluation of the relationship between serum levels of zinc, vitamin B12, vitamin D, and clinical outcomes in patients with COVID-19. *Journal of Medical Virology*, 94(1), 141–146.

Shakoor, H., Feehan, J., Mikkelsen, K., Al Dhaheri, A. S., Ali, H. I., Platat, C., Stojanovska, L. & Apostolopoulos, V. (2021). Be well: A potential role for vitamin B in COVID-19. *Maturitas*, 144, 108–111.

Stipp, M. M. (2020). SARS-CoV-2: Micronutrient optimization in supporting host immunocompetence. *Int J Clin Case Rep Rev*, 2(2), 1–10.

Suksmasari, B. H. T. (2015). Multivitamin supplementation supports immune function and ameliorates conditions triggered by reduced air quality. *Vitamins & Minerals*, 4(2). https://doi.org/10.4172/2376–1318.1000128.

Tang, C. F., Ding, H., Jiao, R. Q., Wu, X. X., & Kong, L. D. (2020). Possibility of magnesium supplementation for supportive treatment in patients with COVID–19. *European Journal of Pharmacology*, 886, 173546.

Te Velthuis, A. J., van den Worm, S. H., Sims, A. C., Baric, R. S., Snijder, E. J., & van Hemert, M. J. (2010). Zn2+ inhibits coronavirus and arterivirus RNA polymerase activity in vitro and zinc ionophores block the replication of these viruses in cell culture. *PLoS Pathogens*, 6(11), e1001176.

Theofilis, P., Sagris, M., Oikonomou, E., Antonopoulos, A. S., Siasos, G., Tsioufis, C., & Tousoulis, D. (2021). Inflammatory mechanisms contributing to endothelial dysfunction. *Biomedicines*, 9(7), 781.

Walls, J., Sinclair, L., & Finlay, D. (2016). Nutrient sensing, signal transduction and immune responses. *Seminars in Immunology*, 28(5), 396–407. https://doi.org/10.1016/j.smim.2016.09.001.

Weiss, G. (2002). Iron and immunity: A double-edged sword. *European Journal of Clinical Investigation*, 32, 70–78.

Wessels, I., Maywald, M., & Rink, L. (2017). Zinc as a gatekeeper of immune function. *Nutrients*, 9(12), 1286.

Wessling-Resnick, M. (2018). Crossing the iron gate: Why and how transferrin receptors mediate viral entry. *Annual Review of Nutrition*, 38(1), 431–458.

Widasari, L., Chalid, M. T., Jafar, N., Thaha, A. R., & Dirpan, A. (2020). The role of multimicronutrients on improving better pregnancy outcomes: A literature review. *A Multifaceted Review Journal in the Field of Pharmacy*, 11(8), 550–553.

World Health Organization (WHO). (2021a). *Naming the Coronavirus Disease (COVID-19) and the Virus that Causes It*. Retrieved March 4, 2021, from https://www.who.int/emergencies/diseases/novel-coronavirus-2019/technical-guidance/naming-the-coronavirus-disease-(covid-2019)-and-the-virus-that-causes-it.

World Health Organization (WHO). (2021b). *Rolling Updates on Coronavirus Disease (COVID-19)*. Retrieved March 4, 2021, from https://www.who.int/emergencies/diseases/novel-coronavirus-2019/events-as-they-happen.

World Health Organization (WHO). (2021c). *Virtual Press Conference on COVID-19 – 11 March 2020*. Retrieved March 4, 2021, from https://www.who.int/docs/default-source/coronaviruse/transcripts/who-audio-emergencies-coronavirus-press-conference-full-and-final-11mar2020. pdf?sfvrsn=cb432bb3_2.

Wu, D., Lewis, E. D., Pae, M., & Meydani, S. N. (2019). Nutritional modulation of immune function: Analysis of evidence, mechanisms, and clinical relevance. *Frontiers in Immunology*, 9. https://doi.org/10.3389/fimmu.2018.03160.

Zhang, L., & Liu, Y. (2020). Potential interventions for novel coronavirus in China: A systematic review. *Journal of Medical Virology*, 92(5), 479–490.

Index

U

V

W

Y

Z